The Cambridge Illustrated Dictionary of Natural History

R. J. LINCOLN
and
G. A. BOXSHALL

Illustrations by Roberta Smith

CAMBRIDGE UNIVERSITY PRESS

CAMBRIDGE
NEW YORK
NEW ROCHELLE MELBOURNE SYDNEY

Published by the Press Syndicate of the University of Cambridge
The Pitt Building, Trumpington Street, Cambridge CB2 1RP
32 East 57th Street, New York, NY 10022, USA
10 Stamford Road, Oakleigh, Melbourne 3166, Australia

© Cambridge University Press 1987
First published 1987

Printed in Great Britain

British Library Cataloguing in Publication Data

Lincoln, Roger J.
    The Cambridge illustrated dictionary of natural history.
    1. Natural history — Dictionaries
    I. Title II. Boxshall, G. A.
    508'.03'21   QH13

Library of Congress Cataloging in Publication Data

Lincoln, Roger J.
    The Cambridge illustrated dictionary of natural history.
    1. Natural history — Dictionaries
    I. Boxshall, Geoffrey Allan. II. Title.
    QH13.L56   1987   508'.03'21   87-8018

ISBN 0 521 30551 9

HO

PREFACE

The words 'natural history' conjure up an immediate vision of the living countryside – wild animals, flowers, birds and butterflies. This popular image of natural history is reflected in the content of this book. It is a dictionary about life on Earth, about the rich diversity of living organisms and their habitats. Mostly, it deals with present day species but it does include many fossil groups as well. This book is a response to the blossoming popular interest in natural history, which has undoubtedly been stimulated through the medium of television. Previously hidden and intimate wonders of the natural world, both microscopic as well as macroscopic, have been revealed in fascinating detail to a huge audience of children and adults alike. At the same time natural history, within the framework of habitat and species' conservation, exploitation of natural resources, pollution, waste disposal and other ecological issues, has become an issue of serious political debate. This enhanced public and political awareness of, and concern for, organisms and habitats is working its way into educational programmes and is apparent in the current popularity of zoological and botanical parks, natural history museums and exhibitions.

The abundant variety of the natural world is hard to comprehend. Already, more than one million species of animals and plants have been described and named. This is but a small part of the whole, as estimates of the total world fauna and flora put the number of species anywhere between 5 and 30 million. Groups with which we are all familiar, such as vertebrates (fishes, amphibians, reptiles, mammals, birds) and flowering plants, contain comparatively few species, numbering only hundreds of thousands. It is the ubiquitous insects that account for by far the greater part. Studying such a prodigious diversity would be impossible without some form of indexing system. Thus, for purposes of study and reference species are ordered into extensive classifications – hierarchical systems of named groups (taxa). These classification systems serve to summarize information at different levels. It should be remembered,

however, that these classificatory terms are artificial, they are only concepts and have no actual existence in nature. A group's identity stems from the fact that its members share features that distinguish them from other such groups. Most systems of classification are designed to correspond with supposed evolutionary history and relationships. A primary aim of this book is to provide a brief explanation for many of these long, Latinized and usually unfamiliar terms. As far as possible, we have incorporated information on the types of organisms contained within the groups, the approximate numbers of species, their habitat preferences and geographical distributions, and some of the morphological features which characterize them and by which they can be recognized. The major taxonomic groups of all living organisms, from microorganisms through to mammals and flowering plants, are included, down to the level of order. For flowering plants, vertebrate animals and the economically more important insect groups, the coverage extends further, to include families. Many common names of organisms or of groups have been incorporated but only as a cross-reference to the family or higher taxonomic category to which they belong. No attempt has been made to include descriptions of individual species. Selected representatives of a wide range of groups are illustrated. These line drawings, over 700 of them, serve to supplement the definitions and should help readers to visualize some of the small, rare forms such as protistans and lower invertebrates.

Several different classification schemes are in use today and a great many others have been used in the past. For us, this was a dilemma which had to be resolved as it was abundantly clear that we could not include everything. Our solution was to produce a composite scheme drawn from a variety of recent sources. In some cases we also incorporated parallel terms from other widely used classifications. Protozoans and other single-celled organisms are a case in point; they have variously been treated as part of the Kingdom Plantae, the Kingdom Animalia and the Kingdom Fungi, and have also been grouped together in their own Kingdom, the Protista. Equivalent terms have been cross-referenced wherever possible.

As with our earlier *Dictionary of Ecology, Evolution and Systematics* (Cambridge University Press, 1982) this new volume is conceived as a working dictionary comprising concise definitions that are adequate to explain the meaning or essential concept of a term, or to briefly outline the characteristics of a group of organisms. The entries are not discursive essays such as would be found in an encyclopaedia.

The new dictionary combines extensive coverage of classification with a broad selection of ecological terms. Ecological studies are the key to our understanding of the complex interrelationships of the natural world. We have sought to cover topics such as feeding, reproduction, population biology, behaviour, soils, ecological energetics, habitats and life histories, which are all part of modern natural history.

R.J.L.
G.A.B.
March 1987

a

aardvark Orycteropodidae *q.v.*

abalone Archaeogastropoda *q.v.*

abduct To move away from the midline; **abduction**; *cf.* adduct.

aberrant 1: Not conforming to type. 2: An individual exhibiting unusual characters due to external environmental influences rather than to genetic factors; **aberration.**

abience An avoidance reaction; withdrawal or retraction from a stimulus; *cf.* adience.

abiogenesis Spontaneous generation; the concept that life can arise spontaneously from non-living matter by natural processes without the intervention of supernatural powers; *cf.* biogenesis.

abioseston The non-living component of the total particulate matter suspended in water; *cf.* bioseston.

abiotic Devoid of life; non-living; *cf.* biotic.

ablation The removal of a surface layer, as of ice by melting or evaporation.

aboospore A spore produced from an unfertilized female gamete.

aborigine The original or indigenous biota of a geographical region; **aboriginal.**

abort To arrest development; **abortion.**

abrasion The process of erosion by rubbing off or wearing away of surface material.

Abrocomidae Rat chinchillas; family containing 2 species of rat-like rodents (Hystricomorpha) that inhabit burrows and crevices in the Andes of South America; stiff hairs project over the nails of the 3 central digits of the hindfeet.

rat chinchilla (Abrocomidae)

abscission The natural process by which two parts of an organism separate; **abscise.**

absenteeism The behaviour shown by animals which nest away from their offspring but visit them from time to time with food, providing minimal parental care.

absolute age The precise geological age of a fossil or rock, usually calculated by means of radiometric dating; *cf.* relative age.

absolute dating A method of geological dating which involves measuring the amount of decay of a radioactive isotope and gives a direct measure of the amount of time that has elapsed since formation of the rocks; *cf.* relative dating.

absolute humidity The actual amount of water vapour present in a mass of air; *cf.* relative humidity.

absorption 1: The process by which one substance (the absorbate) is taken into and incorporated in another substance (the absorbent); **absorb.** 2: In ecological energetics, that part of the total intake of food (consumption) not voided during regurgitation or as faeces.

abundance The total number of individuals of a species in an area, volume, population or community; often measured as cover in plants.

abyssal Pertaining to great depths within the Earth, or to zones of great depth in the oceans or lakes into which light does not penetrate; commonly used in oceanography of depths between 4000 and 6000 m; **abyss**; see marine depth zones.

abyssal plain The more or less flat ocean floor below 4000 m, excluding ocean trenches, having a slope of less than 1 in 1000; see marine depth zones.

abyssobenthic Living on or in the ocean floor in the abyssal zone; see marine depth zones.

abyssopelagic Living in the oceanic water column at depths between 4000 and 6000 m, away from the ocean floor; see marine depth zones.

acacia Mimosaceae *q.v.*

Acanthaceae Zebra plant, shrimp plant; family of Scrophulariales containing about 2500 species of often twining herbs or woody plants commonly accumulating orobanchin and iridoid compounds; mostly tropical in distribution; flowers frequently condensed in cymes; bracts and bracteoles often coloured and sometimes enclosing the flowers; individual flowers having a calyx with 4–5 clefts, 4–5 petals, 2, 4 or 5 stamens; producing a capsular fruit.

zebra plant (Acanthaceae)

Acantharia Class of protozoans (Actinopoda *q.v.*) possessing a skeleton made of radial spines which are joined in the centre, arranged regularly and composed of strontium sulphate; most float passively in surface waters of the open sea and reproduce by the formation of spores.

Acanthobdellida Monotypic subclass of hirudinoidean worms (Annelida) ectoparasitic on freshwater fishes in cold lakes of northern Europe and Alaska; morphology intermediate between oligochaetes and leeches; body with 30 segments each bearing 4 annuli, length 3–35 mm; possessing posterior sucker only.

Acanthocephala Thorny-headed worms, spiny-headed worms; phylum of exclusively parasitic, pseudocoelomate worms found mainly in vertebrate intestinal tracts and typically utilizing an arthropod intermediate host; body divided into retractable proboscis (typically armed with hooks), neck and trunk; lacking a digestive tract; comprising 3 classes, Archiacanthocephala, Eoacanthocephala and Palaeacanthocephala.

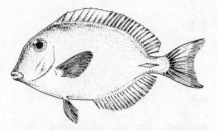

surgeonfish (Acanthuridae)

Acari Mites and ticks; diverse order of small free-living and parasitic arthropods (Arachnida) comprising about 30,000 recognized species although this may represent only a small proportion of the world fauna; body mostly small (80μm–20 mm), compact; head and abdomen fused; mouthparts include chelicerae and subcapitulum; immense variety of form and habit in 3 suborders, Acariformes, Opilioacriformes and Parasitiformes.

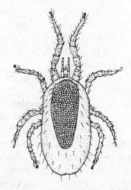

mite (Acari)

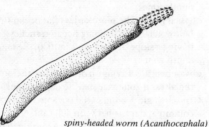

spiny-headed worm (Acanthocephala)

Acanthochitonida Small order of specialized chitons typically with the shell valves completely or partially covered by the peripheral girdle; surface of each valve usually divided into 4 distinct sections; includes the largest known chiton, the gumshoe.

Acanthodii Extinct class of primitive bony fishes (Osteichthyes) known from the Silurian to the Permian; tail fin heterocercal; ganoid scales present; stout spines present in front of fins.

Acanthopodina A suborder of Amoebida *q.v.*; also treated as a class of the protoctistan phylum Rhizopoda.

acanthosoma Late zoeal larval stage of sergestoid shrimps (Decapoda).

Acanthuridae Surgeonfishes; family containing 7 species of tropical marine perciform teleosts most abundant in the Indo-Pacific; body compressed, up to 500 mm in length; dorsal and anal fins long, sometimes with filamentous rays; mostly herbivorous, feeding on algae scraped from coral.

Acariformes Diverse suborder of mites (Acari) comprising 3 subgroups, Astigmata, Oribatei, Prostigmata; includes phytophagous, fungivorous, predatory and parasitic forms; Actinotricha.

acarology The study of mites and ticks; **acarological.**

acarophilous Thriving in association with mites; **acarophile, acarophily.**

acarophytism Symbiosis between plants and mites.

Acarpomyxa Class of rhizopod protozoans comprising small multinucleate plasmodial forms and similar uninucleate forms, usually branching and sometimes forming a network; a test is absent and there are no fruiting bodies or spores as in Eumycetozoa *q.v.*; includes 2 classes, Leptomyxida and Stereomyxida.

accessory chromosome Any chromosome differing from the normal A-chromosomes.

accidental Not normally occurring in a particular community or habitat.

accidental parasite A parasite found associated with an organism which is not its normal host.

Accipitridae Eagles, harriers, hawks, kites, Old World vultures; large and diverse family of raptors (Falconiformes) distributed worldwide except for polar regions; feeding mainly on small vertebrates captured using powerful feet, some feeding on carrion; solitary or gregarious in habits, monogamous, nesting in trees, cliffs or on the ground; contains about 215 species.

acclimation The gradual and reversible adjustment of physiology or morphology as a result of changing environmental conditions; often used with reference to an individual organism in an artificial or experimentally manipulated environment; *cf.* acclimatization, accommodation.

acclimatization The gradual and reversible adjustment of physiology and morphology to changing natural environmental conditions; often used to refer to the changes observed in a species over a number of generations; *cf.* acclimation, accommodation.

acclivous Having a gentle upward slope.

accommodation 1: The capacity of a plant to adapt to changes in the environment. 2: A decrease in response or sensation as a result of repeated stimulation; *cf.* acclimation, acclimatization.

accrescent Increasing in size with age; used of plants that continue to grow after flowering.

accretion 1: Increase in size by the external addition of new material. 2: Deposition of material by sedimentation.

accumulator Any organism that actively concentrates a particular element or compound in its tissues.

acellular Not composed of cells; *cf.* cellular.

Aceraceae Maple; family of Sapindales containing over 100 species of trees and shrubs with mostly palmately lobed or veined leaves; flowers regular, pentamerous; fruit is a samara.

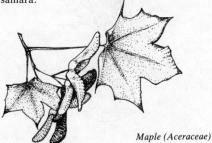

Maple (Aceraceae)

Achariaceae Small family of Violales containing only 3 species of climbing or stemless monoecious herbs without a latex system; confined to South America.

Achatocarpaceae Small family of Caryophyllales comprising 8 species of spiny shrubs or trees found from Texas to Argentina.

achene A small, dry, indehiscent fruit developed from a single carpel containing a single seed.

achromatic Without colour; unpigmented.

A-chromosome Any of the normal chromosomes of a eukaryotic organism.

achroous Without colour, unpigmented.

acid rain Rain with a high level of acidity (low pH) due to pollution by oxides of sulphur and nitrogen produced by oil and coal combustion.

acidic habitat A habitat with a pH less than 7, typically poor in nutrients.

acidophilic Thriving in an acidic environment; **acidophile, acidophily**; *cf.* acidophobic.

acidophobic Intolerant of acidic environments; **acidophobe, acidophoby**; *cf.* acidophilic.

acidotrophic Feeding on acidic food or acidic substrates.

Acipenseridae Sturgeons: family of large (1–5.5 m) freshwater or anadromous fishes widespread in the northern hemisphere; body spindle-shaped bearing rows of large bony scutes; mouth protrusible; teeth absent in adult; swim bladder large; contains 25 species, some exploited commercially; eggs marketed as caviar.

sturgeon (Acipenseridae)

Acipenseriformes Order of primitive bony fishes (Osteichthyes) comprising 2 families, Acipenseridae (sturgeons) and Polyodontidae (paddlefishes).

acme A period of maximum vigour; the highest point attained in phylogenetic or ontogenetic development; **acmic**; *cf.* epacme.

Acochlidioidea An order of small, elongate opisthobranch molluscs found in sandy sediments of marine or sometimes brackish waters; commonly lacking a shell and a head shield; possessing a pair of tentacles; typically discarding their old radula teeth.

Acoela Order of primitive free-living turbellarians found abundantly in or on marine sediments but with a few brackish water or freshwater forms; characterized by their simple construction, the lack of a permanent gut cavity, the lack or weak development of a muscular pharynx, and the presence of a statocyst containing a single statolith.

Aconchulinida Order of Filosa *q.v.*; naked amoebae with true filose pseudopodia (filopodia), that feed on various other microorganisms and algae, either by ingestion or parasitic penetration.

acorn barnacle Thoracica *q.v.*

acorn worm Hemichordata *q.v.*

acoustics The study of sound; **acoustic.**

acquired characters, inheritance of
1: Lamarckism; that changes in use or disuse of an organ result in changes in size and functional capacity and that these modified characters are transmitted to the offspring. 2: Neo-Lamarckism; that characters acquired by organisms as a response to environmental factors are assimilated into the genome and transmitted to the offspring.

acquired trait A character or trait which is a result of direct environmental influences.

Acrania Cephalochordata *q.v.*

Acrasea Cellular slime moulds *q.v.*; treated as a class of rhizopod protozoans.

Acrasiomycetes Cellular slime moulds *q.v.*; treated as a class of Myxomycota in the fungi.

Acrasiomycota Cellular slime moulds *q.v.*; treated as a phylum of protoctistans.

Acrididae Large and diverse family of insects (order Orthoptera) comprising the locusts and true grasshoppers; contains about 7000 species in 1100 genera, including the majority of the economically important orthopterans.

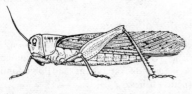

locust (Acrididae)

acridophagous Feeding on grasshoppers; **acridophage, acridophagy.**

Acrochordidae Wart snakes; family of primitive snakes (Serpentes) found in India, Sri Lanka and some Indo-Australian Islands; ridged scales on body do not overlap; nasal openings on top of snout.

acrodendrophilous Thriving in tree top habitats; **acrodendrophile, acrodendrophily.**

acronurus Transparent larval stage of surgeonfish (Acanthuridae).

acropleustophyte A large aquatic plant floating freely at the water surface.

Acrosiphoniales Order of marine and brackish water green algae with a simple or branched uniseriate filamentous body attached to the substrate by rhizoids; cells usually multinucleate with a perforated cylindrical chloroplast.

Acrothoracica Order of free-living, filter-feeding barnacles (Cirripedia) that inhabit burrows in mollusc shells, coral skeleton and limestone in tropical and temperate shallow marine habitats; characterized by reduced calcareous plates and by degenerate complemental males; contains about 50 species.

Acrotretida Order of small inarticulate brachiopods comprising about 25 extant species; pedicle typically short or absent; foramen present.

acrotropism An orientation response resulting in the continued growth of a plant in the direction in which growth originally commenced; **acrotropic.**

actic Pertaining to rocky shores; often used of the zone between high and low tides; intertidal.

Actinelida Order of protozoans (Actinopoda) forming a transitional group between the Acantharia and Heliozoa.

Actiniaria Sea anemones; order of solitary zooantharians comprising about 800 species found in all the seas, from the intertidal zone to abyssal depths; typically lacking a hard exoskeleton; basal end may be an adherent pedal disk or a bulbous structure used for anchoring in soft substrates; oral disk typically circular with surrounding tentacles arranged with a six-fold symmetry.

sea anemone (Actiniaria)

Actinidiaceae Chinese gooseberry; family of Theales containing about 300 species of tanniferous, woody plants, widely distributed in tropics, subtropics and mountains of Asia.

actinobiology The study of the effects of radiation on living organisms; **actinology.**

Actinoceratida Extinct order of nautiloid cephalopod molluscs known from the Lower Ordovician to the Lower Carboniferous; predominantly orthozonic with large siphuncles.

actinomorphy Radial symmetry in flowers.

Actinomyxea Class of Myxozoa *q.v.* containing parasites of aquatic oligochaetes and sipunculans; characterized by an outer spore covering composed of 3 valves which may be drawn out into long processes, and by the presence of 3 polar capsules; treated as a class, Actinomyxida, of the protoctistan phylum Cnidosporida.

Actinophryida Order of protozoans (Heliozoa *q.v.*) containing mostly free cells that lack a skeleton and a centroplast, found in both marine and fresh water.

Actinopoda Superclass of the protozoan subphylum Sarcodina characterized by the possession of radial axopods (consisting of a rigid axial rod surrounded by cytoplasm) amongst the fine pseudopodia; food particles are trapped on the sticky axopods, pulled in towards the capsule and enclosed in a vacuole; most are marine and pelagic; comprising 4 classes, Acantharia, Heliozoa, Phaeodaria and Polycystina; also treated as a distinct phylum of Protoctista.

Actinopterygii Ray-finned fishes; subclass of bony fishes (Osteichthyes) characterized by fins of membranous skin supported by spines or rays; includes the majority of living bony fishes.

Actinotricha Acariformes *q.v.*

actinotroch Planktonic larval stage in the phylum Phoronida *q.v.*

actinula Free-swimming tentaculate larva of some hydrozoan cnidarians, formed from a ciliated planula.

Actinulida Order of free-living Hydrozoa which are solitary and motile but not pelagic and occur in interstitial spaces in marine sands; typically small with a covering of cilia by which they move; exhibiting a general actinula larva-like organization.

active chamaephyte A type of chamaephyte *q.v.* in which the horizontal vegetative shoots persist through unfavourable seasons in a procumbent position.

active process Any process requiring the expenditure of metabolic energy.

actophilous Thriving on rocky seashores; **actophile, actophily.**

actualism The theory that seeks to explain the evolution of the Earth in terms of relatively small scale natural fluctuations or events that have been operative throughout the entire geological history of the Earth and have thus produced changes of great magnitude; uniformitarianism; *cf.* catastrophism.

Aculeata Bees, wasps, ants; diverse group of hymenopteran insects (suborder Apocrita) comprising about 60,000 species many of which are eusocial; ovipositor typically modified as sting for injection of venom, or spread of pheromones or allomones.

acyclic parthenogenesis Reproduction by parthenogenesis alone, in which the sexual phase of an alternation of generations cycle has been lost; *cf.* parthenogenesis.

adaptation 1: The process of adjustment of an individual organism to environmental conditions; adaptability. 2: Process of evolutionary modification which results in improved survival and reproductive efficiency. 3: Any morphological, physiological, developmental or behavioural character that enhances survival and reproductive success of an organism; **adaption.**

adaptiogenesis The production of new adaptations.

adaption Adaptation *q.v.*

adaptive landscape The figurative representation of the fitness of organisms in the form of a topographical map, on which those fit genotypes (species) able to occupy particular ecological niches are depicted as adaptive peaks separated by adaptive valleys representing unfit gene combinations.

adaptive peak A peak on an adaptive landscape *q.v.*

adaptive radiation The evolutionary diversification of a taxon (adaptive type) into a number of different ecological roles or modes of life (adaptive zones), usually over a relatively short period of time and leading to the appearance of a variety of new forms.

adaptive valley A valley in an adaptive landscape *q.v.*

adduct To move towards the midline; **adduction**; *cf.* abduct.

adeciduate Not falling off, or coming away.

adelphogamy Fertilization between two different individuals derived vegetatively from the same parent plant.

adelphoparasite An organism parasitic on a closely related host organism; *cf.* alloparasite.

adelphophagy The fusion of two gametes of the same sex.

adelphotaxy The mutual attraction between spores after extrusion.

Adenophorea Class of nematodes including free-living aquatic and terrestrial forms, and parasites of plants and animals; characterized by the position of the head chemoreceptors (amphids) behind the lips and of the 16 cephalic sensory organs which are found on or behind the lips; external cuticle always 4-layered; comprising 2 subclasses, Chromadoria and Enoplia.

Adephaga Suborder of beetles (Coleoptera) comprising about 34, 000 species including Carabidae (predaceous ground beetles), Dytiscidae (giant water beetles), Gyrinidae (whirligig beetles), the first containing the vast majority of species.

carabid beetle (Adephaga)

adichogamy Simultaneous maturation of male and female reproductive organs of a flower or hermaphroditic organism; **adichogamous**; *cf.* dichogamy.

adience Movement towards a stimulus; an approaching reaction; *cf.* abience.

admiral butterfly Nymphalidae *q.v.*

adnate Closely applied to; growing on; attached along the entire length.

Adoxaceae Family of Dipsacales containing a single species of delicate perennial herbs with a circumboreal distribution.

Adrianichthyidae Family of small (to 200 mm) beloniform teleost fishes comprising only 3 species, restricted to freshwater lakes of the Celebes; body elongate, compressed; head depressed with large jaws; pelvic fins small.

adsorption The adhesion of molecules as an ultra-thin layer on the surface of solids or fluids; **adsorb.**

adspersed Widely distributed; scattered.

adtidal Living immediately below low tide level.

advection The process of transfer by virtue of motion; the transfer of heat or matter by horizontal movement of water masses.

adventitious Accidental; occurring at an unusual site; secondary.

adventive Not native; an organism transported into a new habitat.

adynamandrous Having non-functioning male reproductive organs; **adynamandry.**

adynamogynous Having non-functioning female reproductive organs; **adynamogyny.**

Aegithalidae Long-tailed tits; family containing 7 species of small titmice (Passeriformes) found in America, Eurasia and Java; nest intricate, domed, built in trees and bushes.

aeolation Erosion of a land surface by wind-blown sand and dust; **eolation.**

aeolian Pertaining to the action or effect of the wind; **eolian.**

aeolian deposit Wind-borne soil deposit; *cf.* alluvial deposit, colluvial deposit.

aeon An indefinitely long period of time; **eon.**

Aepyornithidae Elephant birds; extinct family of large, ratite birds standing up to 3 m tall; found in subforest habitats in Madagascar.

aerobic 1: Growing or occurring only in the presence of molecular oxygen; **aerobe, aerobiosis**; *cf.* anaerobic. 2: Used of an environment in which the partial pressure of oxygen is similar to normal atmospheric levels.

aerobiology The study of airborne organisms.

aerochorous Disseminated by wind; **aerochore, aerochory.**

aerogenic Gas-producing; **aerogenesis**; *cf.* anaerogenic.

aerohygrophilous Thriving in high atmospheric humidity; **aerohygrophile, aerohygrophily**; *cf.* aerohygrophobous.

aerohygrophobous Intolerant of high atmospheric humidity; **aerohygrophobe, aerohygrophoby**; *cf.* aerohygrophilous.

aerophilous 1: Pollinated by wind; fertilized by airborne pollen. 2: Thriving in exposed windy habitats; **aerophile, aerophily.** 3: Disseminated by wind.

aerophyte An epiphyte growing on a terrestrial plant and lacking direct contact with soil or water.

aeroplankton Those organisms freely suspended in the air and dispersed by wind; aerial plankton.

aerotaxis The directed movement of a motile organism towards (positive) or away from (negative) an air-liquid interface, or a concentration gradient of dissolved oxygen; **aerotactic.**

aerotolerant Used of anaerobic organisms having the capacity to grow to a limited extent under aerobic conditions.

aerotropism An orientation response to a gaseous stimulus; **aerotropic.**

Aeshnidae Widely distributed family of large, hawker dragonflies (Odonata) which are often marked with blue and green colours; eyes large, almost contiguous across top of head.

aestival Pertaining to the early summer season; *cf.* hibernal, hiemal, serotinal, vernal.

aestivation 1: Passing the summer or dry season in a dormant or torpid state; **estivation, aestivate;** *cf.* hibernation. 2: The manner in which plant structures are folded prior to expansion or opening.

aetiology 1: The branch of science dealing with the study of origins or causes. 2: The demonstrated cause of a disease or trait; causation; **etiology, aetiological.**

Aetosauria Extinct group of thecodontian reptiles known from the Triassic; body rather crocodile-like, armoured with bony plates; up to 3 m long.

Aextoxicaceae Family of Celastrales *q.v.* containing a single tree species native to Chile; flowers borne in axillary racemes, pentamerous; fruit is a drupe.

afforestation The process of establishing a forest in a non-forested area; *cf.* reforestation.

Afrenulata Class of pogonophoran marine worms comprising a single order, Vestimentifera; body devoid of setae; bridle absent from mesosoma.

African subkingdom A subdivision of the Palaeotropical kingdom.

African tulip tree Bignoniaceae *q.v.*

African violet Gesneriaceae *q.v.*

Aftonian interglacial An interglacial period of the Quaternary Ice Age *q.v.* in North America.

agamandroecious Used of a plant having male and neuter flowers in the same inflorescence; *cf.* agamogynoecious, agamohermaphrodite.

agameon A species comprising only non-sexually reproducing individuals.

agamete A mature reproductive cell which does not fuse with another to form a zygote; a non-copulating germ cell; *cf.* gamete.

agamic Without gametes; used of complexes of organisms in which all individuals reproduce asexually.

Agamidae Large, diverse family of arboreal and terrestrial lizards (Sauria) widespread in Oriental and Palaearctic regions from North Africa to Australia; contains about 320 species of diurnal, insectivorous or herbivorous lizards, characterized by a depressed or compressed body, a non-autotomic tail, and by oviparity.

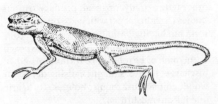

agamid lizard (Agamidae)

Agamococcidiida Primitive order of coccidians comprising only 3 species found in marine annelids; typically without merogony or gamonts in the life cycle.

agamodeme A local interbreeding population of predominantly asexually reproducing individuals; *cf.* deme.

agamogenesis Asexual reproduction.

agamogony Asexual reproduction by development of a new individual from a single cell, following binary or multiple fission, or budding.

agamogynoecious Used of a plant having female and neuter flowers in the same inflorescence; *cf.* agamandroecious, agamohermaphrodite.

agamohermaphrodite Used of a plant having hermaphrodite and neuter flowers in the same inflorescence; *cf.* agamandroecious, agamogynoecious.

agamont The sexual individual or generation producing agametes *q.v.*

agamospecies A species or population comprising only asexually reproducing individuals.

agamospermy Reproduction without fertilization in which embryos and seeds are formed asexually, but not including vegetative reproduction; **agamospermous.**

agamotropic Used of flowers that do not close again once they have opened; *cf.* gamotropic, hemigamotropic.

Agaonidae Fig wasps; family of small phytophagous wasps (Hymenoptera) that live only in the flowers and fruits of figs, and upon which the host is dependent for pollination; adults are sexually dimorphic, the male lacking wings and eyes, and having feeble limbs; contains about 160 species.

Agaricaceae Edible mushrooms; family of typically umbrella-shaped fungi in which the cap breaks cleanly from the stalk, which usually bears a ring; spore masses on gills dark coloured when mature; found on the ground in grassland and open woodland.

Agaricales Large, diverse order of hymenomycete fungi which can be saprobic in a variety of habitats, parasitic on plants and other fungi, or mycorrhizal; produce fruiting bodies (basidiocarps) which are diverse in form but having basidia arranged in a fertile layer (hymenium) which is typically exposed at the time of basidiospore formation and discharge; includes fungus causing dry rot of wood, edible mushrooms, boletes, fly agarics and inkcap mushrooms.

field mushroom (Agaricales)

Agaricogastrales Order of gasteromycete fungi in which the fertile layer (hymenium) persists after maturity; fruiting body typically resembling the edible mushroom.

agaricolous Living on mushrooms and toadstools; **agaricole.**

Agavaceae Century plant, dragon tree, snake plant, yucca; family of Liliales comprising nearly 600 species of coarse herbaceous or arborescent plants with perennial leathery or fleshy leaves; native to warm, mostly arid regions.

age 1: To become old; to attain maturity. 2: The period of time a group or organism has existed. 3: The length of geological time since the formation of a rock either by solidification from a molten state (igneous) or by sedimentation (sedimentary). 4: A period of geological history characterized by a dominant life form, such as the age of fishes. 5: A geological time unit shorter than an epoch.

age of amphibians That period of the Earth's history dominated by amphibians; the Carboniferous and Permian periods.

age of cycads That period of the Earth's history dominated by cycads; the Jurassic period.

age of fishes That period of the Earth's history dominated by fishes; the Silurian and Devonian periods.

age of gymnosperms That period of the Earth's history dominated by gymnosperms; the Mesozoic era.

age of mammals That period of the Earth's history dominated by mammals; the Cenozoic era.

age of man That period of the Earth's history dominated by man; the Quaternary period.

age of marine invertebrates That period of the Earth's history dominated by marine invertebrates; the Ordovician and Cambrian periods.

age of reptiles That period of the Earth's history dominated by reptiles; the Mesozoic era.

age structure The number or percentage of individuals in each age class of a population.

ageing The process of irreversible decline of bodily function and adaptability with time or increasing age.

Agelasida Small order of tetractinomorph sponges widely distributed in shallow tropical and warm temperate waters; having a skeletal reticulum of spongin fibres with protruding short, spined megascleres.

Agelenidae Sheet-web spiders, house spider; family of spiders (Araneae) that construct sheet webs with tubular retreats; body with a flattened thorax, legs long and thin, covered with long hairs; eyes arranged in two short rows.

Ageneiosidae Family of predatory tropical South American freshwater catfishes (Siluriformes) found mostly in slow-moving backwaters; body naked, dorsal fin with stout spine, barbels rudimentary; gas bladder enclosed in bony capsule; contains about 35 species, some important as food, others collected for aquaria.

ageotropism The absence of orientation movements in response to gravity; **ageotropic.**

aggregation A society or group of conspecific organisms which has a social structure and consists of repeated members or units but with a low level of coordination, integration or genotypic relatedness.

aggression A hostile act or threat made to protect territory, the family group or offspring, or to establish dominance.

aggressive mimicry Mimicry *q.v.* in which a predator mimics a non-predatory model in order to deceive its prey.

Agnatha Jawless fishes; class or superclass of aquatic vertebrates lacking true jaws formed from modified gill arches; with poorly developed or absent paired fins; gills present on inner surfaces of gill arches; possess 2 pairs of semicircular canals; contains 2 extant orders, Myxiniformes and Petromyzoniformes, and several fossil groups.

Agnotozoic Proterozoic *q.v.*

Agonidae Poachers; family containing 50 species of small (to 300 mm) bottom-dwelling marine scorpaeniform teleost fishes found on soft sediments from intertidal to 1200 m; body elongate, covered with rows of bony plates.

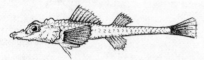

poacher (Agonidae)

agonistic behaviour Social interaction between members of a species, involving aggression or threat and conciliation or retreat.

agonistic buffering The use of infants by adults to inhibit the aggressive behaviour of other adults.

Agonomycetales Small order of hyphomycete fungi characterized by the lack of propagative spores (conidia); reproduction and dispersal is by means of rounded compact masses of hyphae or groups of cells.

agouti Dasyproctidae *q.v.*

agrarian Pertaining to cultivation or cultivated plants.

agrestal Growing on arable land.

Agriochoeridae Extinct family of ruminants (Tylopoda) having a lightly built pig-like body with long tail and clawed feet; found from the Upper Eocene to the Miocene.

agroclimatology The study of climate in relation to the productivity of plants and animals of agricultural importance.

agrology The branch of agriculture dealing with the study of soils.

agronomy The theory and practice of agricultural management, crop production and husbandry.

agrophilous Thriving in cultivated soils; **agrophile, agrophily.**

agrostology The study of grasses; graminology.

Agulhas Current A warm surface ocean current that flows south off the coast of South Africa, derived in part as an extension of the Mozambique Current and Indian South Equatorial Current; see ocean currents.

ahermatypic Pertaining to a non-colonial assemblage, or an individual; used of a coral that lacks symbiotic algae; **ahermatype.**

A-horizon The dark coloured upper mineral horizon of a soil profile, immediately below the O-horizon and comprising some humified organic material as a result of biological activity or cultivation; see soil horizons.

aigialophilous Thriving in beach habitats; **aigialophile, aigialophily.**

aigicolous Living in beach habitats; **aigicole.**

Ailuropodidae Giant panda; family containing single species of large bear-like mammals (Carnivora) restricted to a few mountain forest areas in the south west of China; feeding exclusively on bamboo but habits and biology little known.

giant panda (Ailuropodidae)

aiphyllophilous Thriving in evergreen woodland; **aiphyllophile, aiphyllophily.**

aiphyllus Evergreen; **aiophyllus.**

air capacity That volume of air remaining in a soil after saturation with water.

air porosity The ratio of the volume of air in a given mass of soil to its total volume.

aithalophilous Thriving in evergreen thickets; **aithalophile, aithalophily.**

aitiogenic Used of a movement or reaction induced by an external stimulus.

aitionomic Used of growth patterns and other phenomena imposed by the environment.

aitiotropism Any tropism resulting from an exogenous stimulus; **aitiotropic.**

Aizoaceae Fig marigold, stone plant, hottentot fig; large family of succulent herbs (Caryophyllales) mostly from South Africa and Australia; characterized by adaptations to

stone plant (Aizoaceae)

arid conditions such as fleshy leaves covered with hairs, leaves buried in ground with only the tips showing, or having only 2 leaves; flowers often brightly coloured, with numerous petals.

Ajacicyathida Extinct order of usually solitary archaeocyathids (Monocyathea) known from the Lower and Middle Cambrian.

Akaniaceae Family of Sapindales containing a single species of alkaloid-producing tree native to eastern Australia.

akaryotic Lacking a discrete nucleus; non-nucleated; **akaryote**; *cf.* eukaryotic, prokaryotic.

akinesis Absence or cessation of movement.

akinete A highly resistant, thick-walled, resting spore formed by some blue-green algae during periods of unfavourable environmental conditions.

aktology The study of shallow inshore ecosystems.

Akysidae Family of small (to 20 mm) southeast Asian freshwater catfishes (Siluriformes); comprising 7 species each with a naked body, short dorsal fin with 1–2 spines and 3 pairs of barbels.

Alangiaceae Small family of Cornales containing about 20 species of sometimes thorny woody plants native to the Old World.

alarm call A sound produced by an animal when danger threatens but is still a significant distance away.

alarm pheromone A chemical substance, exchanged by members of a group, that induces a state of alarm or alertness.

alarm reaction The sum of all non-specific responses to sudden exposure to stimuli to which the organism is not adapted.

Alaska Current Aleutian Current *q.v.*

Alaudidae Larks; family containing about 80 species of small terrestrial passerine birds found worldwide in open barren areas and on shores; habits gregarious, migratory, feeding on invertebrates; nest solitary, on ground.

albatross Diomedeidae *q.v.*

albinism The absence of pigmentation in animals; partial albinism in plants is termed variegation.

Albulidae Bonefishes; family containing 4 species of tropical marine teleosts found mainly in shallow coastal waters; body elongate, herring-like, snout produced; feed primarily on benthic molluscs and crustaceans.

Alcedinidae Kingfishers; family of small colourful woodland birds (Coraciiformes) that feed on fishes and other aquatic vertebrates by diving from flight, or on insects caught on the wing; habits solitary, monogamous, non-migratory; nesting in burrows in banks or in tree holes; contains about 90 species distributed worldwide but most diverse in Old World tropics.

kingfisher (Alcedinidae)

Alcidae Auks, puffins; family containing 23 species of diving seabirds (Charadriiformes) found in the northern hemisphere; wings short and rounded offering poor flight but efficient underwater swimming; habits gregarious, migratory; feeding mostly on fishes and crustaceans.

puffin (Alcidae)

Alcyonacea Soft corals; order of colonial octocorals distributed mainly in shallow waters of tropical seas but extending into polar seas; body form varying from encrusting to erect and branching; packing coenenchyme thick and filled with spicules; colony usually without polyps on basal portion.

Alcyonaria Octocorals; subclass of Anthozoa characterized by 8 hollow, marginal tentacles and 8 complete mesenteries dividing the gastrovascular cavity; found mainly fixed to hard substrates in shallow tropical waters but occasionally found anchored in soft sediment and in deep or cold water.

alder Betulaceae *q.v.*

alderfly Sialidae *q.v.* (Neuroptera).

Alepisauridae Lancetfishes; family containing 3 species of large voracious myctophiform teleosts found in the deep sea; body slender, naked and with large sail-like dorsal fin; teeth dagger-like.

Alepocephalidae Slickheads; family containing about 35 species of benthic and pelagic deep-sea salmoniform teleost fishes; body elongate, to 700 mm length; scales may be absent; adipose fin lacking; dorsal and anal fins positioned posteriorly.

Alestidae Tetras; family containing about 100 species of mostly small tropical African freshwater characiform teleost fishes included by some authorities in the Characidae; larger species are predatory, smaller forms usually omnivorous; often kept by aquarists.

aletophilous Thriving on roadside verges and beside railway tracks; **aletophile, aletophily.**

Aleutian Current A warm surface ocean current that flows north and west off the coast of Alaska, derived as a deflection of the North Pacific Gyre; Alaska Current; see ocean currents.

Aleyrodidae Whiteflies; family of small, actively flying insects (Homoptera) which have their forewings and bodies covered with a waxy, white powder; immature stages typically live on undersides of leaves, feeding on plant juices, and are commonly attended by ants as they produce honeydew; contains about 1200 species.

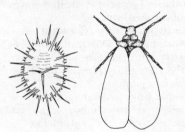

whitefly (Aleyrodidae)

alfalfa Fabaceae *q.v.*

alfonsino Berycidae *q.v.*

algae Eukaryotic algae; an informal assemblage of photosynthetic non-vascular organisms which differ from more advanced plants in their lack of multicellular sexual organs sheathed with sterile cells, and by their failure to retain the embryonic sporophyte within the female organ.

algal bloom An explosive increase in the density of phytoplankton within an area.

algal line The highest continuous line on the shore along which any particular algal species occurs; *cf.* tang line.

algal wash Shoreline drift, comprising mainly filamentous algae.

algicolous Living on algae; **algicole.**

algology The study of algae.

Algonkian Proterozoic *q.v.*

algophagous Feeding on algae; **algophage, algophagy.**

Algophytic Archaeophytic *q.v.*

alien Non-native; a species occurring in an area to which it is not native.

alima Larval stage of stomatopods belonging to the genus, *Squilla*; a type of erichthus larva.

alimentation 1: Feeding, taking in nourishment. 2: Those processes, including precipitation, sublimation and refreezing, that serve to increase the mass of a glacier or snowfield.

Alismataceae Arrowhead, water plantain; cosmopolitan family containing about 75 species of perennial aquatic or marsh-inhabiting herbs.

arrowhead (Alismataceae)

Alismatales Order of Alismatidae, containing 3 families of perennial, aquatic or semiaquatic herbs with alternate leaves clustered near base; flowers usually with 3 green sepals and 3 white petals.

Alismatidae Archaic subclass of monocotyledons (Liliopsida) containing 4 orders of herbs, either aquatic, or of wet places, sometimes lacking chlorophyll and mycotrophic in habit; flowers with 3 sepals and 3 petals or reduced, producing seeds which usually lack endosperm.

alkaline Pertaining to habitats or substances having a pH greater than 7; basic.

alkalinity 1: The properties of an alkali. 2: A measure of the pH of sea water, calculated as the number of milliequivalents of hydrogen ion that is neutralized by one litre of sea water at 20°C.

alkaliphilic Thriving in alkaline habitats; **alkaliphile, alkaliphily**; *cf.* alkaliphobic.

alkaliphobic Intolerant of alkaline habitats or conditions; **alkaliphobe, alkaliphoby**; *cf.* alkaliphilic.

alkaloduric Extremely tolerant of high pH (alkaline) conditions.

alkaloid A nitrogenous and often poisonous organic compound, produced especially by flowering plants; many alkaloids such as nicotine, caffeine, morphine. atropine, cocaine, quinine and strychnine have a pronounced physiological activity in animals.

allele Any of the different forms of a gene occupying the same locus on homologous chromosomes, and which undergo pairing during meiosis, and can mutate one to another; allelomorph; **allelic, allelism.**

allele frequency Gene frequency *q.v.*

allelochemic A secondary substance produced by an organism that has the effect of modifying the growth, behaviour or population dynamics of other species, often having an inhibitory or regulatory effect (allelopathic substance).

allelogenic Producing offspring in broods that are entirely of one sex; **allelogeny**; *cf.* amphogenic, arrhenogenic, monogenic, thelygenic.

allelomorph Allele *q.v.*

allelopathic substance An allelochemic; a waste product, excretory product or metabolite having an inhibitory or regulatory effect on other organisms.

allelopathy The chemical inhibition of one organism by another.

Allen's Law The generalization that the extremities (ears and tails for example) of mammals tend to be relatively shorter in colder climates than in warmer ones.

alligator Alligatoridae *q.v.*

Alligatoridae Alligators, caiman; family containing 7 species of freshwater crocodilians with broad, flat snouts; fourth teeth of lower jaw fitting into pit in upper jaw; often included in the family Crocodylidae.

allocryptic Used of organisms that conceal themselves under a covering of other material, living or non-living.

allochthonous Exogenous; originating outside

and transported into a given system or area; **allochthony**; *cf.* autochthonous.

allochronic Not contemporary; existing at different times; used of populations or species living, growing or reproducing during different seasons of the year; *cf.* synchronic.

allochroic Exhibiting colour variation, or having the ability to change colour.

allochoric Occurring in two or more communities within a given geographical region; **allochore.**

allochemic Any secondary compound produced by plants as part of their defence mechanism against herbivores; acting either as a toxin or digestibility reducer.

allobiosphere That part of the biosphere in which heterotrophic organisms occur but into which organic food material must be transported as primary production does not take place; *cf.* autobiosphere.

allogamy Cross fertilization; **allogamous.**

allogenetic plankton Planktonic organisms transported into an area by movement of the medium, but normally living and reproducing elsewhere.

allogenic 1: Used of factors acting from outside the system, or of material transported into an area from outside. 2: Having different sets of genes; **allogenetic**; *cf.* syngenic.

allogrooming Grooming by one individual of another.

alloiogenesis Alternation of generations; an alternation between sexual and asexual phases in a life cycle.

allokinesis Passive or involuntary movement; drifting; **allokinetic.**

allomaternal A female alloparent *q.v.*

allometric growth Differential growth of body parts resulting in a change of shape or proportion with increase in size; **allometry**; *cf.* isometric growth.

allomixis Cross fertilization; **allomictic.**

allomone A chemical substance produced and released by one species in order to communicate with another species; *cf.* pheromone.

alloparasite An organism parasitic on an unrelated host organism; *cf.* adelphoparasite.

alloparent An individual that assists a parent in the care of its young, either male (allopaternal) or female (allomaternal).

allopaternal A male alloparent *q.v.*

allopatric Used of populations, species or taxa occupying different and disjunct geographical areas; **allopatry**; *cf.* dichopatric, parapatric, sympatric.

allopatric speciation The differentiation and reproductive isolation of populations that are geographically separated.

allopatry Spatial separation; disjunction; **allopatric.**

allopelagic Used of organisms occurring at any depth in the pelagic zone; *cf.* autopelagic.

allophilous Used of a plant that lacks morphological adaptations for attracting and guiding pollinators; **allophily;** *cf.* euphilous.

allopolyploid A polyploid hybrid having chromosome sets derived from two different species or genera; **alloploid;** *cf.* autopolyploid.

allosematic Pertaining to coloration or markings that imitate warning patterns of other typically noxious or dangerous organisms.

allosome Any chromosome or chromosome fragment other than a normal A-chromosome.

allotherm An organism having a body temperature determined largely by the ambient temperature; *cf.* autotherm.

allotopic Used of populations or species that occupy different macrohabitats; **allotopy;** *cf.* syntopic.

allotrophic 1: Obtaining nourishment from another organism; heterotrophic *q.v.* 2: Pertaining to the influx of nutrients into a water body or ecosystem from outside.

allotrophic lake A lake receiving organic materials by drainage from the surrounding land; *cf.* autotrophic lake.

allotropism 1: The mutual attraction of cells, especially gametes; **allotropic.** 2: The condition of a flower having a plentiful supply of readily available nectar.

allotropous Used of unspecialized insect species that are able to feed on a variety of kinds of flowers; *cf.* eutropous.

alluvial deposit A silty deposit transported by water; *cf.* aeolian deposit, colluvial deposit.

Alluvial soil An azonal soil formed from a relatively unmodified recent alluvial deposit in flood plains and deltas.

alluviation The deposition of sediment by a river at any point along its course.

Aloeaceae Aloe, red-hot poker; family of Liliales comprising about 700 species of coarse plants with some secondary growth of the monocotyledonous type, native to Africa, Arabia and some nearby islands; plants from virtually stemless herbs to arborescent with clusters of succulent leaves crowning the branches or at ground level.

aloe (Aloeaceae)

Alopiidae Thresher sharks; widespread family of large (to 6 m), active but inoffensive, pelagic lamniform elasmobranch fishes, comprising 3 species; upper lobe of caudal fin extremely long, used to manoeuvre and stun prey which includes pelagic fishes, cephalopods and crustaceans.

thresher shark (Alopiidae)

alpestrine Living at high altitude above the tree line, commonly used as synonymous with alpine; occasionally used of plants found below the tree line (subalpine plants).

alpine 1: Pertaining to the Alps. 2: Used of habitats and organisms found between the tree line and snow line in mountainous regions.

Alpine Meadow soil An intrazonal soil with a dark coloration, formed under meadow grass above the tree line in alpine regions; typically shallow and stony with thin litter and duff; A-horizon granular and strongly acidic, B-horizon absent, and C-horizon gleyed due to poor drainage.

Alpine Turf soil An intrazonal soil with a dark yellow-brown coloration formed under alpine turf; typically a well-drained soil with thin litter and duff; A-horizon stony and strongly acidic, B-horizon stony and moderately acidic.

Alseuosmiaceae Small family of Rosales containing about 12 species of shrubs native to New Caledonia and New Zealand.

alsocolous Living in woody groves; **alsocole.**

alsophilous Thriving in woody grove habitats; **alsophile, alsophily.**

alternation of generations The alternation of generations having different reproductive processes, typically of sexual (diploid) and asexual (haploid) phases, in the life cycle of an organism.

Altithermal period A postglacial interval (*c.* 7500–4000 years B.P.) characterized by distinctly warmer conditions than at present; *cf.* Anathermal period, Medithermal period.

altitude The vertical distance between a given point and a datum surface, or chart datum; elevation.

altricial Used of offspring or of species that show a marked delay in the attainment of independent self-maintenance; **altrices**; *cf.* precocial.

altruism The situation in which one individual acts to promote or enhance the fitness of an unrelated individual or of other members of a group at the same time reducing its own fitness; **altruistic.**

Alucitidae Small family of moths (Lepidoptera) in which the wings are divided into 6 or more plumes; larvae burrow into plants and may induce formation of galls.

Amanitaceae Fly agarics; family of mushroom-like fungi (Agaricales) in which the cap can be broken cleanly from the stalk; gills not attached to stalk; includes many poisonous forms.

Amaranthaceae Amaranths, cockscomb; family of Caryophyllales containing about 900 species of mostly herbs, with a cosmopolitan distribution in warm regions; typically with regular flowers either solitary or in axillary cymes; flowers with 1 to 5 free stamens and a superior ovary.

Amaryllidaceae Family of monocotyledons usually included in the Liliaceae (Liliales *q.v.*); sometimes separated on the basis of their inferior ovary.

amathophilous Thriving in sandy plains; **amathophile, amathophily.**

amber snail Stylommatophora *q.v.*

ambient Surrounding; background.

ambisexual 1: Having separate male and female flowers on the same plant. 2: Pertaining to both sexes.

ambivalence Behaviour resulting from two incompatible motivations, often taking the form of a mixture of the two motivational tendencies; displacement activity.

ambivorous Feeding on grasses and broad-leaved plants; **ambivore, ambivory.**

Amblycipitidae Family of small freshwater catfishes (Siluriformes) found in fast-flowing rivers of India, Burma and Thailand; body slender, naked; dorsal fins short with weak spine; 4 pairs of barbels present; gas bladder partly enclosed by bony plates.

Amblyopsidae Cavefishes; family containing 6 species of small (to 100 mm) North American freshwater teleosts (Percopsiformes) found in cave systems or shallow coastal swamps; body slender, eyes and pigment absent in cave-dwelling forms; single dorsal fin present, pelvic fins reduced or absent; eggs brooded by female in spacious branchial chamber.

Amblypoda Order including a variety of large, primitive ungulates belonging to the suborders Pantodonta, Dinocerata, Xenungulata and Pyrotheria.

Amblypygi Tailless whip scorpions, whipspiders; order of predatory nocturnal terrestrial arthropods (Arachnida) found mainly in forest litter, in crevices, and beneath stones in tropical and subtropical regions; a few of the 70 species are cavernicolous; cephalothorax and abdomen joined by narrow waist; chelicerae with proximal fang, pedipalps spinose and raptorial, first pair of legs whip-like.

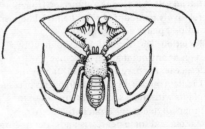

tailless whip scorpion (Amblypygi)

daffodil (Amaryllidaceae)

amaryllis Amaryllidaceae *q.v.*

amathicolous Inhabiting sandy plains; **amathicole.**

Amborellaceae Family of flowering plants containing a single species of evergreen shrub from New Caledonia; uniquely amongst the Laurales the wood has no vessels.

ambulatorial Adapted for walking.

Ambystomatidae Family containing about 30 species of mostly terrestrial, North American salamanders which have internal fertilization and lay eggs in ponds and streams; fore and hind limbs, and lungs present; larvae with long filamentous gills, sometimes neotenic; includes axolotl and mud salamanders.

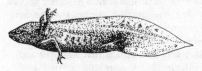

axolotl (Ambystomatidae)

ameiotic parthenogenesis Parthenogenesis *q.v.* in which meiosis has been suppressed so that neither chromosome reduction nor any corresponding phenomenon occurs.

amensalism An interspecific interaction in which one species population is inhibited, typically by toxin produced by the other, which is unaffected.

ametabolous Used of a pattern of development in which the offspring gradually assume adult characters through a series of moults; **ametabolism**; *cf.* hemimetabolous, holometabolous.

ametoecious 1: Used of a parasite having only a single host during its life cycle. 2: Used of a parasite species that is highly host specific; **ametoecius**; *cf.* metoecious.

amictic 1: Pertaining to females that produce unfertilized eggs that develop into offspring of one sex only. 2: Used of a lake that has no period of overturn because it is perennially frozen; *cf.* mictic.

Amiidae Bowfin; primitive family of North American freshwater fishes found in still weedy backwaters; body with cycloid scales, mouth large with strong teeth; caudal fin heterocercal; gas bladder serves as accessory respiratory organ.

Amiiformes Order of primitive bony fishes (Osteichthyes) comprising a single extant family, the Amiidae.

amino acid The basic structural component from which proteins are contructed; each comprising carboxylic acid and amino groups.

amixis Absence or failure of interbreeding.

ammochthophilous Thriving on sand banks; **ammochthophile, ammochthophily.**

ammocoete Freshwater larval stage of a lamprey (Petromyzoniformes).

ammocolous Living or growing in sand; **ammocole.**

Ammodytidae Sand lances; family containing 12 species of bottom-dwelling marine teleost fishes (Perciformes) usually found in large schools over sandy sediments from low water to 150 m; body elongate, head pointed with lower jaw prolonged; teeth and pelvic fins absent.

sand lance (Ammodytidae)

ammonification The release of ammonia from nitrogenous organic material by the action of microorganisms.

ammonite Ammonoidea *q.v.*

Ammonoidea Ammonites; extinct subclass of cephalopod molluscs which typically possessed planispiral septate shells; siphuncle generally ventral; sutures often complex; comprised the largest subclass of cephalopods with over 160 families found extensively from the Devonian to the Upper Cretaceous.

ammonite (Ammonoidea)

ammophilous Thriving in sandy habitats; **ammophile, ammophily.**

amnicolous Living on sandy river banks; **amnicole.**

Amniota Vertebrate group comprising reptiles, birds and mammals characterized by the possession of extra embryonic membranes; *cf.* Anamniota.

amoeba A single-celled eukaryotic organism that is naked and moves by means of pseudopodia; Amoebida *q.v.*

Amoebida Amoebae; order of protozoans (Gymnamoeba) which possess mitochondria

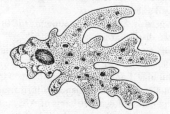

amoeba (Amoebida)

but lack a flagellate phase; includes the amoeba and the entamoebae, which cause amoebic dysentery; subdivided into 5 suborders, Acanthopodina, Conopodina, Flabellina, Thecina and Tubulina, each of which is also treated as a class of the protoctistan phylum Rhizopoda.

Amoebidiales Order of trichomycete fungi containing two genera of freshwater symbiotic forms typically found in the rectum of crustaceans or larval insects but occasionally attached to external surface of host; once classified as protozoans as they lack both cellulose and chitin in their cell walls.

amphiatlantic Occurring on both sides of the Atlantic Ocean.

Amphibia Class of quadruped vertebrates containing frogs, newts, salamanders, toads, caecilians and many fossil groups; skin glandular without epidermal scales, feathers or hairs; tail present primitively but lost in some groups; limbs or girdles reduced or absent in some forms; eggs anamniotic, primitively laid in water with external fertilization; tadpole larvae possess gills and open gill slits; ovoviviparity and viviparity exhibited by some species; contains a single extant subclass, Lissamphibia.

amphibious Adapted to life on land as well as in water; **amphibian.**

amphiblastula Free-swimming larval stage of sponges, possessing some external flagellate cells.

amphicarpogean Producing fruits above ground that are subsequently buried; **amphicarpogenous**; *cf.* hypocarpogean.

amphichromatism The occurrence of different coloured flowers on individual plants in different seasons; **amphichromatic.**

amphicryptophyte A marsh plant with amphibious vegetative parts.

Amphidiscophora Subclass of hexactinellid sponges which are rooted to the substratum by a basal tuft of anchoring spicules.

amphigamy Normal cross fertilization; the fusion of male and female gametes.

amphigean Used of a plant having underground as well as aerial flowers.

amphigenesis Sexual development; the fusion of two dissimilar gametes.

amphigony Sexual reproduction involving cross fertilization.

Amphiliidae Family containing about 50 species of small freshwater catfishes (Siluriformes) found in fast-flowing streams of Africa; body flattened ventrally, sometimes with bony

armature; mouth suctorial with 2–3 pairs of barbels.

Amphilinidea Order of cestodarian tapeworms found in turtles and fishes; typically with a flattened, elongate body and an indistinct anterior holdfast; uterine pore positioned near anterior end of body.

amphimixis True sexual reproduction; the union of male and female gametes; may be either autogamy (inbreeding) or allogamy (outbreeding); **amphimict, amphimictic.**

Amphineura Class of elongate, bilaterally symmetrical molluscs that lack eyes and tentacles; shell, if present, consisting of 7 or 8 overlapping calcareous plates on the dorsal surface; head poorly differentiated; posterior mantle cavity contains anus and gills; comprises the subclasses Polyplacophora and Aplacophora, which are usually treated as separate classes.

Amphinomida Fireworms; order of errant polychaete worms containing about 165 species; pharynx cylindrical, eversible, unarmed; parapodia biramous with simple setae; cirri and branchiae present.

amphioecious Used of a population or species showing broad variable tolerance of habitat and environmental conditions; *cf.* euryoecious, stenoecious.

Amphionidacea Order of eucaridan crustaceans comprising a single planktonic species found in open oceans to a depth of about 1700 m; body shrimp-like with a brood pouch, one pair of maxillipeds, and reduced thoracic legs and gills; feeding limbs and gut non-functional in the adult female.

amphioxus Cephalochordata *q.v.*

amphiphyte A plant able to live either rooted in damp soil above the water level or completely submerged.

Amphipoda Sandhoppers; order of usually small peracaridan crustaceans found primarily free-living in aquatic habitats from freshwater to marine, although some are semiterrestrial or parasitic; body usually laterally

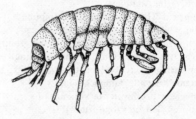

sandhopper (Amphipoda)

compressed, occasionally linear or globular; comprises about 6000 species in 4 suborders: Gammaridea (sandhoppers, scuds, beach fleas), Hyperiidea (marine pelagic forms), Caprellidea (skeleton shrimps, whale lice) and Ingolfiellidea (hypogean forms).

Amphisbaenia Worm lizards; suborder of small, limbless, fossorial reptiles (Squamata) most widely distributed in Africa and South America; body annulated, eyes concealed, external ear openings absent; contains about 135 species in 2 families; may be oviparous or ovoviviparous.

Amphisbaenidae Family of burrowing worm lizards (Amphisbaenia) found mainly in tropical and subtropical America and Africa but also in southwestern Asia and the Iberian Peninsula; contains about 130 species, most feeding on soil arthropods; with small wedge-shaped head used for digging.

amphistylic The condition in which both the hyomandibular and the palatoquadrate arch are involved in the suspension of the lower jaw in fishes; *cf.* autostylic, holostylic, hyostylic.

amphitoky Parthenogenesis in which both male and female offspring are produced.

amphitopic Used of a population or species showing broad, variable tolerance of habitat and environmental conditions; *cf.* eurytopic, stenotopic.

amphitrophic Used of an organism that lives as a phototroph during daylight and as a chemotroph in darkness; **amphitroph.**

Amphiumidae Family containing 3 species of eel-like salamanders (Caudata) from southeastern United States; fore- and hindlimbs greatly reduced, lungs and a single pair of gill slits (spiracles) present; fertilization internal; egg strings laid on mud, guarded by female.

amphogenic Producing male and female offspring in approximately equal numbers; **amphogenous, amphogeny**; *cf.* allelogenic, arrhenogenic, monogenic, thelygenic.

Amphycyonidae Dog-bears; extinct family of large, heavily built dog-like carnivores found in northern hemisphere during the Oligocene.

amplectant Clasping or twining for support.

amplexus Precopulation pairing; the grasping of the female by the male prior to copulation.

amplitude The range of tolerance to environmental conditions of an organism or species; ecological amplitude.

Ampulicidae Digger wasps; family containing about 160 species of primitive hunting wasps found mainly in the tropics; antennae inserted low on face; females hunt and paralyse cockroaches with their stings, using them to provision their nests.

Amynodontidae Extinct family of rhinoceros-like mammals (Ceratomorpha) known from the Eocene to the Miocene; similar to the hippopotamus in size and shape and probably also lived in rivers.

Anabantidae Climbing gouramis; family containing 40 species of freshwater perciform teleost fishes found in Africa, India and Indonesia which can walk on land using the spiny margins of gill opercula; able to utilize atmospheric oxygen through an accessory respiratory structure, the suprabranchial organ.

Anabantoidei Gouramis; suborder of freshwater perciform teleosts comprising about 70 species in 4 families found in Africa, India and southeast Asia.

anabiosis A state of greatly reduced metabolic activity assumed during unfavourable environmental conditions; **anabiotic.**

Anablepidae Four-eyed fishes; family containing 3 species of surface-living insectivorous nocturnal fresh and brackish water cyprinodontiform teleosts from Central and South America; anal fin of male modified as gonopodium for sperm transfer; eyes highly specialized for simultaneous aerial and aquatic vision with 2 pupils and a divided retina.

four-eyed fish (Anablepidae)

anabolism That part of metabolism involving the manufacture of complex substances from simpler substrates with the consequent utilization of energy; **anabolic**; *cf.* catabolism.

Anacanthobatidae Family of deep-water rajiform fishes containing about 15 species; body shape subcircular to rhomboidal; snout long and pointed; tail slender; dorsal fins absent.

Anacardiaceae Poison ivy, poison oak, sumach, smoke tree, mango, pistachio and cashew nut; family of Sapindales containing about 600 species of woody plants often producing allergenic reactions, or poisonous to the touch; often producing waxy or oily seeds; widespread in tropical regions; flowers small and regular, with 3–5 sepals and petals, 3–10 stamens; the fruit is commonly a drupe.

anachoric Living in fissures, holes and crevices; **anachoresis.**

anaclinotropism A tropism resulting in growth or movement in a horizontal plane; **anaclinotropic.**

anaconda Boidae *q.v.*

anadromous Migrating from salt to fresh water, as in the case of a fish moving from the sea into a river to spawn; **anadromy**; *cf.* catadromous, oceanodromous, potamodromous.

anaerobic 1: Growing or occurring in the absence of molecular oxygen; **anaerobe, anaerobiosis**; *cf.* aerobic. 2: Used of an environment in which the partial pressure of oxygen is significantly below normal atmospheric levels; deoxygenated.

anaerobiotic Used of habitats devoid of, or significantly depleted in, molecular oxygen; **anaerobiosis.**

anaerogenic Not producing gas; used of microorganisms that do not produce visible gas during the breakdown of carbohydrate; *cf.* aerogenic.

analogous 1: Pertaining to similarity of function, structure or behaviour due to convergent evolution rather than to common ancestry; **analogue, analogy**; *cf.* homologous. 2: Used of organisms having similar habitats or distributions.

Anamniota Vertebrate group comprising the lower classes (fishes and amphibians) that lack complex extraembryonic membranes and have only a trilaminate yolk sac; *cf.* Amniota.

Anamorpha Subclass of chilopods in which young hatch from egg with only 7 pairs of legs and pass through several moult stages at which trunk segments and limbs are added; adult with 19 trunk segments and 15 pairs of legs; eggs laid singly and not brooded; comprises 2 orders, Scutigerida and Lithobiida.

anaplasia 1: The progressive phase in the development of an organism, prior to attaining maturity. 2: An evolutionary state characterized by increasing vigour and diversification of organisms; **anaplasis**; *cf.* cataplasia, metaplasia.

Anapsida Subclass of Reptilia containing a single living order, Testudines (turtles and tortoises); characterized by a skull which lacks apertures in the temple regions behind the eyes.

Anaptomorphidae Extinct family of tarsier-like primates (Prosimii) with large eyes and a short face; olfactory regions of brain better developed than modern tarsiers; found in Europe and America from the Palaeocene to the Oligocene.

Anarhichatidae Wolffishes; family containing 7 species of bottom-living marine blennioid teleosts (Perciformes), found at moderate depths in the North Pacific and Atlantic; body elongate, to 2.5 m, scales present or absent; dorsal fin spinose, pelvics absent, pectorals well developed.

Anaspidacea Order containing about 15 species of freshwater syncaridan crustaceans found in surface and subterranean waters of the southern hemisphere (Australia, New Zealand and southern Africa).

Anaspidea Sea hares; order of herbivorous opisthobranch molluscs typically found on seaweed in shallow waters; with a small, more or less internal shell and ear-shaped rhinophores on the head; sometimes producing coloured ink which can be expelled from the mantle cavity.

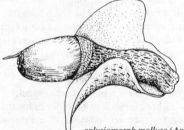

aplysiomorph mollusc (Anaspidea)

Anaspidiformes Extinct order of jawless, fish-like vertebrates (Cephalaspidimorphi); body fusiform, with hypocercal tail and up to 15 pairs of external gill openings; known from the Silurian and Devonian.

anastomosis The union of branches to form a network.

anathermal Pertaining to a period of rising temperature; *cf.* catathermal.

Anathermal period A postglacial interval (*c.* 10 000–7500 years B.P.) characterized by rising temperatures following the last glaciation; *cf.* Altithermal period, Medithermal period.

Anatidae Ducks, geese, swans; diverse family

duck (Anatidae)

containing about 145 species of small to large water birds (Anseriformes), cosmopolitan in freshwater and coastal marine habitats; feed on wide variety of plants and animals; most are monogamous and solitary during breeding; nest on ground or in trees; some important as game or domesticated birds.

anatomical Pertaining to the structure of organisms, especially as revealed by dissection; **anatomy.**

anatropism The inversion of a seed or ovule so that the micropyle is bent downwards; **anatropic.**

anautogenous Used of a female insect that must feed before it can produce mature eggs; *cf.* autogenous.

ancestor A progenitor of a more recent descendent taxon, group or individual; any preceding member of a lineage; **ancestral.**

ancestrula Primary zooid from which a bryozoan colony is formed by budding.

anchialine Maritime; used of coastal salt-water habitats having no surface connection to the sea.

anchor plant Rhamnaceae *q.v.*

anchovy Engraulidae *q.v.*

anchusa Boraginaceae *q.v.*

Ancistrocladaceae Small family of Violales containing about 20 species of shrubs climbing by hooked or twining branch tips, native to tropical Africa and Asia.

ancocolous Living in canyons; **ancocole.**

ancophilous Thriving in canyon forests; **ancophile, ancophily.**

Ancylopoda Chalicotheres; extinct suborder of Holarctic horse-like mammals (Perissodactyla) known from the Eocene to Pleistocene; possessed 3 toes which probably bore claws rather than hooves; long forelimbs possibly used for digging up roots.

Andreaeopsida Granite mosses; small class of ancient bryophytes distributed worldwide but most abundant on rocky substrates in polar or cold temperate regions; characterized by the dehiscence of the spore capsule along longitudinal slits and by the small dark coloured, epilithic, tufted or cushion-like growth form of the gametophyte.

Andrenidae Family containing over 4000 species of generalized, soil- nesting bees (Hymenoptera) which produce a wax-like lining to nest cells; feeding on nectar and pollen as both adults and larvae.

andric Male; *cf.* gynic.

androchorous Dispersed by the agency of man; **androchore, androchory.**

androcyclic parthenogenesis Reproduction in which a series of parthenogenetic generations is followed by the production, by a portion of the population only, of males as a sexual generation *cf.* parthenogenesis.

androdioecious Used of plant species having male and hermaphrodite flowers on separate plants; **androdioecy**; *cf.* gynodioecious.

androecious Used of a plant having male flowers only; *cf.* gynoecious.

androecium The male component of the flower, typically comprising a whorl of stamens surrounding the female component (gynoecium).

androgamy The impregnation of the male gamete by the female gamete.

androgenesis Male parthenogenesis; the development of an egg carrying paternal chromosomes only (patrogenesis) or the production of a haploid plant by the germination of a pollen grain within an anther; **androgenetic**; *cf.* gynogenesis.

androgenous Producing male offspring only; *cf.* gynogenous.

androgynous Having both male and female reproductive organs; hermaphrodite; **androgyne, androgynism.**

andromonoecious Having male and hermaphrodite flowers on the same plant; *cf.* gynomonoecious.

andromorphic Having a morphological resemblance to males; **andromorphy**; *cf.* gynomorphic.

androphilous Thriving in proximity to man; **androphile, androphily.**

androsome Any chromosome found only in a male nucleus.

androspore 1: A male spore; pollen grain. 2: An asexual zoospore giving rise to a dwarf male plant.

anebous Prepubertal; immature.

anecdysis The absence of moulting, or a prolonged intermoult period.

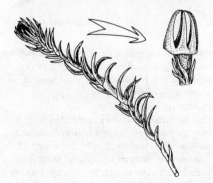

granite moss (Andreaeopsida)

anemochorous Having seeds, spores or other propagules dispersed by wind; **anemochore**, **anemochory**.

anemone Ranunculaceae *q.v.*

anemoneuston Terrigenous organic and inorganic material transported by wind to the surface of water bodies.

anemophilous 1: Dispersed by wind. 2: Pollinated by wind-blown pollen; **anemophile**, **anemophily**. 3: Thriving in sand draws.

anemophobic Intolerant of exposure to wind; **anemophobe**, **anemophoby**.

anemophyte 1: A plant occurring in windswept situations. 2: A plant fertilized by wind-blown pollen.

anemoplankton Wind borne organisms; aerial plankton.

anemotaxis A directed reaction of a motile organism towards (positive) or away from (negative) wind or air currents; **anemotactic**.

anemotropism Orientation response to an air current; **anemotropic**.

aneuploid Having a chromosome number that is more or less than, but not an exact multiple of (euploid), the basic chromosome number.

Angara A continental landmass comprising most of central and northern Asia, formed from Eurasia *q.v.* after the break up of Pangaea *q.v.*; *cf.* Cathaysia, Euramerica.

angel shark Squatinidae *q.v.*

angelfish Chaetodontidae *q.v.*

Angiospermae Flowering plants; the major division of seed plants (Spermatophyta); characterized by the production of seeds fully enclosed within fruits; commonly known as the Magnoliophyta *q.v.*

Angiosperms Flowering plants; Magnoliophyta *q.v.*

anglerfishes Lophiiformes *q.v.*

Anglian glaciation A glaciation of the Quaternary Ice Age *q.v.* in the British Isles with an estimated duration of 90 thousand years; Lowestoft glaciation.

angonekton Short-lived organisms inhabiting temporary pools on rocks, in tree stumps and similar sites.

Anguidae Slow worms; family containing about 75 species of terrestrial and arboreal lizards (Sauria) widely distributed in the Americas, Europe, North Africa and southern Asia; limbs well developed to absent; feeding on invertebrates and small vertebrates; tongue forked; eyelids mobile; tail autotomous; reproduction either oviparous or ovoviviparous.

Anguillidae Freshwater eels; family containing 16 species of small (to 1 m) anguilliform teleost fishes found in North Atlantic, Indian and western Pacific Oceans and adjoining freshwater systems; body cylindrical; dorsal, caudal and anal fins continuous, pectorals well developed; most species migrating to sea to spawn, the leptocephalus larval stage lasting up to 3 years; important in commercial fisheries.

Anguilliformes Order of mostly marine, eel-like, teleost fishes in which the pelvic fins and girdle are absent; body smooth or bearing simple cycloid scales; development includes leptocephalus larval stage; contains 2 suborders, Anguilloidei (freshwater, spaghetti, snipe, moray, conger, cutthroat, arrowtooth, pike conger, duckbill, snake, worm, and longneck eels) and Saccopharyngoidei (deep-sea gulper eels).

anheliophilous Thriving in diffuse sunlight; **anheliophile**, **anheliophily**.

Anhimidae Screamers; family containing 3 species of South American water birds (Anseriformes) found in marshes or wet grasslands; long-legged, short-necked; wings broad with a spur on leading edge; feed on plants; nest on ground near water.

Anhingidae Anhingas, darters; small family of fish-eating aquatic birds (Pelecaniformes), cosmopolitan in tropical and warm temperate wooded lakes, swamps and rivers; feeding by diving from water surface, spearing fishes with long bill; nest in colonies in trees.

anhinga (Anhingidae)

anholocyclic parthenogenesis Reproduction by parthenogenesis alone, as when the sexual reproduction phase has been lost from an alternation of generations cycle; *cf.* parthenogenesis.

anhydrobiosis Dormancy induced by low humidity or by desiccation; **anhydrobiose**.

Aniliidae Pipe snakes; family containing 10 species of subterranean snakes (Serpentes) from South America and southeast Asia; eyes small, not covered by scales; belly scales enlarged; often treated as part of the Typhlopidae *q.v.*

animal Any member of the kingdom Animalia.

animalcule An archaic term for any microscopic organism.

animalculism The belief that the embryo was contained within the sperm cell.

Animalia Kingdom comprising all multicellular eukaryotic organisms exhibiting holozoic nutrition and having capacity for spontaneous movement and rapid motor response to stimulation.

anise Apiaceae *q.v.*

anisogametic Producing gametes of dissimilar size, shape or behaviour; *cf.* isogametic.

anisogamy The fusion of gametes of dissimilar size, shape or behaviour; *cf.* isogamy.

anisomorphic Differing in shape or size; **anisomorphy.**

Anisophylleaceae Small family of Rosales containing about 40 species of trees or shrubs from tropical forests.

anisoploidy The condition of having an odd number of chromosome sets in somatic cells; **anisoploid.**

Anisoptera Dragonflies; suborder of active predatory insects (Odonata) in which the long slender wings are held horizontal at rest; larvae aquatic, body typically broad with depressed abdomen and rectal gills.

Anisozygoptera Small suborder of paleopterous insects (Odonata), comprising a single family and genus, exhibiting transitional characters between damselflies (Zygoptera) and dragonflies (Anisoptera).

Ankylosauria Ankylosaurs; a group of dinosaurs having body heavily armoured with bony plates; head small, teeth reduced and sometimes absent; known from the Cretaceous.

ankylosaur (Ankylosauria)

anlage A rudiment or primordium from which a given organ or structure develops during ontogeny.

Annelida Ringed worms; phylum of bilaterally symmetrical coelomate worms comprising 3 classes, Polychaeta (bristleworms), Oligochaeta (earthworms, pot worms, sludge worms) and Hirudinoidea (leeches); body divided into cylindrical rings (segments) with serially arranged organs; segments may have lateral lobes (parapodia) bearing variety of chaetae and bristles.

polychaete worm (Annelida)

Anniellidae Small family of limbless, burrowing lizards (Sauria) from California, containing 2 species; body length 250 mm; reproduction viviparous.

Annonaceae Pawpaw, custard apple; largest family of the order Magnoliales, containing about 2300 species found mostly in the tropics; flowers often fragrant, each with 3 sepals, 6 petals and many stamens, carpels superior; fruits typically fleshy.

annotinous Used of growth added during the previous year; one year old.

annual Having a yearly periodicity; living for one year; *cf.* biennial, perennial.

annual turnover The total biomass produced in one year.

Anobiidae Furniture beetles, death-watch beetle, woodworm; family of small brown beetles (Coleoptera) in which the head is hooded by the pronotum; typically feeding on dead wood; contains about 1600 species distributed worldwide; larvae C-shaped, white and fleshy with tiny legs; bore into wood and bark; adults emerge from wood leaving typical woodworm holes.

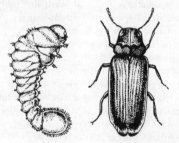

furniture beetle & woodworm (Anobiidae)

Anocheta Superorder of helminthomorphan diplopods (millipedes) comprising a single order, Spirobolida.

anoestrus A non-breeding period; **anoestrous**; *cf.* dioestrus, monoestrous, polyoestrous.

Anomalepidae Family containing 20 species of small (to 300 mm) worm-like, insectivorous, burrowing snakes (Serpentes) from tropical American rainforests.

Anomalodesmata Subclass of marine bivalve molluscs comprising about 500 species, which may be infaunal, burrowing, nestling or attached in habit; commonly found in the deep sea.

Anomalopidae Lanterneye; family of small (to 250 mm) pelagic beryciform teleost fishes, containing 4 species characterized by a well developed light organ associated with the orbit of each eye, apparently used for communication.

Anomaluridae Scaly-tailed squirrels; family containing 7 species of small arboreal mammals (Rodentia) found in tropical and subtropical forests of Ethiopian Region; tail long, hairy with characteristic patch of scales near its base; gliding membranes present between fore- and hindlimbs; feed on fruit, seeds and other plant material.

Anomura Hermit crabs, squat lobsters; infraorder of pleocyematan decapod crustaceans; body crab-like or lobster-like, abdomen extended, bent upon itself or flexed beneath thorax; last thoracic sternite free; some species utilize empty gastropod shells as portable refuges.

Anopla Class of nemerteans characterized by the possession of a separate mouth and proboscis pore; comprising 2 orders, Heteronemertea and Palaeonemertea.

Anoplogastridae Fangtooth; family of small mesopelagic beryciform teleost fishes with a deep body which is robust anteriorly, compressed posteriorly; teeth very large and fang-like.

Anoplopomatidae Sablefish, skilfish; family containing 2 species of bottom-living temperate North Pacific marine scorpaeniform teleosts widespread from shallow water to 1500 m.

Anoplura Sucking lice; order of small (0.5–4.0 mm) wingless hemipterodean insects that live exclusively as ectoparasites of placental mammals; mouthparts specialized for piercing and sucking; abdomen distensible; legs robust, modified for attachment; eggs usually cemented to body hair; contains about 500 species, including the human body louse, *Pediculus*.

Anostomidae Leporins, headstanders; family of small colourful South and Central American freshwater characiform teleost fishes; body elongate; dorsal fin anterior to mid-body; contains about 100 species, some of which adopt an oblique, head down posture (headstanders); popular amongst aquarists.

Anostraca Fairy shrimps; order of sarsostracan crustaceans containing about 180 species of suspension feeders distributed worldwide in ephemeral inland saline or freshwater bodies; characterized by an elongate body with up to 19 pairs of foliaceous limbs; includes the brine shrimp, *Artemia salina*.

fairy shrimp (Anostraca)

Anotopteridae Daggertooth, javelinfish; monotypic family of large (to 1 m) predatory myctophiform teleost fishes found in temperate and subtropical seas to 300 m depth; body very long and slender, naked; jaws strong, bearing large dagger-like teeth; dorsal fin and photophores absent.

anoxic Used of a habitat devoid of molecular oxygen; **anoxicity**; *cf.* normoxic, oligoxic.

Anseriformes Diverse order of water birds comprising 2 families, Anhimidae (screamers) and Anatidae (ducks, geese and swans).

ant Formicidae *q.v.* (Apocrita, Aculeata).

ant lion Myrmeleontidae *q.v.* (Neuroptera).

Antarctic Circumpolar Current A major cold surface current that flows eastwards through the Southern Ocean producing a circumglobal antarctic circulation; West Wind Drift; see ocean currents.

antarctic cod Nototheniidae *q.v.*

Antarctic convergence That region of the oceans of the southern hemisphere (at about 50° to 60° S) where cold antarctic surface water moving northeast meets and sinks below warmer water moving southeast.

Antarctic divergence The region of extensive upwelling close to Antarctica where water from intermediate depths replaces surface water drifting northeast in the Antarctic Circumpolar Current.

Antarctic kingdom One of the six major phytogeographical areas characterized by floristic composition; comprising New Zealand region, Patagonian region and South Temperate Oceanic island region.

Antarctogaea The Australian zoogeographical region, excluding New Zealand and Polynesia.

antbird Formicariidae *q.v.*

anteater Myrmecophagidae *q.v.*

antelope Bovidae *q.v.*

antenatal Before birth; during gestation.

Antennariidae Frogfishes; family containing about 60 species of voracious piscivorous anglerfishes (Lophiiformes) found on rocky substrates or amongst floating weed in shallow tropical and subtropical waters; body swollen, naked or warty, length to 350 mm; mouth oblique to vertical, teeth villiform, pectoral fin robust with elbow-like joint.

antennation The act of touching or probing with antennae.

antepisematic Pertaining to a character or trait aiding recognition and which implies a threat; *cf.* episematic.

anthecology The study of pollination and the relationships between insects and flowers.

anther The apical part of a stamen which produces the pollen or microspores.

antheridium The male sex organ of the lower plants, usually producing numerous motile, flagellate gametes (antherozoids).

anthesis The period of flowering; the time at which flower buds open; the period of maximum physiological activity in plants.

anthesmotaxis The organization and arrangement of the parts of a flower.

Anthocerotales Anthocerotopsida *q.v.*; the hornworts treated as an order of the class Hepaticae.

Anthocerotopsida Hornworts; a class of bryophytes containing a single family of prostrate thalloid plants which may have lobate or membranous margins but lack leaves and ventral scales; rhizoids smooth; sporophyte consists of a bulbous foot, a persistent meristematic zone and a capsule which releases spores by dehiscence.

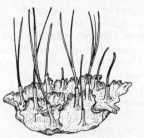

hornwort (Antherocerotopsida)

Anthocyathea Irregulares; extinct class of mostly solitary invertebrates (Archaeocyatha) known from the Cambrian; conical cup often irregular in outline, commonly with 2 porous walls.

anthogenesis The production of male and female offspring by asexual forms.

Anthomedusae Suborder of predominantly marine cnidarians (Hydroida) in which the polyps are typically colonial with a horny rather than calcareous exoskeleton; free medusae tall and bell-shaped with gonads usually on the stomach.

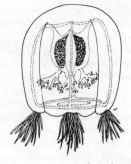

anthomedusa (Anthomedusae)

anthophilous Attracted to, or feeding on, flowers; **anthophile, anthophily.**

Anthophoridae Carpenter bees; large, diverse family containing over 4000 species of long-lived bees (Hymenoptera) which exhibit solitary, communal and primitive social behaviour; typically nesting in hollow stems or wood; producing unlined nest cells.

Anthophyta Magnoliophyta; flowering plants.

anthoptosis The fall of flowers.

anthotaxis The arrangement of the parts of a flower; **anthotaxy.**

anthotropism An orientation response of a flower to a stimulus; **anthotropic.**

Anthozoa Class of marine Cnidaria in which the medusa stage is absent and the polyp reproduces sexually as well as asexually; polyp solitary or colonial and typically benthic; polyp usually cylindrical containing a gastrovascular cavity divided longitudinally by radially arranged mesenteries; dispersal is commonly by means of planula larvae; comprises 3 subclasses, Alcyonaria, Ceriantipatharia and Zoantharia.

anthracobiontic Living or growing on a burned or scorched substratum.

Anthracotheriidae Extinct family of large pig-like mammals (Artiodactyla) known from the Eocene to the Pleistocene; possibly amphibious in habit; may have been ancestral to the hippopotamuses.

anthrageny The decomposition of plant material to form peat.

Anthribidae Coffee-bean weevil; family of beetles (Coleoptera) which are usually covered in a pattern of dark-coloured or white scales or flattened hairs; contains about 2600 species feeding on seeds, fungi, as predators or on crops of economic importance.

anthropic Pertaining to the influence of man; **anthropical.**

anthropocentric Interpreting the activities of

other organisms in relation to human values; **anthropocentrism.**

anthropochorous Having propagules dispersed by the agency of man; **anthropochore, anthropochoric, anthropochory.**

anthropogenesis The descent of man; the origin (phylogeny) and the development (ontogeny) of man.

anthropogenic Caused or produced through the agency of man; **anthropogenous.**

Anthropoidea Anthropoid apes; suborder of primates comprising the monkeys, apes and man; brain large; face mobile with forward-facing eyes, external ears small, lower incisors vertical; one pair of pectoral mammae present; digits usually with nails rather than claws; social behaviour often highly developed.

anthropology The study of man; **anthropological.**

anthropophilous 1: Thriving in close proximity to man; **anthropophile, anthropophily.** 2: Used of biting or blood-sucking insects that feed on man; **anthropophilic.**

Anthropozoic Geological time since the appearance of man.

Anthuridea Small order of isopod crustaceans with slender, elongate bodies.

antibiosis The antagonistic association between two organisms in which one species adversely affects the other, often by production of a toxin; **antibiotic.**

antibody An immunoglobulin serum protein that binds to a specific antigen.

antiboreal Pertaining to cool or cold temperate regions of the southern hemisphere; *cf.* boreal.

anticryptic Used of protective coloration facilitating attack or capture of prey; *cf.* procryptic.

anticyclonic Used of a region of high atmospheric pressure at sea level; also used of a wind system around such a high pressure centre that has a clockwise rotation in the northern hemisphere or an anticlockwise motion in the southern hemisphere; **anticyclone;** *cf.* cyclonic.

antigen A substance which induces formation of antibodies.

antigenic 1: Having the properties of an antigen; **antigenicity.** 2: Exhibiting sexual dimorphism.

Antilles Current A warm surface ocean current that flows north in the western North Atlantic forming the westerly limb of the North Atlantic Gyre; see ocean currents.

Antilocapridae Pronghorn; monotypic family of antelopes (Artiodactyla: Ruminantia) found in open arid habitats of the western Nearctic region; characterized by an unbranched, permanent horn core with the horn sheath branched and shed annually.

pronghorn (Antilocapridae)

Antipatharia Black corals, thorny corals; order of colonial ceriantipatharians found mainly in deeper waters in the tropics and subtropics; characterized by an erect branching form with a dark coloured, horny axial skeleton and small, hardly contractile polyps typically possessing 6 tentacles.

antiphyte That generation of a plant showing alternation of generations in which the gametes are produced.

antirrhinum Scrophulariaceae *q.v.*

antithetic Used of a life cycle exhibiting either an alternation of haploid and diploid generations or an alternation of morphologically dissimilar generations.

antitropism The orientation of growth in a plant so that it twines in the opposite direction from that followed by the sun across the sky; **antitropic;** *cf.* eutropism.

antizoea Early free-swimming larval stage of some stomatopods, possessing uniramous antennules and 5 pairs of biramous thoracic limbs, but lacking pleopods.

Anura Frogs and toads; large and diverse order of Amphibia comprising almost 3000 species

toad (Anura)

in 23 families; body length to 400 mm, tail absent, hind-limbs enlarged for jumping; habitats terrestrial and aquatic, including arboreal and fossorial forms; eggs and larvae (tadpoles) typically aquatic, but reproductive strategies variable, including ovoviviparity and viviparity.

apandrous Lacking, or having non-functional, male reproductive organs; **apandry.**

apatetic coloration A misleading coloration pattern, such as coloration resembling physical features of the habitat; camouflage.

ape Pongidae *q.v*

aperiodicity The irregular occurrence of a phenomenon; **aperiodic.**

Aphaniptera Older name for the Siphonaptera (fleas).

aphaptotropic Not influenced by, or responsive to, a touch or contact stimulus; **aphaptotropism.**

Aphelenchida Order of diplogasterian nematodes containing a variety of predators, fungal feeders, and parasites of higher plants and insects; often classified as a suborder of the Tylenchida.

apheliotropism An orientation response away from sunlight; negative heliotropism; **apheliotropic.**

aphercotropism An orientation response resulting in a turning away from an obstruction; **aphercotropic.**

aphid Aphididae *q.v.*

aphidicolous Living amongst aggregations of aphids; **aphidicole.**

Aphididae Aphids, blackflies, greenflies; family of over 3500 species of pear-shaped, soft-bodied insects (Homoptera) which move slowly over their host plants sucking juices and injecting toxic saliva; typically excreting honeydew or secreting wax for protection; life cycles complex, commonly involving both parthenogenetic and bisexual reproduction; includes many pests of agricultural plants.

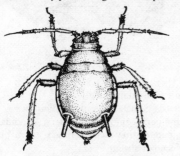

apple aphid (Aphidae)

aphidivorous Feeding on aphids; **aphidivore, aphidivory.**

aphotic Used of an environment or habitat having no sunlight of biologically significant intensity.

aphototaxis 1: The absence of a directed response to a light stimulus in a motile organism; **aphototactic.** 2: Negative phototaxis *q.v.*

aphototropism 1: An orientation response away from light; negative phototropism; **aphototropic.** 2: The absence of an orientation response to light.

Aphragmophora Cosmopolitan order of pelagic chaetognaths characterized by the absence of ventral musculature; contains about 45 species in 3 families.

Aphredoderidae Pirate perch; family of small (to 130 mm), nocturnal, eastern North American freshwater teleost fishes (Percopsiformes) inhabiting well vegetated lowland lakes and ponds; body compressed with single tall dorsal fin and rounded caudal fin; vent located near the throat in adults.

aphydrotaxis 1: The directed reaction of a motile organism away from moisture; **aphydrotactic.** 2: The absence of a directed response to moisture in a motile organism.

Aphytic The period of geological time before the appearance of plant life; *cf.* Archaeophytic, Caenophytic, Eophytic, Mesophytic, Palaeophytic.

Apiaceae Umbels, anise, dill, celery, caraway, carrot, coriander, fennel, parsley, hemlock, parsnip, sea holly, samphire; large, nearly cosmopolitan family containing 3000 species of aromatic, often poisonous herbs, shrubs and trees with flowers typically small and borne in compound umbels; individual flowers are regular and bisexual, and commonly have 5 very small sepals, 5 petals and stamens and an inferior ovary; formerly known as the Umbelliferae.

Apiales Order of Rosidae containing 2 families, Apiaceae and Araliaceae, of woody plants and herbs; Umbellales.

Apicomplexa Phylum of parasitic protozoans characterized by the presence, at one end, of a

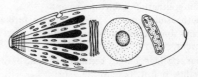

apicomplexan (Apicomplexa)

complex of microtubules, club-shaped rhoptries, polar rings and other structures; parasitic in invertebrates and vertebrates, often intracellular, causing various diseases including malaria; only the male gametes are flagellate; consists of 2 classes, Perkinsea and Sporozoa; also classified as a phylum of Protoctista.

Apidae Cosmopolitan family of hymenopteran insects containing the majority of eusocial bees in 3 well circumscribed subfamilies, Bombinae (orchid bees, bumblebees), Meliponinae (stingless bees), Apinae (honeybees); all construct nests with wax or resin cells, mostly within naturally occurring cavities; none excavates burrows; contains almost 1000 species, many extremely important as pollinators of crop plants.

honey bee (Apidae)

apivorous Feeding on bees; **apivore, apivory.**

Aplacophora Alternative name for the Solenogastres *q.v.* and Caudofoucata *q.v.* (Mollusca), treated as subclass of the class Amphineura.

aplanogamete A non-motile gamete; *cf.* planogamete.

aplanetic Having non-motile spores; **aplanetism.**

Aploactinidae Velvetfishes; family containing 30 species of mostly small western Pacific coastal marine scorpaeniform teleosts; body compressed, covered with spines or prickles; head bearing large blunt spines or papillae.

Aplochitonidae Family containing 4 species of small (to 300 mm), mostly anadromous, fresh- and brackish-water salmoniform teleost fishes from southern South America and southeastern Australia; body elongate, terete, scales sometimes absent; jaw teeth small or absent; adipose fin present.

Aplodactylidae Family containing 5 species of shallow marine perciform teleost fishes, resembling hawkfishes and morwongs, found

primarily in the southern hemisphere around Australia and New Zealand; jaw teeth flattened and recurved for feeding on algae.

Aplodontidae Mountain beaver; monotypic family of burrowing rodents (Sciuromorpha) found in the Rocky Mountains; body heavy, limbs and tail short, claws large.

Aplousobranchia Order of colonial tunicates (Ascidiacea) having a simple branchial sac lacking internal longitudinal folding, and gonad located in loop of intestine; individual zooids small, 1–25 mm in length; colonies formed by various types of budding; zooids may have separate tunics or a common tunic may envelop entire colony.

Apocrita Bees, wasps, ants; suborder containing about 125 000 species of hymenopteran insects, which have a narrow waist between first and second abdominal segments; ovipositor modified as sting in some species; the larvae are grub-like; habits diverse, including free-living, parasitic, parasitoid, solitary and social; 2 divisions recognized, Parasitica (gall wasps, fig wasps, chalcid wasps, and other parasitic wasps), feeding mainly on body fluids of arthropod prey, and Aculeata (bees, wasps, ants) feeding mainly on nectar or honeydew; Clistogastra.

Apocynaceae Oleander, periwinkle, golden trumpet, Chilean jasmine, corkscrew flower; family of Gentianales containing 2000 species of trees, shrubs, herbs or vines producing iridoid compounds and cardiotonic glycosides mostly from warm regions; flowers regular with 4 or 5 joined sepals, 4 or 5 joined petals forming a funnel supporting 5 stamens, and often 2 ovaries containing many seeds.

periwinkle (Apocynaceae)

Apoda Gymnophiona *q.v.*

Apodacea Subclass of holothurians (sea cucumbers) comprising 2 orders, Apodida and Molpadiida, in which the tube feet are greatly reduced or absent.

Apodida Order of littoral to abyssal holothurians (Apodacea) comprising about

215 species typically found under rocks or buried in soft sediment; test worm-like, 50 mm to 2 m in length, smooth or rugose, tentacles simple or pinnate, tube-feet and respiratory trees absent.

Apodidae Swifts; cosmopolitan family containing about 80 species of small aerial birds (Apodiformes) characterized by very fast acrobatic flight; head large, bill small with broad gape; wings elongated, curved; habits gregarious, monogamous, often crepuscular and migratory; usually nest on vertical surface of rock or tree; feed on insects caught on the wing.

Apodiformes Order of mostly small neognathous birds specialized for fast and highly manoeuvrable flight; comprising 2 suborders Apodi (swifts) and Trochili (hummingbirds).

apogamy Reproduction without fertilization; direct production of a plant from the gametophyte by budding, without the formation of gametes; **apogamous.**

apogenic Sterile.

apogeotropism An orientation movement against gravity; negative geotropism; **apogeotropic.**

Apogonidae Cardinalfishes; family containing 170 species of small (to 200 mm) primarily tropical, shallow marine, reef-dwelling teleosts (Perciformes), including also some brackish, freshwater and deep-sea species; body elongate, colourful, with large eyes and 2 dorsal fins; many species are mouth brooders.

apogynous Lacking, or having non-functional, female reproductive organs; **apogyny.**

apomixis Reproduction without fertilization, in which meiosis and fusion of gametes are partially or totally suppressed; **apomict, apomictic.**

apomorphic Derived from and differing from an ancestral condition; **apomorph, apomorphy;** *cf.* plesiomorphic.

Aponogetonaceae Cape pondweed; family of Najadales containing about 40 species of perennial freshwater herbs with secretory canals containing oil or latex; leaves basal and usually with a long petiole and a floating blade; native to the tropics of the Old World and South Africa; flowers borne on spikes, bisexual, with 1–3 perianth segments, usually 6 stamens and 3 carpels.

apophyte A native weed which appears after cultivation.

Aporidea Small order of tapeworms found in the gizzard of ducks, swans and geese; scolex typically consisting of an armed rostellum and simple suckers, external segmentation lacking; mostly protandrous hermaphrodites.

Apororhynchida Order of archiacanthocephalan thorny-headed worms found as parasites in the digestive tract of birds; typically with a large, globular proboscis which has an infolded anterior wall covered with spirally arranged, spine-like hooks.

aposematic coloration Coloration having a protective function; sometimes used in a restricted sense for warning coloration only; **aposematism; aposeme;** *cf.* proaposematic coloration, pseudoaposematic coloration, synaposematic coloration.

apospory The production of a diploid gametophyte from a sporophyte without spore formation; **aposporous.**

Apostomatida Order of hypostomatan ciliates found predominantly as parasites or symbionts on marine invertebrates; typically with an inconspicuous cytostome which is sometimes absent.

apotypic Diverging from the basic or typical form.

appeasement display A display that serves to prevent attack by reducing the opponent's attack drive and by releasing other non-aggressive behaviour.

Appendicularia Class of small (to 5 mm) paedomorphic planktonic tunicates, comprising about 70 species, that retain the larval tail and notochord in the adult; body enveloped within a complex transparent house that serves for food collection; water is forced through the house by movements of the tail, large particles are removed on coarse outer filter and only fine nanoplankton reaches the main inner filter; the houses are frequently discarded and new ones secreted; development is direct; Larvacea.

appetitive behaviour 1: A variable introductory phase of an instinctive behaviour pattern or sequence; *cf.* consummatory act. 2: Behaviour aiding in the satisfaction of a need in response to an internal condition.

apple Rosaceae *q.v.*

apple canker Hypocreales *q.v.*

apricot Rosaceae *q.v.*

Apteronotidae Small family of freshwater catfishes (Siluriformes) from South America; body tapering posteriorly, anal fin well developed.

Apterygidae Kiwis; family of small forest-dwelling ratite birds (Palaeognathae) from

kiwi (Apterygidae)

holly (Aquifoliaceae)

New Zealand comprising 3 species; height up to 0.3 m; legs stout, wings inconspicuous, feathers hair-like, eyes reduced; long sensitive bill used to probe for earthworms and insects; habits nocturnal, cursorial.

Apterygiformes Order of small ratite birds containing a single family, Apterygidae (kiwis).

Apterygota Subclass of primitively wingless insects, most of which are free-living, elongate forms with a direct larval development; formerly included the orders Collembola, Diplura, Protura and Thysanura; now often restricted to the Thysanura with the other groups treated as distinct classes of Hexapoda.

aptosochromatosis Change of colour in birds, without moulting.

aquaculture The cultivation of aquatic organisms.

aquatic Living in or near water; used of plants adapted for a partially or completely submerged life.

aquaticolous Living in water or aquatic vegetation; **aquaticole.**

aquatosere An ecological succession commencing in a wet habitat, leading to an aquatic climax.

aqueous desert An area of the sea floor or the bed of a lake more or less devoid of macroscopic organisms, typically with unstable sediment.

aquifer Permeable underground rock stratum which holds water.

Aquifoliaceae Holly; family of Celastrales containing about 400 species of shrubs and trees, widely distributed but most abundant in the New World; flowers borne in small cymes; petals and stamens usually free; fruit is a small drupe containing 3 or more stones.

Aquilonian region A biogeographical region comprising Europe, Asia north of the Himalayas, Africa north of the Tropic of Cancer and America north of latitude 45° N.

arable Used of land that is cultivated or fit for cultivation.

Aracaceae Aroids; arum, dumb cane, philodendron, skunk cabbage, Swiss cheese plant, jack-in-the-pulpit; family of Arales containing about 1800 species of herbs, scrambling shrubs, climbing vines with aerial roots or free-floating aquatics, often producing latex or mucilage, with numerous tiny, malodorous flowers borne in a spadix; fruit usually a berry.

arum (Aracaceae)

arachnactis Free-swimming tentaculate larva of some anemone-like cerianthids (Cnidaria).

Arachnida Large and diverse class of mainly terrestrial arthropods (Chelicerata) comprising about 65 000 species in 11 extant orders, Scorpiones (scorpions), Uropygi (whip scorpions), Schizomida, Amblypygi (tailless whip scorpions), Palpigradi (micro whip scorpions), Araneae (spiders), Ricinulei, Pseudoscorpiones (pseudoscorpions), Solpugida (wind scorpions), Opiliones (harvestmen), Acari (mites); antennae, wings and compound eyes absent; body divided into prosoma (cephalothorax) and opisthosoma (abdomen); 4 pairs of legs present, abdomen segmented or unsegmented, only rarely bearing appendages; respiration by means of tracheae and/or book lungs.

Araeolaimida Order of chromadorian nematodes found mostly in marine or brackish habitats; characterized by simple annulation of the body, lacking punctations and by the simple spiral form of the chemosensory amphids.

Arales Aroids and duckweeds; order of Arecidae comprising two families of herbs, shrubs or vines, sometimes thalloid and free-floating; the family Lemnaceae is probably derived from the Aracaceae.

Araliaceae Ginseng, ivy; family of Apiales containing about 700 species of mostly woody, sometimes prickly, plants, with small flowers mainly borne in umbels and heads; widespread in warm regions; individual flowers normally with 5 sepals, petals and stamens, an inferior ovary, and producing drupe fruits.

Araneae Spiders; order of terrestrial arthropods comprising about 35 000 recognized species in 3 suborders, Mesothelae, Orthognatha, Lapidognatha; cephalothorax and abdomen joined by slender waist, chelicerae possessing distal fang; pedipalps leg-like, modified in male for sperm transfer; abdomen bearing book lungs and tracheae; usually 3 pairs of spinnerets, the silk serving many functions – used for traps, draglines, ballooning, wrapping prey, construction of domiciles, and defence.

Araneidae Orb-web spiders; family containing about 2500 species of spiders (Araneae) which spin typical orb webs with viscid spirals and many radii; prey caught in web are cut out and wrapped in silk; legs armed with many long spines, chelicerae with many teeth.

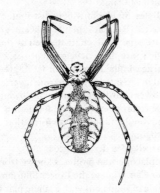

orb-web spider (Araneidae)

araneophagic Feeding on spiders; **araneophage**, **araneophagy**.

Araphideae Order of pennate diatoms (Pennales) lacking a true raphe; (slit-like suture in valve); also treated as a suborder, Araphidineae.

Araucariaceae Family of conifers (Pinatae) containing the Monkey Puzzle tree, native to southern hemisphere excluding Africa; leaves broad to lanceolate; male cone is a catkin, female cone globose, often massive.

Arbacioida Order of littoral to abyssal echinoids (Echinacea) comprising about 25 extant species; test sculptured, dentition stirodont, anus at aboral pole; includes the genus *Arbacia* used widely in laboratory studies of embryology and biochemistry.

arboreal Living in trees; adapted for life in trees.

arboreous desert An area of sparsely scattered trees with little or no vegetation between; desert forest.

arboretum A botanic garden or parkland dominated by trees.

arboricide A chemical that kills trees.

arboricolous Living predominantly in trees or large woody shrubs; **arboricole.**

arboriculture The cultivation of trees.

arborvirus An arthropod-borne virus; **arbovirus.**

arbusticolous Living predominantly on scattered shrubs and perennial herbs with shrub-like habit; **arbusticole.**

Arcellinida Testate amoebae; lobose amoebae having a test with a single opening, found predominantly in freshwater habitats although some occur in soil, litter and salt-water habitats.

Archaean A geological period (*c.* 4500–2400 million years B.P.); a subdivision of the Precambrian; **Archean**; see geological time scale.

Archaebacteria One of the three primary kingdoms (urkingdoms *q.v.*) of living organisms, comprising all methanogenic bacteria; *cf.* Eubacteria, Urkaryota.

Archaeoceti Extinct suborder of whales (Cetacea) containing the most ancient whales, known from the Eocene; porpoise-sized, with an elongate snout and nostrils on top of the skull; hind legs reduced to vestiges not visible externally; fed on fishes.

Archaeocyatha Extinct phylum of reef-forming organisms known from the Cambrian; resembling both sponges and corals in some features but possibly representing a higher grade of organization than the sponges.

Archaeocyathida Order of solitary or colonial Archaeocyatha (Anthocyathea); cup-shaped often irregular in outline; known from Lower and Middle Cambrian.

Archaeogastropoda An order of primitive prosobranch snails comprising about 3000 species of mostly marine and herbivorous species, as well as a few freshwater and terrestrial forms; includes the abalones, limpets, pheasant shells and top shells; shell variably shaped, often with mother-of-pearl internally; anterior mantle cavity typically containing 2 gills; usually with external fertilization.

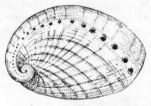

ormer (Archaeogastropoda)

Archaeognatha Order of primitively wingless insects (Apterygota); comprises about 250 species of fusiform insects which run fast and jump by abdominal flexion; body covered with pigmented scales; compound eyes large; abdomen with ventral styles on somites 2 to 9; free-living and nocturnal in habit, feeding on plant material.

archaeophyte A plant taken into cultivation in prehistoric times.

Archaeophytic The period of geological time during the evolution of the earliest plants, the algae; Algophytic; *cf.* Aphytic, Caenophytic, Eophytic, Mesophytic, Palaeophytic.

Archaeopulmonata A small order of primitive pulmonate molluscs found at the edge of the sea; characterized by the type of reproductive system and by the separation of the genital pores.

Archaeornithes Subclass of fossil birds

Archaeopteryx (Archaeornithes)

characterized by a long tail, 3 clawed digits on the forelimbs, small sternum and teeth on both jaws; includes *Archaeopteryx* from the Jurassic period.

Archaeozoic A geological period preceding the Cryptozoic, within the early Precambrian era; sometimes used for the whole of the early Precambrian period; **Archeozoic**; see geological time scale.

archaeozoology The study of animal remains from archaeological sites.

archegonium The female sex organ of many lower plants, bryophytes, pteridophytes and most gymnosperms, which typically comprises a distal neck and a swollen base containing the female gamete.

Archeobatrachia Suborder of primitive frogs (Anura) containing about 80 species in 6 families; eggs and larvae typically aquatic.

archerfish Toxotidae *q.v.*

archetype The hypothetical ancestral type; the earliest common ancestor; **archetypal**, **archetypic**.

Archiacanthocephala Class of parasitic thorny-headed worms found as adults mainly in predatory birds and mammal hosts, and using insects and myriapods as intermediate hosts; characterized by a usually retractable proboscis with hooks arranged in concentric circles.

Archiannelida Group of archaic aquatic annelid worms with narrow bodies, primitive nervous systems and, sometimes, parapodia; formerly treated as a distinct class of Annelida.

archibenthal zone The continental slope; the sea floor from the edge of the continental shelf to the continental rise; **archibenthic zone**; see marine depth zones.

archicleistogamic Having reduced reproductive organs within permanently closed flowers; *cf.* archocleistogamic.

Archidiales Small but widely distributed order of mosses (Bryidae); plants delicate, growing on bare sandy soil; characterized by their unique sporophyte development.

Archigregarinida Primitive order of gregarines found in the intestine of marine invertebrates and lower chordates, multiplying by multiple fission (merogony), gamete formation (gamogony) and sporozoite formation (sporogony).

Archinacelloidea Extinct order of monoplacophoran molluscs known from the Upper Cambrian to the Lower Silurian; shells variable in shape but with a single pair of muscle scars.

archocleistogamic Having mature reproductive organs within permanently closed flowers; *cf.* archicleistogamic.

Archoophora Subclass of exclusively marine flatworms (Turbellaria) comprising the Acoela and Polycladida.

Archosauria Large group of advanced diapsid reptiles including the crocodiles, dinosaurs, pterosaurs and thecodonts; the birds may be descended from archosaurs as their single temporal opening may be derived by fusion from the 2 apertures in the diapsid skull.

Archostemata Small suborder of beetles (Coleoptera) comprising about 30 species in 3 families.

Arcoida Ark shells; an order comprising about 250 species of mainly marine, sedentary pteriomorphian bivalves; characterized by their shell structure.

Arctic brown soil A soil having a shallow profile; acidic crumb-structured yellow-brown A-horizon, B-horizon absent, C-horizon more or less neutral.

Arctic period The earliest period of the Blytt–Sernander classification (*c.* 14 000–10 000 years B.P.) characterized by primarily tundra vegetation and cold late-glacial conditions.

Arctiidae Tiger moths; family of medium-sized colourful nocturnal moths (Lepidoptera) containing about 2000 species of widespread distribution but most abundant in the New World tropics; many produce sounds using a stridulatory organ on the side of the body; the characteristically hairy larvae are known as woolly bears; the larvae of some species are economically important pests causing damage to the foliage of broad-leaved trees.

Arctocyonidae Extinct family of generalized primitive mammals (Condylarthra) known from the Cretaceous to the Eocene; slim bodied with slender limbs, clawed feet, small brain and primitive teeth; included omnivorous and herbivorous forms.

Arctogaea The biogeographical area comprising the Palaearctic, Nearctic, Ethiopian and Oriental regions; *cf.* Megagaea, Neogaea, Notogaea, Palaeogaea.

Ardeidae Herons, bitterns, egrets; family of medium to large wading birds (Ciconiiformes) characterized by long legs, long neck, long broad wings and a strong pointed bill; feed on fishes and other aquatic organisms; solitary or colonial breeders that nest on or off the ground; contains 62 species with cosmopolitan distribution in fresh to marine waters.

heron (*Ardeidae*)

Arecaceae The palms; formerly known as the Palmae; the only family of the order, Arecales *q.v.*

Arecales Palms; order of Arecidae comprising a single family of nearly 3000 species of mostly slender trees with an unbranched trunk and a terminal crown of large evergreen leaves; flowers generally small and insect-pollinated; mainly occurring in tropical and warm temperate regions; includes betel palm, coconut palm and date palm.

Arecidae Diverse subclass of monocotyledons (Liliopsida) comprising about 5700 species in 4 orders; includes herbs, shrubs and trees with alternate leaves and small crowded flowers.

arena An area used for communal courtship display; lek.

arenaceous Derived from or containing sand; having the properties of sand; growing in sand; sandy.

Arenicolidae Lugworm; family of capitellidan polychaete worms found burrowing in soft estuarine and inshore sediments; widely used as fishing bait by anglers.

arenicolous Living in sand; **arenicole.**

areogeography Study of the distributions of plants and animals.

arescent Becoming dry or arid.

Argentinidae Herring smelts, argentines; family containing about 12 species of mesopelagic salmoniform teleost fishes found on outer shelf and upper continental slope; body elongate, to about 500 mm length, adipose fin present.

Argentinoidei Suborder of mostly deep-sea salmoniform teleosts comprising about 110 species in 5 families; including herring smelts, deep-sea smelts, barreleyes, slickheads, and tubeshoulders.

argillaceous Comprising clay-sized particles; having the properties of clay.

argillicolous Living on or in clay; **argillicole.**

argillophilous Thriving in clay or mud; **argillophile, argillophily.**

argodromilic Pertaining to slow-flowing streams.

argotaxis Passive movement due to surface tension; **argotactic.**

Arguloida Only order of Branchiura *q.v.*

arheic region An area in which no rivers arise; *cf.* endorheic region, exorheic region.

Arhynchobatidae Monotypic family of rajiform fishes from New Zealand; single dorsal fin present.

Arhynchobdellae Order of freshwater, amphibious and terrestrial leeches parasitic or predacious on wide range of hosts; proboscis absent, jaws or piercing stylets present; contains about 200 species in 9 families; commonly blood feeders.

arid Used of a climate or habitat having a low annual rainfall of less than 250 mm, with evaporation exceeding precipitation and a sparse vegetation.

Ariidae Family of marine catfishes (Siluriformes) cosmopolitan in tropical and subtropical seas; body naked, dorsal and pectoral fins with stout venomous spine; some forms migrate into estuaries or streams to spawn; eggs brooded orally; some species utilized as food-fish; referred to by some authorities as Tachysuridae.

Aristolochiales Order of aromatic woody vines (Magnoliidae) comprising a single family of about 600 species; flowers often malodorous and lacking petals.

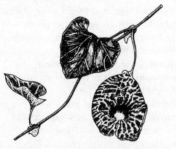

Dutchman's pipe (Aristolochiales)

ark shell Arcoida *q.v.*

armadillo Dasypodidae *q.v.*

armyworms Larvae of noctuid moths (Lepidoptera) that migrate in huge numbers and cause extensive damage to crops.

aroids Aracaceae *q.v.*

arrhenogenic Producing offspring that are entirely, or predominantly male; **arrhenogeny;** *cf.* allelogenic, amphogenic, monogenic, thelygenic.

arrhenotoky Parthenogenesis in which unfertilized eggs produce haploid males and fertilized eggs produce diploid females; **arrhenotokous;** *cf.* thelytoky.

Arripidae Australian salmon; family of primarily coastal marine teleost fishes (Perciformes) containing 2 species confined to Australia and New Zealand; body elongate, cylindrical, to 1 m length; important as sport and as food-fishes.

arrow worm Chaetognatha *q.v.*

arrowhead Alismataceae *q.v.*

arrowroot Taccaceae *q.v.*

arrowtooth eel Dyssomidae *q.v.*

Artamidae Wood swallows; family containing 13 species of small passerine birds found in open countryside and forest habitats of southeast Asia and Australian regions; bill pointed with broad gape; wings long and pointed; habits gregarious, aerial, some species migratory; feed on insects caught on wing.

Arthracanthida Order of acantharians with pyramidal bases on the 20 radial spines and a capsule membrane separating the endoplasm from the ectoplasm.

Arthrochirotida Extinct order of sea cucumbers (Holothuroidea) characterized by an axial skeleton of stout sclerites in the tentacles; known from the Devonian.

Arthrodira Group of primitive fossil fishes (Placodermi); body covered with bony plates and a head shield; known from the Devonian.

Arthropoda Very large phylum of invertebrates that includes the insects, arachnids, crustaceans and many smaller groups, totalling several million species and far outnumbering all other phyla added together; body metamerically segmented, most segments carrying a pair of jointed appendages; at least one pair of limbs specialized as jaws, the others variously modified for locomotion, feeding, respiration or reproduction; body enclosed in chitinous exoskeleton (cuticle) that is periodically moulted to permit growth; size range 80 μm to 3.6 m. Four subphyla recognized, Crustacea, Chelicerata, Uniramia, Pentastomida; classification and status of the Arthropoda is still controversial, component groups are sometimes regarded as distinct phyla retaining the term arthropod as a grade of organization.

Articulata 1: Class of epifaunal brachiopods
(lamp shells) in which the valves are hinged by
ventral valve teeth and dorsal valve sockets;
lophophore usually with skeletal support; gut
blind ending, without functional anus;
contains about 280 species in 2 orders,
Rhynchonellida and Terebratulida, found
from the intertidal zone to depth of 5000 m. 2:
Sole extant subclass of the echinoderm class
Crinoidea comprising the stalked
Millericrinida, Cyrtocrinida,
Bourgueticrinida, Isocrinida and the free-
living Comatulida.

artificial selection Selection by man;
domestication; selective breeding.

Artiodactyla Even-toed ungulates: diverse
order of mostly large herbivorous or
omnivorous terrestrial mammals
characterized by possession of a cloven hoof in
which digits 3 and 4 are dominant and carry
most of the weight (paraxonic); bony horns
often present; stomach simple to complex;
contains 3 suborders, Suiformes (pigs,
peccaries, hippopotamuses), Tylopoda
(camels, llamas), Ruminantia (deer, giraffe,
cattle, sheep, goats).

llama (Artiodactyla)

Ascetospora Phylum of spore-producing
protozoans which are parasitic mostly in
marine invertebrates; spores are multicellular
but lack polar capsules; comprising 2 classes,
Paramyxea and Stellatosporea.

Aschelminthes A loose grouping of
invertebrate groups characterized by the
possession of a pseudocoelom; formerly
treated as a phylum comprising the Rotifera,
Gastrotricha, Kinorhyncha, Nemata and
Nematomorpha but now regarded as a level of
organization and each of its constituent groups
is now classified as a phylum.

Ascidiacea Sea squirts; class of solitary or
colonial, sessile tunicates typically attached
posteriorly, with branchial (inhalent) and
atrial (exhalent) apertures at anterior end;
gonads situated in loop of intestine or in body
wall, heart ventral, direction of blood flow
periodically reversed; many species
accumulate heavy metals such as vanadium
and iron; 3 orders recognized,
Aplousobranchia, Phlebobranchia, and
Stolidobranchia.

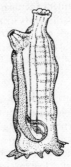

sea squirt (Ascidiacea)

arum Aracaceae *q.v.*

arundineous Having an abundance of reeds;
reedy.

asaccharolytic Used of microorganisms that are
unable to break down carbohydrates, alcohols
or polysaccharides.

Ascalaphidae Owl-flies; family of large, often
strikingly coloured, neuropteran insects which
fly mostly at dusk and at night, catching prey
insects whilst on the wing; predacious larvae
found on soil and in vegetation, feeding by
ambushing prey; contains about 400 species.

Ascaridida Order of exclusively parasitic
rhabditian nematodes found in vertebrate
hosts; generally possessing 3 or 6 lips with an
outer circle of 8 sensilla and paired pore-like
amphids; 2 spinules usually present in the male
and 2 ovaries in the female.

Asclepiadaceae Milkweed, butterfly weed, wax
flower; large family of Gentianales containing
about 2000 species of vines or erect herbs
commonly producing various cardiotonic
glycosides and alkaloids, mostly from warm
regions; flowers typically in cymose
inflorescences; each flower pentamerous,
petals contorted; fruit consists of paired
follicles.

Ascoceratida Extinct order of cephalopods
(Nautiloidea) which have a narrow juvenile
portion and inflated adult portion to the test;
siphuncle about central in position; known
from the Middle Ordovician to the Silurian.

Ascomycota Ascomycotina *q.v.*; treated as a
separate phylum of the kingdom Fungi.

Ascomycotina Sac fungi, Ascomycetes; group
of fungi possessing a minute reproductive

structure (ascus) formed after plasmogamy, between an ascogonium and another cell, produces ascogenous hyphae containing nuclei which fuse only after binucleate cells are walled off within the ascogenous hyphae; meiosis and mitosis usually results in the formation of 8 haploid ascospores inside a sac-like ascus; asexual reproduction by the production of conidiospores is also widespread; classified as a subdivision of true fungi, Eumycota, or as a phylum of the kingdom Fungi under the name Ascomycota; comprises 5 classes, Hemiascomycetes, Discomycetes, Loculoascomycetes, Plectomycetes and Pyrenomycetes.

Ascoseirales Small order of brown algae found only in antarctic and subantarctic waters.

Ascosphaerales Order of plectomycete fungi in which the ascogonium is a multinucleate swollen cell that arose as a lateral branch and the asci are evanescent; ascospores sometimes forming a membrane-enclosed spore ball; contains soil-living and root nodule species as well as parasites of bees.

Ascothoracica Order of ectoparasitic and endoparasitic barnacles (Cirripedia) found on echinoderms and cnidarians from shallow coastal to abyssal waters; characterized by mouthparts modified for piercing and sucking; contains about 75 species; sometimes treated as a distinct subclass.

ascus The specialized cell in Ascomycote fungi in which the haploid ascospores develop.

A-selection Selection operating in a consistently and predictably adverse or resource limited environment and characterized by low population density, low species richness and little interspecific competition.

Asellariales Order of trichomycete fungi which are obligate gut symbionts of marine, freshwater and terrestrial isopods, and of terrestrial insects.

Asellota Water lice; suborder of isopod crustaceans containing about 2000 species widely distributed in freshwater and marine benthic habitats; especially diverse in the deep sea.

asexual 1: Lacking functional sex organs. 2: Used of reproduction not involving the fusion of gametes, including budding, fission, gemmation, parthenogenesis, spore formation and vegetative propagation.

asexual reproduction Reproduction without the formation of gametes and in the absence of any sexual process.

ash Oleaceae *q.v.*

Asilidae Robber flies; cosmopolitan family containing about 5000 species of active predatory flies (Diptera) found commonly in open sunny habitats; anterior legs robust, bearing strong hairs, for gripping prey which are caught in flight; many mimic bees or wasps.

asparagus Liliaceae *q.v.*

aspect 1: The degree of exposure of a site to environmental factors. 2: The seasonal changes in the appearance of vegetation.

aspection The periodic changes in the appearance of vegetation within an area or community; the periods are usually recognized as seasons; **aspectation.**

aspidistra Liliaceae *q.v.*

Aspidochirotacea Subclass of holothurians (sea cucumbers) in which the test has a marked bilateral symmetry, the tentacles are non-retractile, tube-feet are present, and respiratory trees are present or absent; some species have a toxin (holothurin) in the body wall; 2 orders recognized, Elasipodida and Aspidochirotida.

Aspidochirotida Order of mainly shallow water, tropical, holothurians (Aspidochirotacea) comprising about 300 species which have a thick test, 15–30 tentacles, and in which respiratory trees are present; the dried body wall of some species is considered a gastronomic delicacy in Oriental regions (trepang, blêche-de-mer).

Aspidogastrea Subclass of trematodes comprising about 32 species endoparasitic in molluscs, fishes and turtles; characterized by their cylindrical, non-spinous body with a flattened ventral holdfast (haptor); development may involve an intermediate host or may be direct.

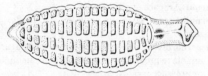

aspidogastrean (Aspidogastrea)

asporogenic Not producing, or developing from, spores; **asporogenous.**

Aspredinidae Banjo catfishes; family of tropical nocturnal fresh- and brackish-water teleosts (Siluriformes) from South America; body depressed, tuberose, broad anteriorly, narrow posteriorly; head armoured with bony plates; 3 pairs of barbels; large species important locally as food-fish; small forms popular amongst aquarists.

ass Equidae *q.v.*

assassin bug Reduviidae *q.v.*

assemblage A group of fossils occurring together in the same stratigraphic level (an assemblage zone).

association 1: A climax plant community characterized by two or more dominant species which have the life form *q.v.* typical of the formation to which the association belongs. 2: Sometimes used in general to indicate a large assemblage of organisms in a particular area, with one or two dominant species, or to refer to any group of plants growing together and forming a small unit of natural vegetation.

assortment The distribution of chromosomes to the germ cells during meiosis.

Astacidea Crayfishes, Norway lobster, lobster; infraorder of pleocyematan decapod crustaceans; first 3 pairs of thoracic legs chelate, uropods well developed.

Asterales Composites; cosmopolitan order of Asteridae containing a single enormous family, Asteraceae, of over 20 000 species of mostly herbs producing various repellent chemicals and latex or resin; flowers borne in centripetally flowering heads and may be perfect, pistillate, staminate or neutral; pollen is released into a tube and pushed out by growth of the style; includes many familiar species, aster, chamomile, chrysanthemum, cineraria, cocklebur, coneflower, dahlia, daisy, dandelion, goldenrod, hawkweed, lettuce, marigold, ragweed, sagebrush, sunflower, star thistle, thistle, tickseed; family formerly known as Compositae.

composite (Asterales)

Asteridae The most advanced subclass of dicotyledonous (Magnoliopsida) plants with various sorts of repellents, which can be distinguished from the majority of other dicotyledons by their sympetalous flowers in which the stamens are equal to the number of,

and alternate to, the corolla lobes, or, in which there are fewer stamens than corolla lobes.

Asterinales Order of mostly tropical loculoparenchemycetid fungi which are typically epiphytes or external parasites of plants.

Asteroidea Starfish, sea stars; subclass of intertidal to deep-sea asterozoan echinoderms comprising about 1500 species in 5 extant orders, Platyasterida, Paxillosida, Valvatida, Spinulosida and Forcipulata; arms typically merging gradually with central disk, 5 to multi-radiate; sexes usually separate, fertilization external, asexual reproduction by disk fission occurring in some species; development may be direct but commonly includes planktonic bipinnaria and brachiolaria larvae; feed as omnivorous scavengers or opportunistic predators, large prey may be digested externally.

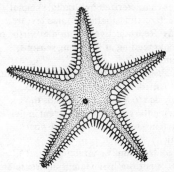

starfish (Asteroidea)

Asterozoa Subphylum of echinoderms comprising a single class, Stelleroidea, and 3 subclasses, Somasteroidea, Asteroidea (sea stars) and Ophiuroidea (brittle stars); body flattened with radiating arms, mouth central and downward facing, anus absent or central aboral.

asthenobiosis A period of reduced metabolic activity.

Astigmata Diverse subgroup of acariform mites (Acari) typically found on or in association with arthropods, vertebrates and other animals, only rarely free-living throughout the life cycle; common on carrion, dung, in nests, and as pests of stored products.

astilbes Saxifragaceae *q.v.*

astogeny 1: The development of a colony by budding. 2: The developmental history of a colonial organism.

Astomatida Order of hymenostome ciliates characterized by their total lack of a mouth (cytostome), found mostly as endosymbionts in the digestive tract of terrestrial or aquatic oligochaetes.

Astrapotheria Order of extinct South American ungulates (Protoungulata) known from the Eocene to Miocene; anterior body robust, hindlimbs slender; possessed elephant-like proboscis.

Astroblepidae Family containing about 35 species of freshwater catfishes (Siluriformes) typically found in fast-flowing montane streams of tropical South and Central America; body depressed, naked, head broad; dorsal and adipose fin often with stout spine; single pair of barbels.

Astronesthidae Snaggletooths, star-eaters; family containing 27 species of small (to 150 mm) luminescent deep-sea stomiiform teleost fishes; body bearing 2 rows of photophores; front teeth fang-like.

Astrophorida Order of tetractinomorph sponges characterized by a radial skeletal architecture and generally coarse texture.

asyngamy Reproductive isolation by virtue of different breeding or flowering seasons; **asyngamic.**

atavism A reversion to an ancestral character state not evident in recent generations, possibly due to a recessive gene or to complementary genes; the aberrant individual is sometimes referred to as a throwback; **atavistic.**

Ateleopodidae Family containing about 10 species of bottom-living lampridiform teleost fishes widespread in tropical and warm temperate seas; body elongate, to 600 mm length, anal fin long and continuous with small caudal; pelvics comprising only 2 filamentous fin rays.

ateliosis Dwarfism; reduced size but with normal proportions.

Atelopodidae Family containing about 30 species of small brightly coloured tree frogs (Anura) found in Central and South America; many are poisonous; most walk rather than hop.

Atelostomata Superorder of irregular

atelopid frog (Atelopidae)

euechinoidean echinoids that burrow in soft sediments; anus displaced from aboral pole to posterior interambulacrum, dental apparatus absent in adult, gills lacking; aboral tube-feet serve respiratory function; 2 orders recognized, Cassiduloida and Spatangoida.

Athalamida Order of naked Granuloreticulosa protozoans, lacking any test or shell.

athalassic Used of waters or water bodies that have not had any connection to the sea in geologically Recent times, all ions in solution are thus derived from the substratum or atmosphere.

Athecanephria Order of pogonophoran marine worms possessing a sac-like anterior body cavity in the tentacular region and lateral ciliated ducts opening to the exterior; spermatophores cylindrical or spindle-shaped.

Atherinidae Silversides, sand-smelts; family containing about 150 species of primarily marine atheriniform teleost fishes widespread in tropical and temperate coastal waters; body fusiform, to 600 mm length, typically almost transparent with silvery lateral stripe.

Atheriniformes Order of small marine, brackish and freshwater teleost fishes widespread in tropical and temperate regions; contains about 180 species in 4 families; including silversides, sand-smelts, rainbowfishes.

athermobiosis Dormancy induced by relatively low temperatures; **athermobiose.**

Atlantic Equatorial Countercurrent A warm surface ocean current that flows east in the tropical Atlantic, forming the northerly part of the equatorial circulation; see ocean currents.

Atlantic Equatorial Undercurrent A strong subsurface ocean current that flows eastwards across the Atlantic at the equator.

Atlantic North Equatorial Current A warm surface ocean current that flows westwards in the equatorial Atlantic and forms the southerly limb of the North Atlantic Gyre; see ocean currents.

Atlantic period A period of the Blytt–Sernander classification q.v. (c. 7500–4500 years B.P.) characterized by oak, elm, linden and ivy vegetation and by the driest and warmest postglacial conditions.

Atlantic South Equatorial Current A warm surface ocean current that flows westwards in the equatorial Atlantic and forms the northerly limb of the South Atlantic Gyre; see ocean currents.

atlas moth Saturniidae q.v.

atmobios Organisms living in the air; aerial organisms.

atmophyte An epiphyte that obtains water by aerial assimilation over its entire surface.

atokous Without offspring; non-reproductive; vegetative; *cf.* epitokous.

atrophic Used of organisms or life cycle stages that do not feed.

atrophy Marked reduction in size and functional significance.

attenuation 1: A reduction in strength or intensity; weakening. 2: The loss of virulence of a pathogenic microorganism.

aubergine Solanaceae *q.v.*

Auchenipteridae Family containing about 50 species of small (to 250 mm) tropical South American freshwater catfishes (Siluriformes); body naked, lateral line often zig-zagged; dorsal fin with single spine; 3 pairs of barbels.

aucuparious Attractive to birds.

aufwuchs A community of aquatic organisms and associated detritus adhering to and forming a surface coating on submerged stones, plants and other objects; fouling organisms.

augur Neogastropoda *q.v.*

auk Alcidae *q.v.*

aulophyte A non-parasitic plant living within a hollow cavity in another plant.

Aulopidae Family of primitive bottom-living myctophiform teleost fishes containing 8 species found in warm coastal waters of the North Atlantic and Pacific to 1000 m depth; body slender, to 600 mm length, dorsal fin high and anteriorly placed; photophores absent.

Aulorhynchidae Tubesnouts, tubenose; family containing 2 species of shallow marine gasterosteiform teleost fishes of the North Pacific that inhabit a variety of rocky, sandy, eelgrass- or algae-covered substrata; body very slender, to 160 mm length, cylindrical, bearing armour of lateral bony scutes; snout elongate, upper jaw protractile.

Aulostomidae Trumpetfishes; family containing 3 species of tropical and subtropical shallow marine gasterosteiform teleost fishes often found around reefs in association with gorgonian corals; body slender, compressed, to 800 mm length; snout tubular; dorsal fin comprising 8–13 separated spines.

auricularia A free-swimming, planktonic larval stage of sea cucumbers characterized by a single, continuous band of cilia looping around the paired lobes of the body.

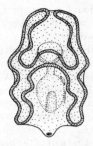

auricularia

Auriculariales Order of fungi (Phragmobasidiomycetes) now mostly included in the order Eutremellales *q.v.*

auroral Pertaining to the dawn.

austral Southerly; pertaining to cool or cold temperate regions of the southern hemisphere; antiboreal; *cf.* boreal.

Australian chat Ephthianuridae *q.v.*

Australian kingdom One of the six major phytogeographical areas characterized by floristic composition; comprising Central, North & East, and South & West Australian regions.

Australian laurel Pittosporaceae *q.v.*

Australian region A zoogeographical region comprising Australia, Maluku (the Moluccas), Sulawesi, Papua–New Guinea, Tasmania, New Zealand and the oceanic islands of the South Pacific; subdivided into Australian, Austro-Malayan, New Zealand and Polynesian subregions; Hawaii and Polynesia are sometimes excluded and regarded as distinct regions.

Australian salmon Arripidae *q.v.*

australopithecines Extinct group of homonoid apes that lived about 5–1.5 million years B.P., known from fossil remains found in Africa, in particular along the East African Rift Valley; they are thought to have walked upright on two legs; facial and skeletal characteristics link the group closely to chimpanzees and gorillas as well as to man; the best known of the australopithecines is Lucy, found at Hadar (Ethiopia) in 1974.

Austrobaileyaceae Family of archaic flowering plants (Magnoliales) comprising a single species of evergreen woody vine, native to northeast Australia; flowers are foul smelling and have poorly differentiated sepals and petals.

autamphinereid An autotrophic amphibious plant.

autecology The ecology of individual organisms or species; **autoecology**; *cf.* synecology.

autephaptomenon Autotrophic organisms
which live attached to a surface;
autophaptomenon.

authigenic Derived from sea water and
precipitated directly onto the sediment rather
than settling through the water column.

autoallogamy The condition of a species in
which some individuals are adapted to cross
fertilization and others to self-fertilization;
autallogamy.

autobiosphere That part of the biosphere in
which energy is fixed by photosynthesis in
green plants; *cf.* allobiosphere.

autochorous Having motile spores or
propagules disseminated by the action of the
parent plant; **autochore, autochory.**

autochthonous Endogenous; produced within a
given habitat, community or system;
indigenous; endemic; native; **autochthony**; *cf.*
allochthonous.

autocoprophagous Used of an organism that
consumes its own faeces; **autocoprophage,
autocoprophagy.**

autodeme A deme *q.v.* composed
predominantly of self-fertilizing individuals.

autoecious Used of a parasite that passes
through the different stages of its life cycle
within the same host individual; **autoecius.**

autogamy 1: The process of self-fertilization or
self-pollination. 2: The fusion of two
reproductive nuclei within a single cell,
derived from a single parent. 3: Reproduction
in which a single cell undergoes reduction
division producing two autogametes which
subsequently fuse. 4: In protistans, the
division of the micronuclei to produce eight or
more nuclei, of which two fuse and give rise to
a new macronucleus.

autogenic Occurring within a given system;
produced by the activities of living organisms
within a system and acting upon that system.

autogenous Used of a female insect that does
not have to feed in order to facilitate
maturation of her eggs; *cf.* anautogenous.

autoicous Having male and female
inflorescences on the same plant; monoecious.

autoinfection The direct reinfection of a host
individual by larval offspring of an existing
parasite.

autolysis The breakdown of a cell by its own
enzymes; **autolytic.**

automictic parthenogenesis Parthenogenesis
q.v. in which meiosis is preserved and diploidy
reinstated either by the fusion of haploid
nuclei within a single gamete or by the
formation of a restitution nucleus.

automimicry Mimicry in which a palatable
morph (the mimic) resembles a non-palatable
morph (the model) of the same species; the
polymorphism with respect to palatability can
arise from a choice of different food plants by
model and mimic.

autonastic Pertaining to growth curvature
promoted by endogenous factors;
autonastism, autonasty.

autoparthenogenesis The development of an
unfertilized egg that has been activated by a
chemical or physical stimulus.

autopelagic Epipelagic; used of planktonic
organisms living continually at the sea surface;
cf. allopelagic.

autophagous Used of precocious offspring
capable of locating and securing their own
food; **autophage, autophagy.**

autophilous Self-pollinated; **autophily.**

autophyte A plant capable of synthesizing
complex organic substances from simple
inorganic substrates, typically by
photosynthesis; **autophytic.**

autoploid Used of an organism with the
characteristic chromosome set of its species.

autopolyploid A polyploid variety resulting
from the multiplication of the chromosome set
of a single species; *cf.* allopolyploid.

autopotamic Pertaining to organisms adapted
to stream environments and which complete
their life cycles in streams; *cf.* eupotamic,
tychopotamic.

autoradiography A photographic technique for
determining the location of a chemical
previously labelled with a radioactive isotope,
which has been incorporated into a cell or
organism.

autosome Any chromosome other than sex
chromosomes.

autostylic The condition in which the
hyomandibular plays no part in the suspension
of the lower jaw in fishes, since the
palatoquadrate arch articulates directly with
the skull; *cf.* amphistylic, hyostylic, holostylic.

autotherm An organism that regulates its body
temperature independent of ambient
temperature changes; *cf.* allotherm.

autotomy Self-amputation of an appendage or
other part of the body.

autotoxin Any substance produced by an
organism that is toxic to itself.

autotrophic 1: Capable of synthesizing complex
organic substances from simple inorganic
substrates; including both chemoautotrophic
and photoautotrophic organisms. 2: Used of
any organism for which environmental carbon

dioxide is the only or main source of carbon in the synthesis of organic compounds by photosynthesis; **autotroph, autotrophy**; *cf.* heterotrophic.

autotrophic lake A lake in which all or most of the organic matter present is derived from within the lake and not from drainage off the surrounding land; *cf.* allotrophic lake.

autotropism The tendency to grow in a straight line unaffected by external factors; **autotropic.**

autoxenous Used of a parasite that passes through the different stages of its life cycle in the same host individual.

auxanometer A simple apparatus for measuring growth (in length) of plant organs.

auxin A plant hormone promoting or regulating growth.

auxotrophic Used of a microorganism having a biochemical deficiency and requiring supplementary growth factors not needed by the wild type; **auxotroph, auxotrophy.**

Aves Birds; class of feathered, bipedal, warm-blooded tetrapods; forelimbs specialized as wings for flight; stout keeled sternum present, pelvic girdle fused to synsacrum; neck elongate; teeth absent; right aortic arch only; reproduction oviparous, fertilization internal; eggs large, yolked, amniotic, shelled; contains 9000 Recent species in a single subclass, Neornithes.

aviculture The practice of keeping birds in captivity for scientific study.

avifauna The bird fauna of an area or period.

avocado pear Lauraceae *q.v.*

avocet Recurvirostridae *q.v.*

axenic Used of a pure culture free from contaminant organisms; *cf.* dixenic, monaxenic.

Axinellida Order of tetractinomorph sponges found mainly in shallow tidal and sublittoral waters; typically branching form with the spicule-enclosing spongin fibres condensed axially or basally.

axolotl Ambystomatidae *q.v.*

aye aye Daubentoniidae *q.v.*

ayu Plecoglossidae *q.v.*

azalea Ericaceae *q.v.*

azara Flacourtiaceae *q.v.*

azoic Without life.

Azoic The earliest geological period; the subdivision of the Precambrian era preceding the Archaeozoic; see geological time scale.

azonal soil A soil lacking a well defined profile; a soil maintained in a permanently immature state by persistent deposition, truncation or erosion; *cf.* intrazonal soil, zonal soil.

azotification Nitrogen fixation.

Azygiida Order of digenetic trematodes in which the miracidium larva possesses a single pair of flame cells (excretory organs); the body of the cercaria is retracted into the tail.

azygospore A spore produced from an unfertilized female gamete.

azygote An organism produced by haploid parthenogenesis.

lapwing (Aves)

b

babbler Timaliidae *q.v.*
baboon Cercopithecidae *q.v.*
Bacillariophyceae Diatoms; large class of microscopic, unicellular chromophycote algae common in marine and freshwater habitats but also found in soil and damp moss; possessing an often ornate siliceous test (frustule) comprising 2 valves; chlorophylls *a* and *c* present with various xanthin pigments; reproduce by binary fission but sexual processes involving the production of haploid gametes are also found; comprises 2 orders, Centrales and Pennales; also treated as a distinct phylum of Protoctista under the name Bacillariophyta.

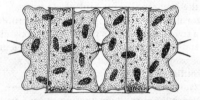

diatom (Bacillariophyceae)

Bacillariophyta The diatoms classified as a phylum of Protoctista; Bacillariophyceae *q.v.*
background radiation Natural radiation, comprising cosmic radiation (elementary particles from outer space) and terrestrial radiation (derived from the decay of naturally occurring isotopes).
backshore The zone of a beach above mean high water; also used for the zone covered only in exceptionally severe storms.
Bacteria A division of moneran microorganisms with rigid or semi-rigid external walls containing peptidoglycans, except in the Mollicutes; divided into 19 groups identified by trivial names with no taxonomic significance, such as Mollicutes, Spirochaetes and Rickettsias; cell division usually occurs by binary fission, mitosis never occurs.
bactericide An agent that destroys bacteria; **bactericidal.**
bacteriochlorophyll A form of chlorophyll found in photosynthetic bacteria.
bacteriology The study of bacteria; **bacteriological.**
bacterioneuston The bacteria at the sea surface; the bacterial component of the neuston.

bacteriophage 1: A virus infesting and usually lysing bacteria; often abbreviated to phage. 2: An organism that feeds on bacteria.
bacteriophagous Feeding on bacteria; **bacteriophage, bacteriophagy.**
bacteriostatic An agent inhibiting the growth of bacteria but not destroying them.
badger Mustelidae *q.v.*
badland An arid or semi-arid area with scanty vegetation and marked surface erosion.
Bagridae Family of primitive Old World catfishes (Siluriformes) found in fresh waters from Africa to southeast Asia; body naked, to 1 m length, dorsal fin short with strong spine; adipose fin usually large; 3 pairs of barbels present; larger species often utilized as food-fish, some small forms popular amongst aquarists.
Balaenidae Right whales; family of large (6–20 m) marine mammals (Mysticeta) comprising 3 species; baleen plates elongate; widespread except for tropical and south polar waters.

right whale (Balaenidae)

Balaenopteridae Rorquals; family of large (9–30 m) baleen whales (Mysticeta) found in northern and southern hemispheres; typically spending summers in high latitudes, migrating to warmer waters for winters; throat and chest region bearing numerous parallel grooves; contains 5 species including blue, humpback and minke whales.
Balanopaceae Small family of Fagales containing about 10 species of dioecious evergreen trees found in the southwestern Pacific region; male flowers borne in catkins, female flowers solitary with scale-like perianths; fruit an acorn-like drupe.
Balanophoraceae Family of Santalales containing about 60 species of fleshy root parasites, lacking chlorophyll and often fungus-like in appearance, widespread in tropical and subtropical regions.
bald An elevated treeless area within a region of forest vegetation, usually grassy.
baleen whales Mysticeta *q.v.*
Balistidae Triggerfishes, filefishes; family containing about 120 species of tropical marine tetraodontiform teleosts; body

compressed with 2–3 dorsal fin spines; in triggerfishes the second spine locks against the first as a trigger, used to wedge the body in crevices.

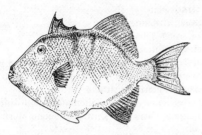

triggerfish (Balistidae)

ballistic Used of dehiscence in which the seeds are catapulted from the fruit.

balm Lamiaceae *q.v.*

balsa Bombacaceae *q.v.*

Balsaminaceae Touch-me-not, impatiens, busy Lizzie; family of Geraniales containing about 450 species of mostly glabrous herbs, mainly from the tropical Old World; flowers bisexual, commonly with 3 or 5 sepals, one forming a spur, 5 petals and stamens, and a superior ovary.

bamboo Bambusaceae *q.v.*

bamboo rat Rhizomyidae *q.v.*

bamboo worm Capitellida *q.v.*

Bambusaceae Bamboos; family of grass-like plants (Liliopsida) most of which have woody stems; may reach heights of 40 m; usually grow in clumps with a large underground rhizome; commonly treated as a subfamily of the Poaceae *q.v.*

banana Musaceae *q.v.*

bandfish Cepolidae *q.v.*

bandicoot Peramelina *q.v.*

Bangiales 1: Order of marine and freshwater red algae, of filamentous or parenchymatous body form. 2: Bangiophycideae *q.v.* treated as a class of the phylum Rhodophyta.

Bangiophycideae Subclass of red algae with thalli that are unicellular or consist of aggregations of cells that are filamentous or parenchymatous but of minimal complexity and in which cell division is predominantly intercalary; cells mostly uninucleate and with a single chloroplast; containing 4 orders Porphyridiales, Bangiales, Compsopogonales and Rhodochaetales; also treated as a class of the phylum Rhodophyta, under the name Bangiales.

banjo catfish Aspredinidae *q.v.*

baobab Bombacaceae *q.v.*

barachorous Having propagules dispersed by their own weight; **barachore, barachory.**

barb Cyprinidae *q.v.*

Barbados cherry Malpighiaceae *q.v.*

barberry Berberidaceae *q.v.*

barbet Capitonidae *q.v.*

Barbeyaceae Family of Urticales containing a single species of small, dioecious tree native to northeast Africa and Arabia.

Barbourisiidae Whalefishes; monotypic cosmopolitan family of small (to 250 mm) bathypelagic beryciform teleosts; body stout and microscopically spinose; mouth very large, teeth minute; small pelvic fins present.

Barclayaceae Small family of water lilies (Nymphaeales) native to region from southeast Asia to New Guinea.

bark beetle 1: Curculionidae *q.v.* 2: Scolytidae *q.v.*

bark louse Psocoptera *q.v.*

barley Poaceae *q.v.*

barnacle Cirripedia *q.v.*

barokinesis A change of linear or angular velocity in response to a change in pressure; **barokinetic.**

barophilic Thriving under conditions of high hydrostatic or atmospheric pressure; **barophile, barophily.**

barotaxis A directed reaction of a motile organism to a pressure stimulus; **barotactic.**

barotropism An orientation response to a pressure stimulus; **barotropic.**

barracuda Sphyraenidae *q.v.*

barracudina Paralepididae *q.v.*

Barrandeoceratida Extinct order of cephalopod molluscs (Nautiloidea) known from the Middle Ordovician to the Middle Silurian; lacking cameral and siphuncular deposits.

barreleyes Opisthoproctidae *q.v.*

barren 1: Incapable of producing offspring; unproductive; infertile; incapable of producing seeds or fruit. 2: Devoid of vegetation, or of fossils.

barrial A mud flat.

barrier Any obstruction to the spread of an organism or to gene flow between populations; an unfavourable biological, climatic, geographical or other factor which prevents successful dispersal and establishment of a species.

barymorphosis Those structural changes in organisms that result from the effect of pressure or weight; **barymorphic.**

basal area 1: A cross-sectional area of a tree determined from the diameter of the trunk at breast height, 1.4 m above ground. 2: The total area of ground covered by trees measured at breast height. 3: The actual surface area of soil covered or occupied by a plant measured close to the ground; ground cover.

Basellaceae Ulluco, Madeira vine; small family of tropical or subtropical herbs with twining or scrambling stems and thickened or tube-bearing rhizomes.

basic Rich in alkaline minerals and typically rich in nutrients as well; alkaline.

basic number The number of chromosomes in a basic chromosome set representing the lowest monoploid number in a polyploid series.

Basidiomycotina Basidiomycetes; subdivision of true fungi (Eumycota); characterized by sexual reproduction involving the formation of connections between hyphal cells so that each contains 2 compatible nuclei which fuse to form a basidium; meiotic division of the basidium produces 4 haploid basidiospores attached to the outside of the basidium; the basidia are arranged on a fruiting body (basidiocarp); comprises 4 classes, Teliomycetes, Phragmobasidiomycetes, Hymenomycetes and Gasteromycetes.

basidium The specialized spore-producing structure of Basidiomycote fungi on the surface of which are borne the haploid basidiospores.

basifuge A plant unable to tolerate basic soils.

basket shell Neogastropoda *q.v.*

basket star Ophiuroidea *q.v.*

basking range The temperature range within which a reptile basking in direct sunlight remains inactive although alert.

basking shark Cetorhinidae *q.v.*

Basommatophora Pond snails, ramshorn snails, freshwater limpets; order of pulmonate molluscs found in fresh water or along the margins of the sea; with unstalked eyes situated at the base of the tentacles and nearly contiguous genital pores.

ramshorn snail (Basommatophora)

basophilous Thriving in alkaline habitats; **basophile, basophily.**

Bassleroceratida Extinct order of straight or curved cephalopod molluscs (Nautiloidea) in which the siphuncle is typically ventral and sutures often lobate; known from the Ordovician.

basslet Grammidae *q.v.*

bass wood Tiliaceae *q.v.*

bat Chiroptera *q.v.*

bat fly 1: Nycteribiidae *q.v.* 2: Streblidae *q.v.*

Bataceae Small family containing 2 species of tropical maritime shrubs.

Batales Small order of Dilleniidae comprising 2 families of trees and shrubs producing mustard oil.

Batesian mimicry The close resemblance of a palatable or harmless species (the mimic) to an unpalatable or venomous species (the model) in order to deceive a predator; *cf.* mimicry.

batfish Ogcocephalidae *q.v.*

bathophilous Thriving in deep-water habitats; **bathophile, bathophily.**

bathyal 1: Pertaining to the sea floor between 200 m and 4000 m; see marine depth zones. 2: Pertaining to the deep sea.

Bathydraconidae Family containing 15 species of antarctic perciform teleost fishes related to the Nototheniidae; body slender, to 160 mm length, depressed anteriorly with prolonged snout; single long dorsal fin present.

Bathyergidae Mole rats; family containing 9 species of small fossorial hystricomorph rodents widespread in the Ethiopian region; body fusiform, limbs short and strong; eyes and ears vestigial; feeding mainly on roots and bulbs.

Bathylagidae Black smelts; family containing 35 species of small (to 250 mm) deep-sea salmoniform teleost fishes resembling herring smelts (Argentinidae); body slender to deep; mouth small; eyes very large.

bathylimnetic Pertaining to the deep-water regions of a lake.

bathylittoral That part of the marine sublittoral zone that is devoid of algae.

Bathymasteridae Ronquils; family containing 7 species of North Pacific, bottom-dwelling, coastal marine teleost fishes (Perciformes); body slender, to 300 mm length, single long dorsal fin present; dorsal and anal fins lacking spines.

bathymetry The measurement of ocean or lake depth and the study of floor topography; **bathymetric.**

Bathynellacea Order of very small (0.5–3.5 mm) blind syncaridan crustaceans containing about 130 species found mostly in subterranean freshwater habitats but occasionally in saline waters; body vermiform, with reduced appendages as an adaptation to interstitial groundwater habitats; distributed worldwide.

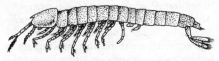

bathynellacean (Bathynellacea)

bathypelagic Living in deep water below the level of light penetration but above the abyssal zone, between 1000 m and 4000 m.

bathyphilous 1: Thriving in lowlands. 2: Thriving in the deep sea; **bathyphile, bathyphily.**

bathyplankton Planktonic organisms which undergo diurnal vertical migration, moving up towards the surface at dusk and down away from the surface at dawn; **bathyplanktonic;** *cf.* epiplankton.

Bathypteroidae Tripod fishes; family containing 8 species of bottom-living deep-sea myctophiform teleosts; body cylindrical, becoming compressed posteriorly, snout depressed; eyes very small, teeth minute; photophores absent; some pelvic, pectoral and caudal fin rays elongate.

Bathysauridae Family of bottom-living deep-sea lizard fishes (Myctophiformes) comprising 3 species which are widespread below about 900 m; body compressed, to 600 mm length, head depressed; teeth recurved, barbed, needle-like; photophores absent.

bathyscaphe A free-moving manned vehicle for deep ocean exploration.

bathysphere A spherical manned chamber suspended from a mother ship and lowered into the deep sea for underwater exploration.

Batoidea Superorder of mainly benthic marine elasmobranch fishes comprising about 450 species in 5 orders, Rhinobatiformes (guitarfishes), Rajiformes (skates and rays), Pristiformes (sawfishes), Torpediniformes (electric rays), Myliobatiformes (eagle, sting, devil rays); a few species epipelagic or freshwater; body more or less flattened dorsoventrally; pectoral and pelvic fins may be greatly expanded, anal absent, dorsal and caudal present or absent.

batology The study of brambles (Rosaceae).

Batrachoididae Toadfishes; family containing about 55 species of mostly marine bottom-dwelling teleosts; body naked or with tiny scales, length to 600 mm; head large, depressed, bearing barbels or flaps; dorsal fin spines may be venomous; photophores present in some species.

Batrachoidiformes Order of bottom-living predatory marine and freshwater teleost fishes comprising a single family, Batrachoididae (toadfishes).

Baventian glaciation An early glaciation of the Quaternary Ice Age *q.v.* in the British Isles.

bay laurel Lauraceae *q.v.*

B-chromosome A supernumary or accessory chromosome differing from normal A-chromosomes in morphology, genetic effectiveness and pairing behaviour.

Bdelloida The sole order of the Bdelloidea *q.v.*

Bdelloidea Digonota; class of bottom-living aquatic rotifers found commonly in fresh water but also in damp moss and soil; body typically vermiform, covered with a glycoprotein cuticle; differentiated into a foot, trunk and head which bears a ciliated retractable corona with 2 trochal disks used for swimming and feeding; reproduction probably parthenogenetic; comprising a single order, Bdelloida.

Bdellonemertea Small order of enoplan nemerteans found as commensals in marine bivalve molluscs; body dorsoventrally compressed, with an unarmed proboscis and a single posterior sucker.

beach The gently sloping strip of land along the margin of a body of water that is washed by waves or tides sufficiently to inhibit all or most plant growth.

beach flea Semiterrestrial amphipod crustacean capable of athletic jumping activity and often abundant along the strandline of the sea shore; sandhopper.

bead tree Elaeocarpaceae *q.v.*

beaked whale Ziphiidae *q.v.*

bean Fabaceae *q.v.*

bean weevil Bruchidae *q.v.*

bean wilt Hypocreales *q.v.*

bear Ursidae *q.v.*

beard worm Pogonophora *q.v.*

beardfish Polymixiidae *q.v.*

beaver Castoridae *q.v.*

bed A unit layer in a sequence of rock strata.

bed load The quantity of rock and debris moved along a stream or river by the flow of water; *cf.* silt load.

bedbug Cimicidae *q.v.*

bedding plane The division plane that separates individual beds or strata of a sedimentary or stratified rock.

bee Apidae *q.v.* (Apocrita, Aculeata).

bee eater Meropidae *q.v.*

bee fly Bombyliidae *q.v.*

beech Fagaceae *q.v.*

Beestonian glaciation A glaciation of the Quaternary Ice Age *q.v.* in the British Isles, with an estimated duration of 100 000 years.

beet Chenopodiaceae *q.v.*

beetle Coleoptera *q.v.*

Begoniaceae Begonia; large family of Violales containing about 1000 species of glabrous or often shaggy haired herbs to soft shrubs, largely tropical in distribution and including many garden ornamentals; leaves usually asymmetrical, flower stems in leaf axils often with branched inflorescences arranged with male flowers proximally and female flowers distally.

begonia (Begoniaceae)

behaviour Any observable action or response of an organism.

Belemnoidea Belemnites; extinct order of cephalopod molluscs (Coleoidea) with an internal shell comprising phragmacone, rostrum and pro-ostracum; known from the Upper Carboniferous to the Eocene.

belemnite (Belemnoidea)

belladonna Solanaceae *q.v.*

bellflower Campanulaceae *q.v.*

Belonidae Needlefishes, garfishes; family containing 25 species of surface-living piscivorous beloniform teleosts widespread in tropical and temperate seas; body elongate, to 2 m length; jaws prolonged bearing sharp teeth; pectoral fins short.

garfish (Belonidae)

Beloniformes Order of primarily surface-living marine and freshwater teleost fishes comprising about 140 species in 5 families; including flying fishes, half-beaks, needlefishes, sauries.

Belontiidae Gouramis; family containing about 30 species of mostly small (to 150 mm) freshwater perciform teleost fishes from west Africa, India and southeast Asia; upper jaw protractile; utilize atmospheric oxygen through accessory respiratory structure, the suprabranchial organ; eggs laid in bubble nest at water surface.

belt A narrow strip or area of vegetation.

belt transect A narrow belt of predetermined width set out across a study area, and within which the occurrence or distribution of plants and animals is recorded.

beluga Monodontidae *q.v.*, white whale.

Benguela Current A cold surface ocean current that flows northwards off the east coast of Africa and forms the easterly limb of the South Atlantic Gyre; see ocean currents.

Bennettitales Order of gymnosperms resembling the cycads, known only as fossils from the Mesozoic era; reproductive structures resembled flowers rather than cones.

benthic Pertaining to the sea bed, river bed or lake floor; benthonic; *cf.* pelagic.

benthogenic Derived from the benthos *q.v.*

benthonic Benthic *q.v.*

benthophyte A plant living at the bottom of a water body or on the bed of a river.

benthopleustophyte Any large plant resting freely on the floor of a lake but capable of drifting slowly with the currents.

benthopotamous Living on the bed of a river or stream.

benthos Those organisms attached to, living on, in or near the sea bed, river bed or lake floor; *cf.* endobenthos, epibenthos, haptobenthos, herpobenthos, psammon, rhizobenthos.

Berberidaceae Barberry; large family of often

barberry (Berberidaceae)

spiny shrubs or herbs belonging to the Ranunculales; containing about 650 species widespread in the temperate northern hemisphere; flowers composed of free parts in 2–7 whorls which vary in morphology; ovary consists of a single carpel; fruit a berry or capsule.

Bergmann's rule The generalization that geographically variable species of warm-blooded vertebrates tend to be represented by larger races in the colder parts of their range than in the warmer parts.

berm The large deposits of dry loose sediment above the high tide line on a beach.

Beroida Small order of pelagic ctenophores with a worldwide distribution; body strongly flattened in the tentacular plane, lacking tentacles and tentacle bulbs.

berry A fleshy indehiscent fruit containing many seeds and surrounded by a tough outer skin derived from the outer fruit wall (epicarp); includes the grape and tomato.

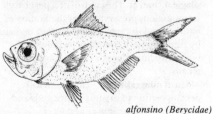

grapes (berry)

Berycidae Alfonsinos; cosmopolitan family containing 10 species of mainly deep-sea beryciform teleost fishes; body oval, compressed, length to 600 mm, scales ctenoid or cycloid; eyes and mouth large, teeth small; dorsal and anal fins robustly spinose.

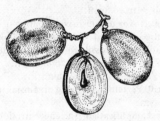

alfonsino (Berycidae)

Beryciformes Diverse order of marine teleost fishes comprising about 160 species in 13 families widespread in benthic and pelagic habitats of coastal waters and continental slopes to about 1300 m depth; body typically oblong, compressed, with relatively large head, mouth and eyes; includes pricklefishes, slimeheads, fangtooth, alfonsinos, lanterneyes, squirrelfishes and whalefishes.

betel nut Piperaceae *q.v.*

Betulaceae Birch, alder; family of Fagales containing about 120 species of trees or shrubs widely distributed in the temperate and cold temperate regions of northern hemisphere; flowers arranged in unisexual catkins; fruit is a samara.

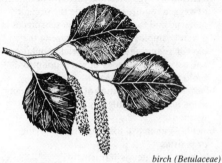

birch (Betulaceae)

B-horizon The upper subsoil horizon with an accumulation of clay, humus, iron and sesquioxides as a result of leaching and translocation from upper layers; see soil horizons.

bibulous Highly absorbent; capable of absorbing water.

bichir Polypteridae *q.v.*

biennial Lasting for 2 years; occurring every 2 years; requiring 2 years to complete the life cycle; *cf.* annual, perennial.

biferous Producing fruit twice a year, or producing two crops per season.

biflorous Flowering in both spring and autumn.

big-bang reproduction Semelparous *q.v.* reproduction.

bigeye Priacanthidae *q.v.*

big-headed turtle Platysternidae *q.v.*

bight A large indentation in the coastline or continental shelf margin (seabight).

Bignoniaceae Flame tree, empress tree, African tulip tree, calabash tree, trumpet creepers; family of Scrophulariales containing about 800 species of mostly woody plants, often vines, with large showy flowers; mainly tropical in distribution, especially tropical America; flowers usually borne in cymose clusters; petals and sepals fused, ovary superior; fruit commonly a capsule containing small winged seeds.

billfish Istiophoridae *q.v.*

billion A thousand million, 10^9; this United States value has been adopted internationally; formerly a value of 10^{12} (a million million) was widely used.

binary fission Division into two parts; the commonest form of asexual reproduction in protistans.

bindweed Convolvulaceae *q.v.*

bioassay Biological assay; the use of an organism for assay purposes.

biocenosis Biocoenosis *q.v.*

biochemical oxygen demand (BOD) The amount of oxygen required to degrade the organic material and to oxidize reduced substances in a water sample; used as a measure of the oxygen requirement of bacterial populations and serving as an index of water pollution; biological oxygen demand.

biochronology The dating of biological events using biostratigraphic or palaeontological methods.

biocide A chemical toxic or lethal to living organisms.

bioclast A single fossil fragment.

bioclastic sediment A sediment composed of broken fragments of organic skeletal material.

bioclimatic law Any climatic rule *q.v.*

bioclimatology The study of climate in relation to fauna and flora.

biocoenology The qualitative and quantitative study of communities of organisms.

biocoenosis A community or natural assemblage of organisms; often used as an alternative to ecosystem but strictly it is the fauna/flora associations themselves, excluding physical aspects of the environment; biocenosis; *cf.* thanatocoenosis.

biodegradable Capable of being decomposed by natural processes.

bioenergetics The study of energy flow through ecosystems; ecological energetics.

bioerosion Erosion resulting from the direct action of living organisms.

biofacies A subdivision of a sedimentary unit based on a distinctive assemblage of fossils.

biogenesis The principle that all living organisms have derived from previously existing living organisms; **biogenetic**; *cf.* abiogenesis.

biogenetic Produced by the activity of living organisms; **biogenic.**

biogenetic law The theory that an individual during its development passes through stages that resemble adult forms of its successive ancestors so that the ontogeny of an individual recapitulates the phylogeny of its group; recapitulation.

biogenic meromixis The mixing of a lake caused by an accumulation in the monimolimnion of salts liberated from the sediment by biological activity; *cf.* crenogenic meromixis, ectogenic meromixis.

biogenous Living on or in other organisms; symbiotic.

biogeochemical cycle The cyclical system through which a given chemical element is transferred between biotic and abiotic parts of the biosphere, as for example in the carbon and nitrogen cycles.

biogeochemistry The study of mineral cycling and of organism–substrate relationships.

biogeographical region Any geographical area characterized by distinctive flora and/or fauna.

biogeography The study of the geographical distributions of organisms, their habitats (ecological biogeography) and the historical and biological factors which produced them (historical biogeography).

biogeosphere That part of the lithosphere *q.v.* within which living organisms can occur; eubiosphere.

bioglyph A trace fossil.

bioherm 1: Any organism contributing to the formation of a coral reef. 2: A mound-like accumulation of fossil remains on the site where the organisms lived; **biohermal.**

biolith A rock of organic origin.

biological Pertaining to living organisms or life processes.

biological amplification The concentration of a given persistent substance by the organisms in a food chain so that the amount of the substance present in the body increases at each successive trophic level.

biological clock An inherent physiological mechanism for measuring time or maintaining endogenous rhythms.

biological control The control of a pest by the introduction, preservation or facilitation of natural predators, parasites or other enemies, by sterilization techniques, by the use of inhibitory hormones or by other biological means.

biological mineralization The biological decomposition of organic compounds and liberation of inorganic minerals; *cf.* immobilization.

biological oxygen demand (BOD) Biochemical oxygen demand *q.v.*

biological races The sympatric populations of a species which differ biologically but not morphologically and in which interbreeding is inhibited by different food or host preferences or behaviour cycles.

biological rhythm A regular periodicity

exhibited by biological processes.

biological species Groups of actually or potentially interbreeding natural populations genetically isolated from other such groups by one or more reproductive isolating mechanisms; biospecies.

biological weathering Changes in soil structure and composition resulting from biotic activity, as part of the soil-forming process.

biology The science of life; the study of living organisms and systems; **biological.**

bioluminescence Light produced by living organisms and the emission of such biologically produced light; **bioluminescent;** *cf.* phosphorescence.

biolysis Death and subsequent tissue degradation of an organism.

biomass Any quantitative estimate of the total mass of organisms comprising all or part of a population or any other specified unit, or within a given area at a given time; measured as volume, mass (live, dead. dry or ash-free weight) or energy (calories); standing crop.

biome A biogeographical region or formation; a major regional ecological community characterized by distinctive life forms and principal plant (terrestrial biomes) or animal (marine biomes) species; see map on page 48.

biometeorology Study of the effects of atmospheric conditions on living organisms.

biometrics The application of statistical methods to biological problems and the mathematical analysis of biological data.

biomineralization The process by which organisms produce skeletal structures containing crystalline or amorphous inorganic substances.

bionomics Ecology; the study of organisms in relation to environment.

biont An individual organism.

biophagous Consuming or destroying other living organisms; **biophage, biophagy;** *cf.* saprophagous.

biophilous Thriving on other living organisms, often used specifically of plant parasites; **biophile, biophily.**

biophysics The application of physics to the study of living organisms and systems.

biophyte A plant that feeds on other living organisms; a parasitic or predatory plant.

biopoiesis The origin of life, including the abiotic synthesis of macromolecular systems and the transformation (eobiogenesis) of these systems into the first living organisms (eobionts).

bioseston Plankton, nekton and suspended organic particulate matter derived from living organisms; the biological component of seston.

biospecies 1: A biological species *q.v.* 2: A species defined primarily on biological characters.

biospeleology The study of subterranean life.

biosphere The global ecosystem; that part of the Earth and atmosphere capable of supporting living organisms.

biostratigraphy The study and classification of rock strata based on their fossil content; stratigraphic palaeontology; **biostratigraphic.**

biostratinomy The study of the relationship between fossils and their environments.

biostrome An accumulation of fossils that are distinctly bedded but do not form a mound-like or reef-like structure (bioherm); a fossil bed having no pronounced topographical relief.

biosynthesis The production of organic compounds by living organisms.

biota The total flora and fauna of a given area.

biotaxis The directed reaction of a motile organism towards (positive) or away from (negative) a biological stimulus; **biotactic.**

biotelemetry Study of the behaviour and activity of organisms using remote detection and transmission equipment; radio tracking.

biothermal Pertaining to the interrelationship of temperature and living organisms.

biotic Pertaining to life or living organisms; caused by, produced by, or comprising living organisms; *cf.* abiotic.

biotic climax A plant community maintained at climax by some biotic factor such as grazing.

biotic community 1: A group of interacting species coexisting in a particular habitat; community. 2: In palaeontology, a group of species frequently found together.

biotic potential The maximum potential rate of increase of an organism or population under ideal conditions; reproductive potential.

biotic pyramid A diagrammatic representation of the successive trophic levels of a food chain in the form of a pyramid of numbers, with primary producers at the base and the sequence of consumers leading to the apex.

biotope 1: The smallest geographical unit of the biosphere or of a habitat that can be delimited by convenient boundaries and is characterized by its biota. 2: The location of a parasite within the host's body.

biotrophic Used of a parasite deriving nutrients from the tissues of a living host; **biotroph, biotrophy.**

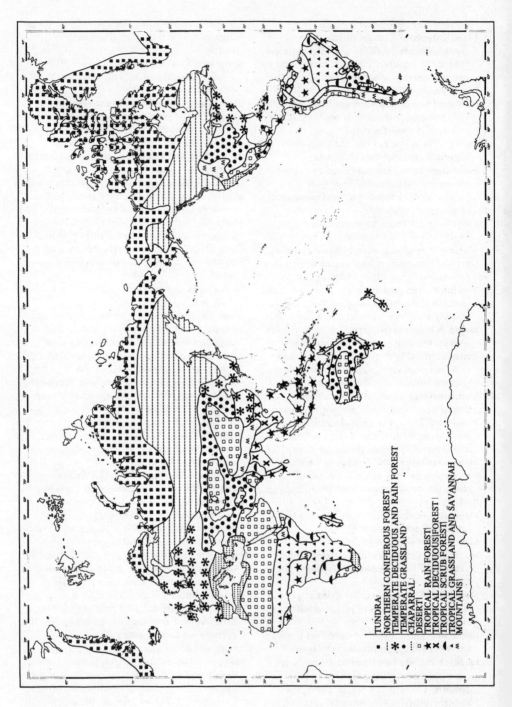

terrestrial biomes (biome)

bioturbation The mixing of a sediment by the burrowing, feeding or other activity of living organisms.

biparous 1: Producing two offspring in a single brood; **biparity**; *cf.* uniparous, multiparous. 2: Having produced only one previous brood.

bipinnaria A free-swimming, planktonic larval stage of starfishes, possessing two separate bands of cilia looping around paired lobes on the body.

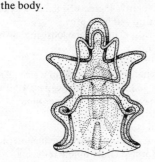

bipinnaria

bipolarity The occurrence of species or forms in both northern and southern hemispheres but not in the intervening equatorial belt; **bipolar.**

Biraphideae Order of pennate diatoms (Pennales) possessing a fully differentiated raphe on each valve; also treated as a suborder, Biraphidineae.

birch Betulaceae *q.v.*

bird of paradise Paradisaeidae *q.v.*

bird-of-paradise flower Strelitziaceae *q.v.*

bird Aves *q.v.*

bird's-nest fungus Nidulariales *q.v.*

birdwing butterfly Papilionidae *q.v.*

birth rate The number of offspring produced per head of population per unit time; natality.

bisect A profile of vegetation and soil showing the natural distribution of growth above and below ground; a cross-section of vegetation and soil as revealed by a trench extending down to the deepest plant roots.

bisexual 1: Used of a population or generation composed of functional males and females; gonochoristic. 2: Used of an individual possessing both male and female functional reproductive organs; hermaphrodite. 3: Used of a flower possessing both male and female reproductive organs; perfect; *cf.* unisexual.

biting midge Ceratopongonidae *q.v.*

bittern Ardeidae *q.v.*

bitterroot Portulaceae *q.v.*

Bivalvia Bivalves; a large class of marine, estuarine and freshwater molluscs which are ciliary or filter feeders on small particles; characterized by a shell of two calcareous valves joined by a flexible ligament along a hinge line; head not differentiated; gills large, ciliated and involved in feeding; also known as Pelecypoda or Lamellibranchia; comprising 5 extant subclasses, Anomalodesmata, Heterodonta, Paleoheterodonta, Protobranchia and Pteriomorpha.

Bivalvulida Order of Myxosporea *q.v.* characterized by having a spore with 2 shell valves.

bivoltine Having two generations or broods per year; *cf.* voltine.

Bixaceae Lipstick tree; small family of Violales containing about 20 species of tropical trees or shrubs which produce a red or orange secretion inside specialized cells or canals.

black coral Antipatharia *q.v.*

black earth Chernozem *q.v.*

black fly Simuliidae *q.v.*

black mildew Meliolales *q.v.*

black mud A terrigenous marine sediment, black in colour, rich in hydrogen sulphide and having a high organic content; typical of poorly ventilated anaerobic basins.

black smelt Bathylagidae *q.v.*

blackberry Rosaceae *q.v.*

blackbird 1: Icteridae *q.v.* 2: Turdidae *q.v.*

blackfly Aphididae *q.v.*

blackwater Water rich in humic acids and with low nutrient concentrations.

bladdernut Staphyleaceae *q.v.*

bladderwort Lentibulariaceae *q.v.*

blastochorous Pertaining to a plant that is dispersed by means of offshoots; **blastochore.**

Blastocladiales Order of chytridiomycete fungi comprising about 50 species of saprophages or microscopic parasites with a fungal thallus differentiated into root-like filaments and reproductive structures, sometimes with incomplete cross walls (pseudosepta): also treated as a class of the protoctistan phylum Chytridiomycota, under the name Blastocladia.

Blastodiniales Small order of ectoparasitic marine dinoflagellates found on invertebrate and fish hosts.

blastogenesis Asexual reproduction by budding or gemmation.

Blastoidea Extinct class of sessile, bud-like echinoderms (Pelmatozoa); with well developed pentamerous symmetry and biserial brachioles; theca composed of 17 major plates arranged in 3 circlets; known from the Silurian to the Lower Permian. (Picture overleaf).

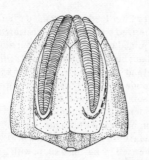

blastoid (Blastoidea)

Blattaria Cockroaches; order of active
orthopterodean insects with a depressed body
and long cursorial legs; forewings hard or
leathery, hindwings membranous, but may be
reduced or absent; typically live on the ground
under stones, in litter and wood debris, in
nests, or in caves; contains about 3700 species,
a few of which are household pests.

cockroach (Blattaria)

Blenniidae Combtooth blennies; family
containing about 275 species of tropical to
temperate, shallow marine, brackish and
freshwater, perciform teleost fishes; body
length to 600 mm, but usually much smaller;
scales absent or vestigial.

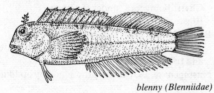

blenny (Blenniidae)

Blennioidei Blennies; suborder of perciform
teleost fishes comprising about 650 species in
15 families; pelvic fins when present in jugular
position; total number of dorsal and anal fin
rays equal to number of vertebrae.

blenny Blennioidei *q.v.*; Tripterygiidae
(threefin blennies); Clinidae (scaled
blennies); Blenniidae (combtooth blennies).
bletting The change in consistency, without
putrefaction, of some fruits.
blight A non-specific name for a variety of plant
diseases, typically caused by a fungus.
blind shark Brachaeluridae *q.v.*
blister beetle Meloidae *q.v.*
bloom An explosive increase in the density of
phytoplankton within an area.
blow off The loss of humus and loose top soil by
wind action.
blowfly Calliphoridae *q.v.*
blowout A small area of land from which all or
most of the soil has been removed by wind
erosion; an excavation in sandy ground
produced by wind action.
blue Lycaenidae *q.v.*
blue mud A terrigenous marine sediment, blue
in colour, formed under mildly reducing
conditions, similar to a green mud.
blue shark Carcharhinidae *q.v.*
blue whale Balaenopteridae *q.v.*
bluebell Liliaceae *q.v.*
blueberry Ericaceae *q.v.*
bluebird Turdidae *q.v.*
bluebottle Calliphoridae *q.v.*
bluefish Pomatomidae *q.v.*
bluegrass Poaceae *q.v.*
blue-green algae Cyanophycota *q.v.*
blue-green bacteria Cyanobacteria *q.v.*
Blytt–Sernander climatic classification A
classification of late glacial and postglacial
climate based on stratigraphic evidence from
plant remains; comprising the Arctic,
Preboreal, Boreal, Atlantic, Subboreal and
Subatlantic periods.
boarfish Caproidae *q.v.*
bobcat Felidae *q.v.*
body louse Anoplura *q.v.*
bog An area of wet peaty substrate rich in
organic debris but low in mineral nutrients,
with a vegetation of ericaceous shrubs, sedges
and mosses.
Bog soil An intrazonal soil with a muck or peat
surface layer and an underlying peat horizon
over an impervious grey-blue clay, formed in
boggy or swampy conditions; typical of
glaciated basins.
Boidae Pythons, anacondas, boas; family of
small to very large (to 10 m) non-venomous
constrictor snakes (Serpentes); left lung and
vestige of pelvic girdle present; habits
terrestrial, fossorial, arboreal or semiaquatic;
feeding on variety of birds and mammals;

reproduction oviparous or viviparous; containing about 60 species with pantropical distribution.

Boletaceae Boletes; family of fungi (Agaricales) which produce typical mushroom-shaped fruiting bodies; fertile spore-producing layer present lining tubes on underside of the cap, replacing the gills; often found in mycorrhizal association with trees.

bolete (Boletaceae)

bolete Boletaceae *q.v.*

bolochorous Having propagules dispersed by propulsive mechanisms; **bolochore, bolochory.**

Bombacaceae Balsa, baobab, kapok tree, silkcotton tree; small family of Malvales containing about 200 species of trees, commonly with light and soft wood and large showy flowers which are typically asymmetrical with 5 sepals, 5 or many stamens and a superior ovary; fruit is a large woody capsule.

baobab (Bombacaceae)

bombadier beetle Carabidae *q.v.*
Bombay duck Harpadontidae *q.v.*
Bombycidae Family of broad robust moths (Lepidoptera) from Asia that comprises about 100 species, including the domesticated silkworm that no longer exists in the wild; commercial silk is derived from the pupal cocoon.

Bombycillidae Waxwings, silky flycatchers; family containing 9 species of passerine birds found in forest and open grassland habitats in the northern hemisphere; bill short and stout, with wide gape; feeds on insects and fruit; typically gregarious and monogamous; construct cup-shaped nests in trees.

Bombyliidae Bee flies; cosmopolitan family containing about 4000 species of nectar-feeding flies (Diptera) with robust hairy bodies; commonly found hovering over flowers in sunny habitats; larvae parasitic on eggs or larvae of other insects.

bonding The formation of a close relationship between two or more individuals; **bond.**

bone bed A stratum rich in fragments of fossil vertebrates.

bonefish Albulidae *q.v.*

bonytongue Osteoglossidae *q.v.*

booby Sulidae *q.v.*

Booidea Superfamily of primitive non-venomous snakes (Serpentes) retaining a moderately developed left lung and vestige of pelvic girdle; contains about 135 species including the large constrictors (boas and pythons).

book louse Liposcelidae *q.v.* (Psocoptera).

Boraginaceae Anchusa, lungwort, forget-me-not; widely distributed family of Lamiales containing about 2000 species of herbs or woody plants commonly provided with unicellular hairs often with mineralized walls; flowers bisexual, usually regular and with 5 sepals, 5 petals forming a tube or funnel, 5 stamens and a superior ovary.

anchusa (Boraginaceae)

Bordeaux mixture A common fungicide, containing copper sulphate and calcium hydroxide.

boreal Pertaining to cool or cold temperate regions of the northern hemisphere; *cf.* antiboreal.

Boreal kingdom One of the six major phytogeographical areas characterized by floristic composition; comprising Atlantic North American, Arctic and Subarctic, Euro-Siberian, Macaronesian, Mediterranean, Pacific North American, Sino-Japanese and Western and Central Asiatic regions.

Boreal period A period of the Blytt–Sernander classification (*c.* 9000–7500 years B.P.), characterized by pine and hazel vegetation and by relatively warm and dry conditions

borer Bostrichidae *q.v.*

Borhyaenoidea Extinct superfamily of carnivorous marsupials (Metatheria) known from the Palaeocene to the Pliocene in South America; the largest attained the size of a wolf.

Boston ivy Vitaceae *q.v.*

Bostrichidae Borers; family of small, dark beetles (Coleoptera); head tucked under rounded thorax; curved larvae are fleshy with tiny head; bore into dead or dying trees, grain, bamboo and other plant material.

botany The study of plants; **botanical.**

botfly 1: Cuterebridae *q.v.*. 2: Oestridae *q.v.*

Bothidae Cosmopolitan family containing about 210 species of pleuronectoid flatfishes in which the eyes are on the left side of the head; body length to 1.5 m.

Bothriocidaroida Extinct order of sea urchins (Echinoidea) known from the Middle Ordovician to the Upper Silurian; test with 2 columns of perforate plates and 1 of imperforate plates.

bottle urchin Spatangoida *q.v.*

bottleneck A sudden decrease in population density with a corresponding reduction of total genetic variability.

bottletree Sterculiaceae *q.v.*

bottom water The dense, deep oceanic water mass formed by the cooling and sinking of surface waters at high latitudes.

bougainvillea Nyctaginaceae *q.v.*

boulder A large sediment particle greater than 256 mm in diameter and too big to be handled easily; see sediment particle size.

boundary layer The layer of fluid adjacent to a physical boundary in which the fluid motion is significantly affected by the boundary and has a mean velocity less than the free stream value.

Bourgueticrinida Order of articulate crinoids in which the stalk is typically slender, cirri are absent, and attachment is by a terminal plate or root-like branches; about 40 extant species in 3 families, found from shelf to 9500 m depth.

Bovidae Cattle, antelopes, gazelles, goats, sheep; diverse family of herbivorous terrestrial ungulates (Artiodactyla: Ruminantia); comprising about 110 species found in wide range of habitats from tundra to tropical forest in Holarctic, Ethiopian and Oriental Regions; now cosmopolitan through domestication; unbranched horns present in males and females of many species; horn core permanent - not shed annually as in deer.

bison (Bovidae)

Bovoidea Superfamily of artiodactyls (Pecora) comprising the pronghorn (Antilocapridae) and the cattle, sheep, antelopes and gazelles (Bovidae); horns are present and covered with keratin (horn) rather than skin.

bowerbird Ptilonorhynchidae *q.v.*

bowfin Amiidae *q.v.*

boxfish Ostraciontidae *q.v.*

boxwood Buxaceae *q.v.*

B.P. Before present.

Brachaeluridae Blind sharks; family containing 2 species of small (to 1 m) bottom-dwelling orectolobiform elasmobranch fishes found on rocky reefs of eastern Australia; body cylindrical, head broad and depressed, nostrils with long barbels; caudal fin lacking ventral lobe.

brachiation Progression in trees by swinging from branch to branch using the forelimbs; **brachiating.**

brachiolaria Late larval stage of some starfish possessing conspicuous larval arms and a glandular adhesive sucker.

Brachionichthyidae Family of little-known Australian anglerfishes (Lophiiformes) resembling the frogfishes (Antennariidae); body small, globose and denticulate.

Brachiopoda Lamp shells; phylum of solitary, unsegmented marine coelomates possessing a bivalve shell and typically attached to the substratum by a stalk (pedicle); a few species are unattached or cemented directly to substratum; ventral valve usually larger than dorsal; body with looped or spiral array of tentacles (lophophore); gut simple. Brachiopods range from intertidal to 7500 m in cool-temperate or cold regions; 350 species in 2 living classes, Inarticulata, Articulata; abundant in Palaeozoic seas with 12 000 fossil species described.

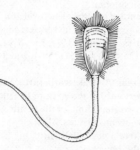

lamp shell (Brachiopoda)

Brachiopterygii Group of bony fishes (Osteichthyes) possessing spinous dorsal finlets, ganoid scales, lobate pectoral fins, a spiracle, and a pair of lung-like air bladders; includes the bichir and reedfishes (Polypteridae *q.v.*)

Brachycephalidae Small family of terrestrial toads (Anura) from southeastern Brazil comprising 2 species; body length to only 15 mm; forelimbs with only 2 functional digits, hindlimbs with 3; development thought to be direct, with terrestrial eggs.

Brachycera Suborder of flies (Diptera) comprising about 69 000 species including soldier flies, horseflies, clegs, robberflies, bee flies, hoverflies, fruitflies, vinegarflies, shore flies, botflies, houseflies, blowflies, fleshflies, warble flies and tsetse flies.

Brachydiniales Small order of marine unicellular dinoflagellates.

Brachypoda Cephalocarida *q.v.*

Brachyura Crabs; largest infraorder of pleocyematan decapod crustaceans containing all true crabs; abdomen permanently flexed beneath carapace.

crab (Brachyura)

brackish Pertaining to water of salinity intermediate between fresh water and sea water; brackish water may be classified according to salinity as mesohaline, oligohaline or polyhaline.

Braconidae Braconid wasps; family of small, parasitic wasps (Apocrita) related to the ichneumons; eggs laid on larvae of other insects which are then parasitized and consumed before pupation.

Bradypodidae Tree sloths; family containing 5 species of arboreal herbivorous Neotropical mammals (Edentata); teeth simple, peg-like; limbs elongate, bearing 2–3 strong claws, used for hanging from branches and clinging to trunks; tail vestigial; feed mainly on leaves.

sloth (Bradypodidae)

bramble shark Echinorhinidae *q.v.*

Bramidae Pomfrets; cosmopolitan family containing 20 species of mostly mid-water oceanic teleost fishes (Perciformes); body deep, compressed, to 800 mm length; body scales spiny or ridged; single long dorsal fin present.

pomfret (Bramidae)

branchicolous Living on the gills of fishes or other aquatic organisms; **branchicole.**

Branchiobdellida Subclass of minute (3–10 mm) hirudinoidean worms (Annelida) ectoparasitic on freshwater crayfish; body with 15 segments each bearing 2 annuli; pharynx with strong jaws; anterior and posterior sucker present; comprises a single family containing 10 genera, formerly classified with oligochaetes.

Branchiopoda Class of small aquatic crustaceans inhabiting mostly inland waters ranging from fresh to hypersaline, occasionally marine; contains about 820 extant species in 4 orders, Notostraca (tadpole shrimps), Anostraca (fairy shrimps), Conchostraca (clam shrimps), Cladocera (water fleas).

Branchiostegidae Tilefishes; family containing 25 species of bottom-living, tropical and warm-temperate, marine teleosts (Perciformes); body elongate, compressed, to 1 m length; head rather square, often with high dorsal crest; single dorsal fin present.

Branchiura Fish lice; class of ectoparasitic crustaceans found on freshwater and coastal marine fishes, or rarely on amphibians; capable of swimming but typically attaching to host by paired suckers, feeding on host's blood and mucus by means of a sucking tubular mouth; contains about 150 species in a single order, Arguloida.

fish louse (Branchiura)

Brandenburg glaciation A subdivision of the Weichselian glaciation *q.v.*

Brassicaceae Crucifers; cabbage family; turnip, honesty, watercress, horse radish, swede, rape, stocks, pepper root, radish, wallflower, candytuft; large family of Capparales containing about 3000 species of mostly herbs

honesty (Brassicaceae)

found mainly in temperate regions; many are cultivated for food or ornament; flowers typically with 4 sepals, 4 petals arranged in a cross, 6 stamens and a superior ovary; many are rich in sulphur.

Braun–Blanquet cover scale A scale for estimating cover of a plant species, comprising 6 categories: + (under 1%), 1 (1–5%), 2 (6–25%), 3 (26–50%), 4 (51–75%), and 5 (76–100%).

Brazil Current A weak, warm surface current that flows south off the coast of Brazil and forms the westerly limb of the South Atlantic Gyre; see ocean currents.

Brazil nut tree Lecythidales *q.v.*

bread mould Mucorales *q.v.*

breadfruit Moraceae *q.v.*

breed 1: To reproduce. 2: To propagate organisms under controlled conditions. 3: A group of organisms related by descent; an artificial mating group having a common ancestor; used especially in genetic studies of domesticated species.

breeding system The mode, pattern and extent to which individuals interbreed with others from the same or different taxa.

Bregmacerotidae Codlets; family containing 7 species of aberrant pelagic warm-water gadiform teleost fishes cosmopolitan in tropical and subtropical seas; first dorsal fin represented by solitary fin ray on top of head; second dorsal fin and anal fin elongate; pelvic fin rays markedly prolonged.

brephic stage A pre-adult stage; a larval stage.

Bretschneideraceae Family of Sapindales containing a single species of deciduous tree producing mustard oil, native to the mountains of western China.

bridal wreath Rosaceae *q.v.*

brill Scophthalmidae *q.v.*

brimstone butterfly Pieridae *q.v.*

brine fly Ephydridae *q.v.*

brine shrimp Anostraca *q.v.*

bristlemouth Gonostomatidae *q.v.*

bristletail Thysanura *q.v.*

bristleworm Polychaeta *q.v.*

brittle star Ophiuroidea *q.v.*

broadbill Eurylaimidae *q.v.*

Bromeliales Bromeliads, Spanish moss, pineapple, puya; order of Zingiberidae comprising a single family, Bromeliaceae, of mostly short-stemmed epiphytic herbs containing silica bodies, with firm, succulent leaves which are narrow, parallel-veined and often spiny-margined, arranged so that rainwater is channelled into a reservoir formed by the sheathing leaf bases; the 2000 species are widely distributed in America, mostly in the tropics.

bromeliad (Bromeliales)

Brontotheriidae Extinct suborder of large, rhinoceros-like mammals (Perissodactyla) which flourished during the Eocene and the Oligocene; limbs short and massive, teeth primitive, skull long and low with small braincase; includes brontotheres and titanotheres.

Bronze Age An archaeological period dating from about 5 000 years B.P. in Europe; a period of human history characterized by the use of bronze tools.

bronze leaf Saxifragaceae *q.v.*

brood 1: The offspring of a single birth or clutch of eggs; any group of young animals that are being cared for together by an adult. 2: To incubate eggs.

brood parasitism The use of a host species to brood the young of another species (the parasite).

Brooksellida The sole order of the class Protomedusae *q.v.*

broom Fabaceae *q.v.*

broticolous Living in close proximity to man, in dwelling houses or other buildings; **broticole.**

brotochorous Having propagules dispersed by the agency of man; **brotochore, brotochory.**

brotula Ophidiidae *q.v.*

Brotulidae Ophidiidae *q.v.*

brown algae Phaeophyceae *q.v.*

brown clay A pelagic sediment comprising an accumulation of volcanic and aeolian dust with less than 30% biogenic material.

Brown Earth Brown Forest soil *q.v.*

Brown Forest soil An intrazonal soil with a mull horizon but no pronounced illuvial layer, derived from calcium-rich parent material under deciduous forest in temperate conditions; typically with negligible litter,

dark brown friable A-horizon and lighter coloured B-horizon; Brown Earth.

Brown Podzolic soil A zonal soil similar to Podzol soil *q.v.* but lacking the highly leached intermediate horizon; typically with moderate layer of litter and duff, dark greyish brown A-horizon and a granular light brown acidic B-horizon.

Brown soil A zonal soil with dark brown to greyish brown, slightly alkaline A-horizon and lighter coloured B-horizon over calcareous material, formed in temperate to cool arid climates.

brown water Fresh water rich in suspended organic matter but having a low nutrient content.

browse 1: To feed on parts of plants; **browser.** 2: The edible plant material within reach of browsing animals.

browse line A line marking the height to which browsing animals have been feeding.

Bruchidae Bean weevils, pea weevils; widespread family of beetles (Coleoptera) with the head usually constricted behind the eyes to form a neck; all feed on seeds, especially of leguminous plants, and several are pests of stored products; contains about 1500 species.

Brunelliaceae Small family of Rosales containing about 50 species of evergreen trees native to tropical America.

Bruniaceae Small family of Rosales containing about 75 species of shrubs native to South Africa.

Brunizem Prairie soil *q.v.*

Brunoniaceae Family of Campanulales containing a single species of perennial herbs storing carbohydrates as inulin and having a special type of surface hair; native to Australia; flowers borne terminally on long heads; 5 sepals united into lobed tube, 5 petals and a superior ovary; fruit is a small nut.

Bryales Large, cosmopolitan order of mosses (Bryidae); mat-forming annual or perennial plants occurring mainly on logs, soil or humus in both stable and disturbed habitats; stems usually erect, simple or sparsely branched, sometimes frondose; spore capsule inclined or pendulous on long stalk.

Bryidae Large subclass of mosses (Bryopsida); terrestrial or freshwater green plants arising from a filamentous branched protonema; stems creeping or erect, simple or variously branched and with a poorly developed conducting system; contains between 8000 and 12 000 species.

bryochore That area of the Earth's surface covered by tundra.

bryocolous Living on or in moss; **bryocole.**

bryology The study of mosses and liverworts; **bryological.**

bryophilous Thriving in habitats rich in mosses and liverworts; **bryophile, bryophily.**

Bryophyta Division of plants exhibiting regular alternation of generations between a haploid, thallose or leafy gametophyte which produces male (antheridia) and female (archegonia) reproductive organs and gametes, and small diploid sporophyte which typically comprises a stalked, spore-bearing capsule and an absorptive basal portion; spores produced by meiosis, germinate either directly or via a protonematal stage into the gametophyte; contains 5 classes, liverworts (Hepaticopsida), hornworts (Anthocerotopsida), true mosses (Bryopsida), granite mosses (Andreaeopsida) and peat mosses (Sphagnopsida).

Bryopsida True mosses; large class of bryophytes characterized by having sporophytes that remain attached to the gametophytes; leafy gametophyte rarely has dimorphous leaves and branching is of various non-dichotomous types; leaves have single growing point and are never bilobed; capsule produced on long-lived, unbranched stalk (seta), and typically opening by an apical operculum; capsule often with basal stomates (epidermal openings controlled by guard cells); spore dispersal often aided by a peristome of one or more rows of teeth; comprises 3 subclasses distinguished mainly by peristome characters, Polytrichidae, Bryidae and Tetraphidae.

moss (Bryopsida)

Bryopsidales Order of mainly marine green algae which are most diverse and abundant in warm waters; possessing a thallus consisting of branched tubes (siphons) which are rarely septate; photosynthetic pigments include the distinctive siphonein and siphonaxanthin;Siphonales; Codiales; also treated as a class of the protoctistan phylum Chlorophyta under the name Bryopsidophyceae.

Bryozoa Moss animals; diverse phylum of small sessile aquatic coelomates that typically form colonies containing from a few to a million individuals (zooids); each zooid has a ciliated tentaculate lophophore around the mouth; skeleton cuticular, calcareous or gelatinous; gut U-shaped with anus opening outside lophophore; colonies formed by budding from primary zooid (ancestrula). Most bryozoans are marine, some brackish and freshwater, attached to hard substrates or algae, rarely on soft sediments; contains about 4000 extant species in 3 classes, Phylactolaemata, Stenolaemata and Gymnolaemata, and abundant fossil forms; formerly known as Polyzoa or Ectoprocta.

bubble shell Cephalaspidea *q.v.*

Bucconidae Puffbirds; family containing about 30 species of small insectivorous Neotropical forest birds (Piciformes) found from Mexico to Paraguay; bill strong, straight or hooked, used to catch insects on the wing; habits solitary, sedentary, monogamous, non-migratory; nest in burrow in bank.

Bucephalus larva Cercaria of the digenean fish parasite *Gasterostomum* possessing a forked tail.

Bucerotidae Hornbills; family of large arboreal birds (Coraciiformes) characterized by long massive bill bearing casque on upper margin; plumage loose, tail and wings elongate, legs short and strong; habits solitary or weakly gregarious, non-migratory, monogamous; feed mostly on fruit; nest in tree hole; contains 45 species in the Old World tropics from Africa to New Guinea.

hornbill (Bucerotidae)

buckbean Menyanthaceae *q.v.*
buckeye Hippocastanaceae *q.v.*
buckthorn Rhamnaceae *q.v.*
buckwheat Polygonales *q.v.*
budding 1: A type of asexual reproduction in which a new individual develops as a direct growth from the body of the parent, and may subsequently become detached. 2: The division of insect colonies; colony fission.
Buddlejaceae Butterfly bush; family of Scrophulariales containing about 150 species of shrubs or trees often with hairs of various sorts; mainly tropical and subtropical in distribution; flowers typically with 4 sepals, petals and stamens, and a superior ovary; sometimes included in the Loganiaceae.

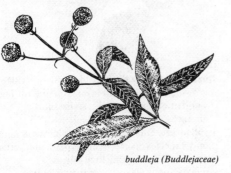

buddleja (Buddlejaceae)

budworm Tortricidae *q.v.*
buffalo fly Simuliidae *q.v.*
buffer 1: A substance which stabilizes the pH of a solution against the addition of acidic or alkaline material. 2: To protect a system from change by external factors; **buffering.**
buffer species A plant or animal acting as an alternative food supply for another organism and thereby buffering the effect of the predator on its normal prey.
buffering The amelioration of environmental conditions by vegetation or topographic features; **buffer.**
Bufonidae Toads; cosmopolitan family containing about 280 species of terrestrial or fossorial amphibians (Anura) typically with a thick, glandular and often warty skin; limbs short, body up to 240 mm length; reproductive habits range from fully aquatic eggs and larvae to terrestrial eggs and direct development.
bug Hemiptera *q.v.*
bulb An underground shoot specialized as a perennating organ formed mainly from fleshy, colourless leaves containing food reserves surrounding the central apical bud and all enclosed in protective scales.

bulbifery Asexual reproduction by the formation of bulbils in the axils of leaves, or in the floral branches.
bulbil A small bulb formed from an aerial bud.
bulbul Pycnonotidae *q.v.*
bullhead Cottidae *q.v.*
bullhead shark Heterodontiformes *q.v.*
bumblebee Apidae *q.v.*
bumblebee bat Craseonycteridae *q.v.*
bunt order A social hierarchy or order of dominance established by butting.
bunting Emberizidae *q.v.*
Buprestidae Jewel beetles; family of metallic coloured, wood-boring beetles (Coleoptera) containing about 15 000 species; adults feeding on nectar, larvae on wood or as leaf miners.

jewel beetle (Buprestidae)

bur reed Sparganiaceae *q.v*
burden The total number or biomass of parasites infecting a given host individual.
Burhinidae Stone curlews; family containing 9 species of medium-sized, long-legged, thick-kneed charadriiform shore birds widespread in dry, open, sandy and stony habitats; habits terrestrial, crepuscular or nocturnal, monogamous, often migratory; feeding on invertebrates and small vertebrates; nesting in scrape on ground.
Burmanniaceae Small family of Orchidales comprising about 130 mostly tropical species of small mycotrophic herbs, sometimes with reduced leaves lacking chlorophyll, otherwise leaves green and clustered at base of stem; fruit a capsule with numerous tiny seeds.
Burramyidae Pygmy phalangers; family containing 9 species of small (to about 250 mm) arboreal marsupials found in Australia, Tasmania and New Guinea; marsupium opens anteriorly; tail somewhat prehensile; habits

nocturnal or crepuscular, mainly insectivorous but also feeding on a variety of plant material and eggs.

burrfish Diodontidae *q.v.*

Burseraceae Gumbo-limbo; pantropical family of Sapindales containing about 600 species of trees and shrubs with prominent resin ducts; flowers small, unisexual; fruit is a drupe or a capsule.

Bursovaginoidea One of 2 orders of gnathostomulids found in marine sands worldwide; characterized by the presence of a female bursa–vagina system, an injectory penis and paired sensory organs on the rostrum; the sperm are not filiform.

bushbaby Galagidae *q.v.*

bush hopper Caelifera *q.v.*

bush rat Octodontidae *q.v.*

bustard Otididae *q.v.*

busy Lizzie Balsaminaceae *q.v.*

butcherbird Cracticidae *q.v.*

Butomaceae Flowering rush; family of Alismatales containing a single species of perennial, emergent aquatic herbs native to temperate Eurasia; inflorescence is a many-flowered umbel; each flower has 6 perianth segments, 9 anthers and 6 follicles containing seeds.

buttercup Ranunculaceae *q.v.*

butterfish Stromateidae *q.v.*

butterfly Lepidoptera *q.v.*

butterfly bush Buddlejaceae *q.v.*

butterfly fish 1: Pantodontidae *q.v.* 2: Chaetodontidae *q.v.*

butterfly ray Gymnuridae *q.v.*

butterfly weed Asclepiadaceae *q.v.*

butterwort Lentibulariaceae *q.v.*

Buxaceae Boxwood; cosmopolitan family of Euphorbiales containing about 60 species of mostly evergreen woody plants; flowers typically very small, each with 4–6 perianth segments and stamens, but often arranged in showy inflorescences, forming shiny black seeds in a fleshy drupe or capsule.

box (Buxaceae)

Buxbaumiales Small order of mosses (Bryidae) growing individually or in mats on soil, rotting wood or on rock surfaces; stems short or absent, sporophytes terminal.

Byblidaceae Small family of Rosales containing only 4 species of insectivorous perennial herbs or half shrubs bearing long oily or mucilaginous glands for trapping insects; sometimes included in the Droseraceae *q.v.*

Byrrhidae Pill beetles; family of small beetles (Coleoptera) containing about 300 species found primarily in temperate regions; larvae occurring in soil, feeding on mosses, liverworts and lichens.

C

C$_3$ plant A plant employing the pentose phosphate pathway of carbon dioxide assimilation during photosynthesis; most green plants belong to this category.

C$_4$ plant A plant employing the dicarboxylic acid pathway for carbon dioxide assimilation during photosynthesis and capable of utilizing lower carbon dioxide concentrations than C$_3$ plants.

cabbage Brassicaceae q.v.

Cabombaceae Small family of aquatic herbs (Nymphaeales) with creeping rhizomes and floating leafy stems; contains 7 species mostly from warm temperate to tropical regions of the . New World.

cacao Sterculiaceae q.v.

cacogenesis The condition of being unable to hybridize.

Cactaceae Cacti; large family of Caryophyllales containing up to 2000 species of mostly spiny, stem-succulents with poorly developed leaves; native to the New World, especially in desert regions; often having large, showy flowers with many petals and stamens and an inferior ovary.

opuntia (Cactaceae)

cadavericolous Inhabiting dead bodies; also used of organisms feeding on dead bodies or carrion; cadavericole.

caddisfly Trichoptera q.v.

caducous Used of structures shed or falling off early.

caecilian Gymnophiona q.v.

Caeciliidae Diverse family containing about 85 species of terrestrial caecilians

(Gymnophiona) widely distributed in the Neotropical region, Africa, India and the Seychelles; body length to 1.2 m; tail absent; scales often present; mostly viviparous but some are oviparous with aquatic larvae.

Caelifera Locusts, short-horned grasshoppers, monkey hoppers, bush hoppers, ground hoppers; suborder of typically diurnal phytophagous orthopteran insects comprising about 11 000 species common in grassland and other vegetation, although some are arboreal or arbusticolous; many engage in swarming and mass migration and some are serious crop pests.

Caenogaea A zoogeographical region incorporating the Nearctic, Oriental and Palaearctic; Cainogaea; Kainogaea; cf. Eogaea.

caenogenetic 1: Of recent origin. 2: Used of transitory adaptations developed in the early ontogenetic stages of an organism; caenogenesis.

Caenolestidae Rat opossums; family of small shrew-like insectivorous marsupials found in forests of western South America; marsupium absent; lower first incisors large.

Caenophytic The period of geological time since the development of the angiosperms in the mid to late Cretaceous; Cenophytic; cf. Aphytic, Archaeophytic, Eophytic, Mesophytic, Palaeophytic.

Caenozoic Cenozoic q.v.

Caesalpiniaceae Senna, royal poinciana, red bud, honey locust, Kentucky coffee tree; large family of Fabales containing about 2200 species of mostly woody leguminous plants widespread in tropical and subtropical regions; flowers usually irregular with 10 or fewer stamens; formerly known as the subfamily Caesalpinioideae of the family Leguminosae.

caespiticolous Living in grassy turf or pastures; caespiticole.

caiman Alligatoridae q.v.

Cainogaea Caenogaea q.v.

Cainotheriidae Extinct family of small, rabbit-like artiodactyls known from the late Eocene and Oligocene of Europe; without gnawing incisors; hindlegs elongate.

Cainozoic Cenozoic q.v.

calabash tree Bignoniaceae q.v.

Calanoida Order of free-living, mainly planktonic copepod crustaceans found at all depths in the sea and in fresh water and inland saline waters; body divided into prosome and urosome behind segment bearing fifth swimming legs; contains about 2500 species, some extremely abundant in surface waters where they can be the main consumers of phytoplankton.

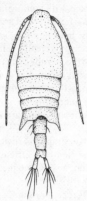

copepod (Calanoida)

Calcarea Class of sponges characterized by a skeleton formed from calcium carbonate spicules laid down as calcite; the spicules may be free or fused; largely restricted to marine waters shallower than 1000 m; comprising 4 subclasses, Calcaronia, Calcinia, Pharetronidia and Sphinctozoa.

calcareous Rich in calcium salts; pertaining to limestone or chalk; growing on or having an affinity for chalky soil.

Calcaronia Subclass of calcareous sponges possessing choanocytes with an apical nucleus associated with the flagellum; producing an amphiblastula larva; comprises 2 orders, Leucosoleniida and Sycettida.

calceolaria Scrophulariaceae *q.v.*

calcicole A plant growing in soils rich in calcium salts; calcipete; **calcicolous**; *cf.* calcifuge.

calcification 1: The deposition of calcium salts in living tissue. 2: The replacement of organic material by calcium salts during fossilization.

calcifuge A plant which is intolerant of soils rich in calcium salts; calciphobe; **calcifugous**; *cf.* calcicole.

Calcinia Subclass of calcareous sponges possessing choanocytes in which the flagellum arises independently of the basal nucleus; releasing larvae at the hollow blastula stage of development; comprises 2 orders, Clathrinida and Leucettida.

calcipete Calcicole *q.v.*

calciphilic Thriving in environments rich in calcium salts; boring into calcareous shells; **calciphile, calciphily.**

calciphobe Calcifuge *q.v.*

calciphyte A plant of soils rich in calcium salts.

calcivorous Used of plants living on limestone or in soils rich in calcium salts; **calcivore, calcivory.**

calcosaxicolous Inhabiting rocky limestone areas; **calcosaxicole.**

California Current A cold surface ocean current that flows southwards along the coast of California, forming the easterly limb of the North Pacific Gyre; see ocean currents.

Californian bluebell Hydrophyllaceae *q.v.*

caliology The study of burrows, nests, hives, tubes and other domiciles constructed by animals.

Callichthyidae Family containing about 130 species of mostly small (to 200 mm) freshwater South American catfishes (Siluriformes); body deep and heavily armoured with bony plates; dorsal and adipose fin with stout spine; 3 pairs of barbels; gas bladder enclosed in bony capsule; aerial respiration may occur through vascularized hindgut.

Callionymidae Dragonets; family containing about 40 species of bottom-dwelling marine teleost fishes (Perciformes); body depressed, to 200 mm length; scales and gas bladder absent; gill openings confined to single aperture on each side of head; sexual dimorphism pronounced; typically live buried in substratum feeding on invertebrates.

dragonet (Callionymidae)

Calliphoridae Bluebottles, blowflies, greenbottles, cluster flies; large family of small to large flies (Diptera) many of which have a metallic body sheen; the maggot-like larvae feed on decaying organic or faecal matter, although some are parasitic; several species are serious pests of livestock or vectors of disease.

Callipodida Order of active carnivorous helminthomorphan diplopods (millipedes) found mostly in dry rocky habitats of the eastern Mediterranean and southern North America, several forms are troglophilic; body elongate, cylindrical, setose; sternal sclerites free, pleural and tergal sclerites fused.

Callitrichaceae Widely distributed family comprising about 35 species of typically aquatic many-branched herbs with small unisexual flowers; each flower without perianth, with a single stamen and a 4-celled ovary.

Callitrichales Order of Asteridae containing 3 small families of aquatic or semiaquatic herbs with reduced flowers.

Callitrichidae Marmosets, tamarins; family containing 14 species of small arboreal insectivorous primates (infraorder Platyrrhini) found in the Neotropical region; body long and slender, squirrel-like, limbs short, tail non-prehensile; locomotion scansorial.

marmoset (Callitrichidae)

Callorhinchidae Family containing 5 species of medium-sized (to 1 m) holocephalan fishes, restricted to shallow waters in temperate regions of the southern hemisphere; body compressed, snout rounded and bearing leaf-like proboscis.

Calmanostraca Subclass of branchiopodan Crustacea comprising the tadpole shrimps (Notostraca).

calobiosis The interrelationship between two species of social insect, in which one lives in the nest of, and at the expense of, the other.

Calobryales Small order of liverworts (Hepaticopsida), widely distributed but individual species typically restricted in range; plants green but lacking secondary pigments; organized into prostrate leafless stolons lacking rhizoids and erect leafy branches with hairs or slime papillae that secrete copious mucilage; female sex organs massive, scattered on stem or sometimes aggregated.

Calycanthaceae Small family of aromatic shrubs or small trees (Laurales) comprising 5 species restricted to North America and China; typically with solitary flowers and ribbon-like stamens; embryo with 2 spirally twisted cotyledons.

Calycerales Order of Asteridae comprising a single family, Calyceraceae, containing about 60 species of herbs with a specialized pollen presentation mechanism; native to Central and South America.

calyptopis Early zoeal larval stage of euphausiacean crustaceans.

Calyptoptomatida Extinct class of marine molluscs known from the Lower Cambrian to the Middle Permian; shell bilaterally symmetrical; aperture usually protected by an operculum possessing paired muscle scars.

calyx The outer whorl of a flower, comprising the sepals.

CAM plant A plant, typically a succulent, which employs the crassulacean acid metabolism for carbon dioxide fixation.

Camallanida Order of spirurian nematodes all of which are parasites utilizing copepod crustaceans as intermediate hosts and various terrestrial and aquatic vertebrates as definitive hosts; possessing simple, uninucleate oesophageal glands.

Camaroidea Extinct order of encrusting graptolites known from the Lower Ordovician; built of autothecae and indistinct stolothecae.

Cambrian The earliest geological period of the Palaeozoic era (*c.* 570–504 million years B.P.); see geological time scale.

Camelidae Camels, llamas; family containing 4 species of herbivorous, terrestrial ungulates (Artiodactyla) native to the Palaearctic and Neotropical regions but now more widespread through domestication; feet with only 2 digits, hooves vestigial; stomach 3-chambered.

camel (Camelidae)

camellia Theaceae *q.v.*

Camerata Extinct subclass of crinoids known from the Lower Ordovician to Upper Permian; thecal plates typically united by fused, rigid polygonal platelets.

Camerothecida Extinct order of calyptoptomatid molluscs known from the Cambrian; operculum lacking.

campanula Campanulaceae *q.v.*

Campanulaceae Campanula, lobelia, bellflower, throatwort, Canterbury bell; large cosmopolitan family containing about 2000 species of mostly herbaceous plants producing latex and storing carbohydrate as inulin; fruit is commonly a capsule or berry; flowers variable but typically with 5 separate sepals, 5 petals, 5 stamens and an inferior ovary bearing 2, 3 or 5 stigmas and carpels.

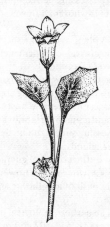

bellflower (Campanulaceae)

Campanulales Order of Asteridae containing 7 families, distinguished by a combination of alternate leaves, an inferior ovary and stamens that are either free or attached at the base of the corolla tube; mostly herbaceous but with some woody species.

Campephagidae Cuckoo shrikes; family containing about 70 species of Old World passerine birds found in forests and bushland habitats from Africa, through Asia to Australia; plumage soft; bill strong and hooked; wings long and pointed; habits arboreal, non-migratory, nesting in trees; feed on fruit and ground insects.

campo A habitat comprising grassy plains with scattered bush and small trees, characteristic of parts of South America.

camptotropism The tendency to return to the natural position if forcibly displaced from it; **camptotropic.**

Canaceidae Surf flies; cosmopolitan family of flies (Diptera) comprising about 60 species found mostly around the seashore; larvae aquatic, often inhabiting intertidal rock pools.

Canaries Current A cold surface ocean current that flows south in the eastern North Atlantic, forming the easterly limb of the North Atlantic Gyre; see ocean currents.

candytuft Brassicaceae *q.v.*

Canellaceae Small family of primitive flowering plants (Magnoliales) found in the tropics; mostly aromatic trees, including wild cinnamon.

Canidae Dogs, foxes; cosmopolitan family of primarily carnivorous terrestrial mammals (Carnivora); habits range from solitary to formation of well-structured social groups; face long; canine teeth large, cheek teeth mostly adapted for crushing, some for cutting (the carnassials); claws blunt, non-retractile; contains about 35 species including dingos, jackals and wolves.

maned wolf (Canidae)

Cannabaceae Cannabis, hops; small family of Urticales containing 4 species of erect or twining herbs, cultivated for fibre (hemp) or psychotrophic drugs (marijuana); small wind-pollinated flowers with pentamerous perianth, 5 stamens and a unilocular ovary.

hops (Cannabaceae)

Cannaceae Cannas; small family of Zingiberales comprising about 50 species of often coarse, leafy-stemmed perennial herbs with silica cells; native to tropical and subtropical regions of the New World; flowers with a single functional stamen bearing a single pollen sac.

cannonball fungi Nidulariales *q.v.*

Canoidea Superfamily of carnivores (Ferae) comprising the dogs (Canidae), bears (Ursidae), raccoons (Procyonidae) and weasels (Mustelidae).

canopy The uppermost continuous stratum of foliage in forest vegetation formed by the crowns of the trees; subdivided into 2 or 3 canopy levels in energy-rich rainforests.

Canterbury bell Campanulaceae *q.v.*

Cantharidae Soldier beetles; family of elongate, parallel-sided beetles (Coleoptera) typically found on flowers and in vegetation, feeding on nectar and pollen; larvae found in soil and litter, primarily predatory; contains about 5000 species distributed worldwide.

cantharophilous Used of plants pollinated by beetles; **cantharophile, cantharophily.**

Cape Horn Current The Antarctic Circumpolar Current in the region of Cape Horn; see ocean currents.

Cape pondweed Aponogetonaceae *q.v.*

caper Capparaceae *q.v.*

capillary water The water held in soil pore spaces and readily available to plants and soil organisms.

Capitellida Order of deposit-feeding polychaete worms comprising about 370 species; body long and cylindrical,; prostomium and peristomium unarmed; pharynx eversible, unarmed; parapodia biramous; includes bamboo worms.

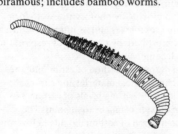

lugworm (Capitellida)

Capitonidae Barbets; family containing about 80 species of pantropical forest birds (Piciformes); body compact, bill short and strong; habits solitary or gregarious, non-migratory, monogamous; feeding mostly on fruit; nest in tree hole or bank burrow.

capitulum A racemose inflorescence in which the flowering shoot is flattened and bears many stalkless flowers (florets) surrounded by a ring (involucre) of bracts; typically found in the composites (Asteraceae).

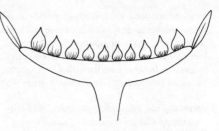

composite flowerhead (capitulum)

Capparaceae Caper; large family of mostly shrubs widespread in warm regions; flowers usually in racemes, with 2 plus 2 sepals, 4 diagonal petals, 4, 6 or more stamens and 2 carpels; capers are the pickled buds of *Capparis spinosa*; Capparidaceae.

Capparales Large order of Dilleniidae consisting of 5 families of herbs to woody plants typically producing mustard oils; including Brassicaceae.

Capparidaceae Capparaceae *q.v.*

caper (Capparaceae)

Caprellidea Suborder of marine amphipod crustaceans comprising the skeleton shrimps (Caprellidae) and whale lice (Cyamidae); caprellids are mostly littoral climbers on algae and hydroids; cyamids are exclusively ectoparasitic on whales and dolphins; Laemodipodea.

caprification The pollination of fig flowers by fig insects (caprifers).

Caprifoliaceae Viburnum, honeysuckle, elderberry, snowberry; family of Dipsacales

honeysuckle (Caprifoliaceae)

containing about 400 species of mostly woody plants mainly from north temperate and boreal regions, and tropical mountains; flowers bisexual and usually in cymose inflorescences; individual flowers usually with 4 or 5 sepals, petals and stamens, and an inferior ovary with many seeds.

Caprimulgidae Nightjars, goatsuckers; family containing about 75 species of small to medium-sized woodland birds (Caprimulgiformes) with cryptic coloration; head large, bill weak with large gape, mouth bearing bristles; habits solitary, largely terrestrial, monogamous, sometimes migratory; feeding on insects caught on the wing.

Caprimulgiformes Small, diverse order of crepuscular or nocturnal, neognathous birds containing 5 families including the frogmouths and nightjars; bill typically weak with a large gape, used to catch insects in flight.

Caproidae Boarfishes; family containing 6 species of small (to 300 mm) strongly compressed, deep-bodied marine zeiform teleost fishes widespread in coastal waters to 600 m depth; body typically reddish, scales ctenoid.

Capromyidae Hutias; family containing 10 species of medium-sized terrestrial and arboreal hystricomorph rodents confined to islands of the West Indies.

capsicum Solanaceae *q.v.*

capsule A dry dehiscent fruit developed from 2 or more many-seeded, fused carpels.

capuchin monkey Cebidae *q.v.*

capybara Hydrochoeridae *q.v.*

Carabidae Predacious ground beetles; tiger beetles, bombadier beetles; family of beetles (Coleoptera) containing over 30 000 species in 1500 genera; mostly active predators found in litter or vegetation; larvae ectoparasitic or predatory, feeding on predigested prey.

Caracanthidae Velvetfishes; family of tiny (to 50 mm) Indo-Pacific marine scorpaeniform teleosts containing only 3 species; body deep, compressed, naked; skin often papillose.

caracaras Falconidae *q.v.*

carambola Oxalidaceae *q.v.*

Carangidae Jacks, pompanos, scads; family of predatory, tropical and temperate, pelagic marine teleost fishes (Perciformes); body variable, compressed to fusiform, length to 1.3 m; 2 dorsal fins present; possessing pair of small spines in front of anal fin; contains about 200 species some popular as game and food-fishes.

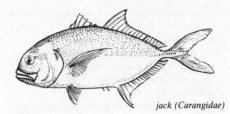

jack (Carangidae)

Carapidae Pearlfishes; family of tropical and subtropical marine and freshwater gadiform teleost fishes; body eel-like, to 300 mm, tail tapering to point; contains 28 species, some inhabiting internal cavities of benthic invertebrates; life cycle can include pelagic vexillifer and benthic tenuis larval stages.

Carapodidae Carapidae *q.v.*

caraway Apiaceae *q.v.*

carbamate A biodegradable organic herbicide.

carbohydrate Any of a group of ketone or polyhydroxy aldehyde compounds including simple sugars (monosaccharides), and the more complex oligosaccharides (such as sucrose) and polysaccharides (such as cellulose and starch) which often serve as energy storage molecules.

carbon-14 A radioactive isotope of carbon having a half-life of 5700 years; used in radiometric dating.

carbon assimilation The incorporation of inorganic carbon from carbon dioxide into organic compounds by photosynthesis.

carbon cycle The biogeochemical cycle of carbon, incorporating the fixation of inorganic carbon dioxide by photosynthesis to form organic complexes and its ultimate return to the atmosphere by processes of respiration and decomposition.

carbon dating A method of estimating the age of archaeological material and subrecent fossils by measuring levels of carbon-14.

carbon dioxide compensation point The concentration of carbon dioxide below which, in a non-limiting light intensity, photosynthesis just compensates for respiration and the value of net photosynthesis is zero.

carbonate compensation depth The oceanic depth at which the rate of solution of carbonate equals the rate of input at that level, typically a depth of about 5000 m.

carbonate dissolution depth The oceanic depth below which the solubility of calcium carbonate increases to the point that dissolution of calcareous shells begins; typically a depth of about 4000 m.

carbonicolous Living on burnt or scorched substrates; **carbonicole.**

Carboniferous A geological period within the Palaeozoic (*c.* 365–290 million years B.P.), divided into Pennsylvanian and Mississippian; see geological time scale.

carbonization An unusual type of fossilization in which organic material has been reduced to a carbon film residue.

carboxyphilic Thriving in carbon dioxide-rich habitats; **carboxyphile, carboxyphily.**

Carcharhinidae Requiem sharks; diverse cosmopolitan family of small to large (0.5–7.5 m), mainly pelagic, carcharhiniform elasmobranch fishes found in shallow water to 400 m; mostly marine but also penetrating tropical rivers and lakes; feed on variety of fishes, marine mammals, reptiles and invertebrates; reproduction typically viviparous with yolk sac placenta; contains about 50 species including tiger, sharp-nosed, grey, lemon and blue sharks; most species reported in shark attacks on people belong to this family.

Carcharhiniformes Large order of mainly marine, benthic or pelagic, coastal to oceanic, littoral to abyssal, galeomorph sharks; snout typically rather elongate, flattened, mouth subterminal; dorsal fins without fin spines; reproduction oviparous or ovoviviparous, to fully viviparous with yolk sac placenta; feed on variety of marine vertebrates and invertebrates; contains about 200 species in 8 families, including cat, hound, weasel, hammerhead, and requiem sharks.

carcinogenic Cancer-producing.

carcinology The study of crustaceans; **carcinological.**

cardinal Cardinalidae *q.v.*

Cardiopteridaceae Family of Celastrales containing only 3 species of glabrous twining herbs producing a milky juice; native from southeast Asia to Australia.

cardinalfish Apogonidae *q.v.*

Cardinalidae Cardinals; family of finches (Passeriformes) with large conical bills, found in woodland in the New World; males often highly coloured; commonly classified in the Emberizidae *q.v.*

Carettochelyidae Monotypic family of large (to 750 mm) predatory freshwater turtles (Testudines: Cryptodira) found in rivers of New Guinea and northern Australia; carapace covered with leathery skin, snout forming proboscis; feet webbed.

Cariamidae Seriemas; family of large terrestrial gruiform birds found in forests and grasslands of central South America; legs strong for fast running, wings short, flight weak; feeding on variety of invertebrates and small vertebrates, including snakes; nest in trees.

Caribbean Current A warm surface ocean current that flows northwest in the Caribbean Sea, derived in part from the Atlantic South Equatorial Current and giving rise to the Florida Current through the Gulf of Mexico; see ocean currents.

caribou Cervidae *q.v.*

Caricaceae Papaya; small family of Violales containing about 30 species of mostly soft-stemmed shrubs or trees typically with an unbranched trunk and terminal cluster of leaves; found in America and Africa; flowers regular, mostly unisexual; fruit is a berry.

Caridea Infraorder of pleocyematan decapod crustaceans including many commercially exploited species of prawns and shrimps; pleura of second abdominal segment overlap those of first and third segments; third thoracic legs not chelate; Eucyphidea.

carnation Caryophyllaceae *q.v.*

Carnivora Diverse order of terrestrial and aquatic carnivorous mammals comprising 10 families grouped in 2 suborders, Caniformia including Canidae (dogs, foxes), Ursidae (bears), Otariidae (sealions), Ailuropodidae (giant panda), Procyonidae (racoons), Mustelidae (weasels, otters), Phocidae (hair seals), and Feliformia including Felidae (cats), Viverridae (civets), Hyaenidae (hyaenas).

carnivorous Flesh-eating; **carnivore, carnivory.**

carnosaur Carnivorous saurischian dinosaur, typically bipedal and powerful, with dagger-like teeth; includes *Tyrannosaurus rex*.

tyrannosaurus (carnosaur)

carotene A type of hydrocarbon, yellow to red pigment found as an accessory photosynthetic pigment in many plant cells.

carp Cyprinidae *q.v.*

carpel The female reproductive structure in flowering plants, that bears and encloses the ovules; typically consisting of a basal ovary, elongate style and terminal, receptive stigma; commonly fused in advanced forms.

carpelotaxis The arrangement of carpels in a flower or fruit.

carpenter bee Anthophoridae *q.v.*

carpet beetle Dermestidae *q.v.*

carpet moth Geometridae *q.v.*

carpet shark Orectolobiformes *q.v.*

carpogenous Growing on or in fruit.

Carpolestidae Extinct family of shrew-sized primates known from the Palaeocene and Eocene; probably herbivorous forms found on forest floors; digits with claws.

carpophagous Feeding on fruit or seeds; **carpophage**, **carpophagy**.

carr A mire containing scrub vegetation; fen woodland.

carrier An organism carrying a disease or infectious agent without showing typical symptoms and which is capable of passing the infection to another individual; *cf.* vector.

carrion Dead putrefying flesh.

carrion beetle Silphidae *q.v.*

carrot Apiaceae *q.v.*

carrying capacity The maximum number of organisms that can be sustained within a given area or habitat.

cartilaginous fishes Chondrichthyes *q.v.*

Carybdeida The sole order of Cubozoa *q.v.*

Caryoblastea Phylum of protoctistans comprising one species, *Pelomyxa palustris*, known from the muddy bottoms of freshwater ponds; characterized by the possession of membrane-bound nuclei and by the lack of the other cytoplasmic organelles characteristic of eukaryotes, except for the 9+2 flagellum, which is intracellular; nuclear division is apparently direct, not involving mitosis.

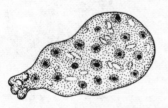

caryoblastean (Caryoblastea)

Caryocaraceae Souari nut; small family of evergreen trees or shrubs confined to tropical America.

Caryophyllaceae Chickweed, pink, carnation, moss campion, sweet William; large family of 2000 species of temperate and warm temperate, northern hemisphere herbs with stems commonly swollen at nodes; flowers typically with 5 sepals, 5 or more petals, 5 or 10 stamens and a superior ovary; producing a capsular fruit with many seeds.

campion (Caryophyllaceae)

Caryophyllales Large well defined order of Caryophyllidae; herbs or woody plants, often succulent and showing C_4 photosynthesis, sometimes with leaves reduced to spines; contains about 10 000 species in 12 families.

Caryophyllidae Subclass of dicotyledons (Magnoliopsida) with a fossil record dating back only to the beginning of the Palaeocene; characterized by their production of betalains but not anthocyanins or by the position of the ovule attachment in their compound ovary; contains 3 orders, Caryophyllales, Polygonales and Plumbaginales.

Caryophyllidea Order of tapeworms in which the posterior part of the body, the strobila, is unsegmented and contains only a single set of reproductive organs and in which the scolex is weakly developed; the life cycle involves freshwater oligochaete worms as intermediate hosts and freshwater teleost fishes as definitive hosts.

caryopsis A dry indehiscent fruit containing a single seed and derived from a single carpel the wall of which fuses with the seed wall during development; typical of members of the Poaceae, including the cereal crops.

cashew nut Anacardiaceae *q.v.*

Cassiduloida Order of irregular, littoral to upper continental slope, tropical to temperate, sea urchins (atelostomata) comprising only about 25 extant species; test rounded, mouth central, aboral ambulacra petal-like.

cassowary Casuariidae *q.v.*

cast A rock formed within a cavity, as in the replacement of a fossil by mineral infiltration.

caste Individuals within a colony belonging to a particular morphological type or age group, or that share a particular behaviour pattern and perform the same specialized function.

casting An object that has been cast off or voided by an organism, such as a faecal pellet or worm casting.

castor bean Euphorbiaceae *q.v.*

Castoridae Beavers; family containing 1 Nearctic and 1 Palaearctic species of semiaquatic mammals (Rodentia) that construct elaborate dams in small streams to form deep ponds; hindlimbs larger than forelimbs, feet webbed; tail broad and spatulate; feed on bark of trees.

beaver (Castoridae)

castration The removal of, or interference with the function of, the testes (geld, emasculation) or ovaries (spey).

casual 1: A plant growing in a community with which it is not normally associated. 2: An alien species that survives for only a short period in the habitat to which it was introduced.

Casuariidae Cassowaries; family containing 3 species of large (to 1.5 m tall) ratite birds (Palaeognathae) found in rain forests of New Guinea and northern Australia; legs stout, powerful, bearing 3 digits; body black, feathers hair-like; upper neck and head naked, coloured blue and red; habits cursorial, polygamous; feed mainly on fruit.

cassowary (Casuariidae)

Casuariiformes Order of ratite birds comprising 2 families, Casuariidae (cassowaries) and Dromaiidae (emus).

Casuarinales Casuarinas, she-oaks; order of Hamamelidae comprising a single family, Casuarinaceae, of about 50 species of evergreen trees bearing slender green branches with minute whorled leaves; native to Australia and adjacent areas.

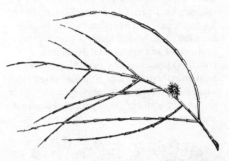

she-oak (Casuarinales)

cat Felidae *q.v.*

cat shark Scyliorhinidae *q.v.*

catabatic Used of winds blowing down a slope, may be warm (foehn wind) or cold (fall wind).

catabolism That part of metabolism involving the degradation of complex substances with the resultant liberation of energy; *cf.* anabolism.

catadromous Migrating from fresh water to sea water, as in the case of fishes moving into the sea to spawn; **catadromy**; *cf.* anadromous, oceanodromous, potamodromous.

catamenial Monthly; menstrual; **catamenia.**

cataplasia An evolutionary state characterized by decreasing vigour; **cataplasis**; *cf.* anaplasia, metaplasia.

catarobic Used of an aquatic habitat in which slow decomposition of organic matter is taking place but in which conditions have not yet become anaerobic.

Catarrhini Infraorder of primates (suborder Haplorhini) comprising 4 families, Cercopithecidae (Old World monkeys), Hylobatidae (gibbons), Pongidae (great apes), Hominidae (humans).

catastrophism The doctrine that fossil faunas were the result of catastrophic changes which had periodically exterminated large numbers of species, so that past cataclysmic geological or climatic events have had a major impact on the course of evolution; *cf.* uniformitarianism.

catathermal Pertaining to a period of decreasing temperature; *cf.* anathermal.

catch crop A quick-growing crop which is cultivated between the rows of a main crop, between the main crops of an ordinary crop rotation or in place of a main crop which has failed.

catchment A natural drainage basin which channels rainfall into a single outflow.

Catenulida Cosmopolitan order of free-living aquatic turbellarians containing about 75 species of slow-moving, elongate and delicate worms; characterized by a simple pharynx and a straight gut, and by sperm lacking flagella.

caterpillar Soft-bodied larva of Lepidoptera (butterflies and moths) and Symphyta (sawflies) possessing 2 types of legs, true segmented thoracic legs, and unsegmented sucker-like false legs (prolegs) typically on the third to sixth and on the terminal abdominal segment.

caterpillar

catfish Siluriformes *q.v*

catfish eel Plotosidae *q.v.*

Cathartidae New World vultures, condors; family containing 7 species of large carrion-feeding birds found in North and South America; head naked, often coloured; wings broad, up to 3.5 m span; habits solitary or gregarious, monogamous; nesting on ground or in tree hollows.

Cathaysia A continental landmass comprising most of China and southeast Asia, formed from the supercontinent Eurasia *q.v.* after the break-up of Pangaea; *cf.* Angara, Euramerica.

catkin A racemose inflorescence in which the main axis is typically elongate bearing many small stalkless unisexual flowers.

catnip Lamiaceae *q.v.*

Catostomidae Suckers; family of freshwater cypriniform teleost fishes found throughout Central and North America, and also in China and Siberia; body deep, subcylindrical to compressed, mouth with thick lips; typically bottom-living, feeding on plankton, insect larvae and detritus; contains about 60 species including buffalo fishes, cultured for food in United States.

cattail Typhaceae *q.v.*

cattle Bovidae *q.v.*

catworm Phyllodocida *q.v.*

Caudata Salamanders, newts; order of tailed Amphibia comprising about 330 species in 8 families, having a primarily holarctic distribution; adults aquatic, terrestrial, occasionally arboreal; forelimbs and hindlimbs usually present and of subequal size; many parts of skeleton cartilaginous; eggs and larvae primitively aquatic; direct development and viviparity also present; some forms neotenic.

Caudofoveata Class of molluscs containing less than 70 species of marine infaunal burrowers which are vermiform and feed on microorganisms and detritus; characterized by the small, cylindrical body covered with a mantle which secretes a chitinous cuticle and embedded imbricating scales, by the bipartite radula, by a reduced gliding surface on the foot, and by the presence of a pair of ctenidia; sometimes treated as an order of the Aplacophora.

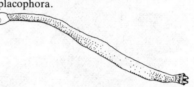

caudofoveate (Caudofoveata)

cauliflory The production of flowers on secondarily thickened plant tissues such as branches or trunks.

caulocarpous Used of a plant producing fruit in each of 2 or more successive years.

Caulophrynidae Family of small (to 200 mm) pelagic, marine anglerfishes (Lophiiformes) comprising 2 species found from depths of 10 m to 1500 m; body of female rounded, naked; mouth large with pointed teeth; pectoral fins fan-like; lure lacking apical bulb; male slender and ectoparasitic on female.

cave cricket Rhaphidophoridae *q.v.*

cavefish Amblyopsidae *q.v.*

cavernicolous Living in subterranean caves or passages; **cavernicole.**

Cavibelonia Order of solenogastres (Mollusca) possessing a mantle with layers of hollow, mainly needle-like, calcareous bodies.

Caviidae Guinea-pigs; family containing 12

guinea pig (Caviidae)

species of small to large terrestrial hystricomorph rodents confined to the Neotropical Region; 4 digits on forelimbs, 3 on hindlimbs; tail vestigial; habits diurnal or nocturnal, herbivorous, living in burrows.

Caytoniales An order of gymnosperms known as fossils from the Jurassic period.

Cebidae New World monkeys; family of primarily arboreal, herbivorous, Neotropical primates; body and limbs elongate, tail long to short, prehensile or not; ischial callosities absent; typically forming family groups; contains about 35 species including capuchin, squirrel, howler, spider, and woolly monkeys.

Cecidomyiidae Gall midges; cosmopolitan family of minute fragile flies (Diptera) containing more than 4000 species; larvae often phytophagous living in galls, others saprophagous or predacious.

gall midge (Cecidomyiidae)

Cecropiaceae Family of Urticales found in tropical regions containing about 300 species of trees, shrubs or woody vines with stilt or aerial roots; sometimes included in the family Urticaceae *q.v.*

Celastraceae Climbing bittersweet, spindle tree; large, cosmopolitan family of usually glabrous woody plants containing about 800 species; small, regular flowers arranged in

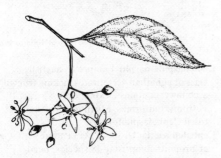

spindle tree (Celastraceae)

cymose inflorescences, each flower typically with 3–5 sepals, petals and stamens, and a superior ovary producing a capsular fruit.

Celastrales Order of Rosidae, containing about 2000 species of alkaloid-producing woody plants or herbs arranged in 11 families.

celery Apiaceae *q.v.*

cellular Composed of cells; pertaining to cells; *cf.* acellular.

cellular slime moulds A group of independently motile, phagotrophic, uninucleate amoeba-like organisms which aggregate into a mass that transforms itself into a spore-producing fruiting body; classified as a phylum of Protoctista (Acrasiomycota), as a class of fungal slime moulds (Acrasiomycetes), or as a class of rhizopod protozoans (Acrasea).

cellular slime mould

Celsius scale (°C) A scale of temperature with the melting point of ice as zero and boiling point of water as 100 degrees at normal atmospheric pressure; equivalent to Centigrade scale.

Cenophytic Caenophytic *q.v.*

Cenozoic A geological era (*c.* 65–0 million years B.P.) comprising the Quaternary and Tertiary periods; the age of mammals; Caenozoic; Cainozoic; Kainozoic; see geological time scale.

cenozoology The study of extant animals.

centi- Prefix meaning hundredth; used to denote unit x 10^{-2}; see metric prefixes.

Centigrade scale (°C) A scale of temperature that takes the melting point of ice as zero and the boiling point of water as 100 degrees; now abandoned in favour of the Celsius scale.

centipede Chilopoda *q.v.*

Centrales Centric diatoms; order of primarily marine algae found in planktonic and epibenthic habitats and in damp terrestrial situations; typically with circular valves which are radially symmetrical; non-motile but with

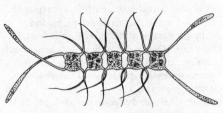

centric diatom (Centrales)

flagellate male gametes; also treated as a class of the phylum Bacillariophyta.

Centrarchidae Sunfishes; family of North American freshwater perciform teleost fishes containing about 30 species; first dorsal fin strongly spined, united with second; tail deep.

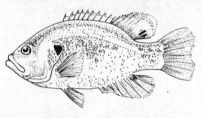

sunfish (Centrarchidae)

Centriscidae Shrimpfishes; family containing 4 species of tropical Indo-Pacific shallow marine gasterosteiform teleost fishes usually found living amongst the body spines of sea urchins; body strongly compressed with armour of thin bony plates; snout elongate; dorsal fin with 3 spines, located at end of body.

Centroceratida Extinct order of cephalopod molluscs (Nautiloidea) in which the sutures are trilobed and the siphuncle subcentral; known from the Lower Devonian to the Upper Jurassic.

Centrohelida Order of Heliozoa characterized by a skeleton composed of siliceous or organic plates or spines and by the presence of a centroplast on which the axopodal microtubules insert.

Centrolenidae Family containing about 60 species of small (to about 30 mm) tropical South American arboreal leaf frogs (Anura); body usually with green coloration; eggs laid on vegetation above water, tadpoles aquatic; feet with pads and extra-digital cartilaginous elements.

Centrolepidaceae Small family of Restionales comprising about 35 species of small, tufted, mostly annual grass-like or moss-like herbs with solid stems and leaves with slender blades; has a widespread but patchy distribution.

Centrolophidae Medusa fishes; family containing about 22 species of pelagic marine stromateoid teleosts (Perciformes); length to 1.4 m, head naked, mouth large; juveniles live in association with jellyfishes.

Centrophrynidae Monotypic family of tiny (to 50 mm) deep-sea anglerfishes (Lophiiformes); female body elongate, slender posteriorly, skin densely spinose; eyes subcutaneous; lure located on snout; male smaller than female, lacking teeth.

Centropomidae Snooks; family of large (to 2 m) inshore marine and brackish water perciform teleost fishes popular as sport and food-fishes; body elongate, compressed, with dorsal fin fully divided.

century plant Agavaceae *q.v.*

Cephalaspidea Bubble shell; order of marine opisthobranch molluscs most of which are bottom-living carnivores, preying on other invertebrates; typically with an external shell, a broad creeping foot and a flattened dorsal head shield used for burrowing; the stomach has grinding gizzard plates.

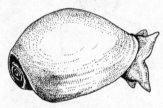

bulloid mollusc (Cephalaspidea)

Cephalaspidiformes Extinct order of fish-like vertebrates known from the Silurian and Devonian; body flattened, head broad; 10 pairs of ventral gill openings present; mouth ventral and lacking jaws.

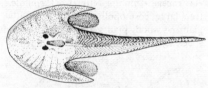

cephalaspid (Cephalaspidiformes)

Cephalaspidomorphi Lampreys, hagfishes; class of agnathan vertebrates lacking true gill-arch jaws; contains 2 extant orders, Petromyzoniformes, Myxiniformes, and the fossil Cephalaspidiformes.

Cephalobaenida Order of pentastome arthropods comprising 18 species; head pointed, hooks arranged in trapezoidal pattern; two families recognized, one parasitic in lizards and snakes, the other in aquatic or semiaquatic birds; secondary insect host known only in the case of the lizard parasites.

Cephalocarida Class of small, primitive crustaceans comprising a single order, Brachypoda, found in intertidal to bathyal fine marine sediments; body elongate, comprising head, 8-segmented thorax with triramous foliaceous limbs, and naked, 12-segmented abdomen; larval development gradual through 13 naupliar and 5 juvenile stages; contains 9 species from coasts of North and South America, southwest Africa and Japan.

Cephalochordata Lancelet, amphioxus; subphylum of invertebrate chordates cosmopolitan in shallow marine sand and gravel habitats; body translucent, to 70 mm length, tapered at both ends, with series of V-shaped muscle blocks; feed by filtering suspended particles from pharyngeal water current produced by gill bar cilia; sexes separate, monomorphic, fertilization external, larvae planktonic; contains about 23 species in 3 genera; Acrania.

lancelet (Cephalochordata)

Cephalopoda Octopus, squid, cuttlefish; a class of marine carnivorous molluscs characterized by the specialization of the head–foot into a ring of arms (tentacles) generally equipped with suckers or hooks, by the funnel of the mantle cavity used for propulsive locomotion, by a powerful parrot-like beak, and by well developed eyes; consists of 2 subclasses, Nautiloidea and Coleoidea.

Cephalotaceae Family of Rosales containing a single insectivorous herb species with some leaves modified as pitchers; found in bog habitats in southwestern Australia; flowers small, hexamerous and borne on a long scape.

Cephalotaxaceae Small, monogeneric family of

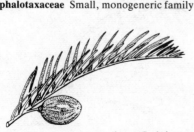

cowtail pine (Cephalotaxaceae)

conifers (Pinatae) native to temperate eastern Asia; foliage yew-like; male cones small, usually globose; female structure erect with a fleshy aril enclosing most of the seed.

Cepolidae Bandfishes; family of bottom-dwelling marine teleosts (Perciformes) found amongst rocks or burrowing in soft sediments; body extremely elongate, compressed, ribbon-like, to 700 mm length; dorsal and anal fins continuous with caudal; 7 species found in Indo-Pacific and eastern North Atlantic.

Ceractinomorpha Subclass of demosponges with a skeleton composed of siliceous spicules and spongin fibres, or spongin fibres alone; variable in body form but commonly branching, found from the intertidal zone down to 7000 m; typically viviparous with incubated parenchymella larvae.

Cerambycidae Long-horned beetles, timber beetles; family of medium-sized, elongate beetles (Coleoptera) containing about 35 000 species in 4000 genera distributed worldwide; adults typically feeding on nectar and pollen; larvae bore into living or dead plant material, including timber.

Ceramiales Large order of predominantly marine algae of filamentous or pseudoparenchymatous body form.

Ceratiidae Sea devils; family of deep-sea anglerfishes (Lophiiformes) comprising 2 species found at depths up to 2000 m; female body robust, skin rough and spinose, length to 1.2 m, lure elongate; males about 60 mm in length, ecto-parasitic on female.

Ceratiomyxomycetidae Subclass of plasmodial slime moulds *q.v.* (Myxomycetes) comprising a single order, Ceratiomyxales, characterized by a fruiting body with hair-like stalks over its surface, each of which bears a single spore at the tip; spores produce flagellate cells which fuse to initiate the plasmodial stage, found migrating through the interstices of wood.

Ceratodontidae Australian lungfish; monotypic family of freshwater fishes found in rivers and reservoirs of northeastern Australia; body compressed, to 1.8 m length, pectoral and pelvic fins broad; the unpaired swim bladder functions as a lung.

Ceratomorpha Suborder of mammals (Perissodactyla) comprising the tapirs and rhinoceroses.

Ceratophyllaceae Small but cosmopolitan family of submersed, rootless perennial herbs (Nymphaeales); flowers small, unisexual; perianth tiny with many narrow lobes, 10–20 stamens; fruit a one-seeded nut.

Ceratopongonidae Biting midges, punkies; cosmopolitan family of minute biting flies (Diptera) comprising about 1200 species that feed on the blood of warm- and cold-blooded vertebrates and are of considerable medical and veterinary importance in some areas as vectors of disease.

biting midge (Ceratopogonidae)

Ceratoporellida Order of sclerosponges known from warm shallow waters in the Caribbean, in which the living sponge tissue forms a thin surface layer over the basal calcareous skeleton and extends down into it in places.

Ceratopsia Horned ornithischian dinosaurs with a large head and extensive bony frill extending back over the neck; jaws beak-like; includes *Triceratops*.

cercaria Tadpole-shaped larval stage of a digenean trematode parasite produced asexually from redia in the molluscan intermediate host; typically with a short free-swimming phase; may infect the definitive host or a second intermediate host.

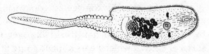

cercaria

Cercidiphyllaceae Family containing 2 species of deciduous, dioecious trees with small wind-pollinated flowers, native to Japan and China; flowers unisexual, with 4 perianth parts; fruit is a follicle.

Cercopidae Froghoppers, spittlebugs; large family of dark-coloured, frog-like, hopping insects (Homoptera) that feed on trees, shrubs and herbaceous plants, and lay eggs which hatch into sedentary nymphs protected from predators and desiccation by a mass of frothy spittle; comprising about 2300 species distributed worldwide.

Cercopithecidae Old World monkeys; diverse family of primarily arboreal primates (infraorder Catarrhini); tail present or absent, non-prehensile; ischial callosities present; habits typically diurnal, herbivorous or omnivorous; most form small to large social groups; family includes baboons, mandrills, langurs and is distributed through the Ethiopian, Oriental and Palaearctic regions.

baboon (Cercopithecidae)

Cercopithecoidea Superfamily comprising the Old World monkeys and their fossil relatives.

ceremony A complex pattern of display behaviour used to promote and maintain social bonds in some groups.

Ceriantharia Tube anemones; order of solitary ceriantipatharians found mostly in warm temperate and tropical seas in both shallow and deep water; typically with an elongate column buried upright in soft sediment to the level of the oral disk; form tube out of mucus and interwoven cnidae into which the animal retracts; lack a pedal disk at the aboral end.

Ceriantipatharia Subclass of anthozoans comprising the black corals (Antipatharia) and the tube anemones (Ceriantharia); typically with weak mesenteric musculature and coupled but not paired mesenteries; sometimes included in the subclass Zoantharia.

cerophagous Wax-eating; **cerophage**, **cerophagy**.

Cerozem Sierozem *q.v.*

Certhiidae Tree creepers; family containing 8 species of small passerine birds widespread in

tree creeper (Certhiidae)

forests and wooded habitats of northern hemisphere and Africa; bill slender, curved, pointed; legs short, feet and claws strong; habits solitary, arboreal, feeding on insects gleaned from tree bark; cup-shaped nests placed in crevice or on tree branch.

Cervidae Deer; family containing about 35 species of herbivorous terrestrial ungulates (Artiodactyla: Ruminantia) widespread in most parts of the world except the Australian region, occupying diverse habitats from Arctic tundra to rainforest; forming small to large herds; some species migratory; antlers are commonly present, at least in males, and are shed and regrown annually; includes reindeer (caribou) and moose (elk).

red deer (Cervidae)

Cervoidea Superfamily of artiodactyls (Pecora) comprising the deer (Cervidae) and giraffe and okapi (Giraffidae).

Cestida Small order of pelagic ctenophores with a worldwide distribution; characterized by a ribbon-like body that is extremely compressed in the tentacular plane and elongated in the stomodeal plane; tentacles are present and are involved in prey capture.

Cestoda Tapeworms; class of exclusively parasitic flatworms found mostly in the intestines of vertebrates as adults and within a vertebrate or invertebrate intermediate host during development; characterized by a body usually consisting of an anterior holdfast (scolex), an unsegmented neck and a chain of

tapeworm (Cestoda)

segments called proglottids formed by serial budding or strobilation; body covered with a syncytial tegument; typically with one or more sets of male and female reproductive systems; shelled embryos are released, and are eaten by the intermediate host within which larval development takes place; comprising 2 subclasses, Cestodaria and Eucestoda.

Cestodaria Small subclass of tapeworms found in primitive fishes and turtles distinguished by the lack of a scolex (an anterior holdfast), an unsegmented body and by the possession of a single set of male and female reproductive organs; the first larval stage, the lycophore, has 10 hooks; comprising 2 orders, Amphilinidea and Gyrocotylidea.

Cetacea Whales, dolphins, porpoises; order of marine mammals comprising 2 extant suborders, Mysticeta (baleen whales) and Odontoceta (toothed whales); body fusiform, forelimbs modified as flippers, hindlimbs absent, tail bearing lateral flukes; nares dorsal; eyes small, pinnae absent; skin essentially hairless with thick layer of subcutaneous fat for insulation.

cetology The study of cetaceans; **cetological.**

Cetomimidae Whalefishes; cosmopolitan family containing 10 species of small (to 150 mm) bathypelagic beryciform teleost fishes; body flabby, fragile, naked; head large, eyes and teeth minute; pelvic fins absent.

Cetopsidae Family of small (to 150 mm) Amazonian catfishes (Siluriformes) comprising 12 species; body robust, cylindrical, naked; gas bladder enclosed in bony capsule; aerial respiration may occur through vascularized hindgut.

Cetorhinidae Basking sharks; family of extremely large (to 13 m) inoffensive, filter-feeding, lamniform elasmobranchs having a widespread temperate distribution, second only to the whale shark as the largest fish; eyes very small, external gill openings extend onto dorsal surface, gill rakers abundant; mouth large; pelagic, feeding on planktonic crustaceans.

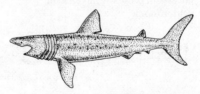

basking shark (Cetorhinidae)

Chacidae Monotypic family of small (to 200 mm) freshwater catfishes (Siluriformes) found in slow-moving lowland streams of India, Burma and Sumatra; body depressed anteriorly, compressed posteriorly, dorsal fin short with single spine; 3 pairs of barbels.

Chaenichthyidae Crocodile icefishes; family containing 15 species of Southern Ocean bottom-dwelling perciform teleost fishes; body naked, to 600 mm length, head large and spinose; blood lacking haemoglobin, oxygen carried in plasma.

Chaetodermatida Alternative name for the Caudofoveata *q.v.*

Chaetodontidae Butterfly fishes, angelfishes; family containing 150 species of colourful shallow-water perciform teleost fishes commonly associated with coral reefs, widespread in tropical and subtropical seas especially of the Indo-Pacific; body deep, compressed, to 600 mm length; snout prolonged, mouth small and protractile, teeth minute; larval stage termed a tholichthys.

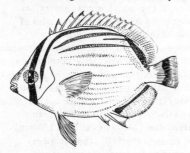

butterfly fish (Chaetodontidae)

Chaetognatha Arrow worms; small phylum of bilaterally symmetrical, free-living, predatory marine coelomates; body slender, head armed with 2–3 pairs of sharp spines around the mouth, translucent trunk bearing lateral and caudal fins; hermaphrodite but reciprocal cross fertilization common. Most chaetognaths are planktonic, a few benthic; may be exceptionally abundant. About 70 species in 2 orders, Phragmophora, Aphragmophora. Evolutionary relationships uncertain, presently linked with deuterostomes.

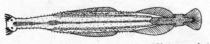

arrow worm (Chaetognatha)

Chaetomiales Order of pyrenomycete fungi comprising about 100 cellulose-decomposing species found in soil, dung and plant debris.

Chaetonotida Order of aquatic and semiaquatic gastrotrichs, of variable body form but lacking pharyngeal pores.

Chaetophoriales Order of predominantly freshwater green algae; typically with thallus of branched filaments composed of uninucleate cells, each containing a single parietal laminate chloroplast; also treated as a class of the protoctistan phylum Chlorophyta.

chaetoplankton Planktonic organisms having spiny outgrowths to reduce their rate of sinking.

Chaetopterida Order of filter-feeding tubicolous polychaete worms that inhabit leathery or parchment-like tubes embedded in soft sediments; body fragile, divided into 3 distinct regions; prostomium indistinct, peristomium collar-like; longitudinal groove produces abundant mucus; contains about 45 species in single family; sometimes included in the order Spionida.

polychaete worm (Chaetopterida)

chaetotaxy The pattern or distribution of bristles, chaetae, setae, spines etc., used as taxonomic characters.

Chaetothyriales Sooty moulds; order of loculoanoteromycetid fungi that are superficial parasites of green plants or saprophytic on plants and other fungi.

chafer Scarabaeidae *q.v.*

Chalastogastra Symphyta *q.v.*

chalazogamy Fertilization in flowering plants in which the pollen tube reaches the nucellus of the ovule by passing through the placenta and the basal region where nucellulus and integuments are fused (the chalaza).

chalcid wasp Chalcididae *q.v.* (Apocrita).

Chalcididae Chalcid wasps; diverse cosmopolitan family of small to large robust parasitic wasps (Hymenoptera) comprising over 1400 species; mostly solitary parasites of other insects, including Lepidoptera, Hymenoptera, Coleoptera, Neuroptera, Diptera.

chalicophilous Thriving on gravel slides; **chalicophile, chalicophily.**

chalicophyte A plant inhabiting gravel slides.

Chalicotheriidae Extinct suborder of perissodactyls known from the Eocene to the Pleistocene; front legs slightly longer than hindlegs.

challenge display A high intensity aggressive

display performed by a male to a conspecific male.

Chamaeleontidae Chamaeleons; family containing about 70 species of arboreal and terrestrial lizards (Sauria) widespread in Africa and Madagascar; length to about 65 mm; eyes highly mobile, can be moved independently; tongue extensible; digits opposable; tail prehensile, non-autotomic; may exhibit very rapid change of colour.

chamaephyte A perennial plant having renewal buds at or just above ground level (up to 250 mm); usually low-growing woody or herbaceous plants common in dry or cold climates and having buds on aerial branches near the ground; sometimes divided into active chamaephyte, passive chamaephyte and suffructicose chamaephyte; see Raunkiaerian life forms.

chamomile Asterales *q.v.*

Chanidae Milkfish; monotypic family of herbivorous Indo-Pacific marine and brackish-water teleost fishes (Gonorhynchiformes); body terete, herring-like, compressed, to 1.7 m length; pelvic fins below dorsal fin, caudal deeply forked; exceptionally fecund; widely cultured in tidal ponds in Indonesia and Philippines.

Channidae Snakeheads; family containing 10 species of piscivorous freshwater teleost fishes (Channiformes) from tropical Africa and eastern Asia; body elongate, to 1.2 m length, cylindrical anteriorly, compressed posteriorly; head broad; dorsal and anal fins elongate; an accessory respiratory organ (suprabranchial organ) is present in the upper part of the gill chamber for aerial respiration.

Channiformes Order of small to large predatory freshwater teleost fishes comprising the Channidae (snakeheads).

Chaoboridae Family of small fragile dipteran insects (flies) which have predacious aquatic larvae that use strong prehensile antennae to catch mosquito larvae and other small arthropods.

Chaparral biome A mild temperate region with a Mediterranean climate of cool moist winters and long dry summers with a vegetation of varied sclerophyllous evergreen shrubs and a diverse fauna; see biomes.

char Salmonidae *q.v.*

Characidae Large family of mostly small South American freshwater characiform teleost fishes comprising over 700 species; the taxonomy of the group is confused and it may include African taxa as well.

Characidiidae South American darters; family containing about 55 species of small (to 100 mm) tropical South American freshwater characiform teleost fishes resembling the true darters, Percidae; mostly benthic, occasionally mid-water; parental care lacking.

Characiformes Characins; diverse order of mostly small, colourful freshwater teleost fishes from South America and Africa; upper jaw protractile; jaws bearing teeth; pharyngeal teeth well developed; pectoral and pelvic fins soft-rayed, spines only rarely present in median fins; pelvics posterior to pectorals; gas bladder simple and connected to inner ear by bony Weberian ossicles; contains over 1300 species in about 15 families; of prime importance in aquarium trade; including tetras, darters, piranhas, pencil fishes, hatchetfishes, leporins, citharinids, and curimatos.

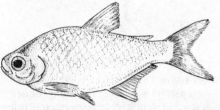

characin (Characiformes)

Charadriidae Plovers; family containing about 65 species of small to medium-sized shorebirds cosmopolitan in open countryside, marshes, and freshwater and marine coastal habitats; legs and bill usually short, wings slender or rounded; habits gregarious, often migratory, monogamous; feed on invertebrates and plant material; nest solitary, in hollow in ground.

plover (Charadriidae)

Charadriiformes Diverse order of mostly small to medium-sized shorebirds and sea birds, comprising 16 families and including oystercatchers, snipes, plovers, sandpipers, avocets, stilts, curlews, gulls, terns, and auks.

Charales The only order of Charophyceae *q.v.*; also treated as a class of the protoctistan phylum Chlorophyta.

Charophyceae Stoneworts; class of fresh- and brackish-water chlorophycote green algae; characterized by a large thallus differentiated into rhizoids, stem and whorls of branchlets; sometimes forming calcareous deposits; comprising a single order, Charales; resemble bryophytes in the presence of a sterile envelope enclosing sex organs.

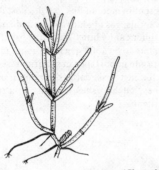

stonewort (Charophyceae)

chart datum The permanently established surface from which depth soundings or tide levels are referenced; formerly referred to low water but more recently to lowest astronomical tide.

chasmochomophyte A plant growing on detritus in rock fissures and crevices.

chasmocleistogamic Used of plants bearing some flowers that open and are pollinated after opening and other flowers that remain unopened and within which self-pollination takes place; **chasmocleistogamy.**

chasmodichogamic Used of plants bearing both cleistogamic and chasmogamic flowers in the same inflorescence; **chasmodichogamy.**

chasmogamy Pollination occurring after the opening of a flower; **chasmogamic, chasmogamous;** *cf.* cleistogamy.

chasmophilous Thriving in rock crevices and fissures; **chasmophile, chasmophily.**

chasmophyte A plant growing on rocks, rooted in detritus and debris in crevices and fissures.

chat 1: Maluridae *q.v.* 2: Turdidae *q.v.*

Chauliodontidae Viperfishes; family containing 6 species of deep-sea stomiiform teleost fishes found over a daytime depth range of 500–3500 m, often migrating towards the surface at night; body slender, to 300 mm length; anterior dorsal fin ray elongate, carrying distal light organ acting as lure; teeth fang-like.

Chaunacanthida Order of acantharians in which the 20 radial spines have loosely articulated, conical or pyriform bases.

Chaunacidae Sea toads; family containing 4 species of anglerfishes (Lophiiformes) widespread on muddy bottoms to depths of 500 m or more; body globose, to 450 mm length; mouth broad oblique.

cheetah Felidae *q.v.*

Cheilodactylidae Morwongs; family of shallow marine perciform teleost fishes widespread in southern hemisphere and northern Pacific; body elongate, slender, to 1.2 m length; single dorsal fin present; pectoral fin rays elongate; contains 15 species; popular in Australia as sport and food-fishes.

Cheilostomata Large order of gymnolaematan bryozoans having diverse colony morphology; may be encrusting, creeping, nodular, fan-shaped, scroll-like or diskoid; zooids typically box-shaped, calcified; polymorphism extensive – autozooids (feeding), gynozooids (female reproductive), androzooids (male reproductive), kenozooids (structural), avicularia (grasping), vibracula (cleaning); cheilostomes are almost entirely marine and comprise the most diverse and successful extant order of bryozoans.

Cheimarrichthyidae Torrentfish; monotypic family of small (to 130 mm) freshwater teleost fishes (Perciformes) found in fast-flowing mountain streams of New Zealand; body depressed anteriorly, dorsal and anal fins elongate, pelvic fins broad and located well in front of pectorals; migrate downstream to breed and spawning may take place in the sea.

Cheirogaleidae Family containing 4 genera of small lemurs (Strepsirhini) in which the eyes are large and the muzzle short.

chelation The process of combining a metal ion with another substance that has the effect of keeping the metal in solution and rendering it non-toxic.

Chelicerata Subphylum of arthropods comprising 3 extant classes, Merostomata (horseshoe crabs), Arachnida (spiders, scorpions, mites), Pycnogonida (sea spiders); body divided into prosoma (cephalothorax) consisting of a prostomium and 6 trunk segments variously amalgamated, and an opisthosoma (abdomen) of up to 12 segments forming an anterior mesosoma and posterior metasoma; antennae absent, first pair of appendages termed chelicerae, second pair pedipalps, and 4 pairs of legs; opisthosomal segments bearing gills, book lungs or spinnerets. Chelicerates are primarily predatory although many mites (Acari) are parasitic.

Chelidae Snake-necked turtles; family containing 30 species of side-necked turtles (Pleurodira) from South America, Australia and New Guinea; head incompletely retractable.

Chelonethida Pseudoscorpiones *q.v.*

Chelonia Alternative name for Testudines *q.v.*

Cheloniidae Sea turtles; family of large (to 1.5 m) marine turtles (Testudines: Cryptodira) comprising 4 species; carapace covered with horny scutes, forelimbs specialized as flippers; distribution pantropical, oceanic.

sea turtle (Cheloniidae)

Chelydridae Snapping turtles; family containing 3 species of aquatic freshwater turtles (Testudines: Cryptodira) with large heads and beaks; carapace bearing horny scutes; found in eastern North America to northwestern South America.

chemoautotrophic Used of those microorganisms that obtain metabolic energy by the oxidation of inorganic substrates, such as sulphur, nitrogen or iron; chemotrophic; **chemoautotroph**; *cf.* photoautotrophic.

chemocline The boundary zone in a lake between the deep stagnant water (monimolimnion) and the overlying region of freely circulating water (mixolimnion).

chemoheterotrophic Chemoorganotrophic *q.v.*

chemokinesis A change of linear or angular velocity in response to a chemical stimulus; **chemokinetic.**

chemolithotrophic Used of organisms that obtain energy from oxidation/reduction reactions and use inorganic electron donors; **chemolithotroph**; *cf.* chemoorganotrophic.

chemonasty A reponse to a diffuse chemical stimulus; a change in the structure or position of an organ in response to a diffuse chemical stimulus; **chemonastic.**

chemoorganotrophic Used of organisms that obtain energy from oxidation/reduction reactions and use organic electron donors; chemoheterotrophic; **chemoorganotroph**; *cf.* chemolithotrophic.

chemostat A steady state laboratory culture apparatus.

chemosynthesis The synthesis of organic compounds using chemical energy derived from the oxidation of simple inorganic substrates; **chemosynthetic**; *cf.* photosynthesis.

chemotaxis The directed reaction of a motile organism towards (positive) or away from (negative) a chemical stimulus; **chemotactic.**

chemotrophic Chemoautotrophic *q.v.*

chemotropism An orientation response to a chemical stimulus; **chemotropic.**

chemozoophobous Used of plants which protect themselves from herbivorous animals by the production of noxious chemical substances (allelochemics); **chemozoophobe.**

Chenopodiaceae Beet, spinach, goosefoot, greasewood; large family of Caryophyllales containing herbs or sometimes woody plants, with stems often succulent, jointed or both; cosmopolitan in distribution but most abundant in dry regions; flowers inconspicuous, borne in dense clusters, variable in morphology; fruits are one-seeded nuts.

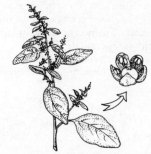

goosefoot (Chenopodiaceae)

cheradophilous Thriving on wet sandbars; **cheradophile, cheradophily.**

cheradophyte A plant living on a sandbar.

Chernozem A zonal soil, typically formed in cool subhumid climates under tall prairie grasses, with a thick black A-horizon and calcium carbonate-rich surface layer, over a lighter intermediate B-horizon and a calcareous C-horizon; black earth.

cherry Rosaceae *q.v.*

chersophilous Thriving in dry wasteland habitats; **chersophile, chersophily.**

chersophyte A plant growing on dry wasteland or in poor shallow soil.

chestnut Fagaceae *q.v.*

Chestnut soil A zonal soil with a thick dark brown, slightly alkaline A-horizon over a lighter coloured B-horizon and a calcareous lower horizon; formed in temperate to cool, subhumid to semi-arid climates beneath mixed tall and short grasses.

chevrotain Tragulidae *q.v.*

chewing louse Mallophaga *q.v.*

Chiasmodontidae Swallowers; family containing 15 species of small (to 200 mm) deep-sea perciform teleost fishes; body slender, black; jaws very large and highly mobile, teeth fang-like; stomach extremely capacious; photophores sometimes present.

chickadee Paridae *q.v.*

chicken Phasianidae *q.v*

chickweed Caryophyllaceae *q.v.*

chigger Any of predatory mites in the family Trombiculidae, some of which bite humans and, in the Oriental region, carry scrub typhus disease.

Chilean jasmine Apocynaceae *q.v.*

Chilopoda Centipedes; class of carnivorous terrestrial arthropods commonly found in soil, leaf litter and decomposing wood; body typically flattened without division into distinct thorax and abdomen, length 6–200 mm; each trunk segment bearing single pair of appendages, first pair specialized as poison fangs; head bearing single pair of antennae, mandibles, 2 pairs of maxillae; eyes present or absent; sexes separate, oviparous, monomorphic; contains about 2500 species in 2 subclasses, Anamorpha and Epimorpha.

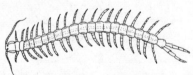

centipede (Chilopoda)

chimaera 1: An organism having tissues of two or more genetic types, formed by mutation, abnormal chromosome segregation, or by artificial grafting; usually used of a plant, rarely of an animal. 2: Holocephalii *q.v.*

Chimaeriformes Chimaeras, rabbitfishes, ratfishes; small order of cartilaginous fishes (Holocephalii) comprising about 35 species in 3 families; body more or less compressed, naked or with placoid denticles; length 0.5–2 m; 4 pairs of gill openings covered by operculum; notochord unsegmented, vertebral centra not developed; reproduction oviparous, eggs encased in horny capsules; widespread bottom-dwelling, sluggish, feeding on benthic fishes and invertebrates.

ratfish (Chimaeriformes)

chimnochlorous Pertaining to plants with thin herbaceous leaves that persist over winter.

chimnophilous Thriving during the winter; used of plants that exhibit maximum development during the winter; **chimonophile, chimonophily.**

chimopelagic Pertaining to deep-water marine organisms which occur in surface waters only during the winter.

chimous Pertaining to winter or to cold climatic conditions.

chimpanzee Pongidae *q.v.*

Chinchillidae Chinchillas, viscachas; family containing 6 species of southern Neotropical herbivorous hystricomorph rodents; habits diurnal or nocturnal, colonial, inhabiting burrows; hindlimbs long with 3 or 4 digits, forelimbs short with 4 or 5 digits.

chinchilla (Chinchillidae)

Chinese gooseberry Actinidiaceae *q.v.*

Chinese lantern Solanaceae *q.v.*

chionophilous Thriving in snow-covered habitats; **chionophile, chionophily;** *cf.* chionophobous.

chionophobous Intolerant of snow-covered habitats; **chionophobe, chionophoby;** *cf.* chionophilous.

chionophyte A plant living in or under snow.

Chirocentridae Wolf herring; monotypic family of very large (to 3.7 m) predatory marine clupeiform fishes found in warm waters of the Indo-Pacific; body long, compressed, teeth large, fang-like; pelvic fins minute.

Chironemidae Kelpfishes; family of marine teleost fishes (Perciformes) comprising 4 species resembling hawkfishes or morwongs, found on rocky coasts of Australia and New Zealand.

Chironomidae Midges; cosmopolitan family of small delicate flies (Diptera) that swarm in vast numbers in damp habitats; larvae typically aquatic feeding on algae, diatoms or detritus, some predacious or parasitic; the blood of some larvae (bloodworms) contains haemoglobin; family comprises about 5000 species in 120 genera.

Chiroptera Bats; order of mostly small nocturnal mammals in which the forelimbs are modified as wings for flight with thin skin membranes extending from sides of body and legs and enclosing elongate digits; most use echolocation for navigation; many are insectivorous, but diet of other species includes fruit, nectar, small vertebrates and blood; some forms hibernate, others are migratory; 2 suborders are recognized, Megachiroptera (1 family, Old World fruit bats), Microchiroptera (16 families).

chiropterophilous Used of plants pollinated by bats; **chiropterophile, chiropterophily.**

chitinolytic Capable of degrading chitin.

chiton Polyplacophora *q.v.*

Chlamydoselachidae Frilled shark; monotypic family of primitive hexanchiform sharks; body elongate, eel-like, to about 2 m length; head broad and depressed, gill slits large; reproduction ovoviviparous; depth range about 100–750 m.

chlamydospore A thick-walled resting spore found in some fungi.

chledophilous Thriving in wasteland habitats and rubbish heaps; **chledophile, chledophily.**

chledophyte A plant growing on rubbish heaps.

Chloramoebales Order of mostly freshwater, unicellular xanthophyceaen algae.

Chloranthaceae Small family of woody or herbaceous plants (Piperales) widely distributed in subtropical and tropical regions; the tiny flowers with 1–3 stamens are borne in compound spikes; fruit is a tiny drupe.

chlorinated hydrocarbon A synthetic contact insecticide, such as DDT, which is relatively resistant to biodegradation and persists in the food web, often accumulating in non-target organisms.

chlorinity A measure of chloride and bromide

ion concentration in sea water (in grams per kilogram, or parts per thousand); used in estimating salinity; (salinity = 1.80655 x chlorinity.)

Chlorococcales Diverse order of non-motile green algae common in freshwater plankton, soil and in subaerial habitats such as snow and ice, occasionally lichenised; typically found as solitary cells, but sometimes forming colonial structures; also treated as a class of the protoctistan phylum Chlorophyta.

Chlorodendrales Order of predominantly marine flagellate Prasinophyceae *q.v.* in which the cells are often non-motile and are enclosed within a pectin-like wall (lorica); occasionally found as symbionts on invertebrates.

Chlorokybales Order of green algae containing a single species found in subaerial habitats in Europe; characterized by a thallus formed from aggregations of cells embedded in a gelatinous matrix.

Chloromonadales Alternative name for the only order of Raphidophyceae *q.v.*

Chloromonadida Raphidophyceae *q.v.* treated as an order of the protozoan class Phytomastigophora.

Chlorophthalmidae Greeneyes; family containing 20 species of bottom-living, mainly deep-sea, myctophiform teleost fishes; body cylindrical becoming compressed posteriorly; teeth small, needle-like; adipose fin conspicuous; photophores absent; hermaphroditic.

Chlorophyceae Large, diverse class of green algae containing over 8500 species characterized by the possession of the photosynthetic pigments chlorophylls *a* and *b*, α, β and γ carotenes and various xanthophylls; typically producing motile zoospores with 2 or 4 flagella arising from an apical pit.

Chlorophycota Green algae; division of eukaryotic algae containing chlorophylls *a* and *b*; comprising 3 classes, Charophyceae, Chlorophyceae and Prasinophyceae; about equivalent in composition to the phyla Chlorophyta and Gamophyta of the kingdom Protoctista.

chlorophyll A photosynthetic pigment reflecting green light and imparting the typical green colour to plants; chlorophyll *a* is found in all autotrophic plants, chlorophyll *b* in land plants, the Chlorophyta and Charophyceae, chlorophylls *c* and *d* occur only in certain groups of algae.

Chlorophyta Phylum of Proctoctista comprising those green algae that form

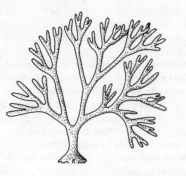

green alga (Chlorophyta)

zoospores or gametes having cup-shaped, grass-green chloroplasts and at least 2 equal anterior flagella; possessing chlorophylls *a* and *b* as well as various carotenoid derivatives; comprising 9 classes; Chlorococcales, Volvocales, Ulotrichales, Oedogoniales, Chaetophorales, Cladophorales, Siphonales, Charales and Prasinophycales.

Chloropidae Frit flies; family containing about 1200 species of flies (Diptera) characterized by a triangular ocellular plate on the dorsal surface of the head; larvae feed on plants; includes several economically important pest species.

Chlorosarcinales Small order of mostly soil-inhabiting, multicellular green algae, exhibiting an advanced type of cell division, desmochisis, in which each daughter protoplast forms a new wall over its entire surface.

Chloroxybacteria Prochlorophycota *q.v.*

Choanichthyes Choanates; group of vertebrates possessing functional lungs, internal and external nares and fleshy, lobe-like paired fins or limbs; comprises the lungfishes (Dipnoi) and the tetrapods.

Choanoflagellida Choanoflagellates; order of mostly unicellular, often stalked

choanoflagellate (Choanoflagellida)

zooflagellates containing about 140 species found in aquatic habitats and characterized by possession of a single emergent flagellum surrounded proximally by a funnel-like collar of tentacles; feed on bacteria extracted by the tentacles from currents set up by the flagellum.

chomophilous Thriving in wastelands and rubbish heaps; **chomophile, chomophily.**

chomophyte A plant growing in detritus and other litter.

Chondrichthyes Cartilaginous fishes; class of carnivorous jawed vertebrates having cartilaginous skeleton that may undergo calcification but not ossification; mostly marine, littoral to abyssal; nostrils and mouth ventral, teeth not fused to jaw, skin usually bearing strong dermal scales; swim bladder absent; reproduction oviparous to viviparous, fertilization internal; eggs develop inside horny cases; feed on a wide range of vertebrates and invertebrates; contains about 800 species in 14 orders in 2 subclasses, Elasmobranchii (sharks and rays) and Holocephalii (rat fishes).

Chondrophora Order of marine colonial Hydrozoa found at or near the ocean surface; lacking a medusa generation they comprise numerous polymorphic zooids attached to a central, chambered, chitinous float; a large feeding gastrozooid is surrounded by reproductive gonozooids which are in turn encircled by stinging dactylozooids; includes the by-the-wind sailor.

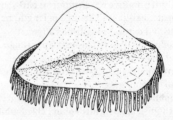

by-the-wind sailor (Chondrophora)

Chondrostei Group of bony fishes (Actinopterygii) comprising the sturgeons (Acipenseriformes) and a number of fossil forms; characterized by a heterocercal tail, spiracle, spiral valve and a partly cartilaginous skeleton.

Chonotrichida Order of hypostomatan ciliates found as ectocommensals on the gills and limbs of aquatic crustaceans; cilia are limited to the atrial area which often forms an apical funnel with the pharynx at the base.

Chordariales Order of brown algae containing small to medium-sized forms characterized by an alternation of gametophyte and sporophyte generations; sporophytes filamentous in construction, sometimes forming large erect thalli composed of compact pseudoparenchymatous filaments.

Chordata Phylum of bilateral coelomate animals characterized by the presence of a notochord, gill slits, dorsal hollow nerve chord and a post-anal tail at some stage of development; 3 subphyla are recognized, Cephalochordata, Tunicata and Vertebrata.

Chordeumatida Order containing about 500 species of small active helminthomorphan diplopods (millipedes); body cylindrical, legs elongate; allomone-producing pores (ozadenes) absent.

Chordodea Order of gordioidan horsehair worms typically possessing cuticular organs called areoles.

C-horizon The lowest layer of a soil profile above the bedrock (R-horizon), comprising mineral substrate with little or no structure; containing gleyed layers or layers of accumulation of calcium carbonate or calcium sulphate in some soils; parent material; see soil horizons.

chorology The description and delimitation of the distributional ranges of taxa.

chresard That portion of the total soil water that is available to plants; *cf.* echard, holard.

Chromadoria Subclass of adenophorean nematodes found free-living in soil, marine or freshwater habitats; typically with a striated or ornamented cuticle and characterized by their looping locomotion rather than the normal sinusoidal movement.

Chromadorida Order of chromadorian nematodes found in marine, freshwater and soil habitats; typically with punctate cuticular ornamentation, the cephalic sensory organs arranged in 1 or 2 whorls at the extreme anterior end of the body.

chromatography A technique for separating and identifying the components in a mixture of compounds; involves passing the dissolved mixture along a piece of filter paper (paper chromatography) or through a column of charged resin (ion-exchange chromatography).

chromatotropism An orientation response to stimulation by light of a particular colour; **chromatotropic.**

Chromophycota A division of eukaryotic algae; a heterogeneous assemblage of 9 classes grouped together mainly on the basis of their possession of chlorophylls *a* and *c* and their

lack of chlorophyll *b*; containing the following classes, Chrysophyceae, Prymnesiophyceae, Xanthophyceae, Eustigmatophyceae, Bacillariophyceae, Dinophyceae, Phaeophyceae, Raphidophyceae and Cryptophyceae.

chromosome A deeply staining nuclear body composed largely of DNA and protein, and comprising a linear sequence of genes.

chromosome complement The actual number of chromosomes in a nucleus.

chromosome deletion A chromosome mutation involving loss of genes.

chromosome insertion A chromosome mutation involving the gain of duplicate genes.

chromosome inversion A mutation involving alteration or reversal of a sequence of genes within a chromosome.

chromosome map A plan of an individual chromosome showing the gene sequence as an arrangement of nucleotide bases.

chromosome translocation A mutation involving the transfer of a segment of one chromosome from one member of a homologous pair to the other.

chromosome pair The two homologous chromosomes that become intimately associated during meiosis and mitosis.

chronistic Pertaining to, or in relation to, time or a time scale.

chronocline A gradual change in a character or group of characters over an extended period of geological time.

chronogenesis The time sequence of occurrence of organisms in stratified rock.

chronospecies 1: A species which is represented in more than one geological time horizon. 2: The successive species replacing each other in a phyletic lineage which are given ancestor and descendant status according to the geological time sequence.

chronotropism An orientation response due to age; used particularly with reference to the movement of leaves in plants; **chronotropic.**

Chroococcales Primitive order of unicellular, blue-green algae which reproduce by binary fission or, rarely, by the formation of nannocytes; contains important primary producers distributed widely in freshwater, soil and marine habitats, and as symbionts in lichens.

chrymosymphily An amicable relationship between ants and lepidopterous larvae, based on the scent produced by the larvae; **chrymosymphile, chrymosymphilous.**

chrysalis A hard shell enclosing an insect pupa.

chrysanthemum Asterales *q.v.*

Chrysididae Cuckoo wasps, gold wasps; family of robust, strongly sculptured, often metallic-coloured, wasps (Hymenoptera) the larvae of which live typically as parasitoids of other insects.

Chrysobalanaceae Coco plum; family of Rosales containing about 450 species of trees or shrubs with a pantropical distribution; flowers irregular, variable; fruit is a drupe.

Chrysochloridae Golden moles; family containing 18 species of small fossorial mammals (Insectivora) found in central and southern Africa; external ears absent; eyes covered with skin; fur has metallic sheen.

golden mole (Chrysochloridae)

Chrysomelidae Leaf beetles, Colorado potato beetle; family of often brightly coloured beetles (Coleoptera) most of which are surface feeders on leaves of flowering plants; some larvae feed on roots or as stem borers; includes numerous agricultural pests amongst the 35 000 known species.

Chrysomonadida Chrysophyceae *q.v.* treated as an order of the protozoan class Phytomastigophora.

Chrysophyceae Class of mainly planktonic chromophycote algae common in temperate and high latitude freshwater bodies; typically unicellular but sometimes forming complex branching colonies; characterized by the possession of 1 or 2 golden yellow plastids containing chlorophylls *a* and *c*, some xanthophylls and carotenoid pigments and by the presence in the life cycle of a stage with 2 unequal flagella, one smooth and one flimmer type; comprises 2 subclasses, Chrysophycidae

chrysophycean alga (Chrysophyceae)

and Dictyochophycidae; also classified as a phylum of Protoctista, under the name Chrysophyta, or as an order of the protozoan class Phytomastigophora under the name Chrysomonadida.

Chrysophycidae Subclass of typically unicellular chrysophyceaean algae characterized by the presence in the life cycle of a silicified resting stage formed within the cell.

Chrysophyta The Chrysophyceae *q.v.* treated as a separate phylum of Protoctista.

Chrysopidae Golden-eyes, chrysopids; the largest family of neuropteran insects many of which produce a noxious odour from glands on the thorax; many larvae feed on plant lice or aphids and some have been used in biological control; contains about 1500 species.

chtonophyte A terrestrial plant obtaining water from the soil through roots.

chylophyte A terrestrial plant rooted on a physically dry and hard substratum.

Chytridiales Chytrids; order of chytridiomycete fungi comprising about 900 species of parasites of various aquatic organisms worldwide; typically with a unicellular thallus within the host cells; sexual reproduction involving fusion of isogametes; also classified as a class of the protoctistan phylum Chytridiomycota, under the name Chytridia.

Chytridiomycetes Class of parasitic and saprobic true fungi found in aquatic and soil habitats; fungal body unicellular to hyphal and generally without cross walls except at reproductive structures; asexual reproduction by production of zoospores bearing a single whiplash flagellum posteriorly; comprises 3 orders, Blastocladiales, Chytridiales and Monoblepharidales; sometimes treated as a phylum, Chytridiomycota, of the Protoctista.

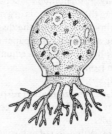

chytrid (Chytridiomycetes)

Chytridiomycota Chytridiomycetes *q.v.*; treated as a phylum of Protoctista.

Chytriodiniales Small order of marine parasitic dinoflagellates found infesting animal and plant hosts.

cicada Cicadidae *q.v.* (Homoptera).

Cicadellidae Leafhoppers; family comprising about 20 000 species of insects (Homoptera) that feed by sucking plant juices; includes many pests of cultivated plants.

Cicadidae Cicadas; family of large insects (Homoptera) the males of which typically have well developed sound-producing organs; eggs laid in twigs and branches of trees; hatching nymphs fall to ground and dig into soil using powerful forelegs; feed on sap of roots; often with long life cycle mostly spent as nymph; contains about 1500 species.

Cichlidae Cichlids; large family of primarily freshwater perciform teleost fishes comprising more than 700 species; body typically perch-like, weakly to strongly compressed with single dorsal fin; feeding habits, reproductive strategies and behaviour extremely variable. Cichlids show immense radiation in African Great Lakes and in South America, with high levels of endemicity; very popular in the aquarium trade; locally important as food-fishes.

Cicindelidae Tiger beetles; family of active brightly-coloured beetles (Coleoptera) found in open sunny situations, the larvae typically inhabiting burrows; adults and larvae feed mainly on small insects.

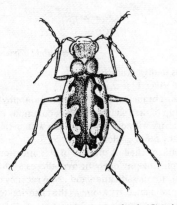

tiger beetle (Cicindellidae)

Ciconiidae Storks; family containing 17 species of large wading birds (Ciconiiformes) with long legs, long straight bills, and long broad wings; feed mostly on fishes, amphibians and insects; solitary or colonial breeders, the nest comprising a large platform of sticks in tree or on cliff; cosmopolitan in tropical and temperate shallow freshwater lakes, marshes, or grassland.

Ciconiiformes Order of medium to large neognathous wading birds comprising 6 families, cosmopolitan in marine and freshwater coastal habitats; characteristically long-necked and long-legged; includes herons, storks, spoonbills and ibises.

Cidaroida Pencil urchins; order of primitive echinoids (Echinozoa) comprising about 140 extant species found mostly at continental shelf and slope depths, rarely littoral or abyssal; body globular with long spines, ambulacral plates small, interambulacral plates large; gills absent; fossil record dating from Palaeozoic.

Ciliophora Ciliates, Infusoria; diverse phylum of protozoans characterized by the possession of two types of nuclei (micronuclei and macronuclei), simple cilia or compound ciliary organelles, and commonly a cell mouth (cytostome) and pharynx (cytopharynx); found free-living in all types of aquatic and terrestrial habitats and as symbionts and parasites on or in a variety of hosts; includes 3 classes, Kinetofragminophora, Oligohymenophora and Polyhymenophora; also treated as a phylum of Protoctista.

Cimicidae Bedbugs; family of flightless bugs (Heteroptera) which suck blood of mammals or birds or, more typically, are predators of other arthropods.

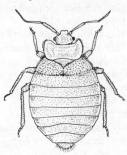

bedbug (Cimicidae)

Cinclidae Dippers; small family containing 5 species of active aquatic passerine birds found along mountain streams and lake shores of Eurasia and western parts of North and South America; habits solitary, non-migratory, swim well underwater using short pointed wings; feed on aquatic invertebrates and small fish; nest in bank or cliff.

cineraria Asterales *q.v.*

Cingulata Infraorder of edentate mammals (Xenarthra) comprising the glyptodonts (Glyptodontidae) and the armadillos (Dasypodidae).

circadian rhythm A biological rhythm having a periodicity of about one daylength (24 hours).

Circaeasteraceae Small family of herbaceous Ranunculales containing 2 species from southeast Asia characterized by their dichotomously veined leaves lacking anastomoses.

circalittoral The lower subdivision of the marine sublittoral zone below the infralittoral zone, dominated by photophilic algae; sometimes used for the depth zone between 100 m and 200 m; see marine depth zones.

circumaustral Distributed around the high latitudes of the southern hemisphere; *cf.* circumboreal.

circumboreal Distributed around the high latitudes of the northern hemisphere; *cf.* circumaustral.

circumneutrophilous Thriving in conditions of about neutral pH; **circumneutrophile, circumneutrophily.**

circumnutation Growth movements of a plant about an axis.

circumpolar Distributed around the north or south polar regions.

circumtropical Occurring throughout the tropics.

Cirratulida Order of burrowing, deposit-feeding, polychaete worms comprising about 245 species in 3 families; included by some authors in the order Spionida.

Cirrhitidae Hawkfishes; family containing 35 species of Indo-Pacific shallow marine perciform teleost fishes; body elongate, to 500 mm length, single dorsal fin present; lower pectoral fin rays enlarged and used for perching on hard surfaces.

Cirripedia Barnacles; class of exclusively sessile marine and brackish water crustaceans typically with calcareous shells enclosing

acorn barnacle (Cirripedia)

modified filtering limbs; contains about 1000 free-living and parasitic species classified in 4 orders, Acrothoracica, Ascothoracica, Rhizocephala and Thoracica.

Cistaceae Small family of Violales containing about 200 species of shrubs or herbs, common in the Mediterranean region; flowers often showy with 3 or 5 sepals, 5 large petals, many stamens and a one-celled ovary; fruit is a capsule.

cistern epiphyte An epiphyte lacking roots and gathering water between the bases of its leaves.

Citharidae Family containing 5 species of small (to 250 mm) pleuronectoid flatfishes comprising 2 subfamilies, one right-eyed (dextral), the other left-eyed (sinistral).

Citharinidae Family containing 80 species of large African freshwater characiform teleost fishes; dorsal fin in mid-body position, adipose fin usually present, lateral line straight; mostly herbivorous or micropredatory; parental care lacking, eggs spread over substratum or vegetation.

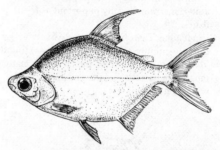

citharinid (Citharinidae)

citrus family Rutaceae *q.v.*

civet Viverridae *q.v.*

clade A branch of a cladogram; a monophyletic group of taxa sharing a closer common ancestry with one another than with members of any other clade.

cladistic method A method of classification employing phylogenetic hypotheses as the basis for classification and using recency of common ancestry alone as the criterion for grouping taxa; cladism; cladistics; *cf.* phenetic method.

Cladocera Water fleas; order of diplostracan crustaceans containing about 450 mostly freshwater species, but with a few marine and brackish forms; characterized by jerky swimming using large antennae; often parthenogenetic; trunk enclosed by carapace, head free.

water flea (Cladocera)

Cladocopina Suborder of marine halocypridan ostracods comprising a single family, Polycopidae, with 40 species, found mainly in shallow waters; efficient swimmers but probably benthic; small body, typically less than 1 mm long.

Cladophorales Order of green algae widely distributed in marine freshwater and brackish water habitats; typically with a thallus comprising uniseriate filaments of multinucleate cells attached by rhizoids; also treated as a class of the protoctistan phylum Chlorophyta.

cladoptosis The periodic shedding of twigs.

Cladoselachiformes Extinct order of sharks (Elasmobranchii) known from the Devonian to the Carboniferous; body elongate; 2 dorsal fins, each with a spine.

clam shrimp Conchostraca *q.v.*

Clariidae Family of Old World freshwater catfishes (Siluriformes) found throughout much of southern Asia and Africa; body eel-like, dorsal fin long, without spines; 4 pairs of barbels; accessory organ present in gill chamber for aerial respiration; contains 100 species, known as walking catfishes as some species may leave the water and move on land for short periods.

classification A process of establishing, defining and ranking taxa within hierarchical series of groups.

clastic 1: Used of sediments that are formed of rock fragments or of clay minerals. 2: Producing or undergoing fragmentation.

clastizoic Used of rock comprising fragmented animal remains.

Clathrinida Order of calcinian sponges which are constructed as a complex system of anastomosing tubes lined with choanocytes, each system terminating in an osculum; widely distributed and ranging from the intertidal to about 850 m.

Clavicipitales Order of pyrenomycete fungi tyically parasitic on grasses, insects, spiders and other fungi; characterized by production of filiform ascospores which are forcibly discharged; includes ergot of rye.

ergot (Clavicipitales)

clay Sediment particles between 0.002 mm and 0.004 mm in diameter and having colloidal properties; sometimes used for all sediment particles less than 0.004 mm in diameter; fine clay; see sediment particle size.

cleg Tabanidae *q.v.*

cleistogamy The condition of having flowers, typically small and inconspicuous, which remain unopened and within which self-pollination takes place; **cleistogamic**, **cleistogamous**; *cf.* chasmogamy.

clematis Ranunculaceae *q.v.*

Clethraceae Small family of Ericales containing 65 species of tanniferous shrubs and small trees usually with stellate hairs; chiefly tropical in distribution; flowers fragrant, bell-shaped, with 5 free imbricate petals, 10–12 stamens and a superior ovary; fruit is a capsule.

click beetle Elateridae *q.v.*

climacteric A critical phase or period in the life cycle of an organism.

climagraph A chart plotting one climatic factor against another, commonly a diagram of mean monthly temperature plotted against mean monthly rainfall or humidity.

climate The long-term average condition of weather in a given area; *cf.* weather.

climatic climax A more or less stable community in which the major factors affecting the vegetation are climatic, typical of zonal soils.

climatic rule Any generalization describing a trend in geographical variation of animals which can be correlated with a climatic gradient; *cf.* Allen's law, Bergmann's rule, Gloger's rule, Rensch's laws.

Climatiiformes Extinct order of fishes (Acanthodii) possessing bony jaws and skeleton, ganoid scales, heterocercal tail and a stout spine in front of each fin.

climatology The study of climate; *cf.* meteorology.

climax A more or less stable biotic community which is in equilibrium with existing environmental conditions and which represents the terminal stage of an ecological succession.

climbing bittersweet Celastraceae *q.v.*

climbing gourami Anabantidae *q.v.*

cline A character gradient; continuous variation in the expression of a character through a series of contiguous populations; **clinal.**

clingfish Gobiesocidae *q.v.*

Clinidae Scaled blennies; family containing 180 species of mostly tropical, intertidal and shallow marine, perciform teleost fishes; body length to 600 mm, but usually much smaller; viviparity common.

clinodeme A deme *q.v.* that forms part of a graded sequence of demes distributed over a given geographical area.

clinotaxis A directed reaction or orientation response of a motile organism to a gradient of stimulation; **clinotactic.**

clinotropism An orientation response to a gradient of stimulation; **clinotropic.**

clisere An ecological succession resulting from a major change in climate.

Clistogastra Apocrita *q.v.*

Clitellata Group of annelid worms possessing a clitellum, comprising the Oligochaeta and Hirudinoidea.

clitochorous Dispersed by gravity; **clitochore, clitochory.**

clod A compacted mass of soil between about 5 mm and 250 mm in diameter.

clone An assemblage of genetically identical organisms derived by asexual or vegetative multiplication from a single sexually derived individual; **clonal;** *cf.* genet, ortet, ramet.

clonodeme A local interbreeding population composed of predominantly vegetatively reproducing individuals; *cf.* deme.

closed community An assemblage of plants which cover the entire ground area of the habitat in which they live, effectively preventing the establishment of other species.

clothes moth Tineidae *q.v.*

clover Fabaceae *q.v.*

clubmoss Lycopodiopsida *q.v.*

Clupeidae Herrings, shads, sardines, menhaden; large family of mostly small planktotrophic teleost fishes (Clupeiformes), typically marine but also found in brackish and freshwater habitats; body compressed to subcylindrical, caudal fin strongly forked; eggs pelagic or benthic; contains about 180 species, many forming large schools; support extremely important commercial fisheries.

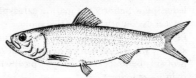

herring (Clupeidae)

Clupeiformes Large order of mostly small planktivorous teleost fishes comprising 3 families, Clupeidae (herrings, shads, sardines, menhaden), Engraulidae (anchovies) and Chirocentridae (wolf herring); swim bladder diverticula connected to inner ear; lateral line absent; no leptocephalus larval stage; contains about 70 genera distributed worldwide; includes many commercially important species.

Clusiaceae Large family of Theales including 1200 species ranging from tropical and woody plants to temperate herbs; produce resinous secretions in canals or chambers and have an oily embryo.

clusium An ecological succession on flooded soil.

cluster fly Calliphoridae *q.v.*

clutch The number of eggs laid at any one time; clutch size.

Clypeasteroida Sand dollars; order of irregular echinoids (Gnathostomata) comprising about 130 species typically found burrowing in soft

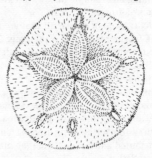

sand dollar (Clypeasteroida)

sediments of shallow tropical or temperate seas; test flattened and disk-like, anus close to posterior margin on adoral surface; tube-feet within petal-like array of aboral ambulacra, serving respiratory function; spines especially abundant on lower surface.

Cneoraceae Family of Sapindales containing only 3 species of glabrous shrubs native to the western Mediterranean region; flowers borne in axillary cymes; flowers tri- or tetramerous; fruit comprising 1–4 small drupes.

Cnidaria Coelenterates; phylum of mostly marine multicellular animals characterized by a body plan that comprises two primary layers, ectoderm and endoderm, separated by a layer of gelatinous connective tissues, the mesoglea, by a basic radial symmetry and by the possession of nematocysts, consisting of an intracellular capsule containing a thread which can be everted in response to a variety of stimuli; two body forms are found, an erect, solitary or colonial, polyp which is attached to the substratum and typically has tentacles surrounding its mouth, and a medusa which is disc-shaped, solitary and pelagic; all are carnivorous, although some may derive nutritive benefit from intracellular symbiotic algae; also known as the Radiata because of their radial symmetry; comprises 4 extant classes, Anthozoa, Cubozoa, Hydrozoa and Scyphozoa.

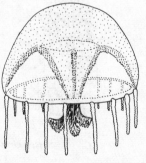

leptomedusa (Cnidaria)

Cnidosporida Phylum of Protoctista containing unicellular organisms that parasitize animals; characterized by the production of a polar filament or thread; equivalent to the protozoan phyla Microspora and Myxozoa.

CO$_2$ compensation concentration The minimum atmospheric carbon dioxide concentration allowing net photosynthesis.

coacervation The aggregation of organic molecules to form particulate matter; **coacervate.**

coaetaneous Of the same age; existing or appearing simultaneously; **coetaneous.**

coati Procyonidae *q.v.*

coat-of-mail shell Polyplacophora *q.v.*

cobble A sediment particle between 64 mm and 256 mm in diameter; large gravel; see sediment particle size.

Cobitidae Loaches; family containing about 145 species of northern hemisphere cypriniform teleost fishes; body terete to vermiform, flattened ventrally, with 3–6 pairs of mouth barbels; gas bladder partly enclosed in bony capsule; primarily nocturnal, feeding on bottom-living invertebrates; some species absorb oxygen from air swallowed through vascularized intestine.

cobra Elapidae *q.v.*

cobra plant Sarraceniaceae *q.v.*

cobweb spider Any of the labidognath spider family Theridiidae, also called comb-footed spiders; comprising about 2000 species including the venomous widow spiders; typically contruct irregular webs in vegetation, amongst stones and litter, and beneath buildings.

cocaine Erythroxylaceae *q.v.*

Coccidia Subclass of sporozoans, mostly parasitic in vertebrates and found throughout the world; mature gamonts are small and typically intracellular; life cycle involves 3 types of multiple fission, merogony, gametogony and sporogony; also treated as a subclass, Coccidiasina, of the protoctistan class Sporozoasida.

Coccinellidae Ladybirds; family of often brightly coloured, rounded beetles (Coleoptera) which are typically predators both as adults and larvae, feeding mainly on other insects including aphids, mealybugs and scale insects; contains about 4500 species, common worldwide.

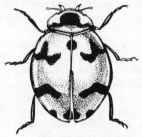

ladybird (Coccinellidae)

coccoid Coccoidea *q.v.* (Homoptera).

Coccoidea Scale insects, coccoids, mealybugs; a group of insect (Homoptera) families in

which the wingless adult females are usually
protected by a scale-like wax covering over the
body; feed on plant juices, and many are
destructive pests of economically important
plants.

coccolith ooze A pelagic sediment comprising
at least 30% calcareous material
predominantly in the form of coccolith
remains. *cf.* ooze.

coccolithophorids Unicellular marine algae
(Prymnesiophyceae) that have the body
embedded in a gelatinous sheath covered with
calcareous plates (coccoliths); common as
fossils.

Coccosphaerales Order of calcareous scale-
covered flagellate algae (Prymnesiophyceae)
found mostly in the marine plankton of warm
seas; includes many fossil species.

cockatoo Psittacidae *q.v.*

cockchafer Scarabaeidae *q.v.*

cockle Veneroida *q.v.*

cocklebur Asterales *q.v.*

cockroach Blattaria *q.v.*

cockscomb Amaranthaceae *q.v.*

coco plum Chrysobalanaceae *q.v.*

cocoon 1:The silken covering around an insect
pupa. 2: A brood chamber spun by some
spiders to receive their eggs.

cod Gadidae *q.v.*

Codiales Bryopsidales *q.v.*

cod icefish Nototheniidae *q.v.*

codlet Bregmacerotidae *q.v.*

codling moth Tortricidae *q.v.*

codon The unit of genetic coding; a triplet of
adjacent nucleotides in DNA or in messenger
RNA, which specifies a particular amino acid;
coding triplet.

coelacanth Latimeriidae *q.v.*

Coelacanthiformes Tassel-finned fishes; order
of bony fishes (Crossopterygii) represented by
the single recent family Latimeriidae, but with
an extensive fossil record from the Devonian
to the Cretaceous; fins supported by movable
lobes.

Coelenterata Coelenterates; this name has been
used to refer to the Cnidaria *q.v.*, and to both
the Cnidaria and the Ctenophora *q.v.*

Coelocheta Superorder of helminthomorphan
diplopods (millipedes) comprising 2 orders,
Callipodida and Chordeumatida; body with
median pleurotergal suture, sterna free;
spinnerets present posteriorly.

Coelomycetes Class of deuteromycotine fungi
containing the conidial states of ascomycotine
fungi and asexual fungi for which no sexual
state is known; most produce closed fruiting

bodies, the nature of which determines the
systematic position; contains 2 orders,
Melanconiales and Sphaeropsidales.

Coelopidae Kelp flies; family of small flies
(Diptera) found in large swarms around
accumulations of kelp; comprising about 20
species, absent from tropical regions and from
South America.

coelozoic Living in the lumen of an organ of the
host individual.

Coelurosauria Group of agile, bipedal
carnivorous dinosaurs (Saurischia) known
from the Triassic to the Cretaceous; up to 3 m
in length.

coelurosaur (Coelurosauria)

coenocarpium A type of multiple fruit with a
fleshy axis, which incorporates the
receptacles, ovaries and floral parts of many
flowers; includes the pineapple.

pineapple (coenocarpium)

coenocline A sequence of communities
distributed along an environmental gradient.

coenosis An assemblage of organisms having
similar ecological preferences.

coenosite An organism that habitually shares
food with another organism.

coenosium A plant community.

coevolution The interdependent evolution of
two or more species having an obvious
ecological relationship, usually restricted to
cases in which the interactions are beneficial
to both species; also used for the evolutionary
interaction between species in which one is

detrimental to the other, such as a parasite and its host, and far wider interactions within an evolving community, such as those between species employing a similar feeding strategy.

coexistence The occurrence of two or more species in the same area or habitat, usually used of potential competitors.

coffee Rubiaceae *q.v.*

coffee-bean weevil Anthribidae *q.v.*

cohort A group of individuals of the same age recruited into a population at the same time; age class.

coition Copulation; sexual intercourse; coitus.

cold blooded Poikilothermic *q.v.*

cold hardiness The tolerance of an organism to low temperatures.

cold monomictic lake A lake having a summer overturn and in which the temperature of the water never rises above 4°C; *cf.* warm monomictic lake.

cold resistance The ability to survive exposure to temperatures below 0°C.

cold temperate zone A latitudinal zone extending between 45° and 58° in both northern and southern hemispheres.

Coleochaetales Small order of aquatic epiphytic or endophytic green algae; thallus typically consisting of cells united or clustered into branched filaments; asexual reproduction involves scaly biflagellate zoospores.

Coleoidea Dibranchiata: the larger subclass of cephalopod molluscs characterized by an internal shell, sometimes lost, by the possession of 8–10 prehensile arms or tentacles often equipped with suckers, by one pair of gills and by spermatophore transfer involving a specially modified pair of arms in the male; comprises 4 orders, Octopoda (octopus), Sepioidea (cuttlefish), Teuthoidea (squid) and Vampyromorpha (vampire squid).

Coleoptera Beetles; large order of holometabolous insects having immense morphological and biological diversity; body typically depressed and heavily sclerotized; forewings modified as rigid elytra covering membranous hindwings; wings may be reduced or absent; development incorporating distinct larval and pupal stages; 4 suborders recognized, Archostemata, Myxophaga, Adephaga and Polyphaga.

Coliidae Mousebirds; small family of gregarious, agile, African forest birds comprising 6 species; wings short and rounded, tail long, feet strong; feed mostly on fruits, seeds and other plant material.

Coliiformes Monofamilial order of small neognathous African forest birds.

collaplankton Planktonic organisms gaining buoyancy from gelatinous or mucous envelopes.

Collembola Springtails; order containing about 2000 species of blind, primitively wingless insects with entognathous biting mouthparts and short antennae; mostly less than 6 mm in length; leaping movement by means of a forked springing organ on the underside of the abdomen; sometimes treated as a distinct class of hexapods.

springtail (Collembola)

colloid 1: A dispersion of one substance within another, having the properties of both a solution and a suspension; the system is composed of two phases, a continuous phase (the dispersion phase) and a discontinuous phase (the dispersed phase) which consists of discrete particles; colloidal system. 2: Very fine sediment particles less than 0.002 mm in diameter; see sediment particle size.

Collothecaceae Small order of mostly sessile rotifers which have an expanded funnel-shaped anterior end and typically live in a gelatinous case, attached to the substrate by a long foot and basal disk.

colluvial deposit Material transported to a site by gravity; as in rock deposits at the base of a scree slope; **colluvium**; *cf.* alluvial deposit, aeolian deposit.

colonist 1: Used of a plant introduced unintentionally as a result of the activities of man, typically a weed of cultivation, and now found only in more or less artificial habitats. 2: An organism which invades and colonizes a new habitat or territory.

colonization 1: The successful invasion of a new habitat by a species. 2: The occupation of bare soil by seedlings or sporelings.

colony Used loosely to describe any group of organisms living together or in close proximity to each other; more precisely it refers to an integrated society in which members may be specialized subunits.

colony fission Multiplication of a colony of social insects by the departure of one or more member groups to establish new colonies whilst leaving the parent colony still viable.

colony odour The particular odour found on the bodies of social insects, providing a basis for colony recognition.

Colorado potato beetle Chrysomelidae *q.v.*

Colpodida Order of vestibuliferan ciliates with highly organized oral ciliature in the vestibulum; mostly found free-living in freshwater or soil habitats.

Colubridae Large and extremely diverse family of snakes (Serpentes); habits terrestrial, fossorial, arboreal and aquatic; feed on wide range of invertebrates and vertebrates; many are venomous; reproduction oviparous and ovoviviparous; contains over 1500 species with an almost cosmopolitan distribution, although largely absent from Australia; teeth usually solid, although grooved in a few venomous species.

Colubroidea Superfamily of morphologically advanced snakes (Serpentes) comprising 4 families that contain the majority of living species; pelvic girdle absent; left lung absent or vestigial; venomous and non-venomous; oviparous and ovoviviparous; contains about 1900 species, worldwide.

colugo Cynocephalidae *q.v.* (Dermoptera).

Columbidae Pigeons, doves; family containing about 300 species of small to medium-sized, terrestrial and arboreal birds found worldwide in woodland, forest and arid areas; habits solitary to gregarious, feeding mostly on seeds and fruit, and nesting in trees, cliffs or on the ground; the young are fed on crop-milk.

pigeon (Columbidae)

Columbiformes Small order of small to medium-sized arboreal and terrestrial birds comprising 2 extant families, Pteroclidae (sandgrouse) and Columbidae (pigeons, doves), and the Raphidae (dodo).

Columelliaceae Small family of Rosales containing 4 species of bitter shrubs or trees native to the northern part of the Andes.

Comatulida Feather stars; order of articulate crinoids comprising about 550 extant species, found from the intertidal to upper slope depths, and abundant on coral reefs; stalk absent in adult but present in early juvenile; cirri present, and may have 40 or more brightly coloured arms.

comb jelly Ctenophora *q.v.*

combfish Zaniolepididae *q.v.*

Combretaceae Indian almond, Rangoon creeper, white mangrove; family of Myrtales containing nearly 400 species of trees or climbing shrubs widespread in tropical and subtropical regions; flowers usually bisexual, regular and with 4 or 5 sepals, 4, 5 or 0 petals, 4, 5, 8 or 10 stamens and an inferior ovary.

combtooth blenny Blenniidae *q.v.*

Comephoridae Oilfishes; family of small (to 200 mm) viviparous freshwater scorpaeniform teleost fishes comprising 2 species confined to Lake Baikal; body slender, covered with thin transparent skin, pectoral and anal fins elongate, pelvics absent.

Commelinaceae Wandering jew, tradescantia, spiderwort; family containing about 700 species of often succulent herbs with stems swollen at nodes, simple alternate leaves, and flowers pollinated by insects although lacking nectaries; widespread in warm regions; flowers bisexual, usually regular, with 3 sepals and petals, 6 stamens and a superior ovary.

purple heart (Commelinaceae)

Commelinales Order of Commelinidae consisting of 4 families of herbs with alternate leaves and flowers lacking nectar but pollinated by insects, producing small abundant seeds.

Commelinidae Subclass of monocotyledons (Liliopsida) comprising 7 orders of mostly herbs with simple, alternate, parallel-veined leaves and usually with insect or wind-pollinated flowers; seed endosperm largely starchy.

commensalism A symbiotic relationship between two species in which one derives benefit from a common food supply whilst the other is not adversely affected; **commensal.**

common carpetweed Molluginaceae *q.v.*

common flax Linaceae *q.v.*

common pokeweed Phytolaccaceae *q.v.*

common rue Rutaceae *q.v.*

communal Pertaining to the cooperation between members of the same generation in nest building but not in care of the brood.

commune A society or group of conspecific organisms which have a social structure and consist of repeated members or modular units with a high level of coordination, integration and genotypic relatedness.

communicable Used of a disease which is readily transferred from one organism to another, usually used of diseases of man and domestic animals.

communication Any action of one organism that modifies the behaviour pattern of another organism.

community Any group of organisms comprising a number of different species that co-occur in the same habitat or area and interact through trophic and spatial relationships.

compatible 1: Capable of cross fertilization. 2: Used of plants having the capacity for self-fertilization.

compensation depth 1: The depth at which primary production and respiration are equal so there is no net production; compensation level; compensation point. 2: The ocean depth at which the rate of solution of calcium carbonate increases to the point that dissolution of calcareous shells begins; typically a depth of about 4000 m, below which calcareous sediments are not found.

competition The simultaneous demand by two or more organisms or species for an essential common resource that is actually or potentially in limited supply (exploitation competition), or the detrimental interaction between two or more organisms or species seeking a common resource that is not limiting (interference competition); **competitor.**

competitive exclusion principle The principle that complete competitors cannot coexist; two species having identical ecological requirements cannot coexist indefinitely.

complemental male A male which lives permanently attached to a female, often small and degenerate except for well developed reproductive organs.

complete flower A flower that has all four parts, pistil, stamen, petal and sepal; *cf.* incomplete flower.

Compositae Asterales *q.v.*

composite species 1: A species comprising two or more subspecies; polytypic species. 2: In palaeontology, a species represented by a group of specimens that were obtained from two or more localities and that are not all of the same geological age.

composites Asterales *q.v.*

compound nest A nest formed by two or more species of social insects in which individuals from the separate colonies may intermingle but in which the broods remain separate.

compression A process of fossilization in which the organic remains have been flattened by overlying strata.

Compsopogonales Small order of tropical and subtropical red algae tyically with parenchymatous thalli which pass through a filamentous stage in early development.

Concentricycloidea Sea daisy; recently discovered class of minute (2–9 mm) deep-sea echinoderms found in submerged driftwood off the New Zealand coast; body circular resembling small jellyfish with spines and tube feet around the periphery; dorsal surface bearing calcareous plates; paired gonads arranged in pentamerous symmetry; mouth, anus and arms absent; unlike other echinoderms the sea daisy has 2 water vascular rings instead of one; the young are brooded in the gonads.

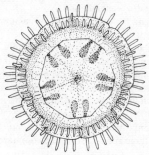

sea daisy (Concentricycloidea)

conch Mesogastropoda *q.v.*

conchology The study of shells; **conchological.**

conchometry The study of the morphometrics of shells.

Conchostraca Clam shrimps; order of diplostracan crustaceans containing about 180 species of detritus and suspension feeders found mostly in freshwater temporary pools; body up to 20 mm in length, enclosed in a bivalved carapace and bearing up to 28 pairs of foliaceous limbs; produce eggs resistant to desiccation.

concolorous Having uniform coloration; **concolour.**

concrescent Coalesced; growing together.

conditioning Modification of the behaviour of an organism such that it responds to a given stimulus with a behaviour pattern normally associated with some other stimulus when the two stimuli have been applied concurrently a number of times.

condor Cathartidae *q.v.*

Condylarthra Extinct order of mammals (Protoungulata) comprising several families of primitive ungulates; somewhat intermediate in form between insectivores and true ungulates; some may have possessed claws but others had hooves and ungulate dentition; known from the Late Cretaceous to the Miocene.

cone shell Neogastropoda *q.v.*

coneflower Asterales *q.v.*

congenital Present at time of birth; used of a condition that has resulted from an embryonic aberration.

conger eel Congridae *q.v.*

Congiopodidae Pigfishes, racehorses; family of little-known marine scorpaeniform teleost fishes found in shallow waters of the southern hemisphere; body deep, compressed, naked; length to 0.8 m; snout prominant.

conglobate To roll up, as in some woodlice and millipedes; **conglobation.**

congregate To collect together or assemble into a group.

Congridae Conger eels; diverse family containing about 100 species of marine anguilliform teleost fishes; body smooth, to 3 m length; dorsal and anal fins long, caudal present or absent, pectorals variable, large to absent; predatory, scavenging or zooplanktotrophic, commonly nocturnal; some large species exploited commercially.

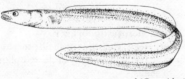

conger eel (Congridae)

conifer Pinatae *q.v.*

Coniferae The largest group of gymnosperms, treated under the alternative name Pinatae *q.v.*

coniophilous Thriving on substrates enriched by dust; used of lichens which thrive when covered with a coat of dust; **coniophile, coniophily.**

conjugation 1: Union of gametes, nuclei, cells or individuals. 2: A type of sexual reproduction during which two cells unite but with only limited genetic exchange between the cells.

Connaraceae Family of Rosales containing about 400 species of often highly poisonous woody plants widespread in tropical regions; flowers pentamerous; fruit usually a one-seeded follicle.

connascent Produced at the same birth.

Conocardioida Extinct order of rostroconch molluscs with an elongate posterior part to the shell; known from the Lower Ordovician to Upper Permian.

Conopodina A suborder of Amoebida *q.v.*; also treated as a class of the protoctistan phylum Rhizopoda.

consanguinity Relationship by descent from a common ancestor.

consecutive hermaphrodite An organism having functional male and female reproductive organs which mature at different times, either the male organs mature first (protandrous hermaphrodite) or the female organs (protogynous hermaphrodite); *cf.* synchronous hermaphrodite.

conservation The planned management of natural resources; the retention of natural balance, diversity and evolutionary change in the environment; *cf.* preservation.

consistence A measure of cohesion of soil particles.

consociation A small climax plant community dominated by one particular species (the physiognomic dominant) which has the life form characteristic of the association.

conspecific Belonging to the same species.

constant plankton The perennial permanent planktonic organisms of an area.

constant species 1: A species invariably present in a given community. 2: A species occurring in at least 50% of samples taken from a given community.

constipated Crowded together.

constrictor snake Boidae *q.v.*

consumer An organism that feeds on another organism or on existing organic matter; includes herbivores, carnivores, parasites and all other saprotrophic and heterotrophic organisms.

consummatory act A behavioural act which

constitutes the terminal phase of an instinctive behaviour sequence; *cf.* appetitive behaviour.

consumption In ecological energetics, the total intake of food or energy by a heterotrophic individual, population or trophic unit per unit time.

contagious disease Any disease transmitted through physical contact.

contact herbicide A herbicide that kills on contact rather than after absorption, such as Paraquat.

contamination The introduction of an undesirable agent such as a pest or pathogen, into a previously uninfested situation.

contiguous Having boundaries that make contact but areas that do not overlap.

continental climate Any climate in which the difference between summer and winter temperatures is greater than the average range for that latitude because of distance from an ocean or sea.

continental drift The theory that the continental landmasses have drifted apart over the course of geological time; *cf.* plate tectonics.

continental island An island that is close to, and geologically related to, a continental landmass, and was formed by separation from the continent; *cf.* oceanic island.

continental rise The gently sloping sea bed from the continental slope to the abyssal plain, with an average angle of slope of about 1°; see marine depth zones.

continental shelf The shallow gradually sloping sea bed around a continental margin, not usually deeper than 200 m and formed by submergence of part of the continent; see marine depth zones.

continental slope The steeply sloping sea bed leading from the outer edge of the continental shelf to the continental rise, with an average angle of slope of about 4° and a maximum of about 20° near the upper margin; see marine depth zones.

continuum A gradual or imperceptible intergradation between two or more extreme values.

Contortae Gentianales *q.v.*

contranatant Swimming, moving, or migrating against the current; *cf.* denatant.

control A parallel experiment or test carried out to provide a standard against which an experimental result can be evaluated.

Conulata Extinct subclass of scyphozoan Cnidaria known from the Middle Cambrian to the Triassic; bodies exhibited tetramerous

symmetry; tentacles present around mouth.

convection A mode of heat transfer within a fluid, involving the movement of substantial volumes of the fluid from one place to another as a result of changes in the density of heated and non-heated areas.

convective rain Rainfall produced by convection, resulting from solar radiation heating the ground; *cf.* frontal cyclonic rain, orographic rain.

convergence 1: Convergent evolution *q.v.* 2: An oceanic region in which surface waters of different origins come together and where the denser water sinks beneath the lighter water, as in the Antarctic convergence; *cf.* divergence.

convergent evolution The independent evolution of structural or functional similarity in two or more unrelated or distantly related lineages or forms that is not based on genetic similarity; *cf.* parallel evolution.

Convolvulaceae Bindweed, morning glory, sweet potato; family of Solanales containing about 1500 species of commonly twining or climbing herbs; nearly cosmopolitan in distribution; flowers regular and usually bisexual, typically with 5 sepals, petals and stamens, and a superior ovary.

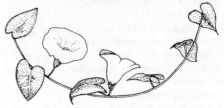

bindweed (Convolvulaceae)

cony Ochotonidae *q.v.*

coot Rallidae *q.v.*

Copepoda Class of mostly small aquatic crustaceans exhibiting a great diversity of form and life history; includes many commensals, parasites of all major animal groups, and free-living forms, abundant in planktonic, benthic and interstitial habitats; primitively with many-segmented antennae, modified for grasping in the male; each pair of swimming legs joined by a rigid plate; contains about 9000 species in 8 orders.

copepodid Larval stage of copepods, possessing functional swimming thoracic legs.

Cope's rule The generalization that there is a trend towards increasing body size within an evolutionary series.

copper Lycaenidae *q.v.*

coppicing Woodland management by regularly cutting back of trees to ground level to encourage adventitious shoot production; *cf.* pollarding.

Coprinaceae Ink caps; family of mushroom-like fungi (Agaricales) in which the spore masses appear dark brown or black; mostly found on dung or rotting wood; gills typically liquefy as spores mature.

ink cap (Coprinaceae)

coprobiont Any animal (coprozoite) or plant (coprophyte) living or feeding on dung; **coprobiontic.**

coprolites Fossilized faecal material.

coprology The study of animal faeces; **coprological.**

coprophagous Feeding on dung or faecal material; **coprophage, coprophagy.**

coprophilous Thriving on or in dung or faecal material; **coprophile, coprophily.**

coprophyte A plant living on dung or faecal material; **coprophytic.**

coprozoite An animal living on or in dung or faecal material; **coprozoic.**

copulate 1: To engage in sexual intercourse; the act of introducing spermatozoa into the female; coition; **copulation, copulatory.** 2: Conjugate; to fuse together, as in the case of gametes.

coquina A deposit of drifted shells.

Coraciidae Rollers; small family of colourful arboreal birds known for their acrobatic tumbling courtship flight; habits solitary, sedentary, aggressive, monogamous, often migratory: feed mainly on insects caught on the wing or by swooping from a perch; nest in hole in bank or tree; widespread in tropical and warm temperate open woodlands of the Old World.

Coraciiformes Diverse order of hole-nesting arboreal birds comprising 10 small families; includes Alcedinidae (kingfishers), Meropidae (bee eaters), Upupidae (hoopoe), Bucerotidae (hornbills) and Coraciidae (rollers); most widespread in tropical regions of the Old World.

coral snake Elapidae *q.v.*

coral-billed nuthatch Hyposittidae *q.v.*

Corallimorpharia Order of solitary or gregarious zooantharians found mainly in tropical, shallow seas often on coral reefs; lacking a hard skeleton and characterized by an adherent pedal disk, radially arranged and non-retractile tentacles, and numerous complete mesenteries.

Cordaitales Extinct order of gymnosperms known from the Carboniferous to the Permian; included trees up to 30 m tall, often with strap-like leaves; cones primitive.

cordillera An entire mountain system, including all subordinate ranges, interior plateaux and basins.

Cordylidae Family of terrestrial lizards containing about 55 species found in xeric rocky and grassland habitats of Africa and Madagascar; mainly insectivorous, sometimes predatory on small vertebrates; reproduction oviparous or ovoviviparous; transverse rows of ossified scales may be elongated into spines.

core A vertical column of bottom sediment.

coriander Apiaceae *q.v.*

Coriariaceae Small family of strongly tanniferous shrubs producing ellagic acid and poisonous terpenoids; contains only 5 species having an interrupted cosmopolitan distribution; sepals imbricate, petals with internal keels, 5–10 free carpels.

Coriolis force The deflecting force of the Earth's rotation that causes a body of moving water to be deflected to the right in the northern hemisphere and to the left in the southern hemisphere.

corkscrew flower Apocynaceae *q.v.*

corm A short, underground stem, specialized as a perennating organ, producing flowers and foliage from one or more buds.

cormidium An assemblage of zooids budded off the coenosarc in some colonial hydrozoan cnidarians (Siphonophora).

cormorant Phalacrocoracidae *q.v.*

Cornaceae Dogwood; family of about 100 species of mostly trees and shrubs, widespread especially in northern temperate regions; flowers with 4, 5, or 0 sepals and petals, 4 and 5 stamens and an inferior ovary.

Cornales Small order of Rosidae containing 4 families of mostly woody plants, Alangiaceae, Cornaceae, Garryaceae and Nyssaceae.

magpie (Corvidae)

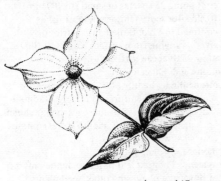

dogwood (Cornaceae)

cornetfish Fistulariidae *q.v.*

corolla The petals, the second outermost whorl of a flower; may be fused into a corolla tube.

Coronata Extinct order of crinoids (Inadunata) in which there was a single circlet of plates beneath the radial plates bearing the arms; plates covering ambulacra; known from the Middle Ordovician to Upper Silurian.

Coronatae Primitive order of Scyphozoa comprising about 25 species of large, dark red or purple jellyfish found mostly in the bathypelagic zone of the world's oceans.

Coronophorales Small order of pyrenomycete fungi found mainly on wood with the mycelium produced within the host; characterized by carbonaceous ascocarps which are completely enclosed.

corrasion Erosion by the mechanical action of a moving agent such as running water, wind or glacial ice.

corridor A more or less continuous connection between adjacent landmasses which has existed for a considerable period of geological time.

Corsiaceae Small family of Orchidales comprising about 9 species of mycotrophic herbs lacking chlorophyll and with leaves reduced to scales and flowers solitary and terminal, restricted to New Guinea and Chile.

corticolous Inhabiting or growing on bark; **corticole.**

Corvidae Crows, jays, magpies; cosmopolitan family containing about 100 species of small to large passerine birds found in wide variety of wooded and open habitats; plumage uniformly dark, or with bold coloration; bill strong, slender; habits gregarious to solitary, non-migratory, often aggressive; feeding on wide range of plant and animal material; nest of sticks in trees or on cliffs.

Corydalidae Dobsonflies; family containing about 200 species of large neuropteran insects with a wingspan usually of 40–160 mm but which do not fly well; subaquatic larvae are voracious predators possessing 8 pairs of simple abdominal gills.

Corylaceae Hornbeam, hazel; family of deciduous trees and shrubs (Fagales) widely distributed in the temperate and cold-temperate northern hemisphere; flowers arranged in unisexual catkins; fruit is a nut; sometimes included within the Betulaceae *q.v.*

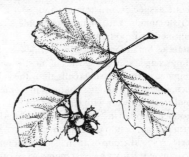

hazel (Corylaceae)

corymb A racemose inflorescence in which lower flower stalks are longer than the upper so the flowers lie in a flattish dome.

corymb

Coryneliales Small order of mostly tropical pyrenomycete fungi which are mostly parasitic on conifers; typically with evanescent asci at maturity.

Corynocarpaceae Family of Celastrales containing only 5 species of trees producing bitter, toxic glucosides; native to Australia and New Zealand.

Coryphaenidae Dolphinfishes; family of predatory, pelagic, teleost fishes (Perciformes) comprising only 2 species, cosmopolitan in tropical oceans; body elongate, compressed, to 1.5 m length; single long dorsal fin present, caudal deeply forked; very popular as game fishes.

coryphilous Thriving in alpine meadows; **coryphile, coryphily.**

coryphophyte A plant inhabiting alpine meadows.

cosmopolitan Having a worldwide distribution, effect or influence; pandemic; ubiquitous.

cosmotropical Circumtropical; occurring throughout the tropics.

Cossurida Order of small (to 15 mm) deposit-feeding polychaete worms that burrow in soft sediments, comprising a single family and about 20 species; prostomium and peristomium naked; pharynx eversible, unarmed; parapodia biramous.

Costaceae Small family of Zingiberales comprising about 150 species of mostly leafy-stemmed herbs with silica cells along the veins, occurring in wet, shady situations in tropical rainforests; flowers with 6 stamens, 1 producing pollen in 2 sacs, the other 5 fused to form a large, petaloid labellum.

coterie A social group of animals which defends a common territory against members of other coteries.

cotinga Cotingidae *q.v.*

Cotingidae Cotingas; family containing about 80 species of neotropical passerine forest birds; male often with bold patterned plumage, some with elaborate courtship display; habits mostly solitary, arboreal, non-migratory, monogamous or polygamous; feed on insects and fruit, and nest in trees or among rocks.

Cottidae Sculpins, bullheads; family containing about 300 species of primarily northern hemisphere marine and freshwater scorpaeniform teleost fishes; body slender to stout, to 750 mm length, often with marked cryptic coloration; head depressed and spinose.

Cottocomephoridae Longwing sculpins; family containing 25 species of small (to 200 mm) freshwater scorpaeniform teleost fishes confined to Lake Baikal and its tributary rivers, ranging from surface waters to 1000 m.

cotton Malvaceae *q.v.*

cottonwood Salicales *q.v.*

Cottunculidae Family of little-known marine sculpins (Scorpaeniformes) having a tadpole-shaped body covered by loose warty skin; head bearing bony processes, mouth large; contains 6 species, widely distributed, mostly in moderately deep water.

cotyledon The first leaf or leaves in the embryo of flowering plants and other seed plants.

Cotylosauria Group of anapsid reptiles which flourished in the late Palaeozoic, becoming extinct in the Triassic.

cougar Felidae *q.v.*

countershading The condition of an animal having a darkly coloured dorsal surface and a lighter ventral surface, that serves to disrupt the silhouette in a gradient of downwelling light.

coural Leptosomatidae *q.v.*

court 1: The group of workers that surround a queen in an insect colony; retinue. 2: That part of a communal display area defended by a particular male.

courtship Any behavioural interaction between males and females that facilitates mating.

cover 1: Plant material, living (vegetative cover) and dead (litter cover), on the soil surface. 2: The area of ground covered by vegetation of a particular plant species; expressed as a scale (Braun–Blanquet scale, Domin scale), or as a percentage.

coverage That part of a sampled area covered by a particular plant species or individual plant canopy.

covert A shelter; a hiding place.

cow killer Mutillidae *q.v.*

cow shark Hexanchidae *q.v.*

cowbird Icteridae *q.v.*

cowrie Mesogastropoda *q.v.*

cowslip Primulaceae *q.v.*

coypu Myocastoridae *q.v.*

crab Brachyura *q.v.*

crab louse Pthiridae q.v.

Cracidae Curassows; family containing about 50 species of medium sized Neotropical tree-living or terrestrial gallinaceous birds (Galliformes) found in forest habitats from Texas to Paraguay.

Cracticidae Butcherbirds; family comprising about 10 species of medium-sized gregarious, arboreal birds found in Australia and Papua

butcherbird (Cracticidae)

New Guinea; bill typically short and stout, wings pointed; feed on insects, small vertebrates and fruit.

cranberry Ericaceae *q.v.*

crane Gruidae *q.v.*

cranefly Tipulidae *q.v.*

cranesbill Geraniaceae *q.v.*

Craniata Those groups of animals possessing a dorsal vertebral column and a bony or cartilaginous skull.

Cranoglanidae Family of Chinese freshwater catfishes (Siluriformes) containing only 3 species; body naked, to 300 mm length, dorsal fin short with single spine; snout depressed bearing 4 pairs of barbels; Cranoglanididae.

crape myrtle Lythraceae *q.v.*

Craseonycteridae Bumblebee bat; monotypic family of small microchiropteran bats from Thailand; tail absent; nose simple with narial pad and open nostrils; second digit partly free; includes the smallest mammal.

crash A precipitous decline in the size of a population; *cf.* flush.

Crassulaceae Sempervivum, houseleek, crassula; large family of Rosales containing about 900 species of succulent herbs or shrubs often with red root tips, commonly with crassulacean acid metabolism *q.v.*; of nearly cosmopolitan distribution and familiar as ornamentals; flowers arranged in cymes, with 3–30 sepals, petals and stamens.

crawfish Palinura *q.v.*

crayfish Astacidea *q.v.*

creationism The belief that all forms of life were created *de novo*, and have undergone little subsequent change; **creationist.**

crèche A group of young animals which have left their nests.

Creediidae Family of small (to 90 mm) Indo-Pacific coastal marine, sand-burrowing, teleost fishes (Perciformes), comprising 3 monotypic genera; body slender, eel-like; mouth protractile; dorsal and anal fins elongate, lacking fin spines.

cremnophilous Thriving on cliffs; **cremnophile, cremnophily.**

cremocarp A dry fruit, developed from 2 fused carpels, that divides into 2 one-seeded units at maturity.

crenic Pertaining to a spring and the adjacent brook water flowing from the spring.

crenicolous Living in springs, or in brook water fed from a spring; **crenicole.**

crenogenic meromixis Mixing in a lake caused by a saline spring delivering dense water into the bottom of a lake displacing the mixolimnion *q.v.*; *cf.* biogenic meromixis, ectogenic meromixis.

crenon The spring-water biotope; *cf.* stygon, thalasson, troglon.

crenophilous Thriving in or near a spring; **crenophile, crenophily.**

Crenuchidae Family containing 3 species of tiny (to 50 mm), colourful, South American freshwater characiform teleost fishes found in parts of the Amazon and Orinoco basins.

Creodonta An extinct order of carnivorous mammals (Ferae) known from the late Cretaceous through to the Pliocene that included mustelid-like and hyaena-like forms; rather small-brained and slow moving; shearing carnassial teeth formed by the molars.

creophagous Carnivorous; used particularly of insectivorous plants; **creophage, creophagy.**

creosote bush Zygophyllaceae *q.v.*

crepitation A defence mechanism in insects in which fluid is explosively discharged.

crepuscular Active during twilight hours; of the dusk and dawn; *cf.* diurnal, nocturnal.

crestfish Lophotidae *q.v.*

Cretaceous A geological period of the Mesozoic era (*c.* 140–65 million years B.P.); see geological time scale.

Cricetidae New World rats and mice, gerbils, hamsters, lemmings, voles; large cosmopolitan family of small (adult length 100–600 mm) terrestrial to arboreal myomorph rodents comprising about 560 species; habits may be fossorial or

deer mouse (Cricetidae)

semiaquatic, often nocturnal, feeding on a variety of plant material and insects.

cricket Gryllidae *q.v.*

cricket wasp Rhopalosomatidae *q.v.*

Cricoconarida Extinct class of molluscs known from the Palaeozoic; shell narrow, with tapering cones of laminated calcium carbonate.

Crinoidea Sea lilies, feather stars; class of shallow- to deep-water echinoderms containing about 625 species in a single extant subclass, Articulata, comprising 5 orders, Millericrinida, Cyrtocrinida, Bourgueticrinida, Isocrinida and Comatulida, widespread but especially abundant in the tropical western Pacific; body globoid, aboral surface forming cup-like calyx, with or without stalk; arms multibranched, fern- or feather-like; most crinoids are non-selective suspension feeders.

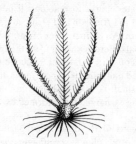

feather star (Crinoidea)

Crinozoa Subphylum of echinoderms comprising about 8 classes only one of which, Crinoidea, survived beyond the Palaeozoic; exhibit radial symmetry; body cup-shaped or globular, enclosed in a plated test; arms or brachioles supporting extensions of feeding ambulacra; also known as Pelmatozoa.

croaker Sciaenidae *q.v.*

crocodile Crocodylidae *q.v.*

crocodile icefish Chaenichthyidae *q.v.*

crocodile shark Pseudocarchariidae *q.v.*

Crocodylia Crocodiles, alligators, gavials; order of amphibious archosaurian reptiles found in freshwater or brackish marsh and river habitats; all are carnivorous, feeding on a variety of mammals, fish and other

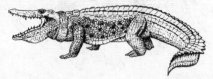

crocodile (Crocodylidae)

vertebrates; reproduction oviparous; contains 22 living species.

Crocodylidae Family of small to very large (up to 7 m) amphibious carnivorous reptiles containing 22 species in 3 subfamilies, Alligatorinae (alligators, caiman, from New World), Crocodylinae (crocodiles, pantropical) and Gavialinae (gavials, India and Burma); may also be regarded as families.

crocus Iridaceae *q.v.*

Cromerian interglacial An interglacial period of the Quaternary Ice Age *q.v.* in the British Isles.

cron A unit of time equal to one million (10^6) years.

cross fertilization The union of male and female gametes from different individuals of the same species; *cf.* self-fertilization.

cross infection The transfer of a disease organism from one individual to another of the same species.

cross pollination Transfer of pollen from one flower to the stigma of a flower on another plant of the same species; *cf.* self-pollination.

crossing The mating of individuals of different breeds, races or strains in order to promote genetic recombination; hybridize; crossbreed.

Crossomataceae Small family of Rosales containing about 9 species of glabrous xerophytic shrubs native to arid parts of North America.

Crossopterygii Subclass of bony fishes (Osteichthyes) comprising the tassel-finned fishes belonging to the Coelacanthiformes and Rhipidistia.

Crotalidae Pit vipers, sidewinder; family containing about 130 species of snakes (Serpentes) that possess heat-sensitive organs just behind the nostrils; feed on small vertebrates; sometimes included in the Viperidae.

croton Euphorbiaceae *q.v.*

crotovina A former animal burrow in one soil horizon that has been filled with organic matter or sediment from another horizon.

crow Corvidae *q.v.*

crowberry Empetraceae *q.v.*

crown The highest part or layer; typically used of the uppermost foliage of a tree.

crown conch Neogastropoda *q.v.*

crown-of-thorns Spinulosida *q.v.*

Cruciferae Crucifers; Brassicaceae *q.v.*

Crustacea Subphylum of arthropod animals including many types of shrimps, lobsters, crabs, barnacles, water fleas, scuds, sand hoppers, slaters, fish lice, whale lice, pill bugs;

found in marine, freshwater, and terrestrial habitats, ubiquitous at all depths from the littoral to ocean trenches; characterized by 2 pairs of antennae but exhibiting an immense diversity of form, life history and habit; contains about 40 000 described species, arranged in 10 extant classes, Cephalocarida, Branchiopoda, Remipedia, Mystacocarida, Tantulocarida, Cirripedia, Copepoda, Branchiura, Ostracoda and Malacostraca.

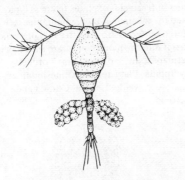

copepod (Crustacea)

crymnion The planktonic organisms of perpetual ice and snow.

crymophilous Thriving in polar habitats; **crymophile**, **crymophily**.

crymophyte A plant of polar regions.

cryochore Those regions of the Earth's surface perpetually covered by snow.

cryoconite Organisms and wind-blown detritus that induce surface melt pits in glaciers.

cryogenic lake A lake formed by local thawing in an area of permanently frozen ground.

cryophilic Thriving at low temperatures; **cryophile**; **cryophily**.

cryophylactic Resistant to low temperatures.

cryophyte 1: A plant growing on ice or snow. 2: A plant thriving at low temperatures.

cryoplankton Planktonic organisms of persistent snow, ice and glacial waters.

cryotropism An orientation response to the stimulus of cold or frost; **cryotropic.**

cryoturbation The physical mixing of soil materials by the alternation of freezing and thawing.

crypsis Concealment.

Crypteroniaceae Small family of Myrtales including 4 species of trees native to India, Malaysia and Philippines; fruit is a capsule.

cryptic Used of coloration and markings that resemble the substratum or surroundings and aid in concealment; *cf.* phaneric.

cryptobiosis The condition in which all external signs of metabolic activity are absent from a dormant organism.

cryptobiotic Used of organisms which are typically hidden or concealed in crevices or under stones; **cryptobios.**

Cryptobranchidae Small family of large (to 1.5 m) stream-dwelling salamanders (Caudata) comprising 3 species, one from eastern North America the others from China and Japan; metamorphosis incomplete, eyelids absent, gills slits present or absent; fertilization external, eggs laid in paired strings; larvae with short gills and caudal fins.

Cryptodira Suborder of testudine reptiles in which the head is withdrawn directly into the shell; contains majority of living taxa, 175 species in 9 families and 3 superfamilies, Trionychoidea (freshwater turtles), Chelonioidea (sea turtles) and Testudinoidea (tortoises, terrapins); distribution mainly northern temperate and tropical.

Cryptodonta Subclass of bivalve molluscs with thin equivalve shells composed of aragonite; lived within the sediment.

cryptofauna The fauna of protected or concealed microhabitats.

cryptogam A lower plant, lacking conspicuous reproductive structures such as flowers or cones; *cf.* phanerogam.

Cryptomonadales Order of flagellate algae (Cryptophyceae).

Cryptomonadida Cryptophyceae *q.v.* treated as an order of the protozoan class Phytomastigophora.

cryptomonads Cryptophyceae *q.v.*

Cryptonemiales Large order of red algae with a thallus that may be encrusting or erect and frondose; sometimes calcareous.

Cryptophyceae Cryptomonads; class of chromophycote algae occurring in freshwater, brackish and marine habitats; consisting of biflagellate unicells, with 2 anterior flagella

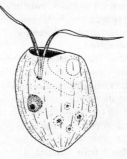

cryptomonad (Cryptophyceae)

inserted in an oral groove and typically possessing phycobilin pigments in addition to chlorophylls *a* and *c*; comprises 2 orders, Cryptomonadales and Tetragonidiales; also treated as a separate phylum of Protoctista, under the name Cryptophyta, and as an order of the protozoan class Phytomastigophora, under the name Cryptomonadida.

Cryptophyta The Cryptophyceae *q.v.* treated as a phylum of Protoctista.

cryptophyte A perennial plant with renewal buds below ground or water level, including geophyte *q.v.*, helophyte *q.v.* and hydrophyte *q.v.*; see Raunkiaerian life forms.

Cryptostigmata Oribatei *q.v.*

Cryptostomata Extinct order of bryozoans in which the colony had a frond-like, reticulate form and a calcite skeleton; individual apertures hidded in a short shaft; known from the Ordovician to the Permian.

Cryptozoic The period of geological time during which only the most primitive life forms are found; a subdivision of the Precambrian era, preceding the Proterozoic period; see geological time scale.

cryptozoic Pertaining to small terrestrial animals (cryptozoa) inhabiting crevices, living under stones, in soil or litter.

cryptozoology The study of cryptozoic organisms.

crystallochorous Dispersed by glaciers; **crystallochore, crystallochory.**

cteinophyte A parasitic plant which destroys its host; cteinotrophic plant.

Ctenocladales Order of branched filamentous green algae in which the filaments are composed of uninucleate cells, each containing a parietal laminate chloroplast; found in marine, brackish and freshwater habitats.

Ctenodrilida Order of minute polychaete worms comprising about 12 species in 2 families; body with few segments; prostomium naked, pharynx globular, eversible, unarmed; asexual reproduction by fragmentation common, some forms protandric hermaphrodites.

Ctenoluciidae Pike-characins; family containing 4 species of predatory, piscivorous, freshwater characiform teleost fishes found in rivers and streams of South America; body elongate, to 1 m length, pike-like; jaws prolonged, dorsal fin posteriorly positioned; adipose fin present.

Ctenomyidae Family containing about 26 species of burrowing hystricomorph rodents found in South America; body compact; legs short, pentadactyl with strong claws; eyes and ears small; soles of feet bordered with bristles.

Ctenophora Ctenophores, sea gooseberries, comb jellies; phylum of carnivorous marine animals comprising about 80 species found in pelagic and benthic habitats but mostly in the open sea; characterized by gelatinous bodies with biradial symmetry and by the possession of mesenchymal muscles, anal openings and 8 rows of fused ciliary plates (ctenes) typically used for swimming; all are bioluminescent; paired tentacles are used to catch prey and pass it into the mouth; most are simultaneous hermaphrodites; divided into 7 orders, Cydippida, Platyctenida, Ganeshida, Thalassocalycida, Lobata, Cestida and Beroida.

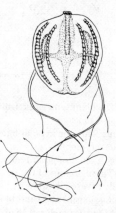

sea gooseberry (Ctenophora)

Ctenostomata Order of gymnolaematan bryozoans in which the colonies are frequently creeping and stolonate, or dense, tufty, encrusting, or frond-like; skeleton membranous, gelatinous, non-calcified; typically marine with some brackish and freshwater forms.

ctetology The study of acquired characters; **ctetological.**

Cubomedusae Cubozoa *q.v.*

Cubozoa Sea wasps, cubomedusae; class of marine pelagic Cnidaria found in all tropical and subtropical waters; typically a tall, colourless medusa with a bell that is square in section and has one or more tentacles at each corner of the umbrella; comprises a single order, Carybdeida; the stings of certain large Australian species are extremely virulent, fast-acting and frequently fatal.

cuckoo Cuculidae *q.v.*

cuckoo shrike Campephagidae *q.v.*
cuckoo wasp Chrysididae *q.v.*
Cuculidae Family containing about 130 species of birds (Cuculiformes) comprising cuckoos and roadrunners; cuckoos found in woodland habitats worldwide, commonly migratory, many Old World species are nest parasites; roadrunners are terrestrial in open arid regions; habits mostly solitary, feeding on invertebrates, small vertebrates, and fruit.

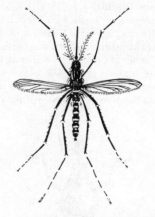

cuckoo (Cuculidae)

Cuculiformes Order of small to large arboreal and terrestrial birds comprising 2 suborders, Musophagi (touracos) and Cuculi (cuckoos).
cucumber Cucurbitaceae *q.v.*
Cucurbitaceae Pumpkin, squash, watermelon, cucumber, gourd, muskmelon and vegetable sponge (*Luffa*); large cosmopolitan family of Violales containing about 900 species of herbaceous mostly climbing or trailing plants commonly with spiral tendrils at nodes; includes mainly cultivated genera; fruits are berries, often very large.
Culicidae Mosquitoes; cosmopolitan family of small dipteran insects (flies) of primary medical and veterinary importance comprising about 3000 species; adults with piercing proboscis for feeding on nectar or blood; larvae aquatic living suspended beneath the surface film; some species transmit pathogens causing malaria, yellow fever, filariasis and dengue.
cull To reduce the number of animals in a breeding group by removing inferior stock or by killing selected animals.
culmicolous Growing on the stems of grasses; **culmicole.**

mosquito (Culicidae)

cultigen An organism known only in cultivation; cultivar.
cultivar A variety of a plant produced and maintained by cultivation.
cultivation The preparation and use of land for crop production.
Cumacea Order of small, tadpole-like peracarid crustaceans containing about 1000 species found in marine bottom sediments.
cumaphyte A surf plant.
cumatophytic Pertaining to structural modifications of plants in response to wave action.
cuniculine Living in burrows resembling those of rabbits.
Cunoniaceae Family of Rosales containing about 350 species of strongly tanniferous shrubs or trees mostly native to the southern hemisphere; flowers small, tetra- or pentamerous; fruit usually a capsule.
cup fungi Discomycetes *q.v.*
Cupressaceae Cypress, juniper; family of evergreen conifers (Pinatae) with leaves scale-like and obscuring the branchlets or occasionally needle-like; mature pistillate cones leathery to woody or berry-like.

cypress (Cupressaceae)

cuprophyte A plant adapted to, or tolerating, high copper levels in the soil; frequently used as an indicator of this particular soil type.

curassow Cracidae *q.v.*

Curculionidae Weevils, bark beetles; family of small to medium-sized beetles (Coleoptera) with a characteristic rostrum on the head; most feed as larvae on rotting wood or on living plants; contains over 50 000 species including many economically important pests such as cotton-boll weevil, grain weevil and rice weevil.

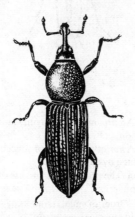

weevil (Curculionidae)

Curimatidae Family of small to medium-sized South American freshwater characiform teleost fishes; body deep, compressed, dorsal fin anterior to mid-body; contains about 130 species; small species popular in aquarium trade, some large forms exploited as food-fish.

curlew Scolopacidae *q.v.*

currant Grossulariaceae *q.v.*

current 1: A non-tidal horizontal movement of the sea. 2: In limnology, a continuous flow of water.

cursorial Adapted for running; running.

cuscus Phalangeridae *q.v.*

Cuscutaceae Dodder; nearly cosmopolitan family of Solanales containing about 150 species of twining parasitic herbs with little or no chorophyll, attached by haustoria to stems of the host; sometimes included in the Convolvulaceae.

cusk eel Ophidiidae *q.v.*

custard apple Annonaceae *q.v.*

Cuterebridae Rodent botflies; family of large robust flies (Diptera), comprising about 70 species confined to the New World, that are serious pests of livestock; the larvae are typically parasitic in the skin of mammals, especially rodents and lagomorphs but also others, including man, and occasionally in birds.

cutlassfish Trichiuridae *q.v.*

Cutleriales Small order of brown algae with a parenchymatous construction and anisogamous sexual reproduction.

cutthroat eel Synaphobranchidae *q.v.*

cuttlefish Sepioidea *q.v.*

cutworm Noctuidae *q.v.*

Cyamidae Whale lice; family of amphipod crustaceans found exclusively as external parasites of whales, dolphins and porpoises.

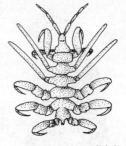

whale louse (Cyamidae)

Cyanastraceae Small family of Liliales comprising only 7 species of perennial herbs arising from a tuberous base, native to forests of tropical Africa; leaves all basal with a broad blade and curved-convergent veins with evident crossveins.

Cyanobacteria Blue-green bacteria; sometimes used as a replacement name for the Cyanophycota *q.v.*

cyanogenic Used of living organisms that produce hydrocyanic acid.

Cyanophyceae The sole class of Cyanophycota *q.v.*, comprising 4 orders, Chroococcales, Nostocales, Pleurocapsales and Stigonematales.

Cyanophycota Blue-green algae; division of photosynthetic moneran microorganisms with an alga-like biology and bacterium-like organization; characterized by the possession of chlorophyll *a* located in single thylakoids not enclosed by membranes and of phycobiliproteins on the thylakoid surface; comprises a single class, Cyanophyceae; also known as the Cyanobacteria.

blue-green alga (Cyanophycota)

cybernetics The study of control and communication in systems, including information systems and feedback control systems.

cycad Zamiaceae *q.v.* (Cycadicae).

Cycadatae Sole extant class of Cycadicae *q.v.*

Cycadicae Cycads; subdivision of Gymnosperms (Pinophyta); evergreen, perennial shrubs or trees with stems that are usually unbranched but thickened by some secondary growth; large pinnate or bipinnate leaves borne spirally near apex of stem; dioecious; male structures arranged like scales on a large cone-like strobilus; female structures, ovules on stalked scales arranged in cone-like strobili; ovule with micropyle pore.

cycad (Cycadicae)

cyclamen Primulaceae *q.v.*

Cyclanthales Fan palms; order of Arecidae consisting of a single family, Cyclanthaceae, of about 180 species of herbaceous perennials, erect shrubs or lianas with alternate leaves comprising a sheath, petiole and expanded blade; fruit juicy and berry-like; confined to tropical America.

cyclic parthenogenesis Reproduction by a series of parthenogenetic generations alternating with a single sexually reproducing generation; *cf.* parthenogenesis.

cyclic succession An ecological succession in which the original community is ultimately restored.

Cyclocystoidea Extinct class of free-living echinoid echinoderms known from the Ordovician to the Devonian; single marginal ring of cuboidal ossicles surrounded a central disk on both dorsal and ventral surfaces.

cyclomorphosis Cyclical changes in form, such as seasonal changes in morphology.

cyclonic Used of a region of low atmospheric sea-level pressure; also of the wind system around such a low pressure centre that has a clockwise rotation in the northern hemisphere and an anticlockwise motion in the southern hemisphere; **cyclone**; *cf.* anticyclonic.

Cyclophyllidea Large order of tapeworms parasitic in amphibians, reptiles, birds and mammals, and utilizing a large variety of intermediate hosts; characterized by a scolex of 4 suckers often with a rostellum armed with hooks, by the presence of a neck, and by a well segmented body with each proglottid containing usually one set of reproductive organs.

Cyclopoida Order of copepod crustaceans containing about 450 species of marine and freshwater planktonic forms and many parasites and commensals; characterized by grasping antennae in the males and by the presence of paired, dorsal egg sacs in the female.

Cyclopteridae Lumpfishes, snailfishes; family containing about 140 species of mostly bottom-living inshore scorpaeniform teleosts, few forms bathypelagic; body globular and tuberose, or elongate and naked; pelvic fins modified as ventral sucker, or absent, pectorals large; utilized as food-fish in some

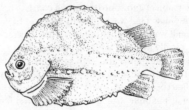

lumpfish (Cyclopteridae)

Cyclorhagida Diverse order of kinorhynchs found mainly in muddy sediments of intertidal estuarine habitats and sandy beaches but also known from deep water; neck typically consisting of a ring of 14–16 plates; cavity of the pharynx round in cross-section; mid-dorsal, lateral and caudal spines present; possessing lateral and ventral adhesive tubes.

Cyclostomata Sole order of bryozoan class, Stenolaemata *q.v.*

Cydippida Primitive order of pelagic ctenophores found from polar to tropical seas; characterized by well developed comb rows and with paired tentacles retractile into tentacle sheaths; usually globular or ovoid in shape.

Cyemidae Monotypic family of bathypelagic snipe eels (Anguilliformes) having short stout dart-like body with short dorsal and anal fins; jaws very slender, upper longer than lower.

cyesis That period between fertilization and birth in animals that produce a single offspring per brood.

cymaphyte A surf plant.

cyme An inflorescence in which the first-opening flower develops at the apex of the central flowering shoot and any further flower buds are formed from lateral growths beneath; cymose inflorescence.

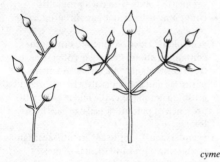

cyme

Cymodoceaceae Family of Najadales comprising about 19 species of glabrous, rhizomatous, submerged marine herbs with small naked, water-pollinated flowers; native to tropical and subtropical sea coasts.

Cynipidae Gall wasps; family of small to very small wasps (Hymenoptera) which, as larvae, typically produce galls on various organs of flowering plants although some are parasites of other insects; containing approximately 2000 species distributed worldwide.

Cynocephalidae Colugos; flying lemurs; family containing 2 species of herbivorous nocturnal arboreal mammals (Dermoptera) found in tropical forests of southeast Asia; possess broad membranes of skin (patangia) extending from sides of neck, between fore- and hindlimbs to the tail, for gliding from tree to tree.

Cynoglossidae Tongue soles; family of tropical marine soleiform flatfishes in which the eyes are on the left side of the head; pectorals and right pelvic fin absent; body length to 300 mm; contains 100 species, many important locally as food-fishes.

colugo (Cynocephalidae)

Cyperaceae Sedges; cosmopolitan family containing about 4000 species of mostly herbs tending to accumulate silica, usually with elongate parallel-veined leaf blades, small and generally wind-pollinated flowers arranged in spikes; includes species such as papyrus.

sedge (Cyperaceae)

Cyperales Order of Commelinidae comprising 2 large and widely distributed families, the Cyperaceae and Poaceae; mainly herbs, sometimes woody, tending to accumulate silica, with small, typically wind-pollinated flowers generally subtended by a chaffy bract; seeds with copious starchy or mealy endosperm.

cyphonautes Ciliated larval stage of bryozoans possessing a triangular bivalve shell; functional digestive system degenerates upon attachment of the larva to the substratum.

cyphonautes

cypress Cupressaceae *q.v.*

Cyprinidae Carps, minnows, barbs; very large family of omnivorous or herbivorous freshwater cypriniform teleost fishes with worldwide distribution except for Australia, New Zealand and South America; body typically moderately compressed, mouth terminal, lips thin; adipose fin absent; contains about 1600 species, many important as food-fish and used widely in pisciculture.

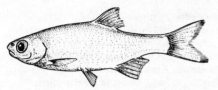

cyprinid (Cyprinidae)

Cypriniformes Order of small to medium-sized freshwater teleost fishes widespread in North America, Eurasia and Africa; upper jaw protractile, jaw teeth absent, pharyngeal bones forming feeding mill; pectoral and pelvic fins soft-rayed, spines rarely present on medium fins; gas bladder simple, connected to inner ear by bony Weberian ossicles; contains about 2000 species in 6 families; important in pisciculture and popular in aquarium trade; includes carps, minnows, barbs, suckers, loaches.

Cyprinodontidae Killifishes, toothcarps, top minnows; family containing about 300 species of small (to 150 mm) mainly freshwater cyprinodontiform teleost fishes widespread in tropical and temperate parts of the New World; body slender, often colourful, upper jaw protrusible; sexual dimorphism may be well developed; reproduction oviparous, fertilization usually external.

Cyprinodontiformes Order of mostly small freshwater, occasionally brackish or coastal marine, teleost fishes widespread in the tropics and subtropics; sexual dimorphism may be well developed; reproduction oviparous to viviparous; comprises about 500 species in 8 families, including killifishes, ricefishes, live-bearers and four-eyed fishes.

cypris Motile planktonic larval stage of barnacles; larval body with 6 pairs of swimming thoracic legs, enclosed within bivalved carapace.

cypsela A dry indehiscent fruit developed from

dandelion fruit (cypsela)

one-seeded inferior ovary so that it incorporates some non-carpellary tissue.

Cyrillaceae Small family of Ericales containing 14 species of glabrous, tanniferous shrubs confined to the warmer parts of the New World; flowers borne in racemes, with 5 sepals and petals, and a superior ovary which develops into a capsule or drupe.

Cyrtocrinida Order of articulate crinoids typically possessing a short stalk, but stalk absent in sole living representative found in shallow waters of the Caribbean.

Cyrtophorida A widespread, mostly marine order of hypostomatan ciliates in which the oral area is occupied by 3 short double rows of kinetosomes derived from the ciliary rows of the left ventral surface.

cyrtopia Late developmental stage of euphausiacean, possessing functional natatory pleopods.

cysticercus Fluid-filled, bladder-like larval stage of cestodes.

cytogamy 1: The fusion of two cells. 2: The fusion of both male nuclei of one of two conjugating organisms with the female nucleus of the other.

cytogenetics The study of chromosomal mechanisms, their behaviour and their effect on inheritance and evolution.

cytokinesis Division of the cytoplasm, usually following meiosis or mitosis; *cf.* karyokinesis.

cytokinin A growth substance which stimulates cell division in plants.

cytology The study of the structure and function of living cells.

cytophagous Feeding on cells; **cytophage, cytophagy.**

cytotaxis 1: The movement of cells in relation to each other; used to describe cells separating (negative cytotaxis) or aggregating (positive cytotaxis); **cytotactic.** 2: The arrangement of cells within an organ.

cytozoic Intracellular; used of an organism living within a host cell.

Cyttariales Small order of discomycete fungi comprising 10 species all of which parasitize southern beeches of the genus *Nothofagus*; producing clustered fruiting bodies in galls on host branches, each consisting of a sterile stroma containing many embedded apothecia.

d

2,4-D 2,4-dichlorophenoxyacetic acid; a translocated hormone weed killer, used to control broad-leaved herbaceous plants, and as a defoliant.

dab Pleuronectidae *q.v.*

Dacnonympha Suborder of lepidopterous insects comprising about 25 species; adult typically possessing reticulate or marbled wing pattern; mandibles functional or reduced and non-biting, maxillae forming small proboscis in some species; larvae mostly leaf or seed miners; pupal mandibles hypertrophied.

Dactylochirotida Order of mainly deep-sea holothurians (Dendrochirotacea) comprising about 35 species that have typically unbranched tentacles and a rigid U-shaped test, and which burrow in soft sediments.

Dactylopteridae Flying gurnards; family of small (to 350 mm) tropical and warm temperate marine, bottom-living, dactylopteriform teleost fishes; head large, spinose and ridged; pectoral fins large with ventral fin rays free; feed on benthic invertebrates and small fishes, resting on the free pectoral fin rays.

Dactylopteriformes Order of marine teleost fishes comprising a single family, Dactylopteridae (flying gurnards).

Dactyloscopidae Sand stargazers; family containing 25 species of small (to 100 mm) New World tropical marine teleost fishes (Perciformes), typically found in shallow water partly buried in sandy sediments; body slender, mouth vertical, eyes dorsal; dorsal and anal fins elongate.

dactylozoid A specialized defensive or protective polyp in colonial cnidarians.

daddylonglegs Tipulidae *q.v.*

daddy-longlegs spider Any of the longlegged spiders of the family Pholcidae (suborder Labidognatha); comprising about 350 species distributed worldwide but most abundant in tropical cave habitats; typically spin a loose cobweb and wrap prey in silk.

daffodil Liliaceae *q.v.*

daggertooth Anotopteridae *q.v.*

dahlia Asterales *q.v.*

daisy Asterales *q.v.*

dam In animal breeding, the female parent; *cf.* sire.

damselfish Pomacentridae *q.v.*

damselfly Zygoptera *q.v.* (Odonata)

dance A highly stylized repetitive movement often associated with courtship.

dance fly Empididae *q.v.*

dandelion Asterales *q.v.*

Daphniphyllales Order of Hamamelidae comprising a single family, Daphniphyllaceae, of about 35 species of trees or shrubs native to eastern Asia and Malaysia; containing a unique type of alkaloid and often accumulating aluminium; flowers typically small, lacking petals.

dark respiration Respiration in photosynthetic plants occurring during the night.

darter 1: Characidiidae *q.v.*; South American darters. 2: Percidae *q.v.* 3: Anhingidae *q.v.*

Darwinism The view of evolution expressed by Charles Darwin; evolution occurring by means of natural selection (the struggle for life) acting on spontaneous variation arising within populations and species, resulting in the survival of the fittest.

Darwin's finch Emberizidae *q.v.*

Dasyatidae Stingrays; family of medium-sized marine, brackish or freshwater myliobatiform fishes; body disk rhomboidal to circular, tail whip-like bearing one or more serrated spines, dorsal and caudal fins absent; habits mainly benthic, feeding on small fishes and bottom invertebrates; contains about 60 species in tropical or warm temperate waters to depths of 100 m.

Dasycladales Mermaid's wine glass; small order of tropical and subtropical marine green algae, characterized by a siphonous, non-septate thallus attached to the substrate by rhizoids and containing a single enormous basal nucleus; includes many fossil forms.

mermaid's wine glass (Dasycladales)

Dasypodidae Armadillos; family containing about 20 species of primarily Neotropical ground-living mammals (Edentata); body protected by armour of dermal bone and scales; teeth peg-like; limbs powerful, bearing strong claws for digging; feed on invertebrates, fruit and carrion.

armadillo (Dasypodidae)

Dasyproctidae Agoutis, pacas; family of Neotropical terrestrial hystricomorph rodents containing about 10 species that excavate burrows and feed on a variety of plant material, including roots, seeds and fruit; hindlimbs elongate for swift locomotion; tail vestigial.

paca (Dasyproctidae)

Dasyuridae Marsupial mice, native cats, Tasmanian devil; family of small to medium-sized (50 mm – 1 m) carnivorous marsupials found in Australia, Tasmania and New Guinea; marsupium opens posteriorly, sometimes absent; tail non-prehensile; habits mainly nocturnal, terrestrial.

Datiscaceae Small family of Violales containing 4 species of perennial herbs to large trees; flowers unisexual, with an inferior ovary; producing capsular fruit with many seeds.

Daubentoniidae Aye aye; monotypic family of arboreal, nocturnal prosimians found only in northern Madagascar; tail long and bushy; all digits except big toe have claws rather than nails; feed on insects and a variety of plant material.

daughter Any of the offspring of a given generation, applicable to both sexes.

Davidsoniaceae Family of Rosales comprising a single species of small tree covered with pungent red hairs on twigs, leaves and fruit; native to northeastern Australia.

day lily Liliaceae *q.v.*

day-neutral A plant in which the flowering response is not dependent on photoperiodism.

DBH Abbreviation of diameter at breast height (1.4 m) of a tree; measured either over the bark or under the bark.

DDT Dichlorodiphenyltrichloroethane; a persistent organochlorine insecticide.

De Vriesianism The belief that evolution in general and speciation in particular are the results of drastic and sudden mutational changes.

dead man's fingers A soft coral of the Alcyonacea *q.v.*

deadly nightshade Solanaceae *q.v.*

deadnettle Lamiaceae *q.v.*

death The total and irreversible cessation of all life processes.

death feint Thanatosis *q.v.*

death rate Mortality; the number of deaths in a population per unit time.

death-watch beetle Anobiidae *q.v.*

debris Accumulated plant and animal remains.

decalcification Removal of calcium carbonate from the soil by leaching.

Decapoda Crabs, lobsters, shrimps; large order of eucaridan crustaceans; carapace fused with all thoracic segments to form a gill chamber above the leg bases; first 3 of the 8 pairs of thoracic legs commonly modified as maxillipeds; comprises 2 suborders, Dendrobranchiata and Pleocyemata, containing about 10 000 recent species found in aquatic and sometimes terrestrial habitats.

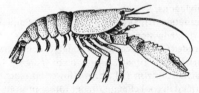

lobster (Decapoda)

decay Organic decomposition during which almost total oxidation of organic substances occurs; *cf.* mouldering, putrefaction.

decennial Pertaining to periods of tens of years; *cf.* millennial, secular.

deci- Prefix used to denote unit x 10^{-1}; see metric prefixes.

deciduous Used of structures that are shed at regular intervals, or at a given stage in development.

decomposer Any organism that feeds by degrading organic matter.

decomposition Metabolic degradation of organic matter into simple organic and inorganic compounds, with consequent liberation of energy; **decomposer.**

decoy A crop of low value cultivated to attract pests away from a more valuable crop.

deductive method A scientific method involving the formulation of theories or hypotheses from which singular statements (predictions) are deduced that can be tested; *cf.* inductive method.

deep scattering layer A more or less well defined layer present in most oceanic waters which reflects sound from echosounding equipment; produced by stratified populations of organisms which scatter sound waves and are recorded on an echosounder as a horizontal layer.

deer Cervidae *q.v.*

deer brush Rhamnaceae *q.v.*

deerfly Tabanidae *q.v.*

defaecate To discharge faeces.

definitive host The host in which a parasite attains sexual maturity; primary host; *cf.* intermediate host.

deflation Erosion of surface layers by wind action.

deflection display Distraction display *q.v.*

defoliant A chemical, such as 2,4-D and 2,4,5-T, that causes a plant to shed its leaves.

deforestation The permanent removal of forest and undergrowth.

Degeneriaceae Family of primitive flowering plants (Magnoliales) containing a single species of large tree native to Fiji; characterized by laminar stamens and 3 or 4 cotyledons.

degradation Breakdown into smaller or simpler parts; reduction of complexity.

Degraded Chernozem soil A zonal soil with very dark brown or black A-horizon over a grey leached horizon and a brown lower horizon; formed in cool climates under forest-prairie transitional vegetation.

degression Regression to a less specialized condition; **degressive.**

dehiscence Spontaneous opening of ripe plant structures to liberate seeds or spores; **dehiscent.**

deimatic behaviour The adoption of an intimidating posture by one animal in order to frighten another.

Deinotherioidea Suborder of extinct elephant-like mammals (Proboscidea) with a long trunk and large downward-curving tusks on the lower jaw; lived from Miocene through to Pleistocene.

deka- Prefix meaning ten, tenfold; used to denote unit x 10; deca-; see metric prefixes.

deleterious 1: Having an adverse effect. 2: Used of a trait which impairs survival, or of a mutation that reduces fitness.

deletion The loss of a segment from a chromosome.

deliquescent Becoming fluid by the uptake of water from air; becoming soft or liquid with age.

delitescence The incubation period of a pathogenic organism.

Delphinidae Dolphins, pilot whale, killer whale; family containing about 30 species of marine toothed whales (Odontoceta) widespread in temperate and tropical oceans and seas; head profile typically showing distinct beak; feed on fishes, squid, crustaceans and, in the case of the killer whale, also on marine birds and mammals.

dolphin (Delphinidae)

delphinium Ranunculaceae *q.v.*

delta A deposit of fine alluvial sediments at the mouth of a river.

deltaic Pertaining to a delta; used of the succession or cycle of processes involved in the formation of a delta.

deme A local interbreeding group; also used loosely to refer to any local group of individuals of a given species; used as a neutral term in combination with qualifying prefixes, including agamodeme, autodeme, clinodeme, clonodeme, ecodeme, endodeme, gamodeme, genodeme, hologamodeme, phenodeme, serodeme, topodeme and xenodeme.

demersal Living at or near the bottom of a sea or lake but having the capacity for active swimming.

demography The study of populations, especially of growth rates and age structure.

Demospongiae Large class of sponges with a skeleton of spongin fibres with or without siliceous spicules; ranging in shape from thin encrusting forms to massive lobed, cylindrical or spherical forms and found at all depths in the sea and in fresh water; sexual reproduction typically involves a dispersal stage, the parenchymella larva, with an outer layer of flagellated cells.

denatant Swimming, moving, or migrating with the current; *cf.* contranatant.

sponge (Demospongiae)

denaturation An alteration in the structural
properties of a protein producing a change in
its biochemical activity.

Dendrobatidae Family containing about 70
species of small (to 50 mm) terrestrial frogs
(Anura) from the New World tropics; skin
secretions often highly toxic, used for poison
darts and arrows; reproduction unique,
amplexus absent, copulation occurring in
horizontal vent-to-vent position; eggs
terrestrial, carried to water on backs of adults.

Dendrobranchiata Suborder of decapod
crustaceans containing about 450 species
found in freshwater, brackish and marine
habitats from the surface to about 2000 m
depth; anterior 3 pairs of legs chelate; gills
with branching primary filaments; eggs
typically hatching as nauplii; includes the
penaeid and sergestid shrimps.

Dendroceratida Order of ceractinomorph
sponges found in the intertidal zone and down
to 650 m from tropical to polar seas; skeleton
typically arising from a basal plate, and with a
branching or reticulate pattern.

Dendrochirotacea Subclass of holothurians (sea
cucumbers) comprising 2 orders,
Dendrochirotida and Dactylochirotida, in
which the tentacles are retractile, and tube feet
and respiratory trees are present.

Dendrochirotida Order of holothurians
(Dendrochirotacea) comprising about 410
species in which the tentacles are
multibranched; found on hard substrate or
burrowing in soft sediments from the littoral
to 1000 m depth.

dendrochore That part of the Earth's surface
covered by trees.

dendrochronology A method of dating using
annual tree-rings; tree-ring chronology.

dendroclimatology The determination of past
climatic conditions from the study of the
annual growth rings of trees.

Dendrocolaptidae Woodcreepers; family
containing about 50 species of small
Neotropical passerine birds found in forests
and woodlands from Mexico through South
America; bill strong and compressed, wings
long; legs short with strong claws; habits
solitary or gregarious, arboreal, non-
migratory, monogamous; feed mostly on
insects and spiders, and nest in tree holes or
crevices.

woodcreeper (Dendrocolaptidae)

dendrocolous Living in, or growing on, trees;
dendrocole.

Dendrogaea The Neotropical region *q.v.*
excluding temperate South America.

dendrology The study of trees.

dendrophagous Feeding on wood;
dendrophage, dendrophagy.

dendrophilous Thriving in trees; living in
orchards; **dendrophile, dendrophily.**

denitrification The release of gaseous nitrogen
or the reduction of nitrates to nitrites and
ammonia by the breakdown of nitrogenous
compounds, typically by microorganisms
when the oxygen concentration is low; on a
global scale thought to occur primarily in
oxygen deficient environments in the oceans.

denizen A plant species found growing in the
wild but considered to have been originally
introduced for cultivation; established alien.

density The number of individuals or
observations within a given area or volume.

density dependence A change in the influence
of an environmental factor (a density-
dependent factor) that affects population
growth as population density changes, tending
to retard population growth (by increasing
mortality or decreasing fecundity) as density
increases or to enhance population growth (by
decreasing mortality or increasing fecundity)
as density decreases.

Dentaliida Order of scaphopod molluscs
characterized by a typically sculptured shell
and by the position of the radula midway along
the body.

Denticipitidae Monotypic family of small
herring-like teleosts restricted to Nigerian

freshwater streams; head covered with small tooth-like denticles.

denudation 1: Erosion of surface material to expose underlying rock. 2: The removal of surface vegetation; **denude.**

deoxygenated Used of an environment depleted in free oxygen; **deoxygenation**; *cf.* oxygenated.

deoxyribonucleic acid DNA *q.v.*

depauperate Impoverished; moribund.

deplumation Moulting in birds.

deposit feeder Any organism feeding on fragmented particulate organic matter in or on the substratum.

depuration The cleansing activity of shellfish effected by immersion in clean water prior to harvesting resulting in the removal of pathogenic organisms and pollutants potentially harmful to man.

Derichthyidae Longneck eels; family of small mesopelagic anguilliform teleosts comprising 2 species; body smooth, eyes large; dorsal, caudal and anal fins continuous; pectorals present.

derived Used of a character or character state not present in the ancestral stock; apomorphic.

Dermaptera Earwigs; order of orthopterodean insects comprising about 1500 species, abundant in tropical regions, preferring cryptic habitats and feeding as scavengers or predators; body length 3–55 mm, posterior cerci modified as large pincers used variously for predation, defence, courtship and grooming; forewings forming short elytra, hindwings membranous; wings frequently reduced or absent.

earwig (Dermaptera)

Dermatemydidae Monotypic family of large (to 600 mm) river turtles (Testudines: Cryptodira) with long necks and webbed feet; found in eastern parts of Central America, from southern Mexico to Honduras.

dermatophyte A fungal skin parasite.

dermatotrophic Living and feeding on skin; **dermatotroph, dermatotrophy.**

dermatozoon An animal parasite of skin.

Dermestidae Carpet beetle, skin beetle; family of small, usually dark coloured beetles (Coleoptera) often found associated with stored products but feeding on a wide range of animal and plant materials; contains about 850 species.

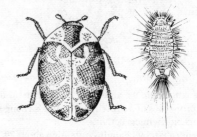

carpet beetle & woolly bear (Dermestidae)

Dermochelyidae Leatherback turtle; monotypic family of very large (to 2.5 m) marine turtles (Testudines: Cryptodira); carapace covered with leathery skin, horny scutes absent; forelimbs modified as flippers; ribs and vertebrae free from carapace; widely distributed in tropical and warm temperate oceans.

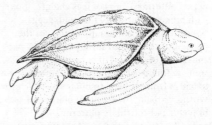

leatherback turtle (Dermochelyidae)

Dermoptera Flying lemurs, colugo; order of arboreal placental mammals of southeast Asia comprising a single family, Cynocephalidae.

desalinization The removal of salts; as in the leaching of saline soils.

descendants 1: All the individuals derived from the sexual union of two individuals. 2: The derivatives of a prototype, ancestor or other source.

desert An arid area with insufficient available water for dense plant growth; more precisely, an area of very sparse vegetation in which the plants are typically solitary and separated from each other by more than their diameter on average.

desert biome Desert: a region of environmental

extremes having less than 250 mm annual
rainfall, high evaporation and low humidity;
typically with a large proportion of barren
ground, sparse vegetation of solitary plants or
small aggregations, and nocturnal animals.

desert forest Arboreous desert *q.v.*

Desert soil A very shallow light coloured zonal
soil overlying calcareous material, formed
under arid conditions; typically with a pale
grey or brown shallow A-horizon, a slightly
darker neutral or alkaline B-horizon and a
calcareous and often compacted lower
horizon; characteristically with sparse
vegetation.

desert wind A dry wind blowing from a desert
region, hot in summer and cold in winter.

deserticolous Living mostly on open ground in
an arid or desert region; **deserticole.**

desertification The development of desert
conditions as a result of human activity or
climatic changes.

desiccation Removal of water; the process of
drying; **desiccate.**

desman Talpidae *q.v.*

Desmarestiales Order of brown algae occurring
in cold waters, especially in the southern
hemisphere; typically with a large thallus and
differentiated holdfast.

brown alga (Desmarestiales)

Desmidoideae Desmids; class of the
protoctistan phylum Gamophyta; typically
comprising a pair of cells joined by a

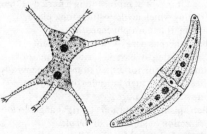

desmids (Desmidoideae)

cytoplasmic isthmus which houses the shared
nucleus; cell walls of cellulose and pectic
substances are often ornate and covered with a
mucilaginous sheath secreted through pores in
cell wall; reproduce asexually by separation of
partners in pair, and sometimes sexually by
leaving shells and conjugating with another to
form a zygote.

desmids Freshwater, unicellular green algae
treated either as a suborder of the
Zygnematales *q.v.* or as a class, the
Desmidoideae, of the protoctistan phylum,
Gamophyta.

Desmocapsales Small order of rare
photosynthetic, unicellular dinoflagellates
occurring in marine and freshwater habitats.

Desmodontidae Vampire bats; usually included
within the family Phyllostomatidae *q.v.*

Desmodorida Order of chromodorian
nematodes found principally in marine
habitats; body cuticle annulate; possessing a
cephalic helmet formed from additional
cuticular layers in the head capsule.

Desmoscolecida Small order of chromodorian
nematodes found mostly in marine habitats;
markedly annulate bodies typically
ornamented with concretion rings, scales or
bristles.

Desmothoracida Order of Heliozoa containing
mostly sedentary organisms in which the adult
stage is usually stalked and enclosed in an
organic lattice; development involves a
flagellated, amoeboid spore.

determinate growth Growth that is limited
during the life span of an individual, so that the
organism reaches a maximum size after which
growth ceases; *cf.* indeterminate growth.

detoxification The biological removal of toxic
substances by conversion into a less toxic or
more readily disposable form.

detriophagous Feeding on detritus;
detriophage, detriophagy.

detritivorous Feeding on fragmented
particulate organic matter; **detritivore,
detritivory.**

detritus Fragmented particulate organic matter
derived from the decomposition of plant and
animal remains; organic debris; **detrital.**

Deuteromycota Deuteromycotina *q.v.*

Deuteromycotina Fungi Imperfecti; an artifical
subdivision of true fungi (Eumycota)
characterized by the absence of a sexual state
but including forms with an ascomycotine-like
or basidiomycotine-like mycelium; a
cosmopolitan group including both saprobic
and pathogenic forms.

Deuterostomia A large group of animals characterized by a radial pattern of cleavage in early embryology; comprising the phyla Chaetognatha, Chordata, Echinodermata, Hemichordata and Pogonophora.

deuterotoky Parthenogenesis in which both male and female offspring are produced; **deuterotokous.**

deutonymph Immature stages of mites (Acari) possessing more complex chaetotaxy and genitalia than preceding protonymph; dcutcronymph.

development The regulated growth and differentiation of an individual, including cellular, tissue and organ differentiation; ontogeny.

Devensian glaciation The most recent glaciation of the Quaternary Ice Age *q.v.* in the British Isles with an estimated duration of about 70 000 years.

devil ray Mobulidae *q.v.*

devil's herb Plumbaginales *q.v.*

Devonian A geological period within the Palaeozoic (*c.* 413–365 million years B.P.); see geological time scale.

dew point The temperature at which liquefaction of a vapour begins.

D-horizon Any layer underlying a soil profile that differs from the parent material of the soil; see soil horizons.

diachronous Belonging to different geological periods.

diacmic Exhibiting two abundance peaks per year; *cf.* monacmic, polyacmic.

Diadematacea Superorder of euechinoidean echinoids comprising 3 orders, Echinothuroida, Diadematoida and Peninoida; anus at aboral pole, dental apparatus and gills usually present.

Diadematoida Order of littoral to abyssal echinoids (Diadematacea), comprising about 50 species, having a flat or spherical fragile test with large hollow spines; may represent conspicuous part of coral reef fauna.

sea urchin (Diademoida)

diadromous Migrating between fresh water and sea water.

diagenesis The chemical and physical processes, in particular compaction and cementation, involved in rock formation after the initial deposition of a sediment.

diageotropism Orientation at right angles to the direction of gravity; typically resulting in growth parallel to the soil surface; **diageotropic.**

diaheliotropism Orientation response of part or all of a plant aligning itself at right angles to the direction of incident sunlight; **diaheliotropic.**

dialect A local geographical variant of behaviour involved in communication.

Dialypetalanthaceae Family of Rosales containing a single tree species from Brazil.

diapause A resting phase; a period of suspended growth or development, characterized by greatly reduced metabolic activity, usually during hibernation or aestivation.

Diapensiales Small order of Dilleniidae containing only 18 species of evergreen perennial herbs and shrubs native to Arctic and northern temperate regions; flowers typically with 5-lobed calyx and corolla, and 5 stamens; fruit capsular.

diaphototaxis An orientation response of an organism or structure at right angles to the direction of incident light; **diaphototactic.**

diaphototropism Orientation at a right angle to the direction of incident light; **diaphototropic.**

Diaporthales Order of pyrenomycete fungi containing many parasites or inhabitants of woody substrates including pathogens causing leaf-spot disease of various plants and chestnut blight; typically with persistent asci.

Diapsida Class of reptiles with two temporal openings in the skull behind the eyes.

diaspore Any part of an organism produced either sexually or asexually that is capable of giving rise to a new individual; propagule.

diastem A gap in the fossil record, denoted by a bedding plane, representing a period of time during which no deposition took place.

diatom Bacillariophyceae *q.v.*

diatomaceous ooze A pelagic sediment comprising at least 30% siliceous material mainly in the form of diatom tests, generally restricted to high latitudes or areas of upwelling; *cf.* ooze.

diatropism Orientation at a right angle to the direction of stimulus; **diatropic.**

Dibamidae Small family of limbless fossorial

lizards (Sauria) comprising 4 species found in the Indo-Pacific region and in Mexico.

Dibranchiata Coleoidea *q.v.*

Dicaeidae Flowerpeckers; family containing about 60 species of small active passerine birds found in variety of forest and open habitats of southeast Asia and Australia; bill slender to stout, occasionally serrated for cutting into flowers, tongue cleft and somewhat tubular distally; habits solitary to gregarious, arboreal, non-migratory; feed on insects, nectar and berries, and nest in bush or in hole in tree or bank.

dicentra Fumariaceae *q.v.*

Diceratiidae Family of small (to 150 mm) anglerfishes (Lophiiformes) found in surface waters of Altantic and Indian Oceans to 100 m depth; contains 4 species known mostly from larval stages.

Dichapetalaceae Family of Celastrales containing about 200 species of often poisonous woody plants with a pantropical distribution; flowers variable, commonly with 5 sepals, 4 or 5 petals, 5 stamens and a superior ovary; fruit usually a drupe.

dichlorvos 2,2-Dichlorovinyl dimethylphosphate (DDVP); an organophosphorus insecticide and acaricide of relatively short persistence.

dichogamy The maturation of male and female reproductive organs at different times in a flower or hermaphroditic organism, thereby preventing self-fertilization; **dichogamic**, **dichogamous**; *cf.* adichogamy.

dichopatric Pertaining to populations or species having geographical ranges separated to the extent that individuals from the two populations never meet and gene flow is not possible; **dichopatry**; *cf.* allopatric, parapatric, sympatric.

dichotomy Division into two parts or categories; **dichotomous.**

dichotypic Dimorphic *q.v.*; used particularly of plants having two types of leaves or flowers.

diclinous Having male and female organs in separate flowers on the same or different plants; **diclinism**, **dicliny**; *cf.* monoclinous.

dicotyledon Magnoliopsida *q.v.*; major group of flowering plants characterized by 2, rarely more, cotyledons in the embryo.

Dicranales Diverse order of mosses (Bryidae) frequently found at high latitudes or altitudes; plants with erect, little-branched gametophytes; apical cell of stem with 3 cutting faces; sporophyte usually borne terminally.

Dicruridae Drongos; family containing 20 species of insectivorous Old World passerine birds found in forest and grassland habitats from Africa to Australia; plumage uniformly dark with some bright feathers; bill strong and hooked; habits solitary or weakly gregarious, arboreal, aggressive; feed on the wing; nest in trees.

Dictyoceratida Order of ceractinomorph sponges with a skeleton lacking spicules, composed of anastomosing spongin fibres which typically vary in size; abundant in shallow tropical and subtropical waters; includes the commercial bath sponges of the family Spongiidae.

Dictyochophycidae Silicoflagellates; subclass of typically unicellular and uniflagellate chrysophyceaen algae, characterized by the possession of a siliceous skeleton; sometimes classified as an order of the protozoan class Phytomastigophora, under the name Silicoflagellida.

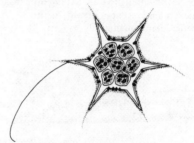

silicoflagellate (Dictyochophycidae)

Dictyoptera An order of insects formerly used to group the cockroaches and mantids, each now treated as distinct orders.

Dictyosiphonales Large order of brown algae which exhibit alternation between a large parenchymatous sporophyte and a microscopic filamentous gametophyte generation, and in which isogamous sexual reproduction occurs.

Dictyostelia Subclass of slime moulds (Eumycetozoa) typically with amoeboid trophic stages with filose pseudopodia and a fruiting body with cellulose walls of the cells and tubes that comprise the stalk; found mostly on humus and in forest soils.

Dictyotales Order of brown algae with a flattened ribbon-like or fan-like body form, occurring in tropical and subtropical waters, possessing uniflagellate male gametes.

(Picture overleaf).

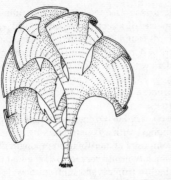

brown alga (Dictyotales)

Dicyemida Order of rhombozoan mesozoans found as parasites in the kidneys (renal system) of cephalopod molluscs.

Didelphidae Opossums; family containing about 70 species of mostly small (to about 1 m) arboreal or terrestrial marsupials widespread in South and Central America with single species extending to Canada; tail prehensile; marsupium present or absent; habits crepuscular to nocturnal.

opossum (Didelphidae)

Didereaceae Small family of spiny, cactus-like shrubs and trees confined to Madagascar.

Didymelales Order containing two species of glabrous, dioecious, evergreen trees occurring only in Madagascar.

diecdysis Continuous moulting, with one ecdysis grading rapidly into the next.

diecodichogamic Used of a population of dichogamic plants in some individuals of which the male flowers mature first while in others the female flowers mature first; **diecodichogamy.**

diel Daily; pertaining to a 24 hour period.

dientomophilous Used of a plant pollinated by two
different insect species and having two kinds of flowers each adapted for one of the insect pollinators; **dientomophily.**

differentiation The process of specialization and the progressive diversification of structure or function; integrated cellular specialization during embryonic development.

diffuse competition Simultaneous interspecific competition between numerous species each having a small degree of niche overlap with other species.

diffusion 1: Passive movement of molecules in solution from a region of high concentration to a region of low concentration. 2: The intermixing of particles as a result of movement caused by thermal agitation.

digametic Having two kinds of gametes, one producing male offspring and the other female; **diagamety.**

Digenea Flukes; subclass of trematodes containing nearly 7000 species most of which are endoparasites in vertebrates, especially fishes; possessing 1 or 2 muscular suckers for attachment to the host, and a digestive system composed of a mouth, pharynx, oesophagus and intestinal caeca; most are hermaphrodites; life cycles are indirect involving one or more intermediate hosts in which a series of larval stages, such as cercariae, metacercariae, rediae and sporocysts develop; the cercarial stage usually has a brief free-living phase.

digenetic fluke (Digenea)

digenesis The regular alternation of sexual and asexual generations.

digenetic Pertaining to a symbiont requiring two different hosts during its life cycle; *cf.* monogenetic, trigenetic.

digger wasp 1: Ampulicidae *q.v.* 2: Sphecidae *q.v.*

digitigrade Walking with only the digits in contact with the ground.

digoneutic Producing two broods per season or year; **digoneutism**; *cf.* monogoneutic, polygoneutic, trigoneutic.

digonic Producing male and female gametes in different gonads of the same individual;

digony; *cf.* syngonic.

Digonota Bdelloidea *q.v.*

dilatent Used of a sediment that becomes more solid as a result of agitation or pressure; **dilatency**; *cf.* thixotropic.

dill Apiaceae *q.v.*

Dilleniaceae Snake vine; family of mostly woody and tanniferous evergreen plants, usually containing a flavonol (myricetin); mostly tropical or subtropical, particularly in the Australian region; flowers typically showy with many stamens and a superior ovary.

dillenia (Dilleniaceae)

Dilleniales Order of woody or herbaceous plants mostly without alkaloids; comprising 2 families, Dilleniaceae and Paeoniaceae.

Dilleniidae Large subclass of dicotyledons that produce various kinds of repellents, often tanniferous and with mustard oils in some groups, but poor in alkaloids; flowers usually regular and with separate petals, pistil usually compound with 2 or more carpels; contains over 25 000 species arranged in 13 orders.

sundew (Dilleniidae)

diluvial 1: In the geological time scale, the present time. 2: Pertaining to a flood.

Dimargaritales Order of zygomycete fungi, all but one of which are obligate parasites on other fungi, of the order Mucorales.

dimictic Used of a lake having two seasonal overturn periods of free circulation, with accompanying disruption of the thermocline; *cf.* mictic.

dimonoecious Used of an individual plant having hermaphrodite, male, female and neuter flowers.

dimorphic 1: Pertaining to a population or taxon having two genetically determined, discontinuous morphological types; **dimorphism**, **dimorphy**; *cf.* monomorphic, polymorphic. 2: Having two different types of flower on a single plant; dichotypic; *cf.* monomorphic.

Dinamoebales Order comprising a single species of marine non-photosynthetic dinoflagellate.

dingo Canidae *q.v.*

Dinocerata Extinct suborder of mainly North American ungulates known from the late Palaeocene to Eocene; some with horn-like bony swellings on the head.

Dinocloniales Small order of marine filamentous dinoflagellates.

Dinoflagellata Dinoflagellates; a group of common, typically marine, planktonic organisms which are usually heterotrophic but occasionally phototrophic; characterized by 2 flagella, one typically lying in an equatorial groove round the cell, the other directed posteriorly; most with rigid test comprising cellulose plates encrusted with silica; treated as a protozoan group (a subphylum of Mastigophora), as a phylum of Protoctista, or a class of Chromophycote algae, the Dinophyceae.

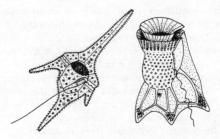

dinoflagellates (Dinoflagellata)

Dinophilida Order of interstitial marine and brackish-water polychaete worms comprising a single family and about 20 species; body transparent, segmentation indistinct; prostomium and peristomium fused, pharynx eversible, parapodia absent; copulate by hypodermic injection.

Dinophyceae Dinoflagellates; Dinoflagellata *q.v.* treated as a class of Chromophycote algae, comprising 4 subclasses, Dinophycidae, Ebriophycidae, Ellobiophycidae and Syndiniophycidae.

Dinophycidae The largest subclass of dinoflagellates (Dinophyceae) in the algal classification, comprising 15 orders of predominantly unicellular free-living, ectoparasitic or symbiotic forms; characterized by the longitudinal flagellum lying in a posteriorly directed groove and by the possession of the pigments chlorophyll *a*, c_2 and peridinin.

Dinophysiales Order of thecate dinoflagellates widely distributed in marine plankton.

Dinornithidae Moas; extinct family of large emu-like ratites, up to 3 m tall; confined to New Zealand but died out during the last few centuries; habits herbivorous and ground-nesting.

moa (Dinornithidae)

Dinornithiformes Moas; order of large ratite birds from New Zealand; comprising the Dinornithidae, now extinct.

dinosaur Heterogeneous assemblage of diapsid reptiles belonging mainly to two groups of archosaurs, Ornithischia and Saurischia; related to the crocodiles and birds; flourished during the Mesozoic era but became extinct at the end of the Cretaceous period.

Diodontidae Porcupine fishes, burrfishes; family containing 15 species of shallow-water tropical marine pufferfishes

porcupine fish (Diodontidae)

(Tetraodontiformes); length to 0.9 m, body covered with sharp spines, jaw teeth fused to form parrot-like beak; typically inflate body with water when threatened.

dioeciopolygamy The presence of both unisexual and hermaphrodite individuals in the same species.

dioecious Used of plants or plant species having male and female reproductive organs on different individuals; unisexual; **diecious**, **dioecy**; *cf.* monoecious, trioecious, trimonoecious.

dioestrus The quiescent period between breeding periods in polyoestrous animals; **dioestrous**; *cf.* anoestrus, monoestrous, polyoestrous.

Diomedeidae Albatrosses; family containing 13 species of large oceanic sea birds (Procellariiformes) found in Southern and Pacific Oceans, specialized for dynamic soaring flight close to water surface; wings long and narrow (up to 3.5 m span); feed on fishes and squid; breed in colonies on islands.

albatross (Diomedeidae)

Dioncophyllaceae Small family of Violales containing only 3 species of lianas or shrubs, climbing by hooked or cirrhose leaf tips; all native to rainforests of tropical Africa.

Dioscoreaceae Yams; family of Liliales comprising about 600 species of twining-climbing, herbaceous vines or erect herbs arising from a tuber or rhizome; largely tropical in distribution; flowers usually small and inconspicuous with an inferior ovary.

dioxin Tetrachlorodibenzoparadioxin (TCDD); a highly toxic and environmentally persistent product of the manufacture of 2,4,5-T.

Dipentodontaceae Family of Santalales containing a single small tree species native to southern China and Burma.

Diphyllidea Small order of tapeworms parasitic in rays, characterized by a scolex of 2 spoon-shaped bothridia lined with minute spines, and an apical organ or rostellum.

diplanetic Having two motile stages during a single life cycle; **diplanetism**; *cf.* monoplanetic, polyplanetic.

dipleurula Bilaterally symmetrical ciliated larval stage of echinoderms.

diplobiont 1: A plant flowering twice in a single season. 2: An organism exhibiting a regular alternation of haploid and diploid generations during the life cycle; *cf.* haplobiont.

Diplocheta Large and diverse superorder of helminthomorphan diplopods (millipedes) comprising 3 orders, Spirostreptida, Julida and Siphoniulida; body cylindrical, 5–300 mm in length, body segments possessing 2 tergal and 2 sternal sclerites, no pleural element apparent.

Diplogasteria Subclass of secernentean nematodes containing free-living microbe feeders and parasites of plants and insects; external cuticle ornamented with annulations which may also be traversed by longitudinal striae; oesophagus with a muscular corpus divided into 3 regions, the middle one usually bearing valves; comprises 3 orders, Aphelenchida, Diplogasterida and Tylenchida.

Diplogasterida Order of mostly free-living diplogasterian nematodes which includes predators, omnivores, bacterial feeders and fungal feeders, as well as a few parasites of insects; characterized by the absence of an axial spear in the stoma.

diplohaplontic Used of an organism exhibiting a regular alternation of haploid and diploid generations during its life cycle.

diploid Having a double set of homologous chromosomes, typical of most organisms derived from fertilized egg cells; *cf.* ploidy.

Diplomonadida Order of small, mostly parasitic, zooflagellates with 1 or 2 nuclei, no mitochondria and 1–4 flagella, one of which is recurrent; reproducing by binary fission.

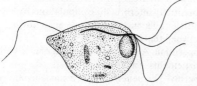

diplomonad (Diplomonadida)

Diplomystidae Family containing 2 species of small (to 250 mm) primitive South American freshwater catfishes (Siluriformes); body smooth with short dorsal and anal fins; adipose fin large; maxillary teeth present.

diplont 1: The diploid stage of a life cycle. 2: An organism having a life cycle in which the direct products of meiosis are haploid and act as gametes; *cf.* haplont.

diplontic life cycle A life cycle characterized by a diploid adult stage producing haploid gametes by meiosis, the zygote forming by fusion of a pair of gametes.

diplophyte A diploid plant; *cf.* haplophyte.

Diplopoda Millipedes; class of mostly small terrestrial arthropods having coalesced double trunk segments typically bearing 2 pairs of legs; body often mineralized and specialized for enrollment into a ball, coil or spiral; many produce toxic secretions (allomones) from glands in trunk; sexes separate, reproduction oviparous, fertilization internal; young typically hatch with 3 pairs of legs and pass through a series of anamorphic growth stages; contains about 10 000 species in 3 subclasses, Penicillata, Pentazonia and Helminthomorpha.

Diplostraca Subclass of branchiopodan crustaceans comprising the clam shrimps (Conchostraca) and the water fleas (Cladocera); body contained within a valvate carapace.

Diplura Cosmopolitan order containing about 660 species of small, slender, blind, whitish insects found in damp soil, under logs and stones; mouthparts enclosed by head capsule.

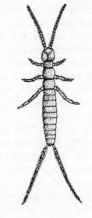

dipluran (Diplura)

Dipneusti Dipnoi *q.v.*

Dipnoi Lungfishes; subclass of lobe-finned bony fishes (Osteichthyes, Sarcopterygii); Dipneusti.

Dipodidae Jerboas; family containing about 25 species of small herbivorous myomorph rodents found in arid habitats of central Palaearctic and northern Ethiopian regions; hindlimbs elongate, adapted for bipedal hopping locomotion; inhabit burrows by day; typically hibernate during winter.

dipper Cinclidae *q.v.*

Diprotodontia Order of small to large herbivorous, terrestrial, arboreal and fossorial marsupials confined to the Australasian region; comprises 7 families, Phalangcridac (phalangers, cuscuses), Burramyidae (pigmy phalangers), Petauridae (gliding phalangers), Macropodidae (kangaroos, wallabies), Phascolarctidae (koala), Vombatidae (wombats) and Tarsipedidae (honey possum).

Dipsacaceae Teasel, scabious; family of mostly herbs with flowers typically grouped into dense cymose heads; comprises about 270 species native to Africa and Eurasia.

Dipsacales Order of Asteridae containing 4 families of herbs or woody plants producing iridoid compounds and various alkaloids, with opposite or whorled leaves; most of its 1000 species are contained in the Caprifoliaceae, Dipsacaceae and Valerianaceae.

Diptera True flies; order of holometabolous insects comprising about 150 000 species in 2 suborders, Nematocera and Brachycera; forewings membranous, hindwings modified as minute club-like balancing organs (halteres); head typically large and mobile with well developed compound eyes; mouthparts adapted for sucking or lapping, occasionally for piercing; feed on nectar, plant and animal secretions, blood or decomposing organic matter; larvae typically cylindrical, lacking true thoracic limbs, exhibiting enormous diversity of feeding strategies; flies are of great medical and veterinary importance as disease vectors, although many are also beneficial as parasitoids and pollinators.

Dipterocarpaceae Family of Theales; comprising usually evergreen trees especially abundant in Malaysian tropical rainforest.

directive coloration Surface markings which divert the attention or attack of a predator to the non-vital parts of the prey body.

directive species A species that attracts a predator, of which it is not a normal prey item, to an area rich in prey species.

Diretmidae Spinyfins; family containing 10 species of little-known deep-sea beryciform teleost fishes; body deep and compressed, almost circular in outline, to 400 mm length, scales ctenoid; eyes extremely large, teeth minute; dorsal and anal fins elongate, lacking spines.

disclimax A disturbed climax; an ecological succession maintained below climax by rapid expansion of introduced species, climatic instability, fire, grazing or by the activities of man.

Discoglossidae Family containing about 10 species of primitive frogs (Anura) found in Europe, North Africa, and temperate regions of Asia to Japan, also Borneo and Philippines; tongues disk-like.

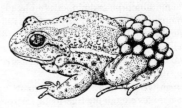

midwife toad (Discoglossidae)

discolorous Having a non-uniform coloration.

Discomycetes Cup fungi; class of ascomycotine fungi typically producing cup-shaped fruiting bodies (apothecia) which open at maturity to expose a fertile layer or tuft consisting of asci interspersed with sterile filaments (paraphyses); comprises mostly saprobic forms with some parasitic, mycorrhizal and lichenized forms as well; contains 7 orders, Cyttariales, Helotiales, Medeolariales, Ostropales, Pezizales, Phacidiales and Tuberales.

discontinuity A marked interruption in an otherwise continuous sequence of variation, populations or events.

Discosorida Order of fossil nautiloid cephalopod molluscs found from Middle Ordovician to Upper Devonian.

diserotization Interference with, or inhibition of, copulation by low temperatures.

disjunct Distinctly separate; used of a discontinuous range in which one or more populations are separated from other potentially interbreeding populations by sufficient distance to preclude gene flow between them.

disoperation A symbiosis in which one or both participants is adversely affected.

dispermy The penetration of a single ovum by two spermatozoa at the time of fertilization; **dispermic;** *cf.* monospermy, polyspermy.

dispersal 1: Outward spreading of organisms or

propagules from their point of origin or release. 2: The outward extension of a species' range, typically by a chance event.

dispersion 1: The pattern of distribution of organisms or populations in space. 2: The non-accidental movement of individuals into or out of an area or population, typically a movement over a relatively short distance and of a more or less regular nature; *cf.* migration.

displacement activity The performance of a behavioural act outside the particular functional context of behaviour to which it is normally related.

display A behaviour pattern or signal which conveys information.

disruptive coloration A distinct coloration which delays recognition of the whole animal by attracting the attention of the observer to certain elements of the colour pattern.

dissemination Scattering or spreading, as of infectious agents, seeds, spores or other propagules; **disseminate.**

disseminule A disseminated propagule; a seed, fruit, spore or other structure modified for dispersal.

dissogony Sexual maturation at two separate stages of a life cycle, with an intervening period during which no gametes are produced.

dissophyte A plant with xerophytic leaves and stems, and a mesophytic root system.

distillation A mode of fossilization involving removal of volatile organic matter, leaving a carbon residue.

distraction display An elaborate pattern intended to attract the attention of an aggressor away from other more vunerable members of a group.

distrophyte A plant living in firm moist soil.

ditch grass Ruppiaceae *q.v.*

ditokous Producing two offspring per brood; *cf.* monotokous, oligotokous, polytokous.

Ditrysia Largest suborder of lepidopteran insects comprising about 136 000 species including many moths and all butterflies; adult mouthparts variable; wings occasionally reduced, especially in females; genitalia of female with separate copulatory opening on abdominal sternite 8, and egg-laying aperture on 9–10; larvae extremely diverse ecologically and morphologically.

ditypic Having two distinct morphs; exhibiting marked sexual dimorphism; used of characters possessing two alternative states: **ditypism.**

diurnae Pertaining to day-flying insects; *cf.* nocturnae.

diurnal 1: Active during daylight hours; *cf.* crepuscular, nocturnal. 2: Lasting for one day only.

diurnal rhythm A biological rhythm having a periodicity of about one day length (24 hours); circadian rhythm.

diver Gaviidae *q.v.*

divergence A zone of oceanic upwelling where deep water rises and spreads out over the surface, as in the Antarctic divergence; *cf.* convergence.

diversionary display Behaviour intended to attract the attention of a predator away from the more vulnerable members of a group.

diversity 1: The absolute number of species in a community; species richness. 2: A measure of the number of species and their relative abundance in a community; low diversity refers to few species or unequal abundances, high diversity to many species or equal abundances.

division 1: Separation into parts; fission. 2: A rank in the hierarchy of plant classification; the principle category between Kingdom and class.

dixenic Used of a mixed culture of one organism together with two other species; *cf.* axenic, monaxenic.

dixenous Used of a parasite utilizing two host species during its life cycle; **dixeny**; *cf.* heteroxenous, monoxenous, oligoxenous, trixenous.

Dixidae Family of small frail dipteran insects (flies) possessing long legs and reduced mouthparts; contains about 150 species with cosmopolitan distribution but most frequent in temperate regions.

dizygotic Used of twins derived from two separate fertilized eggs; fraternal (twins); *cf.* monozygotic.

DL$_{50}$ A measure of drought lethality, the degree of dryness that causes injury to 50% of a population.

DNA Deoxyribonucleic acid; the primary genetic material of a cell; a polymer of the

monarch butterfly (Ditrysia)

nucleotides adenine, guanine, cytosine and thymine, typically containing two polynucleotide chains in the form of a double helix; the sequence of nucleotide pairings in the chains is the basis of the genetic code; DNA molecules are the largest biologically active molecules known.

dobsonfly Corydalidae *q.v.* (Neuroptera).

DOC Dissolved organic matter (carbon), expressed as grams of carbon per litre.

dock Polygonales *q.v.*

dodder Cuscutaceae *q.v.*

dodo Raphidae *q.v.*

dog Canidae *q.v.*

dog-bear Amphycyonidae *q.v.*

dogfish Any of the small sharks belonging to the families Squalidae *q.v.*, Scyliorhinidae *q.v.* and Carcharhinidae *q.v.*

Dogger A geological epoch of the middle Jurassic period (*c.*. 175–160 million years B.P.).

dogwhelk Neogastropoda *q.v.*

dogwood Cornaceae *q.v.*

doliolaria Barrel-shaped larva of some crinoids and holothurians possessing 3–5 ciliated girdles.

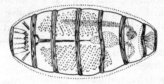

doliolaria

Doliolida Order of pelagic tunicates (Thaliacea) that exhibit an alternation of solitary, sexually reproducing, and compound, asexually produced, forms; zooids small, up to only 5 mm length, inhalent opening anterior, exhalent aperture posterior, body transparent with circular muscles producing locomotory water currents; free-living tailed larva develops into primary asexual oozooid from which colony is produced by budding; colony zooids polymorphic, including gastrozooids (feeding), phorozooids (structural) and gonozooids (sexual, hermaphrodite).

Dollo's law The general principle that evolution is irreversible and that structures and functions once lost are not regained.

dolphin 1: Delphinidae *q.v.*, marine dolphins. 2: Platanistidae *q.v.*, freshwater dolphins.

dolphinfish Coryphaenidae *q.v.*

domestication The adaptation of plants and animals for life in intimate association with man.

domicile A home, nest, burrow, tube, den or other refuge; *cf.* hibernaculum.

domicolous Living in a tube, nest or other domicile; **domicole.**

Domin scale A scale for estimating cover and abundance of a plant species, comprising 11 categories: + (single individual), 1 (very few individuals), 2 (sparsely distributed, less than 1% cover), 3 (frequent but less than 4% cover), 4 (4–10% cover), 5 (11–25% cover), 6 (26–33% cover), 7 (34–50% cover), 8 (51–75% cover), 9 (76–90% cover) and 10 (91–100% cover).

dominance 1: The extent to which a given species predominates in a community because of its size, abundance or coverage, and affects the fitness of associated species. 2: The tendency of the phenotypic traits controlled by one allele (dominant) to be expressed over traits controlled by other alleles (recessives) at the same locus; *cf.* recessiveness. 3: Physical domination initiated and sustained by aggression or other behavioural patterns of an individual.

dominance hierarchy A social order of dominance sustained by aggressive or other behaviour patterns.

dominant 1: The highest ranking individual in a dominance hierarchy; alpha. 2: An organism exerting considerable influence upon a community by its size, abundance or coverage. 3: An allele which determines the phenotype of a heterozygote; *cf.* recessive.

dominule A dominant organism in a microhabitat.

Donatiaceae Family of Campanulales containing only 2 species of dwarf herbs with spirally arranged leaves; native to South America, New Zealand and Tasmania: sometimes included in the Stylidiaceae *q.v.*

Donau glaciation An early glaciation of the Quaternary Ice Age *q.v.* in the Alpine area, with an approximate duration of 260 000 years, tentatively subdivided into 4 separate glacial periods.

donkey Equidae *q.v.*

door snail Stylommatophora *q.v.*

Doradidae Large family of mainly crepuscular tropical South American freshwater catfishes (Siluriformes); body partly covered with spinose bony plates, dorsal and anal fins short, dorsal bearing single spine; 3 pairs of barbels: some species capable of aerial respiration through vascularized intestine; utilized locally as food-fish and also popular in aquarium trade.

dormancy 1: A state of relative metabolic quiescence, such as aestivation, cryptobiosis, diapause, hibernation and hypobiosis. 2: A state in which viable seeds, spores or buds fail to germinate under favourable conditions; *cf.* quiescence.

dormouse Gliridae *q.v.*

dory Zeidae *q.v.*

Dorylaimida Large order of enoplian nematodes found free-living in soil and freshwater habitats and as ectoparasites of plants; most free-living forms are predators or algal-feeders; characterized by the formation of the anterior tooth or a hollow axial spear by a special cell in the anterior part of the oesophagus.

Dothideales Order of loculoparenchemycetid fungi that are parasitic or saprobic in plant tissues; some are occasionally lichenized.

dottyback Pseudochromidae *q.v.*

dove Columbidae *q.v.*

dove shell Neogastropoda *q.v.*

dove tree Nyssaceae *q.v.*

downy mildew Peronosporales *q.v.*

dragon tree Agavaceae *q.v.*

dragonet Callionymidae *q.v.*

dragonfish 1: Stomiatidae *q.v.* 2: Melanostomiatidae *q.v.* 3: Idiacanthidae *q.v.*

dragonfly Anisoptera *q.v.* (Odonata).

Drepanididae Hawaiian honeycreepers; family containing 23 species of small arboreal passerine birds confined to forest habitats of the Hawaiian Islands; plumage often colourful; bill variable indicating the wide spectrum of feeding strategies; tongue occasionally tubular and frilled; feed on invertebrates, seeds, fruit, nectar and other plant material.

driftfish Nomeidae *q.v.*

drimophilous Thriving in salt basins or alkaline plains; **drimophile, drimophily.**

drive A motivating, impelling internal physiological condition.

dromaeosaurid A group of theropod dinosaurs with relatively large brains; known from the Cretaceous.

Dromaiidae Emus; family containing 2 species of large (to 1.8 m tall) Australian ratite birds (Palaeognathae) found in grassland and open forest; legs long, powerful, bearing 3 digits; neck partly naked; habits cursorial, gregarious, monogamous; feeding on fruit, plant material and insects.

Dromornithidae Group of large subfossil ratite birds from Australia.

dromotropism An orientation response in

emu (Dromaiidae)

climbing plants that results in spiral growth; **dromotropic.**

drone The functional male of social Hymenoptera that plays no part in brood maintenance.

drongo Dicruridae *q.v.*

Droseraceae Sundews, Venus' fly-trap; cosmopolitan family of insectivorous perennial herbs catching insects either by an active trap or by irritable, mucilage-tipped hairs.

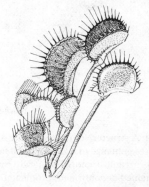

Venus' fly trap (Droseraceae)

Drosophilidae Vinegar flies, fruit flies; cosmopolitan family containing about 1500 species of small brownish yellow flies (Diptera) with red eyes that feed on decaying and fermenting plant material; species of

fruit fly (Drosophilidae)

Drosophila have been the subject of extensive genetic and evolutionary studies both in the laboratory and in the field.

drosphilous Pollinated by dew; **drosophilous, drosphile, drosphily.**

drought 1: A prolonged chronic shortage, usually of water. 2: A period without precipitation during which the soil water content is reduced to such an extent that plants suffer from lack of water.

drought-evading Used of plants sensitive to drought which can survive dry periods by the production of desiccation-resistant seeds or structures.

drought resistance The capacity to withstand periods of dryness, including both desiccation avoidance and desiccation tolerance.

drum Sciaenidae *q.v.*

drupe A fleshy indehiscent fruit containing one or more seeds, each enclosed within a hard endocarp; includes plum, peach, avocado.

peach (drupe)

dubiofossil A structure or object having a marked resemblance to a known fossil but of which the organic or inorganic nature cannot be ascertained or confirmed; problematic fossil.

duck Anatidae *q.v.*

duckbill eel Nettastomatidae *q.v.*

duck-billed platypus Ornithorhynchidae *q.v.*

Duckeodendraceae Family of Solanales containing a single species of tall tree native to the Amazon basin.

duckweed Lemnaceae *q.v.*

duff A product of litter decomposition; incompletely decomposed organic matter in which the original structure is no longer discernible.

Dugongidae Dugong; monotypic family of large nocturnal, herbivorous, aquatic mammals (Sirenia) found along tropical coasts of the Indian Ocean and western Pacific; incisor and cheek teeth present in adults but absent in

older specimens, being replaced by horny plates; tail fluke with median cleft. The recently extinct Steller's sea cow from the Bering Sea is sometimes included in this family.

dulosis Symbiosis in which workers of an ant species capture the brood of another species and rear them as slaves; **dulotic.**

dumb cane Aracaceae *q.v.*

dung beetle 1: Geotrupidae *q.v.* 2: Scarabaeidae *q.v.*

dung fly Scatophagidae *q.v.*

Duplicidentata Older name for the Lagomorpha *q.v.*

Durvillaeales Small order of kelp-like brown algae found only in the southern hemisphere.

dusty wing Neuroptera *q.v.*

Dutch elm disease Eurotiales *q.v.*

dwarfism The condition of being stunted, much smaller than normal; having restricted growth; **dwarf.**

dysanthous Pertaining to the condition of being fertilized by pollen from another plant; **dysanthic.**

dyschronous Pertaining to plant species that do not overlap in their flowering periods.

dysgenesis The condition of infertility between hybrids which are themselves cross-fertile with the parental stocks; **dysgenetic.**

dysgenic Pertaining to, or having, the capacity for decreasing the fitness of a race or breed; *cf.* eugenics.

dysphotic zone The zone of intermediate light intensity in a water body, with insufficient light for photosynthesis but sufficient light for behavioural responses; **disphotic zone**; *cf.* euphotic zone, photic zone.

Dyssomidae Arrowtooth eels; family containing 8 species of small (to 250 mm), littoral to bathyal bottom-living anguilliform teleost fishes; body smooth, eyes and gill openings small, dorsal, caudal and anal fins continuous, pectorals present or absent; snout ridged and grooved ventrally.

dystric Pertaining to unhealthy soils; *cf.* eutric.

dystrophic 1: Used of freshwater bodies rich in organic matter mainly in the form of suspended plant colloids and larger plant fragments, but having low nutrient content; dystrophic lake; *cf.* eutrophic, mesotrophic, oligotrophic. 2: Used of species characteristic of dystrophic habitats.

dyticon An ooze-inhabiting community.

Dytiscidae Family of powerful, predatory beetles (Coleoptera) found in various aquatic habitats both as adults and larvae; feeding on

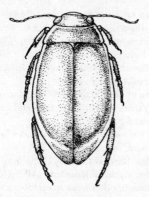

diving beetle (Dytiscidae)

insect larvae, molluscs and small aquatic
vertebrates; contains about 3000 species in 120
genera.

e

eagle Accipitridae *q.v.*

eagle ray Myliobatidae *q.v.*

earless monitor Lanthanotidae *q.v.*

earth star Lycoperdales *q.v.*

earthworm Lumbricina *q.v.* (Oligochaeta).

earwig Dermaptera *q.v.*

East Australia Current A warm surface ocean current that flows southwards off the east coast of Australia, forming the western limb of the South Pacific Gyre; see ocean currents.

East Indian pitcher plant Nepenthaceae *q.v.*

ebb current The tidal current associated with a receding tide; *cf.* flood current.

ebb tide A receding tide; *cf.* flood tide.

Ebenaceae Ebony, persimmon; family containing about 450 species of trees or shrubs with hard dark wood; mostly tropical in distribution; flowers typically unisexual, with fused sepals and petals and a superior ovary; fruit usually a berry.

persimmon (Ebenaceae)

Ebenales Order of Dilleniidae containing 5 families of trees or shrubs, Ebenaceae, Lissocarpaceae, Sapotaceae, Styracaceae and Symplocaceae.

ebony Ebenaceae *q.v.*

Ebriida Order of colourless, marine zooflagellates with 2 flagella and an internal siliceous skeleton; also regarded as a subclass of algae, Ebriophycidae *q.v.*

Ebriophycidae Small subclass of marine phagotrophic dinoflagellates possessing an internal siliceous skeleton; mostly fossil forms but two extant species in order Ebriales.

ecad A plant or animal form produced in response to particular habitat factors, the characteristic adaptations not being heritable; a habitat form.

Eccrinales Order of trichomycete fungi comprising obligate gut symbionts of various crustaceans and uniramians.

ecdemic Foreign; non-native; *cf.* endemic.

ecdysis The act of shedding or moulting the outer exoskeleton or cuticle.

ecesis The pioneer stage of dispersal to a new habitat; successful invasion and establishment by colonizing plants.

echard That part of soil water not available for plant use; *cf.* chresard, holard.

Echeneidae Remoras; family containing 8 species of marine teleost fishes (Perciformes) that attach to other aquatic vertebrates, such as sharks, swordfishes, rays, turtles and whales, by means of a large sucking disk on the flattened dorsal surface of the head, derived from a modified anterior dorsal fin.

echidna Tachyglossidae *q.v.*

Echimyidae Rock rats, spiny rats; family containing about 40 species of small northern Neotropical terrestrial and arboreal hystricomorph rodents; mostly herbivorous, inhabiting burrows or crevices in rocks.

Echinacea Superorder of intertidal to abyssal regular echinoids (Euechinoidea) comprising 5 orders, Salenoida, Phymosomatoida, Arbacioida, Temnopleuroida and Echinoida, and including most of the familiar sea urchins; oral membrane encircled with 5 pairs of ambulacral plates, body spines solid.

Echinodera Kinorhyncha *q.v.*

Echinodermata Phylum of exclusively marine, radially symmetrical (pentamerous), unsegmented, solitary marine coelomates that possess a diverse array of skeletal spicules, plates, spines and larger ossicles composed of calcium carbonate (calcite), and a unique water vascular system the outgrowths of which form tube feet and tentacles used variously for attachment, locomotion and gas exchange; sexes are separate, fertilization normally external, development direct or with

starfish (Echinodermata)

planktonic larvae; asexual reproduction by fission common; 3 extant subphyla recognized, Crinozoa (feather stars, sea lilies), Asterozoa (sea stars), Echinozoa (sea urchins, sea cucumbers); echinoderms have an extensive fossil record.

Echinoida Order of intertidal to lower continental slope echinoids comprising about 65 extant species; test lacking sculpturing, to 300 mm diameter, dentition camarodont; gonads of some species considered a gastronomic delicacy.

Echinoidea Sea urchins, heart urchins, sand dollars; class of echinozoan echinoderms characterized by a hollow globular body formed by fused skeletal plates (test) bearing stiff articulating spines; oral pole facing downwards, anus on upper surface at aboral pole but moved into interambulacral position in bilaterally symmetrical, irregular, burrowing echinoids; a complex buccal apparatus, Aristotle's lantern, is usually present; echinoids are exclusively benthic, feeding by grazing, scavenging or ingestion of sediment; development includes planktonic echinopluteus larva; about 950 extant species in 2 subclasses, Perischoechinoidea and Euechinoidea; rich fossil record comprising about 5000 recognized species.

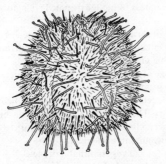

sea urchin (Echinoidea)

echinopluteus The pluteus *q.v.* larval stage of a sea urchin.

Echinorhinidae Bramble shark; family containing 2 species of heavy-bodied squaliform sharks; length to 3.5 m, skin rough bearing thorn-like spines; dorsal fins small, posteriorly positioned, anal fin absent; habits benthic, sluggish; depth range to about 900 m.

Echinorhynchida Large order of palaeacanthocephalan thorny-headed worms found commonly as parasites of fishes and, occasionally, of amphibians and reptiles.

Echinosteliales Small order of slime moulds

(Myxogastromycetidae) with stalked sporangia usually less than 0.5 mm in height, and bearing light-coloured spores; most of the 18 species are restricted to bark of living trees and vines.

slime mould (Echinosteliales)

Echinosteliida The slime mould order Echinosteliales *q.v.*, treated as an order of the protozoan group Myxogastria.

Echinostomida Order of digenetic trematodes possessing one pair of flame cells (excretory organs) in the miracidia larva; cercariae with abundant cystogenous glands, forming cysts on vegetation or in secondary intermediate host.

Echinothuroida Order of deep-sea echinoids (Diadematacea) comprising about 50 living species having a light flexible test, up to 300 mm diameter, that often collapses when removed from water; some spines armed with poison glands; development direct.

Echinozoa Subphylum of echinoderms characterized by a globular compact body with 5 meridional ambulacra that radiate from the oral pole and converge at the aboral pole; body may be elongate (sea cucumbers) or compressed (sand dollars) along oral/aboral axis, some species showing secondary bilateral symmetry; two extant classes, Echinoidea (sea urchins, heart urchins, sand dollars) and Holothuroidea (sea cucumbers).

Echiura Spoonworms; small phylum of soft bodied, unsegmented, coelomate, marine worms; body divided into 2 regions, a sac-like

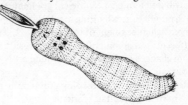

spoonworm (Echiura)

or cylindrical trunk with an extensible ciliated proboscis used in respiration and food collection, and a smooth, annulate or papillose trunk bearing a pair of setae that assist in burrowing and anchorage; inhabit burrows in soft sediments, rock crevices, empty shells or tests, feeding on detritus; contains about 140 species in 3 orders, Echiuroinea, Xenopneusta and Heteromyota; also known as Echiurida.

Echiuroinea Order of echiuran marine worms in which the longitudinal muscles of the trunk body wall are between the outer circular and inner oblique layers; blood system closed; up to 7 pairs of nephridia; contains about 130 species found from the intertidal to abyssal depths.

echiurus Trochophore larva of sipunculan worm.

Echmatocrinea Extinct order of pelmatozoan echinoderms known from the Middle Cambrian; possessing 8–10 short arms and an irregular attachment holdfast.

echolocation The perception of objects using high-frequency sound waves, used by some animals for navigation and orientation within the environment.

echosounding The use of sound waves (acoustic signals) to measure depth of water, plot the bottom profile, or locate submerged objects and dense aggregations of relatively small organisms.

eclipse plumage Dull inconspicuous plumage of birds which alternates with a much brighter breeding plumage.

eclosion The emergence of an adult insect from the pupal case; sometimes also used for the hatching of an egg.

ecochronology The dating of biological events using palaeoecological evidence.

ecocide Any toxic substance that penetrates and kills an entire biological system.

ecoclimate The immediate climate of an individual organism; microclimate.

ecoclimatology The study of plants and animals in relation to climate; bioclimatology.

ecocline 1: A more or less continuous character variation in a sequence of populations distributed along an ecological gradient, with each population exhibiting local adaptation to its particular segment of the gradient. 2: The differences in community structure resulting from changes in slope aspect around a mountain or ridge.

ecodeme A local interbreeding group occurring in a particular habitat; *cf.* deme.

ecogeographical rule Any generalization describing a trend of geographical variation correlated with environmental conditions; climatic rule.

ecography Descriptive ecology.

ecological amplitude The range of a given enviromental factor over which an organism or process can function; range of tolerance.

ecological efficiency The efficiency of transfer of energy from one trophic level to the next.

ecological isolation The absence of interbreeding between populations occurring in the same area because of ecological barriers.

ecological longevity The average life span of an individual in a given population and under stated conditions.

ecological niche The concept of the space occupied by a species, which includes both the physical space as well as the functional role of the species; conceptualized as a multidimensional hypervolume which defines the biological space occupied by a species, and which is unique to the species; niche; *cf.* fundamental niche, realized niche.

ecological pyramid A model of the trophic structure of a community which takes the form of a pyramid of numbers, biomass or energy, in which producers form the base of the pyramid and successive levels represent consumers of higher trophic levels; food pyramid.

ecological race A local race of a species having conspicuous adaptive characters correlated with a given habitat type.

ecological succession The gradual process of progressive community change and replacement, leading towards a stable climax.

ecology The study of the interrelationships between living organisms and their environment; **ecological.**

econ A local vegetational unit; **eca.**

ecoparasite A parasite restricted to a specific host or to a small group of related host species.

ecophene All the naturally occurring phenotypes produced within a given habitat by a single genotype.

ecophenotype A phenotype exhibiting non-genetic adaptations associated with a given habitat, or to a given environmental factor.

ecoproterandry The maturation of staminate flowers before pistillate flowers.

ecoproterogyny The maturation of pistillate flowers before staminate flowers.

ecospecies A group of populations having the capacity for free exchange of genetic material without loss of fertility or vigour, but having a lesser capacity for such exchange with members of other ecospecies groups.

ecosphere Biosphere *q.v.*

ecostratigraphy The study and classification of stratified rocks according to their mode of origin or to the environment at the time of deposition; **ecostratigraphic.**

ecosystem A community of organisms and their physical environment interacting as an ecological unit.

ecotone The boundary or transitional zone between adjacent communities or biomes.

ecotope A particular habitat type within a larger geographical area.

ecotype A locally adapted population; a race or infraspecific group having distinctive characters which result from the selective pressures of the local environment; ecological race.

Ectocarpales Order of brown algae, typically of small size with a tufted filamentous growth form; commonly found in the intertidal zone, growing on rocks or on other algae.

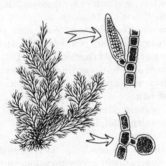

brown alga (Ectocarpales)

ectocommensal A commensal symbiont that lives on the external surface of its host; *cf.* endocommensal.

ectogenic meromixis The mixing of a lake as a result of an external event or agency; as in the influx of saline water into a freshwater lake producing a density stratification; *cf* biogenic meromixis, crenogenic meromixis.

ectogenous Arising or originating outside the organism or system; **ectogenesis**; *cf.* endogenous.

Ectognatha A superorder of insects possessing well developed, protruding mouthparts; comprises the Thysanura and the Pterygota.

ectoophagous Used of an insect larva that hatches from an egg deposited on or near a supply of host eggs and feeds upon them; **ectoophage**, **ectoophagy**; *cf.* endoophagous.

ectoparasite A parasite that lives on the outer surface of its host; *cf.* endoparasite.

ectophagous Feeding on the outside of a food source; **ectophage**, **ectophagy**; *cf.* endophagous.

ectophloedic Growing on the outside of bark; *cf.* endophloedic.

ectophyte A plant parasite living on the outer surface of another organism; **ectophytic**; *cf.* endophyte.

ectopic Occurring in an abnormal place or manner.

Ectoprocta Bryozoa *q.v.*

ectosymbiosis A symbiosis in which both symbionts are external, neither lives within the body of the other; **ectosymbiont**; *cf.* endosymbiosis.

ectothermic 1: Poikilothermic; having a body temperature determined primarily by the temperature of the environment; cold-blooded; **ectotherm**, **ectothermal**. 2: Pertaining to a chemical reaction releasing heat energy; *cf.* endothermic.

ectotrophic Obtaining nourishment externally without marked penetration into the food source; **ectotroph**; *cf.* endotrophic.

ectotropism Orientation movement away from the central axis; **ectotropic**; *cf.* endotropism.

edaphic Pertaining to, or influenced by, the nature of the soil.

edaphology The study of soils, particularly with reference to the biota and man's use of land for plant cultivation.

edaphon The soil flora and fauna; those organisms living in the interstitial water and pore spaces of soil.

edaphotropism An orientation response to a soil water stimulus; **edaphotropic.**

Edentata Order of New World placental mammals in which incisor and canine teeth are absent; premolars and molars absent or peg-like and lacking enamel; contains 3 families, Myrmecophagidae (anteaters), Bradypodidae (tree sloths) and Dasypodidae (armadillos).

Ediacarian period The period of geological time immediately preceding the Cambrian (*c.* 700–570 million years B.P.); the last of the Precambrian periods.

edible frog Ranidae *q.v.*

edible mushroom Agaricales *q.v.*

edible snail Stylommatophora *q.v.*

edobolous Having seeds dispersed by propulsion by means of turgor pressure; **edobole.**

Edrioasteroidea Group of fossil echinoderms known from mid-Cambrian to Lower

edrioasteroid (Edrioasteroida)

Carboniferous; upper (oral) surface with 5 ambulacral grooves radiating from a central mouth; test flexible; anus interambulacral; lower (aboral) surface concave in some forms perhaps forming a sucker for temporary fixation.

eel Anguillidae *q.v.*, freshwater eels.

eelgrass Zosteraceae *q.v.*

eelpout Zoarcidae *q.v.*

eelworm Nematoda *q.v.*

Eemian interglacial The interglacial period preceding the Weichselian glaciation in northern Germany and Poland; see Quaternary Ice Age.

EEZ Exclusive economic zone *q.v.*

effete Functionless; no longer fertile or no longer functioning as a result of age.

effluent The discharge of industrial or urban waste material into the environment; the outflow from a lake or river.

egesta In ecological energetics, that part of consumption expelled as faecal material or regurgitated, and thus not absorbed.

egestion Defaecation; the voiding of faecal or regurgitated matter.

egg plant Solanaceae *q.v.*

egret Ardeidae *q.v.*

E-horizon A pale coloured layer below the A-horizon of a soil profile, having a lower organic content than the layers above and below; a zone of high eluviation; see soil horizons.

Elaeagnaceae Russian olive, oleaster, silver berry, sea buckthorn; small family of Proteales containing about 50 species of often thorny trees or shrubs commonly with root nodules containing nitrogen-fixing bacteria; flowers unisexual or bisexual, often arranged in inflorescences (racemes).

Elaeocarpaceae Lantern tree, bead tree; family of Malvales containing about 400 species of trees and shrubs, widespread in tropics and subtropics except Africa; typically possessing racemes or panicles of small flowers with 4–5 sepals and petals, many stamens and a superior ovary.

elaioplankton Planktonic organisms utilizing oil droplets for buoyancy.

elaphocaris Early zoeal larval stage of sergestoid shrimps (Decapoda).

Elaphomycetales Small order of plectomycete fungi which are mycorrhizal associates of trees forming underground fleshy ascocarps which produce odours attractive to small mammals.

Elapidae Cobras, mambas, kraits, coral snakes; family of small to medium-sized, highly venomous snakes (Serpentes) that feed mainly on cold-blooded vertebrates; contains about 170 species grouped in 2 subfamilies, one primarily terrestrial and oviparous, the other aquatic marine and ovoviviparous; widespread in Australia and Indo-Pacific region.

Elasipodida Order of deep-sea holothurians (Aspidochirotacea) comprising about 110 species which have a fragile, often gelatinous, test that may bear dorsal processes; respiratory trees absent; some species swim, others walk on sea bed using enlarged ventral tube feet; may be locally abundant comprising up to 95% of the biomass.

Elasmobranchii Sharks and rays; subclass of cartilaginous fishes (Chondrichthyes); body elongate and subcylindrical to markedly flattened; 5–7 pairs of gill openings, no gill operculum; upper jaw not fused to cranium; teeth numerous; comprises 4 extant superorders, Squalomorphii, Squatinimorphii, Batoidea and Galeomorphii.

Elateridae Click beetles, wireworms; widely distributed family of common elongate beetles (Coleoptera) which can jump while making a clicking noise; includes predators, plant feeders and saprophages all of which digest food externally and take in liquids; plant-feeding larvae found in soil known as wireworms.

Elatinaceae Small, cosmopolitan family of Theales, containing about 40 species of herbs growing in shallow-water or wet habitats; flowers tiny, solitary or clustered in leaf axils, and with 3–5 sepals and petals.

Elbe glaciation A glaciation of the Quaternary Ice Age *q.v.* in northern Germany and Poland, with an estimated duration of 100 000 years.

elderberry Caprifoliaceae *q.v.*

electric catfish Malapteruridae *q.v.*

electric eel Electrophoridae *q.v.*

electric ray Torpedinidae *q.v.*

electrophoresis A technique for separating mixtures of organic molecules, based on their different rates of travel in an electric field.

Electrophoridae Electric eel; monotypic family of large (to 2.4 m) predatory South American catfishes (Siluriformes); body elongate, cylindrical; dorsal, caudal and pelvic fins absent, anal fin extends most of body length; produce powerful electric shocks to stun prey, or for defence, sufficient to immobilize a large mammal.

electrotaxis A directed reaction of a motile organism in response to an electric field; **electrotactic.**

electrotropism An orientation in response to an electric field; **electrotropic.**

Eleotridae Gudgen; family containing about 150 species of small teleost fishes (Perciformes) distributed worldwide; body round, head blunt with eyes located almost above jaws; 2 dorsal fins present.

elephant Elephantidae *q.v.*

elephant bird Aepyornithidae *q.v.*

elephant fish Mormyridae *q.v.*

elephant shrew Macroscelididae *q.v.*

Elephantidae Elephants; family of massive herbivorous terrestrial mammals (Proboscidea) comprising 2 surviving monotypic genera, one Oriental (Indian elephant), the other Ethiopian (African elephant); upper incisors forming large tusks of solid dentine (ivory); only 2 functional cheek teeth present at one time; nasal region greatly prolonged into prehensile trunk; includes the extinct mammoth.

elephant (Elephantidae)

elfin forest A forest of higher elevation in warm moist regions, characterized by stunted trees with abundant epiphytes; *cf.* krummholz.

elimination In ecological energetics, any loss of biomass or energy by a population or trophic unit, including losses due to mortality, predation, emigration and moulting.

elittoral The zone of the sea bed below the sublittoral, extending to the limit of light penetration; sometimes used for the sea bed below 40 m.

elk Cervidae *q.v.*

Ellesmeroceratida Extinct order of primitive cephalopods found from Upper Cambrian to Lower Silurian.

Ellobiophycidae Subclass of marine dinoflagellates comprising a single order, Thalassomycetales, of ectoparasitic forms infecting arthropods and polychaetes.

elm Ulmaceae *q.v.*

elodea Hydrocharitales *q.v.*

Elopidae Tenpounders; family containing 4 species of tropical marine teleost fishes (Elopiformes) found mainly in shallow coastal waters; body subcylindrical, superficially herring-like, to 1.2 m length, mouth terminal; highly prized as game fish by anglers.

Elopiformes Small order of mainly marine teleost fishes; body rounded to slender and compressed, pelvic fins abdominal, caudal deeply forked; leptocephalus larval stage present; comprises about 7 species in 2 families, Elopidae (tenpounders) and Megalopidae (tarpons).

Elsterian glaciation A glaciation of the Quaternary Ice Age *q.v.* in northern Germany and Poland, with an estimated duration of 90 000 years.

elutriation A method of removing interstitial organisms from a sediment sample by continuous flushing with water.

eluvial layer The leached upper layer of a soil profile.

eluviation The translocation of suspended or dissolved soil material by the action of water; usually the removal of substances in solution is termed leaching *q.v.*

elver Juvenile stage of the common eel (Anguillidae).

emasculation The removal of the male reproductive organs or the inhibition of male reproductive capacity; geld; *cf.* spay.

Emballonuridae Sac-winged bats; family of small insectivorous microchiropteran bats comprising 40 species widespread in tropical and subtropical regions of both Old and New World.

Emberizidae Buntings, cardinals, Darwin's finches, sparrows; diverse family containing about 320 species of passerine birds found in a variety of forest, grassland and open habitats of the New and Old World; coloration cryptic or bold; bill short, conical, robust; habits solitary to gregarious, arboreal to terrestrial, monogamous to polygamous, migratory or not; feed on seeds, other plant material, and insects; nest on or off the ground.

Embiidina Embioptera; webspinners; order of orthopterodean insects comprising about 200 species, most common in tropical forest habitats; inhabit extensive galleries or labyrinths of silk on bark, litter, moss, lichens or within the soil; body slender, legs short; females and some males wingless.

webspinner (Embiidina)

Embioptera Embiidina *q.v.*

Embiotocidae Surfperches, sea perches; family of North Pacific coastal marine teleost fishes (Perciformes); body deep, compressed, to 450 mm length; single dorsal fin present, anal fin of male modified as intromittant organ; reproduction viviparous.

Embrithopoda Extinct order of subungulates known only from the lower Oligocene; contains *Arsinoitherium*, a large form with massive limbs and 2 pairs of horns.

embryo The developing organism between fertilization and hatching or birth, contained within egg membranes or the maternal body.

Embryobionta Embryophytes; a subkingdom of plants; predominantly terrestrial plants which exhibit well developed alternation of generations in which the sporophyte is initially parasitic on the gametophyte; produce spores by meiosis as part of the sexual life cycle but asexual reproduction does not involve spore formation; photosynthetic pigments include chlorophylls *a* and *b*, β-carotene is the main carotene and lutein the main xanthophyll; most have specialized conducting tissues (xylem and phloem) and epidermal openings (stomata) through which gaseous exchange takes place; *cf.* Thallobionta.

embryotrophy The process of nutrition of a mammalian egg before implantation at which point it becomes dependent upon the maternal blood supply.

emergent 1: An aquatic plant having most of the vegetative parts above water. 2: A tree which

reaches above the level of the surrounding canopy; *cf.* submergent.

emersed Pertaining to a plant structure that projects above the water surface; *cf.* submersed.

emersion zone 1: The uppermost part of the eulittoral zone of a lake which is above water level for most of the year. 2: That part of the seashore covered only by extreme high tides.

emigration The movement of an individual or group out of an area or population; **emigrant**; *cf.* immigration.

emmenophyte An aquatic plant devoid of any floating parts; **emmophyte.**

emperor Lethrinidae *q.v.*

emperor butterfly Nymphalidae *q.v.*

emperor moth Saturniidae *q.v.*

Empetraceae Crowberry; small family of Ericales containing 5 species of evergreen shrubs with small inconspicuous flowers, often borne in clusters; fruit a drupe.

Empididae Dance flies; cosmopolitan family containing over 3000 species of small predatory flies (Diptera) commonly found in swarms in damp wooded habitats.

empirical Based upon direct observation and experience rather than theory or preconception.

empress tree Bignoniaceae *q.v.*

emu Dromaiidae *q.v.*

emu bush Myoporaceae *q.v.*

Emydidae Terrapins; large family of aquatic, semiaquatic and terrestrial turtles (Testudines: Cryptodira) having a depressed carapace and webbed feet; contains about 80 species, widespread in tropical and temperate regions especially of North America and the Oriental region.

enantiobiosis Inhibition of, or interference with, one species or population by the action of another.

enation A small outgrowth from plant stem or leaf.

enaulophilous Thriving in sand dunes; **enaulophile, enaulophily.**

enaulophyte A sand dune plant.

Encalyptales Small order of mosses (Bryidae) found mostly in montane and tundra habitats in the northern hemisphere; small dull plants forming tufts on soil generally in calcareous areas.

encasement theory The theory that the embryo is contained within the sperm or egg cell and that it unfolds during development; preformation theory.

Enchytraeidae Pot worms; family of small,

pot worm (Enchytraeidae)

primarily freshwater, oligochaete worms; sometimes included in the order Haplotaxida, or placed in a separate order, Prosotheca.

endangered species A species threatened with extinction.

endemic Native to, and restricted to, a particular geographical region; **endemicity**, **endemism**; *cf.* ecdemic.

endobenthic Living within the sediment; boring into a solid substratum; *cf.* epibenthic, hyperbenthic.

endobenthos Organisms living within the sediment on the sea bed or lake floor; infauna; endobenthic; *cf.* benthos.

endobiontic Used of organisms that live within the substratum; **endobiotic.**

Endoceratida Extinct order of nautiloid cephalopods known only from the Ordovician.

endocommensal A commensal symbiont that lives inside its host; *cf.* ectocommensal.

endodeme A local interbreeding group composed of predominantly inbreeding but dioecious individuals; *cf.* deme.

endoectothrix Growing in or on hair.

endogamy 1: Inbreeding; sexual reproduction between closely related individuals. 2: Pollination of a flower by pollen from another flower on the same plant; self-pollination; *cf.* exogamy.

endogenous Arising or originating from within the organism or system; growing on the inside; **endogenetic**; *cf.* ectogenous, exogenous.

endogenous clock Any physiological system which shows an inherent sustained periodicity.

Endogonales Order of zygomycete fungi containing a single family; many enter endomycorrhizal associations with higher plants.

endolithic Growing within a rock or other hard inorganic substratum; *cf.* epilithic.

endolithophytic Pertaining to plants that penetrate rock or other hard inorganic substratum; **endolithophyte**; *cf.* epilithophytic, exolithophytic.

endomitosis The doubling of chromosomes within a nucleus which does not subsequently divide.

endomixis Self-fertilization in which male and female nuclei from one individual fuse; **endomictic.**

Endomycetales Yeast; order of hemiascomycete fungi containing mostly saprophytic forms but also some parasites, includes many of the yeasts used in fermentation processes; often unicellular forms reproducing asexually by fission and budding.

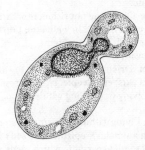

yeast (Endomycetales)

endomycorrhiza A mycorrhiza in which the fungal hyphae penetrate the root cells of the host plant.

endoophagous Used of an insect larva that hatches from an egg deposited within a host egg and feeds upon the contents of that single egg; **endoophage**, **endoophagy**; *cf.* ectoophagous.

endoparasite An internal parasite which lives within the organs or tissues of its host; endosite; entoparasite; *cf.* ectoparasite.

endopelic Used of aquatic organisms that live within the sediment; *cf.* epipelic.

endophagous Feeding from within the food source; **endophage**, **endophagy**; *cf.* ectophagous.

endophloedic Living or occurring within the bark; *cf.* ectophloedic.

endophyllous Living or growing in leaves; *cf.* epiphyllous.

endophyte A plant living within another plant; **endophytic**; *cf.* ectophyte.

Endoprocta Entoprocta *q.v.*

endopsammon The microscopic biota inhabiting sand and mud.

Endopterygota Holometabola *q.v.*

endorheic region An area in which rivers arise but do not reach the sea as they are lost in closed basins or dry courses; *cf.* arheic region, exorheic region.

endosymbiosis Symbiosis in which one symbiont lives within the body of the other; *cf.* ectosymbiosis.

endosymbiosis theory That the three classes of intracellular organelles (mitochondria, basal bodies/flagella/cilia, photosynthetic plastids) of eukaryotes evolved from free-living prokaryotic ancestors through a series of endosymbiotic relationships.

endothermic Warm-blooded; maintaining a body temperature largely independent of the temperature of the environment; homoiothermic; **endotherm, endothermal**; *cf.* ectothermic.

endotrophic Obtaining nourishment internally, as of a mycorrhiza *q.v.* in which the fungus penetrates the host root system; **endotrophism**; *cf.* ectotrophic.

endotropism Orientation movement towards the central axis; **endotropic**; *cf.* ectotropism.

endoxylic Living or growing within wood; *cf.* epixylic.

endozoic 1: Living within or passing through the body of an animal; *cf.* epizoic. 2: Used of a method of seed dispersal in which seeds are ingested by an animal and later voided in the faeces.

endozoochorous Dispersed by the agency of animals, typically after passage through the gut; **endozoochore, endozoochory**; *cf.* epizoochorous.

energetics The study of energy transformation within a community or system; ecological energetics.

energy flow The passage of energy into and out of an organism, population or system; the passage of energy through the different trophic levels of a food chain.

Engraulidae Anchovies: family of small (to 200 mm) plankotrophic teleost fishes (Clupeiformes) cosmopolitan in tropical to temperate seas; typically marine but also found in brackish and freshwater habitats; body fusiform, silvery, tip of snout projecting over large mouth; caudal fin strongly forked; eggs pelagic; many species support important commercial fisheries, primarily for fish meal and bait.

enhalid Used of plants that grow in saltings or in loose soil under salt water.

Enopla Class of nemerteans in which the mouth and proboscis typically share a common pore and in which the mouth is anterior to the brain; comprising 2 orders, Bdellonemertea and Hoplonemertea.

Enoplia Large subclass of adenophorean nematodes containing over 3000 species including the majority of nematodes found in the marine environment as well as many freshwater and soil-inhabiting, free-living forms, and parasites; most are predatory or omnivorous but herbivorous forms are known; cuticle typically smooth; chemosensory organs on head (amphids) non-spiral; exhibit a great diversity of form and structure which makes this assemblage difficult to characterize.

Enoplida Order of free-living enoplian nematodes found in marine, fresh- and brackish-water habitats, feeding as predators, omnivores or on algae; possessing pouch-like chemoreceptors (amphids) with an external slit-like or ellipsoid opening which are positioned behind the lips on the head; other head sensory organs arranged in 3 whorls.

Ensifera Suborder of typically nocturnal orthopteran insects containing about 8500 species including crickets, katydids and long-horned grasshoppers; although widespread ensiferans tend to prefer tropical and arboreal habitats.

Enterogona Order of ascidiacean tunicates comprising 2 suborders sometimes treated as distinct orders, Aplousobranchia and Phlebobranchia.

Enteropneusta Class of mainly littoral, vermiform, hemichordates comprising about 70 species found in soft sediments amongst algal holdfasts and stones; body soft and cylindrical, 250 mm – 2.5 m in length, devoid of appendages, divided into 3 regions, proboscis, collar and trunk; mouth anteroventral in collar; pharynx bearing U-shaped gill slits supported by skeletal elements; coelom greatly reduced; development direct or via tornaria larva.

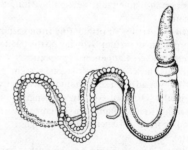

acorn worm (Enteropneusta)

enterozoon A parasitic animal living within the gut of its host.

Entodiniomorphida Order of vestibuliferan ciliates in which the vestibular and body ciliature is reduced to tufts or bands, found commonly as endocommensals in herbivorous mammals.

Entognatha A rarely used superorder of insects comprising the Collembola, Diplura and

Protura; characterized by mouthparts contained within an invagination of the head.

entomochoric Dispersed by the agency of insects; **entomochore, entomochory.**

entomogamous Used of flowers pollinated by insects; entomophilous; **entomogamy.**

entomogenous Living in or on insects.

entomography Description of an insect or of its life history.

entomology The study of insects; **entomological.**

entomopathogenic Causing disease in insects.

entomophagous Feeding on insects; insectivorous; **entomophage, entomophagy.**

entomophilous Pollinated by, or dispersed by the agency of, insects; **entomophile, entomophily.**

Entomophthorales Order of zygomycete fungi comprising 2 families, one saprobic, the other containing many obligate pathogens of plants and animals; characteristically releasing spores by forcible dehiscence; mycelium coenocytic in young stages becoming septate with age; also treated as a class of the phylum Zygomycota.

entomophyte Any fungus growing on or in an insect.

entomosis A disease caused by an insect parasite.

Entomostraca Obsolete name formerly used to group all the lower crustaceans together as distinct from the Malacostraca.

entomotaxy The preservation and preparation of insects for study.

entophagous Feeding from within a food source; **entophage, entophagy.**

entophytic Used of a plant (an entophyte) growing within another plant.

Entoprocta Small phylum of sessile, solitary or colonial, marine (rarely freshwater) invertebrates typically possessing a cylindrical stalk and rounded calyx bearing a horseshoe-shaped ring of ciliated tentacles; gut U-shaped; hermaphroditic, larvae planktotrophic or lecithotrophic; intertidal to 500 m, attached to algae, stones, shells or other surfaces. Relationships remain obscure; sometimes classified as a separate phylum of

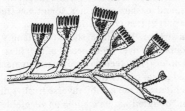

entoproct (Entoprocta)

acoelomates sometimes as a subphylum of the Bryozoa; Endoprocta; Kamptozoa.

entozoic Living or passing through the body of an animal.

entrainment Coupling or synchronization of a biological rhythm to an external time source.

entropy A measure of randomness or disorder in a system.

environment The complex of biotic, climatic, edaphic and other conditions which comprise the immediate habitat of an organism.

enzootic A disease occurring in a given animal species within a limited geographical area.

enzyme A protein molecule specialized to catalyse a biological reaction.

Eoacanthocephala Class of thorny-headed worms found as adults mainly in fish hosts, but also in amphibians and reptiles, usually utilizing crustaceans as intermediate hosts; characterised by a retractable proboscis with radially arranged hooks and by the absence of a protonephridial system; comprising 2 orders Gyracanthocephala and Neoechinorhynchida.

eobiogenesis The transformation of prebiotic macromolecular systems into the first living organisms (eobionts).

eobionts The earliest living organisms developed from prebiotic macromolecular precursors.

Eocene A geological epoch within the Tertiary period (*ca.* 54-38 million years B.P.); see geological time scale.

Eocrinoidea Extinct class of cystoid-like echinoderms (Pelmatozoa) found from the Lower Cambrian to Middle Silurian.

Eogaea A zoogeographical region incorporating Africa, South America and Australasia; *cf.* Caenogaea.

Eognathostomata An assemblage of irregular echinoids containing forms drawn from the orders Holectypoida and Pygasteroida.

eolation Aeolation *q.v.*

eolian Aeolian *q.v.*

eon 1: An indefinitely long period of geological time; aeon. 2: A unit of time equal to 10^9 years.

Eophytic The period of geological time during which the Algae were abundant; *cf.* Aphytic, Archaeophytic, Caenophytic, Mesophytic, Palaeophytic.

eosere A major ecological succession within the climatic climax of a geological period; the ecological succession of vegetation of an era or eon.

Eozoic The Precambrian era *q.v.*; sometimes used to refer to the earliest part of the Precambrian only.

epacme The period in phylogenetic or ontogenetic development of a group or organism just prior to the point of maximum vigour or adulthood; *cf.* acme.

Epacridaceae Family of Ericales containing about 400 species of glabrous shrubs or small trees found mostly in Australia, New Zealand and the East Indies; flowers often fragrant, with usually 5 free sepals, 5 petals joined to form a tube, 5 stamens and a 5-celled ovary.

epedaphic Pertaining to climatic factors or conditions.

epeiric sea A shallow sea covering part of a continental landmass, typically less than 200 m in depth; sometimes also used to include pericontinental seas *q.v.*; epicontinental sea.

ephaptomenon Organisms which are adnate or attached to a surface; *cf.* planomenon, rhizomenon.

ephebic Pertaining to the adult stage, between the juvenile and old age; *cf.* neanic.

Ephedridae Subclass of Gneticae; low evergreen shrubs with slender branches and scale-like leaves; male and female cones in clusters of 2 or 4 at the stem nodes; comprises single family, Ephedraceae, native to arid and desert regions of the Americas and Eurasia.

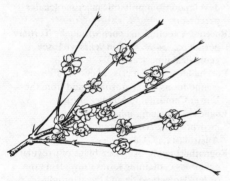

ephedra (Ephedridae)

ephemeral Lasting for only a day; short-lived or transient, as in an organism that grows, reproduces and dies within a few hours or days, or a flower that lasts for a day or less.

Ephemerida Ephemeroptera *q.v.*

Ephemeroptera Mayflies; order of paleopterous insects comprising about 2000 species in 20 families; adults non-feeding, living from 2 to 72 hours, commonly less than 1 day; mouthparts vestigial, compound eyes often sexually dimorphic (large in males, small in females); forewings triangular, distally fluted; hindwings present or absent, up to half length of forewings; male forelegs modified

for grasping female in flight; larvae (nymphs, naiads) aquatic in fresh water, feeding on detritus, diatoms, and microscopic epiphytes, some carnivorous; mayflies unique in having a winged subimago lasting up to 2 days; Ephemerida.

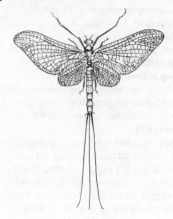

mayfly (Ephemeroptera)

Ephippidae Spadefishes; family containing 15 species of tropical and subtropical reef-dwelling perciform teleosts; body deep, compressed, to 1 m length; mouth protrusible; 2 dorsal fins present.

spadefish (Ephippidae)

Ephthianuridae Australian chats; family of small, brightly coloured passeriform birds with decurved bills; feed on insects caught on ground; nest in low bushes; sometimes treated as part of family Maluridae.

Ephydridae Brine flies, shore flies; family containing over 1000 species of tiny flies (Diptera) that feed on decaying plant material and faecal matter in aquatic or semiaquatic habitats; larvae inhabit diverse fresh, saline and polluted waters feeding on vegetable matter, microorganisms and insect larvae.

ephydrogamous Having water-borne pollen grains transported at the water surface; or being pollinated by such pollen; *cf.* hyphydrogamous.

epibenthic Living on the sea bed or on the lake floor; *cf.* endobenthic, hyperbenthic.

epibenthos The community of organisms living at the surface of the sea bed or lake floor; *cf.* benthos.

epibiontic Living attached to another organism; **epibiotic, epibiont.**

Epicaridea Suborder of mainly marine isopod crustaceans found exclusively as parasites of other crustaceans; adult female often extremely modified, dwarf male retaining isopod-like form; larvae commonly parasitic on planktonic crustaceans.

epiclysile Pertaining to the tide pools of the upper shore.

epicolous Living attached to the surface of another organism but without benefit or detriment to the host; **epicole.**

epicontinental sea Epeiric sea *q.v.*

epidemic Used of a disease affecting a high proportion of the population over a wide area; *cf.* epiphytotic, epizootic.

epidemiology The study of factors affecting the spread of diseases in populations.

epidendric Pertaining to epiphytes growing on trees and shrubs.

epifauna The total animal life inhabiting a sediment surface or water surface; *cf.* infauna.

epigamic Used of any character that serves to attract or stimulate individuals of the opposite sex during courtship, other than essential structures and behaviour of copulation; **epigamous.**

epigean Living or growing at or above the soil surface.

epigenetic Pertaining to the interaction of genetic factors and developmental processes, through which the genotype is expressed in the phenotype.

epigenetics The study of the causal mechanisms of development.

epigeotropism An orientation response of a plant producing growth across the surface of the soil; **epigeotropic.**

epigyny The arrangement of the parts of a flower so that the sepals, petals and stamens are inserted above the ovary (an inferior ovary); *cf.* hypogyny, perigyny.

epilimnion The warm upper layer of circulating water above the thermocline in a lake; **epilimnetic;** *cf.* hypolimnion.

epilithic Growing on rocks or other hard inorganic substrata; *cf.* endolithic.

epilithophytic Pertaining to a plant growing on the surface of stones, rocks and other hard inorganic substrata: **epilithophyte;** *cf.* endolithophytic.

epilittoral 1: The zone on the seashore above the intertidal zone, which is influenced by the effects of salt spray; spray zone. 2: The zone surrounding a lake, lying above water level and uninfluenced by spray.

epimeletic Used of social behaviour patterns in animals relating to the care of other individuals; *cf.* etepimeletic.

Epimorpha Subclass of chilopods in which young hatch with full complement of trunk segments and legs, at least 25 segments and 23 pairs of legs; egg clusters brooded by female; comprises 2 orders, Geophilida and Scolopendrida.

epimorphosis A form of development in arthropods in which all larval forms are suppressed or passed within the egg prior to hatching and the juvenile hatches with the adult morphology.

epinasty Downward curvature of a plant structure due to differential growth of upper and lower surfaces; **epinastic.**

epinekton Organisms attached to actively swimming (nektonic) forms but which are incapable of independent movement against water currents; **epinektonic.**

epineuston Organisms living in the air on the surface film of a water body; **epineustonic;** *cf.* hyponeuston.

epipelagic zone The upper oceanic zone extending from the surface to about 200 m; see marine depth zones.

epipelic Used of those aquatic organisms moving over the sediment surface or living at the sediment/water interface; *cf.* endopelic.

epiphenomenon An event which occurs together with another but which is not causally linked and has no effect upon it.

epiphloedic Growing on the surface of bark.

epiphloeophyte A plant living on the surface of bark.

epiphyllous Used of an epiphyte (an epiphyll) growing on a leaf; *cf.* endophyllous.

epiphyte 1: A plant growing on another plant (the phorophyte) for support or anchorage rather than for water supply or nutrients. 2: Any organism living on the surface of a plant.

epiphytology The study of the nature and ecology of plant diseases.

epiphytotic Pertaining to an epidemic disease amongst plants; *cf.* epidemic, epizootic.

epiplankton 1: Planktonic organisms living within the surface 200 m (the epipelagic zone); **epiplanktonic**; *cf*. bathyplankton. 2: Organisms living attached to larger planktonic organisms or to floating objects.

epipleuston Organisms which move over the surface film of a water body with most or all of their bodies above the water; **epipleustonic.**

Epipolasida Order of demosponges exhibiting a radial structure; appeared first in the Cambrian.

epipsammon Organisms living on the surface of a sandy sediment or on the surface of the sand particles; episammic.

epirhizous Growing on the surface of roots.

episematic Pertaining to a character that aids in recognition, as in some colour markings; **episematism**; *cf*. antepisematic, proepisematic, pseudepisematic.

epitokous Reproductive; having or producing offspring; *cf*. atokous.

epixylic Living or growing on wood; **epixylous**; *cf*. endoxylic.

epizoic 1: Living attached to the body of an animal; used of a non-parasitic animal that lives attached to the outer surface of another animal; **epizoite, epizoon**; *cf*. endozoic. 2: Dispersed by attachment to the surface of an animal.

epizoochorous Dispersed by attachment to the surface of animals; **epizoochore, epizoochory**; *cf*. endozoochorous.

epizoon An organism living attached to the body of an animal; epizoite.

epizootic Pertaining to an epidemic disease in animals; *cf*. epidemic, epiphytotic.

epoch 1: A major interval of geological time; a subdivision of a period. 2: An event or time which marks the beginning of a new phase of development.

Equatorial Current A warm surface ocean current that flows westwards in the tropical Pacific Ocean; see ocean currents.

equatorial tide The tides that occur at intervals of about 2 weeks when the moon is over the equator.

Equidae Horses, zebras, asses, donkey; family of large terrestrial mammals (Perissodactyla) comprising 7 species, now cosmopolitan as result of domestication; all are herbivorous, grazing primarily on grasses, living in small to large herds; Przewalski's horse from central Asia is the only extant wild horse; the group also contains the quagga, an extinct wild ass from southern Africa.

equinoctial 1: Used of a plant which has flowers

quagga (Equidae)

which open and close at particular times during the day; horological.

equinoctial tide A tide of high amplitude occurring when the sun is at or near the equinox.

equinox Either of the two occasions each year when the sun crosses the equator, producing day and night of equal duration; **equinoctial.**

Equisetophyta Horsetails, scouring rushes; division of terrestrial plants primarily found in moist muddy habitats; stems creeping underground and producing erect annual or perennial branches; stems hollow and grooved; leaves minute, whorled into sheaths around stem and non-photosynthetic; sporangia borne in terminal cones on stem; homosporous; comprises single order (Equisetales) and genus of about 20 species; also known as Sphenopsida.

horsetail (Equisetophyta)

era Any of the major intervals of geological time; the sequence of eras being Precambrian, Palaeozoic, Mesozoic and Cenozoic; see geological time scale.

eremacausis The process of humus formation by the oxidation of plant matter.

eremic Pertaining to deserts or sandy regions.

eremobiontic Living in desert regions.

Eremolepidaceae Family of Santalales

containing 12 species of hemiparasitic shrublets producing thickened haustoria on branches of trees; found in tropical America.

eremology The study of deserts.

eremophilous Thriving in desert regions; **eremophile, eremophily.**

eremophyte A desert plant.

Erethizontidae New World porcupines; family containing 8 species of arboreal hystricomorph rodents found primarily in the northern Neotropical region but also ranging into the Nearctic; body covered with strong barbed spines; habits nocturnal, feeding on leaves, fruit, and a variety of plant material.

New World porcupine (Erethizontidae)

Eretmophoridae Cosmopolitan family of little-known deep-water gadiform teleost fishes; 1–3 dorsal and 1–2 anal fins; to 600 mm body length with photophores present in some species; gas bladder with connection to inner ear.

erg A sandy desert.

ergasiaphyte A plant introduced by man for cultivation.

ergasiapophytic Used of plants that colonize cultivated fields.

ergatandromorphic Used of social insects in which worker and male characters are blended; **ergatandromorph**; *cf.* ergatogynomorphic.

ergatandrous Having worker-like males, as in some social insects; *cf.* ergatogynous.

ergatogynomorphic Used of social insects in which worker and female characters are blended; **ergatogynomorph**; *cf.* ergatandromorphic.

ergatogynous Having worker-like females, as in some social insects; **ergatogyne**; *cf.* ergatandrous.

ergatomorphic Resembling a worker.

ergonomics The quantitative study of work, performance and efficiency.

ergot Clavicipitales *q.v.*

Ericaceae Heath, heather, madrone, mountain laurel, sourwood, rhododendron, azalea, blueberry, cranberry; large family of mostly mycorrhizal shrubs often growing in acid soils; widespread in cool temperate to subtropical regions, containing about 3500 species many of which are cultivated; flower morphology variable but often with a 4- or 5-lobed corolla, 4–5 sepals, 5–10 stamens and an ovary with 2–5 cells.

heath (Ericaceae)

Ericales Order of Dilleniidae containing about 4000 species of typically mycotrophic, tanniferous plants; leaves often ericoid (small, firm and evergreen); includes 8 families, the largest of which is Ericaceae.

ericetal Growing on heathland or moorland.

erichthus Larval stage of stomatopods derived from the earlier antizoea or pseudozoea larva.

ericophyte A plant growing on heathland.

Erinaceidae Hedgehogs, moon rats; family containing 14 species of small terrestrial mammals (Insectivora) in which dorsal part of body is covered with long hairs or spines; habits nocturnal or crepuscular, some species hibernate; feed on variety of invertebrates, carrion and fruit; widely distributed in Ethiopian, Palaearctic and Oriental regions.

hedgehog (Erinaceidae)

Eriocaulales Order of Commelinidae comprising a single family of herbs, Eriocaulaceae, growing in wet places mainly in warm regions; having narrow grass-like leaves crowded at base and very small flowers pollinated by insects or wind.

ermine Mustelidae *q.v.*

erosion Wearing away; weathering; the removal of the land surface by water, ice, wind or other agencies; **erode.**

erpoglyph A fossil worm cast.

Errantia Errant polychaetes; a group name formerly used for all motile polychaete worms; *cf.* Sedentaria.

erratic 1: Pertaining to unattached organisms that are moved around by physical agencies. 2: A rock fragment transported into an area from outside, found either incorporated into the sediment or lying free; glacial erratic.

erucivorous Feeding on caterpillars; **erucivore, erucivory.**

erumpent Protruding or bulging out; breaking or bursting through.

Erysiphales Powdery mildews; order of pyrenomycete fungi found widely through temperate and tropical regions as an external hyaline mycelium growing over the surface of the stems, buds, leaves and fruit of flowering plants; producing small spherical fruiting bodies on the mycelium.

Erythrinidae Trahiras; family containing 5 species of predatory South American freshwater characiform teleost fishes found in stagnant weedy pools; body cylindrical, to 1 m length; teeth large; adipose fin absent; gas bladder functioning as accessory respiratory organ.

Erythroxylaceae Family of Linales containing about 200 species of glabrous trees and shrubs often producing tropane alkaloids (including cocaine); mostly tropical New World distribution; includes Andean species, *Erythroxylum coca*, from which cocaine is extracted.

escape A cultivated plant which has become established in the wild.

escatophyte A plant of a climax community.

Escherichia A genus of rod-shaped enterobacteria comprising one species, *E. coli*, found naturally in the gut of mammals and widely studied by biochemists and bacteriologists.

Eschrichtiidae Grey whale; monotypic family of large (up to 15 m) marine mammals (Mysticeta) found in the northern Pacific Ocean; dorsal fin absent; baleen plates short.

esculent Edible.

Esocidae Pikes; family containing 5 species of predatory piscivorous freshwater salmoniform teleost fishes found mostly in weedy lakes and

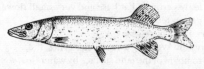

pike (Esocidae)

ponds; confined to Arctic and temperate parts of the northern hemisphere; body elongate, to 1.6 m length, snout prolonged; prized as sport fishes.

Esocoidei Suborder of freshwater salmoniform teleost fishes comprising 2 families, Esocidae (pikes) and Umbridae (mudminnows); dorsal and anal fins positioned posteriorly, adipose fin absent.

essential element A chemical element which is essential to the life of an organism.

estivation Aestivation *q.v.*; **estivate.**

Estrildidae Waxbills, mannikins; family containing about 125 species of small usually colourful passerine birds found in variety of forest, grassland and open arid habitats of Africa, southern Asia and Australia; bill short, conical; habits gregarious, terrestrial to arboreal, non-migratory; feed on seeds and insects; nest domed, constructed from loose mass of grass in a tree.

estrus Oestrus *q.v.*

estuary Any semi-enclosed coastal water, open to the sea, having a high freshwater drainage and with marked cyclical fluctuations in salinity; usually the mouth of a river.

etaerio A cluster of fruits formed from the unfused carpels of a single flower, as in the drupe cluster of a blackberry.

etepimeletic Used of social behaviour patterns in young animals serving to elicit care from adults; *cf.* epimeletic.

Ethiopian region A zoogeographical region comprising that part of the African continent south of the Sahara, tropical Arabia, Madagascar and neighbouring islands; subdivided into East African, Malagasy, South African and West African subregions; Madagascar and its neighbouring islands are sometimes excluded and regarded as a separate region; Palaeotropical region.

ethnobotany Study of the use of plants by the races of man.

ethnology Study of the character, history and culture of the races of man; ethnography.

ethnozoology Study of the use of animals and animal products by the races of man.

ethological isolation The absence of interbreeding between members of different populations because of behavioural differences that preclude effective mating.

ethology The study of animal behaviour; **ethological.**

etiolation Abnormal growth of a green plant when grown in darkness; such plants are pale yellow due to the absence of chlorophyll and the stems are typically elongate bearing small leaves.

etiology Aetiology *q.v.*

Eubacteria One of the three primary kingdoms (urkingdoms *q.v.*) of living organisms; comprising all typical bacteria; *cf.* Archaebacteria, Urkaryota.

eubiosphere That part of the biosphere in which the physiological processes of living organisms can occur; comprising the allobiosphere and autobiosphere.

eucalyptus Myrtaceae *q.v.*

Eucarida Diverse superorder of eumalacostracan crustaceans comprising about 10 000 species in 3 orders, Amphionidacea, Euphausiacea (krill) and Decapoda (crabs, lobsters and shrimps).

eucaryotic Eukaryotic *q.v.*; **eucaryote.**

Eucestoda Tapeworms; subclass of cestodes containing all the tapeworms of medical and veterinary importance; all are parasites of vertebrates except for a single genus which matures in freshwater oligochaetes; body typically segmented with one or more sets of reproductive organs in each proglottid; 6 hooks present on the first larval stage, the oncosphere.

Eucoccidiida Order of coccidians exhibiting 3 types of multiplication in the life cycle, merogony, gametogony and sporogony; host cells, from both vertebrates and invertebrates, are invaded by sporozoites which are ingested by the host or injected by vectors; includes *Plasmodium*, which causes malaria in man.

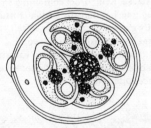

coccidian (Eucoccidiida)

Eucommiales Order containing a single species of tree native to montane forests of western China; dioecious; producing a milky juice; fruit is a samara.

Euconjugatae Class of the protoctistan phylum Gamophyta; comprising generally filamentous forms exhibiting true conjugation during which filaments lie side by side and grow conjugation tubes through which the amoeboid male gamete can pass into the female filament.

Eucryphiaceae Small family of Rosales containing 6 species of gum-producing, evergreen trees or shrubs native to Australia, Tasmania and Chile; flowers large and white, with 4 sepals and petals, many stamens and winged, fleshy seeds.

Eucyphidea Caridea *q.v.*

Euechinoidea Subclass of echinoids (Echinozoa) containing the vast majority of living species and exhibiting great diversity of form; 14 orders recognized in 4 superorders, Diadematacea, Echinacea, Gnathostomata and Atelostomata.

euephemerous Used of flowers that have a single opening period; **euphemeral.**

eugenics The science of breeding; the application of genetic principles to the improvement of the hereditary qualities of a race or breed.

Euglenales Order of euglenoid flagellates with one emergent and one internal flagellum; including *Euglena*.

Euglenamorphales Order of parasitic euglenoid flagellates infesting the rectum of amphibian tadpoles.

Euglenida Euglenophycota *q.v.* treated as an order of the protozoan class Phytomastigophora.

euglenoids Euglenophycota *q.v.*

Euglenophycota Euglenoids; a division of mostly freshwater flagellates comprising about 1000 species, most of which are unicellular, distributed worldwide; characterized by an anterior locomotory flagellum and a pre-emergent flagellum associated with the eyespot; most are photosynthetic green algae containing chlorophylls *a* and *b* in their chloroplasts, but many lack chlorophyll and are osmotrophic or phagotrophic; also classified as a phylum of Protoctista under the name Euglenophyta, and as an order of the protozoan class Phytomastigophora, under the name Euglenida.

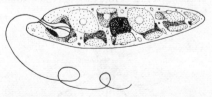

euglenoid (Euglenophycota)

Euglenophyta Euglenoids; the Euglenophycota *q.v.* treated as a phylum of Protoctista.

Eugregarinida Order of parasitic gregarines found mainly in the intestinal tract of annelid and arthropod hosts and which do not reproduce asexually in their hosts by multiple fission but produce gametes or sporozoites.

euhaline Living only in saline inland water bodies.

euhydatophyte Any fully submerged plant that never produces aerial structures.

euhydrophilous Thriving submerged in fresh water; **euhydrophile**, **euhydrophily**.

Eukaryotae Superkingdom of organisms possessing cells with an organized nucleus surrounded by a nuclear envelope, and paired chromosomes containing DNA which are recognizable during mitosis and meiosis; also characterized by elaborate cytoplasmic organelles such as cilia or flagella, the latter exhibiting a universal 9+2 microtubular structure; comprising either 2 kingdoms, Plantae and Animalia, or 4 kingdoms, Protoctista, Fungi, Plantae and Animalia, according to which classification is followed.

eukaryotic Used of organisms the cells of which have a discrete nucleus separated from the cytoplasm by a membrane, with DNA as the genetic material and with defined cytoplasmic organelles; eucaryotic; **eukaryote**; *cf.* prokaryotic.

eulittoral 1: The shore zone between the highest and lowest seasonal water levels in a lake; often a zone of disturbance by wave action. 2: The intertidal zone of the seashore; see marine depth zones.

Eumalacostraca The largest subclass of higher crustaceans (Malacostraca) comprising 13 orders grouped into 4 superorders, Peracarida, Eucarida, Syncarida and Pancarida.

Eumenidae Solitary wasps, mason wasps, potter wasps; diverse family containing about 3000 species of wasps (Hymenoptera) which usually construct nests consisting of a few cells in burrows in soil, twigs or wood; each cell is provisioned, predominantly with lepidopteran or beetle larvae, before the egg is laid; adults feed on nectar.

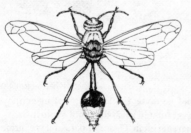

potter wasp (Eumenidae)

Eumetazoa Metazoans; the subkingdom of animals characterized as being multicellular and at the tissue and organ levels of organization as compared to sponges which are multicellular but at the cellular level of organization; comprising 30 phyla.

Eumycetozoa Plasmodial slime moulds *q.v.*, treated as a class of rhizoid protozoans, comprising subclasses Protostelia, Dictyostelia and Myxogastria.

Eumycota True fungi; a division of the kingdom Plantae; a group of ubiquitous heterotrophic organisms with a thallus either in the form of a mycelium of branched thread-like hyphae, or sometimes unicellular; cell walls usually chitinous, occasionally with cellulose; obtain nourishment as parasites or by secretory enzymes that dissolve insoluble food externally before absorption; store food as glycogen or lipid; comprising five subdivisions based on type of reproduction, Mastigomycotina, Zygomycotina, Ascomycotina, Basidomycotina and Deuteromycotina.

Eunicida Order of errant, pelagic, tubicolous and parasitic polychaete worms comprising about 800 species in 10 families; body length up to 2 m; pharynx eversible with well developed jaws; parapodia distinct; one aberrant group occurs in gill chamber of decapod and isopod crustaceans.

eupelagic Used of planktonic organisms found only in open oceanic water away from the influence of the sea bed.

Euphausiacea Krill; order containing 85 species of pelagic shrimp-like eucaridan crustaceans; typically marine oceanic forms often undertaking extensive diurnal vertical migrations and aggregating into vast swarms; filter feeders with exposed gills not covered by carapace; often with photophores.

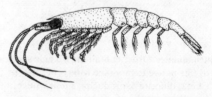

krill (Euphausiacea)

euphemeral Used of flowers that open and close within a single day.

euphilous Used of a plant or flower that has morphological adaptations for attracting and guiding a specialized pollinator; **euphile**, **euphily**; *cf.* allophilous.

Euphorbiaceae Euphorbia, castor bean, poinsettia, croton; family containing about 7500 species of generally woody, sometimes succulent or herbaceous plants, often with a milky or coloured latex and characterized by features of the female organs; seeds usually oily and often poisonous; includes many economically important species such as the sources of Para rubber, castor oil and cassava.

Euphorbiales Order of Rosidae containing 4 families of woody or sometimes herbaceous plants often producing alkaloids and milky or coloured juice; dominated by the very large family Euphorbiaceae.

euphotic zone The surface zone in the sea or large lake with sufficient light penetration for net photosynthesis to occur; *cf.* dysphotic zone, photic zone.

euphototropism Orientation response of plant structures to maximize incident illumination; **euphototropic.**

euploid A polyploid with an exact multiple of the basic chromosome number; **euploidy;** *cf.* aneuploid.

Eupomatiaceae A distinctive family of primitive flowering plants (Magnoliales) comprising a single genus of small trees or herbs confined to New Guinea and eastern Australia; flowers typically large, lacking both sepals and petals, and possessing laminar stamens.

eupotamic Pertaining to an aquatic organism thriving in both flowing and standing fresh water; *cf.* autopotamic, tychopotamic.

Eupteleaceae Family containing two species of deciduous shrubs or small trees possessing small wind-pollinated flowers lacking sepals and petals; native to Japan, China and Assam.

Eupteriomorphia A superorder of Pteriomorphian bivalves comprising the Pterioida (pearl oysters), Limoida (file shells) and Ostreoida (oysters and scallops).

Euramerica A continental landmass comprising Europe and North America formed from Eurasia *q.v.* after the break up of Pangaea; *cf.* Angara, Cathaysia.

Eurasia The northern supercontinent formed by the break up of Pangaea in the Mesozoic (*c.* 150 million years B.P.), comprising North America, Greenland, Europe and Asia, excluding India; Laurasia; *cf.* Gondwana.

euroky The ability to tolerate a wide range of environmental conditions.

Eurotiales Heterogeneous order of plectomycete fungi occurring in a wide variety of habitats including soil, dung and organic debris; includes numerous forms of economic

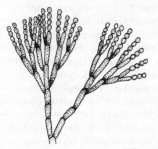

penicillium (Eurotiales)

importance, in causing food spoilage (*Aspergillus*), in antibiotic production (*Penicillium*), for contaminating jet aircraft fuel tanks (*Hormoconis resinae*) and in causing diseases of trees (Dutch elm disease).

eurotophilous Thriving in or on leaf mould; **eurotophile, eurotophily.**

Euryapsida Extinct group of reptiles characterized by a single upper temporal opening behind the eye; includes the plesiosaurs, nothosaurs and placodonts; known from the Permian through to the end of the Mesozoic.

eurybaric Tolerant of a wide range of atmospheric or hydrostatic pressure; *cf.* stenobaric.

eurybathic Tolerant of a wide range of depth; *cf.* stenobathic.

eurybenthic Living on the sea or lake bed over a wide range of depth; *cf.* stenobenthic.

eurybiontic Used of an organism tolerating a wide range of a particular environmental factor; *cf.* stenobiontic.

eurychoric Widely distributed; **eurychorous;** *cf.* stenochoric.

euryhaline Used of organisms that are tolerant of a wide range of salinity; *cf.* holeuryhaline, oligohaline, polystenohaline, stenohaline.

euryhydric Tolerant of a wide range of moisture levels or humidity; *cf.* stenohydric.

euryhygric Tolerant of a wide range of atmospheric humidity; *cf.* stenohygric.

euryionic Having or tolerating a wide range of pH; *cf.* stenoionic.

Eurylaimidae Broadbills; small family containing about 15 species of passerine birds found in the Old World tropical forests of Africa and Asia; head large, bill broad, flattened and hooked; feed on variety of insects, small vertebrates or fruit; nest typically long and pendulous, positioned over water.

eurylumic Tolerant of a wide range of light intensity; *cf.* stenolumic.

euryoecious Tolerant of a wide range of habitats and environmental conditions; *cf.* amphioecious, stenoecious.

euryphagous Utilizing or tolerant of a wide variety of foods or food species; **euryphage, euryphagy**; *cf.* stenophagous.

Eurypharyngidae Monotypic family of bizarre deep-water anguilliform teleost fishes (gulper-eels), widespread in temperate seas and locally quite numerous; body elongate, mouth enormous, teeth small; tail tapering.

euryphotic Tolerant of a wide range of light intensity; *cf.* stenophotic.

Eurypterida Water scorpions; extinct group of merostomes, often of large size; body length up to 3 m; known from the Ordovician to the end of the Palaeozoic.

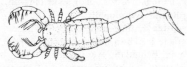

water scorpion (Eurypterida)

Eurypygidae Sunbittern; monotypic family of heron-like gruiform birds found in wooded stream and pond habitats of the Neotropical region; bill, neck and legs long; wings broad; habits solitary, monogamous; feed on aquatic arthropods and nest in trees.

eurythermic Tolerant of a wide range of temperatures; **eurytherm, eurythermous**; *cf.* stenothermic.

eurythermophilic Tolerant of a wide range of relatively high temperatures; **eurythermophile**; *cf.* stenothermophilic.

eurytopic 1: Tolerant of a wide range of habitats; physiologically tolerant; **eurytopy**. 2: Having a wide geographical distribution; *cf.* amphitopic, stenotopic.

eurytropic Used of organisms exhibiting a marked response or adaptation to changing environmental conditions; **eurytropism**; *cf.* stenotropic.

euryxenous Used of a parasite that tolerates a wide range of host species; *cf.* stenoxenous.

eusexual Used of organisms that display a regular alternation of karyogamy and meiosis; *cf.* parasexual.

eusocial Used of a social group in which members are fully integrated and cooperate in caring for young, with non-reproductive individuals assisting those involved in producing offspring, and in which different generations contributing to colony labour overlap; *cf.* presocial.

euspory The formation of spores by normal meiotic divisions; **eusporic.**

eustatic Pertaining to worldwide changes in sea level, such as would be produced by melting of continental glaciers, but excluding relative changes in level resulting from local coastal subsidence or elevation.

Eustigmatida Eustigmatophyceae *q.v.* treated as an order of the protozoan class Phytomastigophora.

Eustigmatophyceae Small class of aquatic, unicellular chromophycote algae characterized by their flagellate zoospores which have a large extraplastidial eyespot and a single yellowish green chloroplast containing chlorophylls a, c_1, c_2 and e; also classified as a separate phylum of Protoctista under the name Eustigmatophyta, and as an order of the protozoan class Phytomastigophora.

Eustigmatophyta The Eustigmatophyceae *q.v.* treated as a phylum of Protoctista.

Eusuchia Sole extant suborder of Crocodylia.

Eutaeniophoridae Tapetail; family containing 3 species of pelagic lampridiform teleost fishes widespread in tropical seas; body elongate, to 70 mm length, naked; dorsal and anal fins close to caudal, pelvics fan-like; caudal fin prolonged as ribbon-like filament.

Eutardigrada Order of freshwater and terrestrial (rarely marine) tardigrades comprising about 200 species; head without appendages or cirri; legs bearing compound claws.

eutelegenesis The improvement of breeding stock by artificial insemination.

Eutheria Placentals; infraclass of live-bearing (viviparous) mammals (Theria) in which embryos are nourished by a placenta; uteri separate or fused, vagina single.

eutraphent Used of an aquatic plant typical of water bodies with high nutrient concentrations; *cf.* mesotraphent, oligotraphent.

Eutremellales Jelly fungi; order of phragmobasidiomycete fungi in which the fruiting body has a gelatinous texture when wet; some species are edible; some are saprobic on wood, others as parasitic.

Eutreptiales Order of euglenoid flagellates with both flagella emergent.

eutric Pertaining to healthy or fertile soils; *cf.* dystric.

eutrophic 1: Having high primary productivity; pertaining to waters rich in the mineral nutrients required by green plants. 2: Used of a lake in which the hypolimnion becomes depleted of oxygen during the summer by the decay of organic matter sinking from the epilimnion; *cf.* dystrophic, mesotrophic, oligotrophic.

eutrophication Over-enrichment of a water body with nutrients, resulting in excessive growth of organisms and depletion of oxygen concentration.

eutrophy The pollination of particular kinds of flowers only by certain specialized insects.

eutropism The orientation of growth in a plant so that it twines in a direction that follows the passage of the sun across the sky; **eutropic**; *cf.* antitropism.

eutropous Used of specialized insect species adapted to feed on particular kinds of flowers; *cf.* allotropous.

evanescent Transient; quickly fading.

evaporation Loss of moisture in the form of water vapour.

evapotranspiration The actual total loss of water by evaporation from soil and from water bodies, and transpiration from vegetation, over a given area with time.

evening primrose Onagraceae *q.v.*

evergreen A plant, typically a tree or shrub, that has leaves all year round, and sheds them more or less regularly through all seasons.

Evermannellidae Sabretooths; family containing 6 species of small (to 150 mm) pelagic myctophiform teleost fishes; body compressed, naked, skeleton fragile; head and jaws large, teeth fang-like; eyes telescopic in some forms; photophores absent.

evolution 1: Any gradual directional change; unfolding. 2: Any cumulative change in the characteristics of organisms or populations from generation to generation.

evolutionary biology The integrated science of evolution, ecology, behaviour and systematics.

evolutionary tree A branching diagram in the form of a tree representing inferred lines of descent; phylogenetic tree.

evolve To change, produce or emit; to undergo gradual directional change; **evolution**.

exa (E) Prefix, used to denote unit x 10^{18}; see metric prefixes.

exaration Erosion by the action of ice.

excitability Sensitivity; the capacity of a living organism to respond to a stimulus.

exclusive economic zone (EEZ) The 200 mile wide marine zone under the jurisdiction of the coastal state.

excreta That part of the energy assimilated by an organism that is removed from the body as secretion, excretion or exudation.

excretion The process of eliminating waste material from the body; **excretory**; *cf.* secretion.

exendotrophic Used of a flower pollinated by pollen from another flower of the same or different plant.

Exobasidiales Small order of parasitic hymenomycete fungi containing 20 species found on stems, leaves and buds of host plants; producing an intracellular or intercellular mycelium which can induce gall formation.

exobiotic Living on the outer surface of another organism or on the surface of the substrate; **exobiont.**

Exocoetidae Flying fishes; family of surface-living marine beloniform teleost fishes widespread in open oceanic waters; body elongate, to 450 mm length, subcylindrical; mouth small, teeth reduced or absent; pectoral and pelvic fins markedly enlarged, caudal deeply forked; contains 50 species, well known for ability to jump out of water and glide considerable distances on the expansive paired fins.

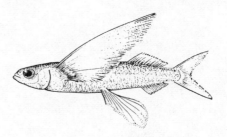

flying fish (Exocoetidae)

exogamy 1: Outbreeding; sexual reproduction between individuals that are not closely related. 2: Pollination of a flower by pollen from a flower on a different plant; cross pollination; *cf.* endogamy.

exogenous Originating from outside the organism or system; due to, or triggered by, external environmental factors; *cf.* endogenous.

exolithophytic Pertaining to plants that grow on the surface of rock or other hard inorganic substrata; **exolithophyte**; *cf.* endolithophytic.

Exoporia Suborder of lepidopteran insects comprising about 520 species, the vast majority of which are in one family, Hepialidae (ghost or swift moths); adult with small or much reduced proboscis; larvae typically burrowing into plant stems and roots.

exorheic region An area in which rivers arise and from which they flow to the sea; *cf.* arheic region, endorheic region.

exotic Not native; alien; foreign; an organism or species that has been introduced into an area.

exotropism An orientation movement of lateral organs away from the main axis; **exotropic.**

exovation The act or process of hatching.

expanding Earth hypothesis That the Earth has increased its volume and hence crustal surface area as a result of global expansion from about 80% of the present dimensions during the last 180–200 million years.

expatriate An individual carried out of its normal range to another area in which it can survive but is unable to reproduce.

expression The manifestation of a character or trait; **expressivity.**

exsiccate To dry up; to remove moisture; **exsiccation.**

exsiccation Desertification; the development of desert conditions as a result of human activity or of climatic changes.

extant Existing or living at the present time; *cf.* extinct.

extinct No longer in existence; no longer living; *cf.* extant.

extinction The disappearance of a species or taxon from a given habitat or biota.

extirpate To remove surgically; to destroy totally; to pull up by the roots; **extirpation.**

extraneous Existing or originating outside the habitat, community or system.

extrinsic Existing or having its origins outside an individual, group or system.

extrorse dehiscence Spontaneous opening of ripe fruit from the inside outwards; *cf.* introrse dehiscence.

exudativorous Feeding on gum and other exudates from trees; **exudativore, exudativory.**

exude To ooze out, or diffuse out; **exudate, exudation.**

exuvium The cast exoskeleton left by an arthropod after moulting.

f

Fabaceae Pea, bean, soybean, peanut, lupin, lentil, wisteria, laburnum, gorse, broom, sweet pea, alfalfa, clover, vetch; large family of over 10 000 species widespread in cold temperate and tropical regions containing many species of agricultural importance; typically leguminous forms with root nodules containing nitrogen-fixing bacteria and legume seed pods; characterized by butterfly-like (papilionaceous) flowers in which 1 petal is large and erect at the back of the bloom, 2 lateral petals form the wings and the other 2 petals are joined along their lower margins forming a keel; formerly known as the subfamily Papilionoideae of the family Leguminosae.

Fabales Legumes, Leguminosae; large order of Rosidae containing about 14 000 species in 3 widely distributed families; herbs, shrubs, trees or vines often with root nodules containing nitrogen-fixing bacteria, the fruit a typical dry legume, usually dehiscent along both sutures; includes the families Caesalpiniaceae, Fabaceae and Mimosaceae which were formerly treated as subfamilies of the family Leguminosae.

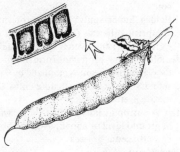

lupin pod (Fabales)

Facetotecta A group of barnacle-like crustaceans known only from their larval stages; also known as y-nauplius, y-cyprid.

facies The general appearance or aspect of an individual, population or community; also the general aspect or appearance of sedimentary rocks or fossils.

facilitation Enhancement of the behaviour or performance of an organism by the presence or actions of others.

facultative Contingent; assuming a particular role or mode of life but not restricted to that condition; used of organisms having the facility to live, or living, under atypical conditions; *cf.* obligate.

facultative anaerobe An organism that normally grows anaerobically but which is able to grow under aerobic conditions.

facultative parthenogenesis The process by which some eggs develop parthenogenetically if not fertilized.

faeces Excrement; waste material voided through the anus; **feces, faecal.**

Fagaceae Beech, chestnut, oak; large family of trees or shrubs with small flowers usually arranged in catkins and wind-pollinated; fruit usually a nut.

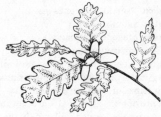

oak (Fagaceae)

Fagales Order of Hamamelidae containing 3 families of strongly tanniferous trees and shrubs with inconspicuous flowers often arranged in catkins.

Fahrenheit scale (°F) A scale of temperature that has the freezing point of water at 32°and the boiling point of water at 212°.

fairy fly Mymaridae *q.v.*

fairy shrimp Anostraca *q.v.*

Falconidae Falcons; cosmopolitan family containing about 60 species of small to medium-sized raptors (Falconiformes); wings usually long and slender, flight rapid; capture prey using powerful feet; one group (caracaras) long-legged, cursorial, feeding on carrion; habits solitary, monogamous, nesting in trees, cliffs or on the ground.

falcon (Falconidae)

Falconiformes Birds of prey; diverse order of small to large diurnal raptors, typically with powerful clawed feet and strong hooked bill; contains about 285 species in 5 families, including vultures, hawks, eagles, falcons, osprey, secretary bird.

Falkland Current A cold surface ocean current that flows northwards off the east coast of Argentina, originating in part as a northerly deflection from the Antarctic Circumpolar Current; see ocean currents.

fall wind A strong cold wind descending from an elevated plateau or glacier.

false cat shark Pseudotriakidae *q.v.*

false conch Neogastropoda *q.v.*

false fruit Pseudocarp *q.v.*

false moray eel Xenocongridae *q.v.*

false scorpion Pseudoscorpiones *q.v.*

false toadflax Santalaceae *q.v.*

false vampire Megadermatidae *q.v.*

family 1: A group comprising parents, offspring and others closely related or associated with them. 2: A category comprising one or more genera or tribes of common evolutionary origin, more or less separated from other such groups by a marked gap; a rank within the hierarchy of taxonomic classification.

family tree A diagrammatic representation of the lineage of a family or group.

fan palm Cyclanthales *q.v.*

fangtooth Anoplogastridae *q.v.*

fanworm Sabellida *q.v.*

far red light Solar radiation *q.v.* in the spectral range 700–800 nm.

farinaceous Containing or comprising flour; starchy.

fat A triacylglycerol that is solid at room temperature.

fauna 1: The entire animal life of a given region, habitat or geological stratum; **faunal**, **faunistic**. 2: A faunal work; *cf.* flora.

faunal province A zoogeographical subregion containing a distinct fauna which is more or less isolated from other regions by barriers to migration.

faunation The assemblage of animal species in a particular area; *cf.* vegetation.

faunistics The study of all or part of the animal species of a particular locality or region.

faunula The animal population of a small geographical area or microhabitat.

feather star Comatulida *q.v.* (Crinoidea).

featherback Notopteridae *q.v.*

feces Faeces *q.v.*; **fecal**.

fecund Producing offspring; fruitful; proliferating.

fecundity The potential reproductive capacity of an organism or population, measured by the number of gametes or asexual propagules; *cf.* fertility.

feedback Those elements of a control system or homeostatic mechanism linked by reciprocal influences, such that the source or input is continuously modified by the product of the process in order to maintain the necessary degree of constancy; a feedback loop may be negative and have a stabilizing or inhibitory effect or positive and have a disruptive or facilitative effect.

Felidae Cats; cosmopolitan family containing about 35 species of small to large carnivorous mammals including the cheetah, cougar (puma), ocelot, lynx, bobcat, lion, tiger, leopard, and jaguar; most are terrestrial and arboreal.

cheetah (Felidae)

fell A bare rocky hillside or mountain slope.

fell-field A type of tundra ecosystem having sparse dwarfed vegetation and flat, very stony soil.

Feloidea Superfamily of carnivorous mammals comprising the cats (Felidae), mongooses (Viverridae) and hyenas (Hyaenidae).

female The egg-producing form of a bisexual or dioecious organism; symbolized by ♀ *cf.* male.

femto- (f) Prefix used to denote unit x 10^{-15}; see metric prefixes.

fen A eutrophic mire, with a winter water table at ground level or above, usually dominated by herbaceous grasses.

fennel Apiaceae *q.v.*

Ferae Superorder of placental mammals (Ferungulata) represented by the order Carnivora but also containing the extinct order Creodonta.

feral Used of a plant or animal that has reverted to the wild from a state of cultivation or domestication.

fermentation The enzymatically controlled anaerobic breakdown of organic substrates.

fern Filicophyta *q.v.*

ferret Mustelidae *q.v.*

fertile Producing or having the capacity to

produce offspring, fruit, pollen or abundant growth.

fertility The actual reproductive performance of an organism or population, measured as the actual number of viable offspring produced per unit time; birth rate; *cf.* fecundity.

fertilization The union of a male and a female gamete to form a zygote, but sometimes used more generally for the act of insemination, impregnation or pollination.

Ferungulata Diverse group of placental mammals comprising the Artiodactyla, Perissodactyla, Carnivora and the primitive ungulates; sometimes classified as a cohort of the Eutheria, sometimes regarded as an artificial assemblage of unrelated forms.

Feyliniidae Family containing 4 species of limbless lizards (Sauria); body cylindrical; head flattened with tiny eyes under transparent scales; feed mainly on termites.

fidelity The degree of restriction of a plant species to a particular situation, community or association; assessed according to a five point scale: 5, exclusive ; 4, selective; 3, preferential; 2, indifferent; 1, strange.

field capacity The amount of water retained by a previously saturated soil when free drainage has ceased; *cf.* chresard, echard, holard.

field layer The horizontal vegetation stratum in woodland comprising all herbaceous plants, mosses and lichens, as well as woody plants not more than 500 mm in height.

fig Moraceae *q.v.*

fig marigold Aizoaceae *q.v.*

fig shell Mesogastropoda *q.v.*

fig wasp Agaonidae *q.v.* (Apocrita).

file shell Limoida *q.v.*

filefish Balistidae *q.v.*

Filicales Largest order of ferns (Filicopsida), containing an estimated 9000 species; producing only one kind of spore (homosporous); gametophytes green, free-living and usually heart-shaped.

Filicophyta Ferns; division of vascular plants

fern (Filicophyta)

which reproduce by spores produced in sporangia borne on the leaves, usually in clusters (sori); stems mostly creeping, sometimes erect or trunk-like, leaves with branching veins, roots wiry to fleshy; gametophytes free-living, green prothalli bearing sex organs on lower surface; contains between 12 000 and 15 000 species classified in 3 subdivisions, Filicopsida, Ophioglossopsida and Marattiopsida; formerly known as Pteridophyta.

Filicopsida Subdivision of ferns (Filicophyta); characterized by specialized sporangia having thin walls and opening by means of a patch or annulus of thickened cells; fronds unroll circinately; comprises 3 orders, Filicales, Marsileales and Salviniales.

polypodium (Filicopsida)

Filosa Class of rhizopod protozoans; amoeboid organisms with filose pseudopodia which are often branching but not reticulate; comprises 2 orders, Aconchulinida and Testaceafilosida.

Filospermoidea One of 2 orders of gnathostomulids found especially in marine sands of the North Sea; characterized by a filiform sperm and by the absence of a female bursa–vagina system and of paired sensory organs from the rostrum.

filter feeder Any animal that feeds by filtering suspended particulate organic matter from water.

fimbricolous Growing in or on dung; **fimbricole.**

fimetarious Pertaining to dung; **fimetarius.**

fimicolous Growing in or on dung; **fimicole.**

finch 1: Fringillidae *q.v.* 2: Emberizidae *q.v.* (Darwin's finches).

fingernail clam Veneroida *q.v.*

fir Pinaceae *q.v.*

fire climax A more or less stable plant community, the structure and composition of which is dependent on regular burning.

fire coral Milleporina *q.v.*

firebrat Thysanura *q.v.*

firefly Lampyridae *q.v.*

firewheel tree Proteaceae *q.v.*

fireworm Amphinomida *q.v.*

firnification The process by which newly fallen snow becomes granular and compacted; **firn.**

fishes A loose assemblage of cold-blooded limbless vertebrates possessing fins and a series of gills either side of the pharynx; typically with scales and paired pectoral and pelvic fins; once accorded the status of a class (Pisces).

fish louse Branchiura *q.v.*

Fissiculata Extinct order of blastoid echinoderms known from the Silurian to the Permian.

Fissidentales Cosmopolitan order of small mosses (Bryidae) commonly found in moist tropical and subtropical forests; leaves distinctive, divided into dorsal and ventral lamina.

fission Asexual reproduction or division by splitting into two (binary fission) or more (multiple fission) parts.

fissiparous Used of organisms that reproduce by division of the body into two or more equal parts.

Fissipedia Suborder of placental carnivores comprising the Canoidea, Feloidea and Miacidae.

fennec fox (Fissipedia)

Fistulariidae Cornetfishes, flutemouths; family of tropical shallow marine gasterosteiform teleost fishes containing 4 species, found on reefs and sea grass beds; body very slender, depressed, naked, to 2 m length; snout tubular; dorsal fin without spines, caudal bearing 1 or 2 elongate fin rays.

fitness A measure of the contribution of a given genotype to the next generation relative to that of other genotypes.

fjord A deep narrow inlet of the sea between steep slopes or mountains; fiord.

Flabellifera Suborder of free-living and parasitic isopod crustaceans widespread in marine, brackish and freshwater habitats; body typically flattened, the limbs simple or prehensile, and the uropods forming a tail fan

with the pleotelson; includes the wood-boring gribble that may cause extensive damage to boat hulls and submerged timbers.

Flabelligerida Order of sluggish deposit feeding polychaete worms, comprising about 160 species in 3 families; found burrowing in soft sediments from littoral to abyssal depths; body long and papillose, prostomium reduced and fused to peristomium; pharynx non-eversible, unarmed; parapodia biramous.

Flabellina A suborder of Amoebida *q.v.*; also treated as a class of the protoctistan phylum Rhizopoda.

Flacourtiaceae Azaras, fried egg tree; large family of Violales containing about 800 species of often cyanogenic trees and shrubs found mostly in tropical regions; characterized by generally alternate, leathery leaves and by flowers with 2–15 sepals, 0–15 petals, many stamens and a superior ovary.

azara (Flacourtiaceae)

Flagellariaceae Family of glabrous, solid-stemmed, cyanogenic herbs tending to accumulate silica, with alternate leaves well distributed along the stem and small trimerous flowers; producing seeds with mealy endosperm containing starch grains; native to the tropical regions of the Old World.

flame tree 1: Bignoniaceae *q.v.* 2: Sterculiaceae *q.v.*

flamingo Phoenicopteridae *q.v.*

flark A local wet area of sparse, weakly peat-forming fen vegetation interspersed with drier areas or features.

flash colour The bright contrasting colour located on concealed surfaces of some animals; thought to attract attention when presented to a predator, thus promoting concealment of the animal when the colour is suddenly withdrawn.

flatfish Pleuronectiformes *q.v.*

flathead Platycephalidae *q.v.*

flatworm Platyhelminthes *q.v.*

F-layer Fermentation layer; the upper or humus layer of a soil profile, exhibiting initial stages of plant decomposition; see soil horizons.

flea Siphonaptera *q.v.*

flesh fly Sarcophagidae *q.v.*

flocculation The aggregation of fine particles in the dispersed phase of a colloid.

floccule An aggregation of microorganisms or particles in a liquid; **flocculate.**

flood current The tidal current associated with an incoming or flood tide; *cf.* ebb current.

flood plain The area of lowland along a water course that is subject to periodic flooding and sediment deposition.

flood tide Incoming tide; *cf.* ebb tide.

flora 1: A floral work; a published work describing the plant life of an area. 2: The plant life of a given region, habitat or geological stratum; **floral, floristic**; *cf.* fauna.

Florida Current A warm surface ocean current derived from the Caribbean current system, that sweeps past Florida from the Gulf of Mexico and flows northeastwards into the Atlantic; see ocean currents.

Florideae The Florideophycideae *q.v.* treated as a class of the phylum Rhodophyta.

Florideophycideae Subclass of red algae with thalli that are obviously filamentous or pseudoparenchymatous and in which cell division is mostly restricted to apical cells of constituent filaments; gametangia well differentiated; contains 6 orders, Nemaliales, Cryptonemiales, Gigartinales, Rhodymeniales, Palmariales and Ceramiales; also treated as a class of the phylum Rhodophyta, under the name Florideae.

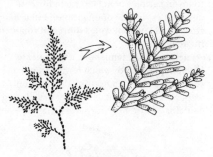

coralline alga (Florideophycideae)

floristics The study of species composition of vegetation, or of the plant species of a particular locality or region.

florology The study of the production and development of plant communities.

florula 1: The plant species of a small area; a local flora. 2: An assemblage of plant fossils

from a single stratum or group of adjacent strata; the floral assemblage of a small zone.

Flosculariaceae Small, diverse order of rotifers containing free-swimming planktonic and sessile forms; characterized by a circumapical girdle on the corona, divided into trochal and cingular parts, and by the lack of toes on the foot.

flounder Pleuronectidae *q.v.*

flower The specialized structure in flowering plants (Magnoliophyta) concerned with sexual reproduction; typically with the female organs (gynoecium) in the centre, surrounded by the male organs (androecium), petals (corolla) and sepals (calyx).

flowering The maturation of the floral organs and the expansion of their envelopes; florification.

flowering plant Magnoliophyta *q.v.*

flowering rush Butomaceae *q.v.*

flowerpecker Dicaeidae *q.v.*

fluke Digenea *q.v.*

flumineous Pertaining to running water.

flush An explosive increase in the size of a population; *cf.* crash.

flushing The deposition of dissolved substances in the upper layers of a soil profile by the action of water; *cf.* leaching.

flutemouth Fistulariidae *q.v.*

fluvial Pertaining to rivers and river action.

fluviatile Inhabiting rivers and streams; fluvial.

fluvioglacial Pertaining to streams flowing from a glacier, or to deposits laid down from glacial streams; glaciofluvial.

fluviology The study of rivers.

fluviomarine Inhabiting rivers and the sea.

fluvioterrestrial Inhabiting streams and the surrounding land.

fluviraption Erosion by running water or wave action.

fly Diptera *q.v.*

fly agaric Amanitaceae *q.v.*

flycatcher 1: Muscicapidae *q.v.* 2: Tyrannidae *q.v.* (tyrant flycatchers). 3: Bombycillidae *q.v.* (silky flycatchers).

flying fish Exocoetidae *q.v.*

flying fox Pteropodidae *q.v.*

flying gurnard Dactylopteridae *q.v.*

flying lemur Cynocephalidae *q.v.*

flyway A migration route of birds.

foam flower Saxifragaceae *q.v.*

foehn wind A warm dry wind on the leeward side of a range of mountains.

foliar Pertaining to leaves.

folicaulicolous Living attached to leaves and stems; **folicaulicole.**

folicolous Living or growing on leaves; **folicole.**

folivorous Leaf-eating; **folivore, folivory.**

follicle A dry dehiscent fruit developed from a single carpel and containing many seeds; splits along one suture only to release seeds on ripening.

fontaneous Pertaining to a freshwater spring.

food chain A sequence of organisms on successive levels within a community, through which energy is transferred by feeding; energy enters the food chain during fixation by primary producers (mainly green plants) and passes to the herbivores (primary consumers) and then to the carnivores (secondary and tertiary consumers).

food chain efficiency The efficiency of transfer of energy from one level to the next in a food chain.

food pyramid A graphical representation of the food relationships of a community, expressed quantitatively as numbers, mass or total energy at each trophic level, with the producers forming the base of the pyramid and successive levels representing consumers of higher trophic levels.

food web The network of interconnected food chains of a community.

footballfish Himantolophidae *q.v.*

forage 1: To search for food; **foraging**. 2: The plant material actually consumed by a grazing animal; *cf.* herbage.

foraminiferal ooze Pelagic calcareous sediment comprising more than 30% calcium carbonate predominantly in the form of foraminiferal tests; *cf.* ooze.

Foraminiferida Foraminiferans; order of rhizopod protozoans which are predominantly marine in distribution; possessing a mineralized, agglutinated or organic shell or test which encloses the amoeboid body; tests varied in form, often with chambers; a pseudopodial net is usually formed through one or more openings in the test and is used in locomotion and in feeding; includes a large number of fossil taxa; also treated as a distinct phylum of Protoctista.

forb A broad-leaved herbaceous plant.

forbicolous Living on broad-leaved plants; herbicolous; **forbicole.**

forbivorous Feeding on broad-leaved plants; **forbivore, forbivory.**

Forcipulata Large order of intertidal to deep-sea asteroidean echinoderms comprising 4 extant families and including many of the most familiar species of sea stars; typically with 5 arms but may be up to 50, arms rounded in cross-section, pedicellariae pincer-like, tube feet with suckers; largest species may exceed 1 m in diameter and a weight of over 10 kg.

foredune A dune ridge, more or less stabilized by plant colonization.

foreshore 1: The zone between mean high water and mean low water on a beach; also used of the upper intertidal zone that is covered only by exceptional spring tides. 2: The strip of land forming the margin of a lake.

forest A relatively large area of closely canopied trees; a small stand of trees may be termed a grove.

Forficulidae Large family of earwigs (Dermaptera) including the common European earwig.

forget-me-not Boraginaceae *q.v.*

form 1: The essential shape and structure of an organism or group. 2: Any minor variant or recognizable subset of a population or species; morph.

Formicariidae Antbirds; family containing about 240 species of small Neotropical passerine forest birds; habits solitary, non-migratory; feed on insects, some species following armies of ants to catch flushed insects; nest in tree or on ground; bill usually strong and hooked, wings short and rounded, flight weak.

Formicidae Ants; family of small aculeate Hymenoptera all of which form perennial societies founded in nests made in rotten wood, crevices, plant cavities or in soil; nest contains one or more fertile, egg-laying

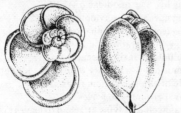

foraminiferans (Foraminiferida)

ants (Formicidae)

queens, hundreds to millions of neuter
workers, and seasonally produced fertile
males (drones) which fertilize queens during
mass nuptial flights; feed as predators of
arthropods, as scavengers, or on nectar and
honeydew; contains about 14 000 species.

forsythia Oleaceae *q.v.*

fossil An organism, fragment, impression or
trace of an organism preserved in a rock; may
be either a body fossil (such as a bone or shell)
or a trace fossil (such as a burrow, track or
imprint); *cf.* recent.

fossil fuel A fuel, such as coal, petroleum and
natural gas, derived from the fossilized
remains of organisms.

fossiliferous Used of a rock, sediment or
horizon containing fossils.

fossorial Adapted for digging or burrowing.

fouling An assemblage of organisms growing on
the surface of floating or submerged man-
made objects, that increases resistance to
water flow or otherwise interferes with the
desired operation of the structure.

Fouquieriaceae Ocotillo; small family of
Violales containing 11 species of woody or
fleshy succulent shrubs found in desert regions
from Mexico to southwestern United States.

four-eyed fish Anablepidae *q.v.*

fox Canidae *q.v.*

foxglove Scrophulariaceae *q.v.*

fractured zone In geology, a zone along which
displacement has occurred; major fractures
are found at right angles to the mid-oceanic
ridges.

Frankeniaceae Small family of Violales
containing about 50 species of halophytic
herbs or shrubs commonly with salt-excreting
glands on the leaves; best developed in the
Mediterranean region and Middle East.

Frankfurt glaciation A subdivision of the
Weichselian glaciation *q.v.*

frass Faecal matter and other fine animal debris
found on the soil surface.

fraternal twins Dizygotic *q.v.* twins.

free-living Living independently of any host
organism.

freemartin A sterile female twin, partially
converted towards a hermaphrodite condition
by some influence of its male twin.

freesia Iridaceae *q.v.*

free-tailed bat Molossidae *q.v.*

freeze drying A method of preserving biological
material by dehydration from the frozen state
under high vacuum.

Fregatidae Frigate birds; small family
containing 5 species of tropical sea birds

frigate bird (Fregatidae)

(Pelecaniformes) with long slender wings and
long pointed tail; feed on fishes and squid
caught in strong hooked bill during flight;
frigate birds do not land on water; breed in
colonies along coast, nesting in bushes with
single egg per clutch.

Frenulata Class of pogonophoran marine
worms possessing a bridle on the mesosoma;
contains nearly 100 species in 2 orders,
Athecanephria and Thecanephria.

fresh water Water having 2 parts per thousand,
or less, of dissolved salts.

fried egg tree Flacourtiaceae *q.v.*

frigate bird Fregatidae *q.v.*

fright cry The sudden loud call made by a
startled animal; distress call.

frigid climate A climate in which a permanently
frozen soil surface is covered more or less
continuously with ice and snow; polar climate.

frigid zone That part of the Earth's surface
within the polar circles.

frigideserta Tundra; the open communities of
cold arctic or alpine regions.

frigofugous Intolerant of cold conditions;
frigofuge.

frigophilic Thriving in cold environments;
frigophile, frigophily.

frilled shark Chlamydoselachidae *q.v.*

fringe tree Oleaceae *q.v.*

Fringillidae Finches; cosmopolitan family

finch (Fringillidae)

containing about 120 species of small to medium-sized passerine birds found in variety of forest, woodland and open habitats; bill short, conical, robust; habits solitary to gregarious, arboreal to terrestrial, many species migratory; feed mainly on seeds, flowers and buds; cup-shaped nest placed on or off the ground.

frit fly Chloropidae *q.v.*

fritillary Nymphalidae *q.v.*

frog Anura *q.v.*

frog shell Mesogastropoda *q.v.*

frogfish Antennariidae *q.v.*

froghopper Cercopidae *q.v.* (Homoptera).

frogmouth Podargidae *q.v.*

front The boundary zone between two air or water masses differing in properties, such as density, pressure, temperature or salinity.

frontal-cyclonic rain Rainfall produced by major cyclonic eddies and discontinuities; of particular significance in the equatorial rain belt and in temperate maritime regions; *cf.* convective rain, orographic rain.

fructicolous Living on or in fruits; **fructicole.**

fructification Any spore-bearing or seed-bearing structure, such as the aerial fruiting body of a fungus.

frugivorous Feeding on fruit; **frugivore, frugivory.**

fruit The structure that is formed from the ovary wall in flowering plants as the enclosed seed or seeds mature.

fruit bat Pteropodidae *q.v.*

fruit fly 1: Tephritidae *q.v.* 2: Drosophilidae *q.v.*

frutescence The time of maturity of a fruit.

fruticolous Living or growing on shrubs; **fruticole.**

Fucales Wracks; order of brown algae having a parenchymatous body form in which apical growth occurs; often with vesicles in the thallus for buoyancy; gametophyte generation reduced, or represented only by gametes; includes *Sargassum* weed.

fuchsia Onagraceae *q.v.*

fucivorous Feeding on seaweed; **fucivore, fucivory.**

fugacious Evanescent; lasting for a short time; falling early from the parent plant.

fugitive species A species which is always excluded locally under interspecific competition, but which persists in newly disturbed habitats by virtue of its high dispersal ability.

fulmar Procellariidae *q.v.*

Fumariaceae Dicentras; family of Papaverales containing about 400 species of herbs, generally with secretory cells but lacking latex; flowers irregular, typically with 2 sepals, 2–4 petals, 6 stamens and a superior ovary; widely distributed in northern temperate regions but also in South Africa.

Funariales Cosmopolitan order of mosses (Bryidae); small annual or ephemeral plants growing on disturbed soils, usually with erect stems bearing terminal sporophytes; leaves with single costa (longitudinal rib).

functional morphology Interpretation of the function of an organism or organ system by reference to its shape, form and structure.

fundamental niche The entire multidimensional space that represents the total range of conditions within which an organism can function and which it could occupy in the absence of competitors or other interacting species; *cf.* realized niche.

fungi An assemblage of ubiquitous heterotrophic, non-vascular organisms either with a thallus in the form of a mycelium of branched thread-like hyphae, or unicellular, and which obtain nourishment either saprophytically by secreting enzymes to dissolve insoluble organic food externally before absorption, or as parasites; cell walls are usually chitinous, occasionally with cellulose; classified either as a division, the Eumycota *q.v.*, of the kingdom Plantae or as a kingdom, the Fungi *q.v.*

Fungi The fungi treated as a separate kingdom of eukaryotic organisms and characterized by their lack of flagella at all stages of the life cycle; the Mastigomycotina *q.v.* are classified in the phylum Protoctista under this scheme; comprises 4 phyla, Zygomycota, Ascomycota, Basidiomycota and Deuteromycota.

Fungi Imperfecti Deuteromycotina *q.v.*

fungicide A chemical used to kill or control fungi.

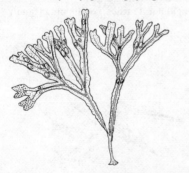

bladder wrack (Fucales)

fungicolous Living in or on fungi; **fungicole.**

fungistasis Inhibition of fungal growth.

fungivorous Feeding on fungi; **fungivore, fungivory.**

fungus gnat Mycetophilidae *q.v.*

fur seal Otariidae *q.v.*

furcilia Late zoeal larval stage of euphausiacean crustaceans.

furiotile lake Any partially disjunct body of water, that connects with the main stream only during high water.

Furipteridae Thumbless bats; small family of bats (Chiroptera), found in tropical America, in which the first digit is reduced.

Furnariidae Ovenbirds; diverse family containing about 220 species of small to medium-sized Neotropical passerine birds that construct a large domed nest of mud and sticks variously placed on the ground, in a tree, or in a hole or burrow; habits solitary or gregarious, arboreal or terrestrial; feeding mostly on insects and other invertebrates.

furniture beetle Anobiidae *q.v.*

ovenbird (Furnariidae)

g

Gadidae Cods; family of primarily northern hemisphere predatory gadiform teleost fishes of outstanding commercial importance; body length to 1.8 m, 1–3 dorsal and 1–2 anal fins present, pelvics jugular; gas bladder lacking connection to inner ear; contains about 55 species, all but one marine, most abundant in Atlantic Ocean.

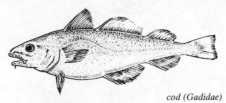

cod (Gadidae)

Gadiformes Order of primarily marine teleost fishes comprising about 500 species in 11 families, including cods, hakes, rattails, eelpouts, pearlfishes and cusk eels; body usually cod-like, lacking fin spines; barbels present; swim bladder typically without connection to inner ear.

Gadilida Order of scaphopod molluscs characterized by a typically smooth shell and by the position of the radula in the anterior third of the animal.

Gaian hypothesis That the development of living organisms on a planet brings about major modifications in the chemical and physical conditions on the planet and that subsequently the organisms themselves mediate the climate and major biogeochemical cycles.

Galagidae Galagos, bushbaby; family containing 7 species of small agile African arboreal prosimians; hindlimbs longer than forelimbs, used for leaping among the branches; ears large, tail elongate; feed on variety of small animals and plant material; sometimes included in the Lorisidae.

Galaxiidae Jollytails; family containing 45 species of southern hemisphere freshwater or anadromous salmoniform teleost fishes from Australia, southern Africa and South America; body slender to stout, to 700 mm in length; scales and adipose fin absent.

Galbulidae Jacamars; family of small insectivorous Neotropical forest birds (Piciformes) with long straight bill used to catch insects on the wing; habits solitary, monogamous, non-migratory; nesting in holes in bank; contains about 15 species distributed from Mexico to Brazil.

Galeomorphii Large superorder of small (250 mm) to large (12 m) neoselachian elasmobranch fishes; body mostly subcylindrical; usually 2 dorsal, anal and caudal fins present, and 5 pairs of gill openings; the dominant group of living sharks containing about 250 species in 4 orders, Heterodontiformes (bullhead sharks), Orectolobiformes (carpet sharks), Lamniformes (mackerel and thresher sharks), and Carcharhiniformes (requiem sharks).

Galeropygoida Extinct order of irregular echinoids with a small central peristome lacking gill slits; known only from the Jurassic.

gall Swelling or abnormal growth in plants typically produced in response to microbial or fungal infection, or to attack by insects, mites and other invertebrates.

gall midge Cecidomyiidae *q.v.*; gall gnat.

gall wasp Cynipidae *q.v.* (Apocrita).

gallery forest A narrow strip of forest along the margins of a river in an otherwise unwooded landscape.

gallicolous Living in galls; **gallicole.**

Galliformes Order of primarily terrestrial gallinaceous (fowl-like) birds comprising 6 families and including turkeys, chickens, grouse, pheasants, peafowl and guinea fowl; wings short for brief explosive flight, bill short for pecking seeds, insects and plant material; some species important as game and domesticated birds.

Indian jungle fowl (Galliformes)

gallinule Rallidae *q.v.*

galliphagous Feeding on galls; **galliphage, galliphagy.**

gallivorous Feeding on galls; **gallivore, gallivory.**

galvanotaxis A directed reaction of a motile organism to an electric current; **galvanotactic.**

galvanotropism An orientation response to an electric current; **galvanotropic.**

gamete A mature reproductive cell (usually haploid) which fuses with another gamete of the opposite sex, to form a zygote (usually diploid); the male gametes are known as sperms (spermatozoa) and the female gametes as eggs (ova); *cf.* agamete.

gametic number The number of chromosomes in the nucleus of a gamete, usually half the somatic chromosome number.

gametogamy The fusion of a male and female gamete to form a zygote; **gametogamic.**

gametogenesis The process of gamete production; **gametogenetic.**

gametophyte The haploid sexual phase of a plant which produces gametes usually by mitotic division; the haploid gametophyte is typically formed by meiotic division of a diploid sporophyte; haplophyte.

gametropism Orientation movements of plant structures immediately before or after fertilization; **gametropic.**

gamic Fertilized.

Gammaridea Largest of the 4 suborders of amphipods (Crustacea), comprising about 4500 species, widespread in marine, brackish, freshwater and occasionally terrestrial habitats, and having diverse morphology and habits, but excluding parasitism.

gamodeme An assemblage of individuals forming a relatively isolated naturally interbreeding population; *cf.* deme.

gamogenetic Pertaining to the formation of gametes; produced by the union of gametes; used of reproduction by union of gametes; sexual; **gamogenesis.**

gamogony Formation of gametes by sporogony.

gamone A substance released by one gamete and serving to attract another.

gamophase The haploid phase of a life cycle; *cf.* zygophase.

Gamophyta Phylum of Protoctista comprising the conjugating green algae; possess chlorophylls *a* and *b* in complex chloroplasts, usually aligned down the long axis of the cell; flagellated stages absent; reproducing asexually by fission of vegetative cells, and sexually by the production of amoeboid gametes that fuse to form a zygote; comprises 2 classes, Desmidoideae and Euconjugatae; *cf.* Zygnematales.

gamosematic Pertaining to coloration, markings or behaviour patterns that assist members of a pair to locate each other.

gamotropic Used of flowers that alternate between open and fully closed; *cf.* agamotropic, hemigamotropic.

gamotropism An orientation response of gametes to one another; the mutual attraction of gametes; **gamotropic.**

Ganeshida Small order of pelagic ctenophores occurring in the waters of southeastern Asia in which the body is compressed in the tentacular plane but not expanded into oral lobes.

gannet Sulidae *q.v.*

gar Lepisosteidae *q.v.*

garden four-o'clock Nyctaginaceae *q.v.*

garden snail Stylommatophora *q.v.*

gardenia Rubiaceae *q.v.*

garfish Belonidae *q.v.*

Garryaceae Small family of Cornales containing 14 species of dioecious evergreen trees and shrubs with small wind-pollinated flowers arranged in unisexual catkins; native to North America and the Greater Antilles.

gasoplankton Planktonic organisms which make use of gas-filled vesicles or sacs for buoyancy.

Gasteromycetes Cosmopolitan class of basidiomycotine fungi in which the basidia and basidiospores mature within the fruiting body; contains over 700 mostly terrestrial and saprobic species in 11 orders, Protogastrales, Hymenogastrales, Podaxales, Gauteriales, Agaricogastrales, Melanogastrales, Phallales, Lycoperdales, Tulostomatales, Sclerodermatales and Nidulariales.

earthstar (Gasteromycetes)

Gasteropelecidae Hatchetfishes; family of tiny (to 70 mm) surface-living, insectivorous, South and Central American freshwater

flying hatchetfish (Gasteropelecidae)

characiform teleost fishes; body deep and strongly compressed, pectoral fins long and broad; containing 7 species, popular amongst aquarists; these fishes are capable of true flight by flapping expanded pectoral fins.

Gasterophilidae Horse botflies; family containing about 45 species of flies (Diptera) in which the larvae are parasitic in the gut of ungulates, in particular horses, elephants and rhinos.

Gasterosteidae Sticklebacks; family containing 8 species of gasterosteiform teleost fishes; body naked or with armour of bony plates, length to 200 mm; dorsal fin with 3–16 strong isolated spines; upper jaw protractile; mostly freshwater but including some marine or anadromous species; males exhibit nest building and territorial behaviour.

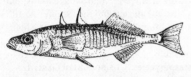

stickleback (Gasterosteidae)

Gasterosteiformes Order of mostly small to medium-sized, marine, brackish and freshwater teleost fishes comprising about 220 species in 11 families; characterized by small mouth on prolonged tubular snout; body usually armoured with dermal plates, caudal fin rounded, pelvics abdominal; including sand eels, tubesnouts, sticklebacks, sea moths, snipefishes, pipefishes, seahorses, shrimpfishes, trumpetfishes and cornetfishes.

Gastraxonacea Small order of colonial octocorals known only from shallow coastal waters off southeastern Australia; possessing a single axial polyp giving rise indirectly to lateral polyps towards the apex.

Gastromyzontidae Homalopteridae *q.v.*

Gastropoda Snails; a large class of aquatic, terrestrial or parasitic molluscs comprising nearly 35 000 species; characterized by a body which has primitively undergone torsion so that the anus opens above the head; body consisting of head, foot and visceral mass covered, at least in part, by a single-valved calcareous shell which is typically spirally coiled; mouth cavity containing a radula comprising backwardly directed teeth arranged in rows; may have separate sexes or be hermaphroditic; contains 3 subclasses, Prosobranchia, Opisthobranchia and Pulmonata.

Gastrotricha Phylum of small, aquatic metazoans found interstitially in loose sediments, on the surface of compacted sediments and as episymbionts on floating vegetation and benthic organisms; characterized by the presence of a true cuticle secreted by the epidermis and lining the foregut and hindgut as well as covering the body, and by oblique, longitudinal and circular muscles, a terminal mouth, a triradiate pharynx, subterminal anus and solenocytic protonephridia (excretory organs); typically hermaphroditic but usually with cross fertilization involving internal fertilization; comprises 2 orders, Chaetonotida and Macrodasyida.

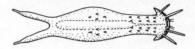

chaetonid (Gastrotricha)

gastrozooid A feeding polyp in colonial cnidarians.

Gauteriales Order of gasteromycete fungi comprising about 20 species occurring as saprophytes on forest soil; produce subglobular or kidney-shaped underground fruiting bodies.

Gavialidae Gavial, gharial; family of crocodilians containing the Indian gharial, found in the Ganges, Indus and Brahmaputra river systems; snout long and narrow, teeth all similar; feed on fish; commonly included in the family Crocodylidae.

Gaviidae Divers, loons; family containing 4 species of specialized aquatic birds; legs located posteriorly for efficient swimming and diving, but locomotion on land clumsy; sexes similar; feed mostly on fishes; breed on freshwater lakes and ponds of northern Holarctic, overwintering along the coast.

Gaviiformes Order of neognathous birds comprising a single family Gaviidae (divers).

gazelle Bovidae *q.v.*

gecko Gekkonidae *q.v.*

geese Anatidae *q.v.*

Geissolomataceae Family of Celastrales containing a single species of xeromorphic evergreen shrub native to Cape Province of South Africa.

geitonogamy Fertilization between different flowers on the same plant; **geitonogamic**; *cf.* xenogamy.

Gekkonidae Geckos; diverse family containing about 730 species of arboreal and terrestrial lizards (Sauria); body often depressed with delicate granular skin, but may be rounded or compressed bearing tubercles or plate-like scales: tail usually autotomic, prehensile in some species; habits mostly nocturnal, insectivorous; reproduction oviparous.

geld The removal of, or interference with the activity of, the testes of a male; emasculation; *cf*. spay.

gelicolous Living in geloid soils having a crystalloid content of between 0.2 and 0.5 parts per thousand; **gelicole**; *cf*. halicolous, pergelicolous, perhalicolous.

geloid soil Soil type characterized by low salt content, weak solutions, strong colloidal properties and a crystalloid content of less than 0.5 parts per thousand; *cf*. haloid soil.

gemma A bud or outgrowth capable of developing into an independent organism.

gemmation A type of vegetative reproduction found in some mosses and liverworts in which a specialized group of cells (a gemma) capable of independent development is produced; usually referred to in animals as budding.

gemmule Internal bud formed during asexual reproduction in sponges.

Gempylidae Snake mackerels; family containing 20 species of scombroid teleost fishes (Perciformes) found mostly in moderately deep waters (to 1200 m); body tuna-like, length to 1.8 m, jaw teeth very large; dorsal and anal fins with finlets, pelvics small or absent; feed on squid, crustaceans and other fishes.

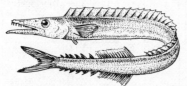

snake mackerel (Gempylidae)

gene The basic unit of inheritance, comprising a specific sequence of nucleotides on a DNA molecule that has a specific function and occupies a specific locus on a chromosome; alternative forms of a gene are known as alleles.

gene flow The exchange of genetic factors within and between populations by interbreeding or migration.

gene frequency The proportion of one allele to the total of all alleles at the same locus in the gene pool.

gene locus The position of a gene on a chromosome.

gene mutation A point mutation; any heritable change in a single gene.

gene pool 1: The total genetic material of a freely interbreeding population at a given time. 2: All the genes at a given locus in a population in a given generation.

genealogy 1: The study of ancestral relationships and lineages. 2: An ancestor–descendant lineage.

genecology The study of intraspecific variation and genetic composition in relation to environment.

generation 1: Formation; production. 2: All the individuals produced within a single life cycle.

generation time 1: The average duration of a life cycle between birth and reproduction. 2: The mean period of time between reproduction of the parent generation and reproduction of the following generation.

generative parthenogenesis The development of a haploid organism from a female gamete that has undergone a meiotic reduction division but has not been fertilized.

genesiology The study of reproduction; **genesiological.**

genet 1: A unit or group derived by asexual reproduction from a single original zygote, such as a seedling or a clone; *cf*. ortet, ramet. 2: Viverridae *q.v.*

genetic Pertaining to the genes.

genetic code The biochemical basis of heredity; the sequence of nucleotide base pairs on the DNA polynucleotide chain which encodes the genetic information; within the code 3 successive nucleotide base pairs (a codon) code for a single amino acid.

genetic drift The occurrence of random changes in the gene frequencies of small isolated populations, not due to selection, mutation or immigration.

genetic engineering Experimental alteration of the genetic constitution of an individual.

genetic system The organization of the genetic material and the reproductive strategy of a species.

genetics The science of heredity and variation.

genodeme A local interbreeding population characterized by genotypic features; *cf*. deme.

genome The minimum set of non-homologous chromosomes required for the proper functioning of a cell; the basic (monoploid) set of chromosomes of a particular species; the gametic chromosome number.

genopathic disease A disease resulting from a genetic condition; congenital disease.

genotype The hereditary or genetic constitution

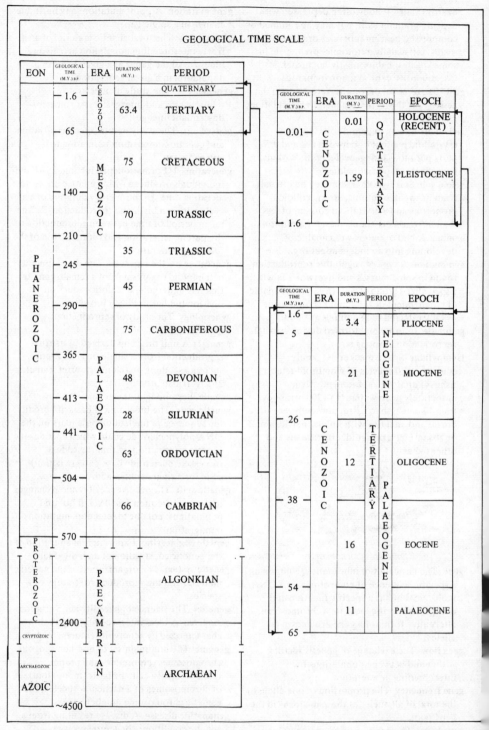

GEOLOGICAL TIME SCALE

EON	GEOLOGICAL TIME (M.Y.) B.P.	ERA	DURATION (M.Y.)	PERIOD
P H A N E R O Z O I C	1.6	C E N O Z O I C		QUATERNARY
			63.4	TERTIARY
	65			
		M E S O Z O I C	75	CRETACEOUS
	140		70	JURASSIC
	210			
	245		35	TRIASSIC
		P A L A E O Z O I C	45	PERMIAN
	290		75	CARBONIFEROUS
	365		48	DEVONIAN
	413		28	SILURIAN
	441		63	ORDOVICIAN
	504		66	CAMBRIAN
	570			
P R O T E R O Z O I C		P R E C A M B R I A N		ALGONKIAN
CRYPTOZOIC	2400			
ARCHAEOZOIC				ARCHAEAN
AZOIC	~4500			

GEOLOGICAL TIME (M.Y.) B.P.	ERA	DURATION (M.Y.)	PERIOD	EPOCH
0.01	C E N O Z O I C	0.01	Q U A T E R N A R Y	HOLOCENE (RECENT)
1.6		1.59		PLEISTOCENE

GEOLOGICAL TIME (M.Y.) B.P.	ERA	DURATION (M.Y.)	PERIOD	EPOCH
1.6	C E N O Z O I C	3.4	N E O G E N E	PLIOCENE
5		21		MIOCENE
26		12	T E R T I A R Y — P A L A E O G E N E	OLIGOCENE
38		16		EOCENE
54		11		PALAEOCENE
65				

geological time scale

of an individual; the genetic material of a cell, usually referring only to the nuclear material; *cf.* phenotype.

gentian Gentianaceae *q.v.*

Gentianaceae Cosmopolitan family containing about 1000 species of often mycorrhizal herbs, commonly accumulating iridoid compounds; flowers regular and perfect with 4–5 sepals, 4–5 petals in the form of a bell or funnel, 4 or 5 stamens, a superior ovary and, commonly, a capsule fruit with many small seeds.

Gentianales Order of Asteridae containing about 5500 species of plants commonly producing iridoid compounds or alkaloids; of the 6 families, the Apocynaceae and Asclepiadaceae are the largest; Contortae.

genus A category in biological classification comprising one or more phylogenetically related and morphologically similar species, forming the principal category between family and species; **genera.**

geoaesthesia The capacity of a plant to perceive and respond to gravity.

geobenthos 1: The sum total of all terrestrial life. 2: That part of the bottom of a stream or lake not covered by vegetation; *cf.* phytobenthos.

geobiology The study of the biosphere.

geobiont An organism spending its whole life in the soil; **geobiontic.**

geobios The total life of the land; that part of the Earth's surface occupied by terrestrial organisms; **geobiontic;** *cf.* halobios, hydrobios, limnobios.

geobotany Plant biogeography; the study of plants in relation to geography and ecology.

geocarpy Ripening of fruits underground.

geochronology The science of dating and the study of time in relation to the Earth   history as revealed by geological data.

geocolous Living in the soil for part of the life cycle; **geocole.**

geocryptophyte A plant having perennating organs or renewal buds below the soil surface.

geodiatropism Orientation at right angles to gravity; **geodiatropic.**

geodyte A ground-living organism.

geographical race A race which is geographically separated from other populations of the same species.

geographical variation Any differences between spatially separated populations of a species.

geological time scale See diagram opposite.

geology Study of the structure, processes and chronology of the Earth.

Geometridae Carpet moths, loopers, inch worms; large cosmopolitan family of small to large moths (Lepidoptera) comprising over 20 000 species including many serious pests of trees and shrubs such as the cankerworm and winter moth; caterpillar larvae often resemble twigs and move with a characteristic looping locomotion.

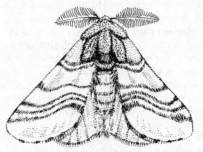

brindled beauty (Geometridae)

Geomyidae Pocket gophers; family containing about 40 species of small New World fossorial mammals (Rodentia) that derive their name from the cheek pouches that open to the outside near the mouth; legs short and strong with large claws for digging; feed primarily on tubers and roots.

pocket gopher (Geomyidae)

geonastic Growth curvature towards the ground; **geonasty.**

geonyctitropism Orientation movements in plants during darkness in response to gravity; **geonyctitropic.**

geoparallotropism An orientation movement of an organ or structure to bring it parallel to the soil surface; **geoparallotropic.**

geoperception The capacity of an organism to perceive and respond to gravity.

geophagous Feeding on soil; deriving nutrients from soil or the sediment; **geophage, geophagy.**

Geophilida Diverse order containing about 650 species of small to large (to 200 mm) epimorphan chilopods found mainly in soil and forest litter; some species inhabit intertidal zone; ocelli absent; trunk elongate, slender, composed of at least 33 segments, legs short and small; genital apparatus exposed.

Geophilomorpha Geophilida *q.v.*

geophilous 1: Thriving or growing in soil; **geophile**, **geophily**. 2: Used of plants that fruit below the soil surface.

geophyte A perennial plant having perennating organs or renewal buds, such as corms or rhizomes, buried well below the soil surface; *cf.* Raunkiaerian life forms.

geoplagiotropism Orientation at an oblique angle to the soil surface; **geoplagiotropic.**

geosere The series of climax formations in an area throughout geological time.

Geosiridaceae Family of Orchidales comprising a single species of small, mycotrophic herbs lacking chlorophyll and with alternate, reduced scale-like leaves; native to Madagascar and other Indian Ocean islands.

geotaxis A directed response of a motile organism towards (positive) or away from (negative) the direction of gravity; **geotactic.**

geothermal Pertaining to heat derived from the Earth's interior.

geotropism An orientation response to gravity; **geotropic.**

Geotrupidae Dung beetles; family containing about 300 species of robust, dark coloured, shiny beetles (Coleoptera); eggs laid in tunnels beneath dung which is carried down into tunnel to feed larvae.

geoxene An organism that becomes a temporary member of the soil fauna.

Geraniaceae Geranium, pelargonium, cranesbill; family containing about 700 species of herbs and shrubs often producing aromatic oils in glandular hairs; widespread in temperate and warm temperate regions; flowers variable but commonly with 5 sepals and petals, 5, 10 or 15 stamens and a superior ovary bearing a simple style with 5 stigmas.

Geraniales Order of Rosidae containing 5 families of mostly herbs or sometimes shrubs, some producing mustard oils; including the large families Geraniaceae and Oxalidaceae.

geranium Geraniaceae *q.v.*

geratology The study of decline and senescence of populations.

gerbil Cricetidae *q.v.*

germ cell A gamete *q.v.* or an agamete *q.v.*

germ line The lineage of generative cells that give rise to the gametes.

germinal Pertaining to, or influencing, the germ cells.

germination The commencement of growth of a propagule or bud.

germule A unit of colonization or migration.

gerontic Pertaining to the later stage of phylogeny or ontogeny.

Gerreidae Mojarras; family containing 40 species of bottom-living coastal marine teleost fishes (Perciformes) widespread in warm seas, and often found in mangrove habitats; body compressed, to 400 mm length, snout pointed, jaws highly protrusible; single dorsal fin present.

Gerridae Pond skaters; family of aquatic hemipteran insects, comprising about 450 species, typically found swimming actively on the surface of flowing as well as static water; legs and body are water-repellent and the insects are supported by surface tension.

Gesneriaceae African violet, gloxinia, hot water plant; family of Scrophulariales containing about 2500 species of mostly herbs, often with unequal cotyledons; widespread in tropical regions; flowers usually bisexual and irregular, with 5 sepals and a tubular corolla with 5 lobes, 2, 4 or 5 stamens and normally a superior ovary.

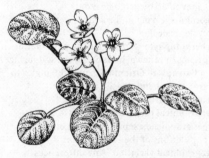

African violet (Gesneriaceae)

gestation The period of development of an embryo within the uterus of a viviparous animal, from conception to birth; **gestate.**

gharial Gavialidae *q.v.*

geranium (Geraniaceae)

G-horizon A gley horizon.

ghost flathead Hoplichthyidae *q.v.*

ghost moth Hepialidae *q.v.*

giant clam Veneroida *q.v.*

giant panda Ailuropodidae *q.v.*

gibberellin A type of growth substance produced in plants which is capable of affecting growth or development.

Gibberichthyidae Family of small (to 120 mm) mesopelagic beryciform teleost fishes comprising a single genus and 2 species; head cavernous, teeth setiform; dorsal and anal fins preceded by series of strong spines, caudal bearing 5–7 procurrent spines; juvenile (kasidoron stage) with characteristic elongate pelvic appendage.

gibbon Hylobatidae *q.v.*

giga- (G) Prefix used to denote unit x 10^9; see metric prefixes.

Gigantactinidae Family containing 18 species of piscivorous deep-sea anglerfishes (Lophiiformes); body elongate, slender, to 400 mm length, lure whip-like; males free-living, lacking teeth.

gigantism The condition of being much larger than normal, or of exhibiting excessive growth; often associated with polyploidy; *cf.* nanism.

Gigantorhynchida Order of archiacanthocephalan thorny-headed worms found as parasites of birds and mammals and utilizing insects as intermediate hosts; characterized by a proboscis shaped like a truncated cone, which bears rooted hooks anteriorly and rootless spines basally, and by the absence of protonephridial glands (excretory organs).

Giganturidae Family containing 5 species of deep-sea salmoniform teleost fishes; body naked, silvery, to 150 mm in length; eyes large and tubular, pectoral fins fan-like; pelvic fins and girdle, and several other bony structures absent.

Giganturoidei Monofamilial suborder of deep-sea salmoniform teleost fishes with silvery bodies and large tubular eyes.

Gigartinales Large order of red algae of variable body form, including crustose, discoid, erect and frondose forms.

gila monster Helodermatidae *q.v.*

ginger Zingiberaceae *q.v.*

Ginglymostomatidae Nurse sharks; circumtropical family containing 4 species of small to large (to 4 m) orectolobiform elasmobranch fishes (carpet sharks) found in shallow continental coastal waters; feed on

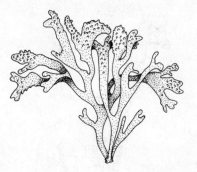

red alga (Gigartinales)

crustaceans, cephalopods and small fishes; occasionally offensive to man.

Ginkgoatae Ginkgo, maidenhair tree; large fossil class containing a single extant species native to China but now widely cultivated; tall resiniferous trees with deciduous fan-shaped leaves borne in clusters on spur shoots; male strobili clustered on axillary spur shoots, female structure stalked with 2 apical ovules.

maidenhair tree (Ginkgoatae)

ginseng Araliaceae *q.v.*

Gipping glaciation Wolstonian glaciation *q.v.*

Giraffidae Giraffe, okapi; family of herbivorous ungulates (Artiodactyla: Ruminantia) comprising 2 genera and species confined to the Ethiopian region; limbs and neck elongate; stump-like horns present; the giraffe inhabits open savannah, forming small herds; the okapi is solitary in tropical forests of Congo.

Girellidae Family containing 15 species of coastal marine perciform teleost fishes found in the Pacific and eastern Atlantic Oceans; body deep, compressed, to 500 mm in length; feed mainly on algae and eel grass.

glacial Pertaining to those geological intervals characterized by cold climate conditions and advancing ice sheets and caps; **glaciation.**

giraffe (Giraffidae)

glacial marine deposit A marine sediment having a significant terrigenous component derived from transportation by icebergs.

glacial relict A species that has survived from Pleistocene faunas and floras, typically in a restricted location or habitat (a Pleistocene refuge *q.v.*).

glaciofluvial Used of sediments transported by ice and deposited from the meltwaters of a glacier.

gladiolus Iridaceae *q.v.*

glareal Growing on dry exposed and typically gravelly ground; **glareous.**

Glareolidae Pratincoles; family containing 16 species of tern-like charadriiform shore birds found in dry sandy and stony habitats, from the Mediterranean through southern Asia to Australia.

pratincole (Glareolidae)

glass sponge Hexactinellida *q.v.*

Glaucosomatidae Glaucosomidae *q.v.*

Glaucosomidae Pearl perches; family of western Pacific marine teleost fishes

(Perciformes) comprising a single genus and 5 species typically found on deep offshore reefs; body stout, deep, with large head and mouth; length to 600 mm; some members important as game and food-fishes.

gley A soil type which is subject to periodic waterlogging because of a poorly permeable C-horizon.

gley horizon A soil horizon characterized by the deposition of iron and manganese compounds; G-horizon.

gleying In waterlogged soils, the removal of iron and manganese compounds from anaerobic surface layers of a soil to deeper layers where they are precipitated under oxidizing conditions; upper layers tend to be dull grey, deeper layers intensely mottled.

Glires Cohort of placental mammals comprising the rodents and the lagomorphs; possess incisors that continue to grow throughout life and are worn away by gnawing.

Gliridae Dormice; family containing about 20 species of small arboreal nocturnal myomorph rodents found in the western Palaearctic and Ethiopian regions; feed mostly on fruit, seeds and insects; many hibernate during winter.

dormouse (Gliridae)

globigerina ooze A pelagic sediment found over extensive regions of the ocean floor, comprising more than 30% calcium carbonate in the form of foraminiferan tests of which *Globigerina* is the dominant genus; *cf.* ooze.

Globulariaceae Family of Scrophulariales containing about 300 species of heath-like herbs and shrubs producing iridoid compounds, native to Africa and western Eurasia; flowers often with a 5-lobed corolla tube, 4 stamens and a superior ovary.

glochidium Bivalved veliger larva of freshwater mussels (Unionoida) specialized for parasitic attachment to the body surface or gills of fishes.

Gloeodiniales Order of freshwater dinoflagellates containing a single species which occurs epiphytically in peat bog habitats.

Gloger's rule The generalization that among warm-blooded animals those races living in

glochidium

warm and humid areas are more heavily
pigmented than those in cool dry areas;
pigments are typically black in warm humid
environments, red and yellow in dry areas, and
generally reduced in cool areas.

Glomerida Order of stout-bodied
oniscomorphan diplopods (millipedes)
containing about 200 species, widespread in
the northern hemisphere.

Glomeridesmida Order of small blind tropical
diplopods (millipedes) comprising about 30
species in a single family; body 22-segmented,
non-conglobating; both sexes with 36 pairs of
legs, the last pair in the male modified as
telopods for clasping; gut simple and straight.

Glossinidae Tsetse flies; small family
containing about 20 species of biting flies
found in tropical Africa; adults feed on the
blood of vertebrates and are of primary
medical and veterinary importance as vectors
of sleeping sickness and nagana.

tsetse fly (Glossinidae)

glow worm Lampyridae *q.v.*
gloxinia Gesneriaceae *q.v.*
glycogen A polysaccharide used for food
storage, found in animals, some fungi,
bacteria and blue-green algae.
glycophyte A plant thriving in a soil of low salt
concentration, typically less than 0.5% sodium
chloride; *cf.* halophyte.

Glyptodontidae Extinct family of large
armadillo-like mammals found in the New
World from the Eocene to the Pleistocene.
gnat Nematocera *q.v.*
gnatcatcher Polioptilidae *q.v.*
Gnathiidea Suborder of isopod crustaceans in
which the adults are non-feeding benthic
forms and the juveniles (praniza larvae) are
parasitic.
Gnathobdellida Order of mainly parasitic or
carnivorous leech-like worms (Hirudinea)
possessing 3 pairs of jaws and a pharynx.
Gnathostomata 1: The jawed vertebrates; a
superclass of vertebrates comprising fishes
(other than lampreys and hagfishes),
amphibians, reptiles, birds and mammals. 2:
Superorder of euechinoidean echinoids in
which the anus is displaced from the aboral
pole to an interambulacral position and the
test has a secondary bilateral symmetry; 2
orders recognized, Holectypoida and
Clypeasteroida.
Gnathostomulida Jaw worms; a phylum of free-
living, bilaterally symmetrical microscopic
worms found worldwide in marine sands which
are rich in organic detritus, at depths ranging
from the intertidal to several hundred metres;
characterized by the monociliated skin
epithelium and by the specialized muscular
pharynx possessing paired jaws and an
unpaired cuticular basal plate; anus absent;
typically hermaphrodite; comprises 2 orders,
Bursovaginoidea and Filospermoidea.

jaw worm (Gnathostomulida)

gnesiogamy Cross fertilization between two
individuals of the same species.
Gneticae Subdivision of gymnosperms
(Pinophyta) comprising evergreen shrubs,
lianas or small trees; leaves scale-like to broad;

staminate and pistillate structures in compound strobili and cones; includes 3 widely divergent subclasses, Ephedridae, Gnetidae and Welwitschiidae, each containing a single family.

Gnetidae Liana; subclass of gymnosperms comprising a single family, Gnetaceae, of evergreen climbing stems or small trees found in tropical and humid regions; large paired leaves, with reticulate venation; male cones elongate, female cone scales fused into a cup-like structure around the axis.

goat Bovidae q.v.

goatfish Mullidae q.v.

goatsucker Caprimulgidae q.v.

Gobiesociformes Order of mostly bottom-living, shallow marine teleost fishes comprising about 100 species in a single family, Gobiesocidae (clingfishes).

Gobiesocidae Clingfishes; family containing about 100 species of intertidal and shallow marine teleost fishes (Gobiesociformes); body flattened ventrally, to 300 mm in length but usually much smaller, head broad with large mouth; pelvic fins forming ventral sucker; body scales and swim bladder absent.

Gobiidae Gobies; large family containing about 1000 species of small (to 100 mm) bottom-dwelling perciform teleost fishes; 2 dorsal fins present, pelvic usually combined to form ventral sucker; often found in association with other fishes or invertebrates; internal fertilization and aerial respiration exhibited by some species.

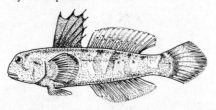

goby (Gobiidae)

Gobioidei Gobies, wormfishes; diverse suborder of marine, brackish and freshwater perciform teleost fishes comprising about 1000 species in 2 families; pelvic fins may form ventral sucker; 2 dorsal fins typically present; lateral line and gas bladder usually absent.

goblin shark Mitsukurinidae q.v.

goby Gobiidae q.v.

gold wasp Chrysididae q.v.

golden mole Chrysochloridae q.v.

golden trumpet Apocynaceae q.v.

golden-eye Chrysopidae q.v.

goldenrain tree Sapindaceae q.v.

goldenrod Asterales q.v.

Gomphotheriidae Extinct family of long-jawed mastodons (Proboscidea) known from the Miocene through to the Pleistocene.

Gomortegaceae Family of aromatic evergreen trees (Laurales) containing a single species native to central Chile; characterized by epigynous flowers arranged in racemes.

Gondwana The southern supercontinent formed by the break up of Pangaea q.v. in the Mesozoic (c. 150 million years B.P.); comprising the present South America, Africa, Arabia, Australia, Antarctica, India and New Zealand; cf. Eurasia.

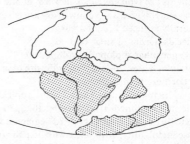

Gondwana

Goniactinida Sole extant order of somasteroidean echinoderms represented by a single living species found on shallow sandy beaches of western Central America.

gonochorism The fusion of male and female gametes produced from separate unisexual individuals; sexual reproduction.

gonochoristic 1: Used of individuals of a single sex; unisexual. 2: Used of populations having male and female individuals in the same population; **gonochoric, gonochorism.**

gonogenesis The formation of germ cells by meiotic division.

Gonorhynchidae Sand eel, mousefish; monotypic family of Indo-Pacific marine teleost fishes (Gonorhynchiformes); body very slender, cylindrical, to 600 mm length; snout pointed and prolonged, with single barbel, mouth ventral; pelvic and dorsal fins posteriorly positioned.

Gonorhynchiformes Small order of mainly Old World freshwater, brackish water and marine teleost fishes comprising about 15 species in 4 families, including milkfish and sand eels.

Gonostomatidae Lightfishes, bristlemouths; family containing about 60 species of luminescent meso- or bathypelagic stomiiform teleosts; body elongate, fragile, naked or with large cycloid scales, mouth and gill openings large; teeth thin and sharp; adipose fin usually present.

gonozoid A reproductive polyp in colonial cnidarians.

Goodeidae Family containing 40 species of small (to 200 mm) viviparous freshwater cyprinodontiform teleost fishes confined to the highlands of Mexico; resemble killifishes (Cyprinodontidae) except male anal fin modified as gonopodium for sperm transfer.

Goodeniaceae Family of Campanulales containing about 300 species of often poisonous perennial herbs storing carbohydrate as inulin; widely distributed in tropical and subtropical regions especially in Australia; flowers typically zygomorphic, with a split corolla tube and an inferior ovary.

goose Anatidae *q.v.*

goose barnacle Thoracica *q.v.*

gooseberry Grossulariaceae *q.v.*

goosefish Lophiidae *q.v.*

goosefoot Chenopodiaceae *q.v.*

Gordea Order of gordioidan horsehair worms characterized by the absence of cuticular organs called areoles.

gordian worm Nematomorpha *q.v.*

Gordioida Class of freshwater or semiterrestrial horsehair worms the juveniles of which are found as internal parasites of a variety of insects; characterized by the lack of lateral rows of swimming hairs and by the presence of paired reproductive organs; comprises 2 orders, Chordodea and Gordea.

Gorgonacea Sea fans, sea whips; order of arborescent colonial octocorals found most commonly in warm shallow waters, especially in association with coral reefs in the Indo-Pacific; characterized by their firm axial skeleton formed from horny proteinaceous material.

gorilla Pongidae *q.v.*

gorse Fabaceae *q.v.*

gourami 1: Belontiidae *q.v.* 2: Anabantidae *q.v.* (Anabantoidei).

gourd Cucurbitaceae *q.v.*

grade A level of organization and adaptation.

graft To induce the union of a tissue (a graft) from one organism with that of another by artificial means or to transplant tissue from one site to another on the same organism; in zoology the two elements are referred to as donor and recipient or host, and in botany the part being introduced is the scion and the recipient is the stock.

graft hybrid A hybrid produced by grafting two dissimilar plants.

grallatorial Adapted for wading.

Gramineae The grass family; now known as the Poaceae *q.v.*

graminicolous 1: Growing on grasses. 2: Used of an animal spending most of its life in a grassy habitat; **graminicole.**

graminivorous Feeding on grass; **graminivore, graminivory.**

graminology The study of grasses.

Grammatidae Grammidae *q.v.*

Grammicolepidae Small family of little-known zeiform teleost fishes widespread in tropical and temperate regions of Pacific and Atlantic Oceans; body compressed, to 250 mm in length, bearing extremely long thin scales forming vertical ridges.

Grammidae Basslets; family containing 9 species of small (to 80 mm) colourful tropical marine perciform teleost fishes commonly found in caves and crevices to 60 m depth; differ from sea basses in having an incomplete or absent lateral line; dorsal fin elongate with 11 fin spines.

Grammistidae Soapfishes; family containing 20 species of coastal marine perciform teleosts; body stout, compressed, to 300 mm length, skin covered with a toxic mucus secretion as protection from predators; mouth large and thick-lipped; single dorsal fin present.

granite moss Andreaeopsida *q.v.*

granivorous Feeding on seeds; **granivore, granivory.**

granule A sediment particle between 2 mm and 4 mm in diameter; small gravel; see sediment particle size.

Granuloreticulosa Class of rhizopod protozoans possessing delicate, anastomosing pseudopodia of continuously streaming, minutely granular cytoplasm; comprising 2 orders, Athalmida and Foraminiferida.

grape vine Vitaceae *q.v.*

grapefruit Rutaceae *q.v.*

Graptolithina Graptolites; a class of extinct colonial marine hemichordates in which individual polyps lived in chitinous tubes

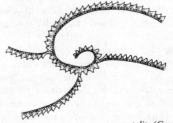

graptolite (Graptolithina)

arranged in single or double rows along characteristic branches (stipes); mainly epiplanktonic, sometimes sessile; known from the Cambrian to the Carboniferous.

grass 1: Poaceae *q.v.* 2: Restionaceae *q.v.*

grass tree Xanthorrhoeaceae *q.v.*

grasshopper Tettigoniidae *q.v.* (Orthoptera)

grassland An area of vegetation dominated by herbaceous grasses, sometimes used for any herb-dominated vegetation.

gravel Sediment particles between 2 mm and 256 mm in diameter; sometimes used for a class of smaller particles between 2 mm and 4 mm in diameter; see sediment particle size.

graveolent Possessing a strong or offensive odour.

gravid Carrying eggs or young; ovigerous; pregnant.

graviperception The perception of gravity.

gravitational water Water which drains by gravity through the soil and which is readily available to soil organisms and plants.

grayling Salmonidae *q.v.*

grazing Feeding on herbage, algae or phytoplankton, by consuming the whole food plant or by cropping the entire surface growth in the case of herbage.

greasewood Chenopodiaceae *q.v.*

great ape Pongidae *q.v.*

Great Interglacial period The major interglacial period in the middle of the Quaternary Ice Age *q.v.*

great white shark Lamnidae *q.v.*

grebe Podicipedidae *q.v.*

green alga Chlorophycota *q.v.*

green mud A terrigenous marine sediment, green in colour, formed under mildly reducing conditions and containing iron in the ferrous state, and some organic matter.

greenbottle Calliphoridae *q.v.*

greeneye Chlorophthalmidae *q.v.*

greenfly Aphididae *q.v.*

greenhouse effect The tendency towards increasing temperature of the lower layers of the atmosphere caused by an increase in atmospheric carbon dioxide which, together with water vapour, absorbs radiated heat more efficiently than it absorbs the incident solar radiation of short wavelengths.

greenling Hexagrammidae *q.v.*

Gregarinia Gregarines; sporozoans which usually live in the intestine or body cavity of invertebrates and lower chordates; the mature stages often appear segmented and commonly attach to each other, a process known as syzygy; also classified as the subclass

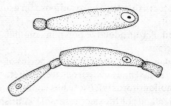

gregarine (Gregarinia)

Gregarinasina of the protoctistan class Sporozoasida.

gregarious Tending to aggregate actively into groups or clusters.

grenadier Macrouridae *q.v.*

gressorial Adapted for walking.

Grey Brown podzolic soil A zonal soil with a thin dark litter layer over a grey-brown, moderately acidic A-horizon and an illuviated lower horizon; formed in humid temperate climates on young land surfaces, typically glacial deposits under deciduous forest.

Grey Desert soil A soil type similar to a Sierozem *q.v.*, but having a calcareous surface layer; formed in warm temperate arid climates.

grey mullet Mugilidae *q.v.*

grey shark Carcharhinidae *q.v.*

grey whale Eschrichtiidae *q.v.*

Grey Wooded soil A woodland soil having a deep litter and duff layer, a deep weakly acidic brown to grey A-horizon, and a deep neutral to moderately acidic B-horizon.

Greyiaceae Wild bottlebrush; small family of Rosales containing 3 species of soft-wooded shrubs or trees confined to South Africa; flowers brilliant scarlet, with 5 small sepals and 5 showy petals.

gribble Wood-boring isopod crustacean (Flabellifera *q.v.*) that burrows into boat hulls, wharf piles and other submerged timbers, and may cause extensive damage.

Grimmiales Order of mainly xerophytic mosses (Bryidae), which are widely distributed but more common in the northern hemisphere; small dark plants usually growing as mats or tufts on rock surfaces.

grooming The act of cleaning the body surface to remove foreign matter and parasitic organisms, by licking, nibbling or other directed behaviour.

gross primary production The total assimilation of organic matter by an autotrophic individual, population or trophic unit; **gross primary productivity**; *cf.* net primary production.

gross production 1: Total assimilation of

organic matter by an individual population or trophic unit; *cf.* net production.

gross secondary production The total assimilation of organic matter or energy by a primary consumer individual, population or trophic unit; **gross secondary productivity**; *cf.* net secondary production.

Grossulariaceae Currant, gooseberry; cosmopolitan family of Rosales containing about 300 species of sometimes spiny shrubs and trees; flowers usually with 5 sepals, petals and stamens, and with an inferior ovary which develops to a juicy berry.

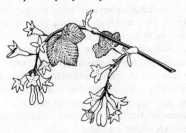

flowering currant (Grossulariaceae)

ground beetle, darkling Tenebrionidae *q.v.*

ground beetle, predacious Carabidae *q.v.*

ground hopper Caelifera *q.v.*

ground sloth Megalonychoidea *q.v.*

ground water All the water that has percolated through the surface soil into the bedrock.

ground-storey The lowest layer of vegetation in a stratified woodland or forest community, comprising small trees and shrubs, herbs and plant debris.

Ground-Water Laterite soil An intrazonal soil with hardpans rich in iron and aluminium, formed immediately above the water table.

Ground-Water Podzol soil An intrazonal soil with a mat of organic surface material over a thin acid humus layer, underlain by a pale grey leached horizon and a dark brown hardpan lower horizon, formed in humid cool to tropical climates under conditions of poor drainage with forest vegetation.

grouper Serranidae *q.v.*

grouse Tetraonidae *q.v.*

growth Increase in size, number or complexity.

growth form The characteristic appearance of a plant under a particular set of environmental conditions.

grub Soft-bodied, legless (apodous) larval stage characteristic of some dipterous, coleopterous and hymenopterous insects.

Grubbiaceae Small family of Ericales, containing 3 species of shrubs confined to the Cape Province of South Africa.

Gruidae Cranes; family containing 15 species of large long-legged wading birds found in wet and marshy habitats throughout the world, excluding South America; bill typically long and straight, wings broad, legs strong; habits gregarious, monogamous, migratory, feeding on small animals; nest made of vegetation, solitary, on ground or in shallow water.

Gruiformes Diverse order of small to large terrestrial or aquatic birds comprising 12 families and including cranes, rails, coots, bustards and several other small groups.

gallinule (Gruiformes)

grumusol A soil rich in clay that expands and contracts under conditions of high or low water content.

grunt Haemulidae *q.v.*

Gryllidae Crickets; large family of orthopterodean insects (Ensifera), mostly with box-like forewings bent down at the sides of the body; ovipositor cylindrical; many species have well developed sound producing mechanisms for auditory communication.

Grylloblattaria Rock crawlers; order of wingless orthopterodean insects comprising 13 species in a single family, found in cold habitats such as glacier margins, ice caves and upland forest; body depressed, poorly pigmented and weakly sclerotized; eyes small or absent; habits primarily nocturnal and omnivorous.

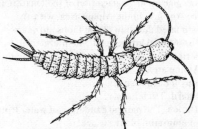

rock crawler (Grylloblattaria)

Gryllotalpidae Mole crickets; small family of large, burrowing orthopterodean insects (Ensifera) with fossorial front legs, reduced forewings and no ovipositor; all produce sound by stridulation.

guano An accumulation of sea bird droppings rich in phosphates and nitrates.

gudgen Eleotridae q.v.

guest An animal living and/or breeding within the nest, domicile or colony of another species.

Guiana Current A warm surface ocean current that flows northwestwards to link the South Equatorial Current with the Caribbean Current; see ocean currents.

guild A group of species having similar ecological resource requirements and foraging strategies, and therefore having similar roles in the community.

Guinea Current A warm surface ocean current that flows southwards off the west coast of North Africa, derived in part from the Canary Current; see ocean currents.

guinea fowl Numididae q.v.

guinea-pig Caviidae q.v.

guitarfish Rhinobatiformes q.v.

Gulf Stream The warm surface ocean current derived from the Antilles Current and the Florida Current that flows northeastwards into the Atlantic and forms the northern limb of the North Atlantic Gyre; see ocean currents.

gull Laridae q.v.

gulper-eel 1: Saccopharyngoidei q.v. 2: Monognathidae q.v. 3: Eurypharyngidae q.v.

gum tree Myrtaceae q.v.

gumbo-limbo Burseraceae q.v.

gumivorous Feeding on gum and other exudates from trees; **gumivore, gumivory.**

gunnel Pholididae q.v.

Gunneraceae Family of Haloragales containing about 50 species of terrestrial perennial herbs often with very large leaves, occurring mainly in southern hemisphere; flowers typically small and borne in large panicles.

Günz glaciation A glaciation of the Quaternary Ice Age q.v. in the Alpine area, with an estimated duration of 100 thousand years.

Günz–Mindel interglacial An interglacial period of the Quaternary Ice Age q.v. in the Alpine area.

guppy Poeciliidae q.v.

gurnard Triglidae q.v.

guttation The natural exudation of water from an uninjured plant surface.

guyot A submarine seamount rising from the abyssal plain.

Gymnamoeba Amoebae; group of ubiquitous, naked, lobose amoebae lacking any obvious test and moving by protoplasmic flow; feed by engulfing with pseudopodia or by invagination of the leading edge; reproducing mainly by binary fission; including many cosmopolitan species occurring in all aquatic habitats and in soil; classified in 3 orders, Amoebida, Schizopyrenida and Pelobiontida.

Gymnarchidae Monotypic family of primitive African piscivorous freshwater teleost fishes; body eel-like, to 0.9 m in length, without pelvic, anal and caudal fins; dorsal fin long; capable of producing strong electric fields; eggs laid in large floating nest.

Gymnascales Order of common and widely distributed plectomycete fungi typically found in soil, dung and animal debris, such as skin, hair and horn; mostly saprophytic but occasionally pathogenic; fruiting bodies (ascocarps) mostly sessile, spherical and lacking ostiole; asci evanescent; comprises over 60 species in 2 families and includes the causative agent of ringworm in man.

Gymnodiniales Order of freshwater and marine dinoflagellates including both photosynthetic and heterotrophic forms.

Gymnolaemata Diverse class of bryozoans comprising 2 orders, Ctenostomata and Cheilostomata; zooids polymorphic, cylindrical or flattened, lophophore circular, epistome absent; colonies variable, encrusting or erect, calcified or soft.

Gymnophiona Caecilians; order of limbless worm-like subterranean amphibians comprising about 150 species in 5 families; adults mostly terrestrial, sometimes aquatic; body elongate, length to 1200 mm, limbs and girdles absent; eyes small, often covered with bone; left lung rudimentary; fertilization internal; some species oviparous, others viviparous; pantropical in distribution; Apoda.

Gymnosomata Pteropods; cosmopolitan order of oceanic opisthobranch molluscs which are active swimmers and spend their lives in the plankton feeding mainly on other invertebrates; shell lacking; possessing lateral swimming fins which are separate from the foot.

gymnosperm Pinophyta q.v.

Gymnostomata Subclass of ciliates (Kinetofragminophora) typically with a uniform covering of simple cilia and a limited amount of ciliature on the body surface associated with the cell mouth (cytostome); reproducing generally by simple binary fission; widely distributed in a variety of habitats.

Gymnotidae Knifefishes; family of nocturnal predatory Central and South American freshwater catfishes (Siluriformes); body elongate, compressed, to 1.4 m in length; dorsal, adipose and pelvic fins absent, anal well-developed; produce weak electric fields for detecting prey; contains 40 species sometimes classified in 3 separate families, Apteronotidae, Rhamphichthyidae and Gymnotidae.

knifefish (Gymnotidae)

Gymnuridae Butterfly rays; family containing 12 species of medium to large, tropical to warm temperate, benthic marine myliobatiform elasmobranch fishes; body disk much broader than long, tail short and slender with or without serrated spine and dorsal fin; dorsal body surface lacking denticles.

gynandromorph An individual of mixed sex; a sexual mosaic having some parts genotypically and phenotypically male and others female; gynander.

gynic Female; *cf.* andric.

gynodioecious Used of plants or plant species having female (pistillate) and hermaphrodite (perfect) flowers on separate plants in a population or species; **gynodioecy**; *cf.* androdioecious.

gynoecious Used of a plant having female flowers only; *cf.* androecious.

gynoecium The central, female component of a flower, typically consisting of one or more carpels, which may be separate or fused.

gynogamete A female gamete; egg; ovum.

gynogenesis The process in which an egg develops parthenogenetically after the egg has been activated by sperm or pollen; pseudogamy; *cf.* androgenesis.

gynogenous Producing female offspring only; *cf.* androgenous.

gynomonoecious Having female (pistillate) and hermaphrodite (perfect) flowers on the same plant; **gynomonoecy**; *cf.* andromonoecious.

gynomorphic Having a morphological resemblance to females; **gynomorphy**; *cf.* andromorphic.

gynopleogamy The condition of a plant species having three sexual forms, with either perfect flowers, pistillate flowers or staminate flowers.

gypsophilous Thriving on chalk or gypsum-rich soils; **gypsophile, gypsophily.**

gypsophyte A plant inhabiting chalk or gypsum-rich soils.

Gyracanthocephala Order of eoacanthocephalan thorny-headed worms found as parasites of marine and freshwater fishes and utilizing crustaceans as intermediate hosts; characterized by the possession of trunk spines.

gyre A circular or spiral system of movement, characteristic of oceanic currents, and other systems of water movement.

Gyrinidae Whirligig beetles; family containing about 700 species of beetles (Coleoptera) found commonly on the surface of still freshwater; capable of detecting prey organisms falling onto water surface; larvae are bottom-living predators with abdominal gills.

whirligig beetle (Gyrinidae)

Gyrinocheilidae Family containing 3 species of small (to 250 mm) herbivorous freshwater cypriniform teleost fishes found in Borneo and southeast Asia; body elongate, mouth suctorial, lips fleshy, pharyngeal teeth absent; the 2 pairs of gill openings serve as inhalent and exhalent apertures for respiratory water current.

Gyrocotylidea Small order of cestodarian

cestodarian (Gyrocotylidea)

tapeworms found as parasites of marine fishes; characterized by a flattened, elongate body often with crenulated margins, a simple anterior holdfast, and a ventral, pre-equatorial uterine pore.

Gyrostemonaceae Small family of Batales containing about 17 species of trees or shrubs producing mustard oil; native to Australia; sometimes included in the Phytolaccaceae.

gyttja Sedimentary peat comprising predominantly plant and animal residues precipitated from standing water.

h

habilines Group of hominids known from fossil remains found in East Africa, that lived about 2–1.5 million years B.P.; from the larger brain size, facial and dental features they are thought to be closer to man than to the australopithecine apes, and have been placed in the same genus, as *Homo habilis,* at the base of the lineage leading directly to man.

habit The external appearance, aspect or growth form of an organism; **habitus.**

habitat The local environment occupied by an organism.

habituation The simplest form of learning in which the reduction or loss of a reponse to a stimulus occurs as a result of repeated stimulation.

habitus The characteristic form and appearance of an organism.

hackberry Ulmaceae *q.v.*

hadal zone Ocean depths below 6000 m; the trenches and canyons of the abyssal region; ultra-abyssal zone; see marine depth zones.

Hadean Early Precambrian; the geological period prior to the Archaeozoic (*c.* 3400 million years B.P.); see geological time scale.

Hadromerida Large order of tetractinomorph sponges characterized by a uniform spiculation of monaxonid megascleres, found in all seas from the intertidal zone down to 4000 m; includes the burrowing sponges of the family Clionidae.

hadrosaur An ornithischian dinosaur which was probably amphibious; up to 9 m in length, duck-billed and with webbed feet; known from the Upper Cretaceous.

haematobium An organism living in blood; **haematobic.**

haematocytozoon A parasite living within a blood cell; *cf.* haematozoon.

haematophagous Feeding on blood; **haematophage, haematophagy.**

haematophyte Any member of the blood flora.

Haematopodidae Oystercatchers; family containing 7 species of large charadriiform shore birds, cosmopolitan on tropical to temperate sandy and rocky shores and coastal marshes; bill characteristically long, robust, bright red; habits gregarious, monogamous; feed on bivalve molluscs; nest solitary, in hollow on ground.

oystercatcher (Haematopodidae)

haematothermal Warm-blooded; **haematothermic.**

haematozoon An animal parasite living free in the blood; *cf.* haematocytozoon.

Haemodoraceae Kangaroo-paw; family of Liliales comprising less than 100 species of perennial geophytic herbs often with a characteristic red pigment in the roots and rhizomes; leaves all basal, consisting of a sheathing base and parallel-veined, sword-shaped blade; mostly southern hemisphere in distribution; flowers perfect, with 6 perianths and 3 or 6 stamens, borne as panicled inflorescences.

kangaroo-paw (Haemodoraceae)

haemoparasite Any parasite inhabiting the blood of its host.

haemophagous Feeding on blood; **haemophage, haemophagy.**

haemotrophic Obtaining nutrients from blood.

Haemulidae Grunts; family containing 175 species of tropical and subtropical coastal marine, perciform teleost fishes usually found over shallow reefs or soft bottoms; the vernacular name refers to their ability to produce sounds by grinding their pharyngeal teeth to resonate the gas bladder; body oblong compressed, to 800 mm in length; mouth small with thick lips; single dorsal fin present; Pomadasyidae.

hagfish Myxiniformes *q.v.*

hair seal Phocidae *q.v.*

hair-streak Lycaenidae *q.v.*

hairyfish Mirapinnidae *q.v.*

hake Merlucciidae *q.v.*

half-beak Hemiramphidae *q.v.*

Half-Bog soil An intrazonal soil with a dark brown-black peaty surface layer over a grey mottled mineral layer, formed in humid, cool to tropical climates under conditions of poor drainage, with grasses, sedges or forest vegetation.

half-life 1: A measure of radioactive decay; the time taken for the level of radioactivity to reduce by one-half. 2: The time taken for an individual or biological system to eliminate one half of a given substance introduced into it.

halibut Pleuronectidae *q.v.*

halic Pertaining to saline conditions.

Halichondrida Order of ceractinomorph sponges, widely distributed and occurring from the intertidal zone down to 2500 m; typically with an encrusting to massive body form and a skeleton of diactinal or monactinal megascleres, or both.

halicolous 1: Living in haloid soils *q.v.* having a crystalloid content between 0.5 and 2 parts per thousand; *cf.* gelicolous, pergelicolous, perhalicolous. 2: Used of a plant living in a habitat with a high salt content; **halicole.**

Halictidae Sweat bees; large family of small to medium-sized bees (Hymenoptera) that feed on pollen and nectar both as adults and larvae and often lick perspiration from human skin; mostly solitary but also exhibiting primitive social nesting behaviour.

haliplankton Marine or inland saltwater planktonic organisms.

halobiont A marine organism or an organism living in a saline habitat; **halobiontic.**

halobios The total life of the sea; that part of the Earth's surface occupied by marine organisms; halibios; **halobiontic;** *cf.* geobios, hydrobios, limnobios.

halocline A salinity discontinuity; a zone of marked salinity gradient.

Halocyprida Order of marine ostracod Crustacea comprising 200 species in 2 suborders, Cladocopina and Halocypridina; found from littoral to abyssal zones in planktonic and benthic habitats.

Halocypridina Suborder containing about 160 species of marine halocypridan ostracods which are typically planktonic, from surface waters down to depths of 4000 m.

haloid soil A soil type characterized by high salt content, concentrated solutions and with a crystalloid content of more than 0.5 parts per

thousand; *cf.* geloid soil.

halolimnetic 1: Pertaining to salt lakes. 2: Halolimnic *q.v.*

halolimnic Used of marine organisms adapted to live in fresh water; **halolimnetic.**

halomorphic Used of an intrazonal soil having an accumulation of salts.

halophilous Thriving in saline habitats; **halophile, halophily;** *cf.* halophobic.

halophobic Intolerant of saline habitats; **halophobe;** *cf.* halophilous.

halophreatophyte A plant utilizing saline ground water.

halophyte A plant living in saline conditions; a plant tolerating or thriving in an alkaline soil rich in sodium and calcium salts; *cf.* glycophyte.

haloplankton Marine or inland salt-water plankton.

Haloragaceae Parrot's feather; family of aquatic, amphibious or marsh-inhabiting herbs, common in southern hemisphere; flowers typically solitary, inconspicuous and wind-pollinated, with inferior ovary.

Haloragales Order of Rosidae containing 2 families of often aquatic herbs, Haloragaceae and Gunneraceae.

Halosauridae Halosaurs; family containing about 14 species of small (to 250 mm) bottom-living notacanthiform teleost fishes widespread and locally common on continental slopes; body scales large, snout spatulate and prolonged; sensory canals cavernous.

halosere An ecological succession commencing in a saline habitat.

Hamamelidaceae Witch hazel, sweet gum; family of tanniferous shrubs or trees often possessing stellate hairs and producing woody capsular fruit; flowers usually in a spike, with 4 or 5 sepals and petals, and up to 14 stamens.

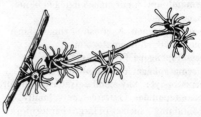

hamamelis (Hamamelidaceae)

Hamamelidae Subclass of dicotyledons (Magnoliopsida) typically with small inconspicuous flowers, often lacking petals, which are usually wind-pollinated.

Hamamelidales Order of woody plants comprising 5 relict families, Cercidiphyllaceae, Eupteleaceae, Hamamelidaceae, Myrothamnaceae and Platanaceae.

hammada A rocky desert.

hammer oyster Pterioida *q.v.*

hammerhead shark Sphyrnidae *q.v.*

hammerjaw Omosudidae *q.v.*

hamster Cricetidae *q.v.*

Hanguanaceae Family of Liliales containing a single species of robust perennial herbs found in moist or wet habitats; native to Malaysia and Sri Lanka.

hapaxanthous Having a single flowering phase during the life cycle.

haplobiont 1: A plant flowering once per season. 2: An organism not exhibiting a regular alternation of haploid and diploid generations during the life cycle; *cf.* diplobiont.

haplodiploidy The genetic system found in some animals in which males develop from unfertilized eggs and are haploid, and females develop from fertilized eggs and are diploid; **haplodiploid, haplodiplont.**

haploid Having only a single set of chromosomes; having the gametic chromosome number; **haploidy**; *cf.* ploidy.

haploid parthenogenesis The development of a haploid individual from a female gamete that has undergone meiotic reduction division but has not been fertilized.

haplometrosis The founding of a colony of social insects by a single fertile female; **haplometrotic.**

haplont 1: The haploid phase of a life cycle. 2: An organism having a life cycle in which meiosis occurs in the zygote to produce the haploid phase; *cf.* diplont.

Haplopharyngida Small order of free-living turbellarians found interstitially in marine sands, characterized by a simple proboscis opening just beneath the anterior end of the body and by the presence of a permanent but weakly differentiated dorsal anal pore.

haplophase The haploid stage of a life cycle.

haplophyte A haploid plant; gametophyte; *cf.* diplophyte.

Haplorhini Suborder of primates comprising 3 infraorders, Tarsii (tarsiers), Platyrrhini (New World monkeys) and Catarrhini (Old World monkeys, apes); nostril without lateral slit, rhinarium non-glandular, lower jaw medially fused or unfused.

Haplosclerida Order of ceractinomorph sponges including both marine and freshwater forms; characterized by a reticulate skeleton comprising triangular, rectangular or polygonal meshes formed of spicules joined with spongin.

Haplosporea Class of parasitic protozoans (Sporozoa) found in invertebrate hosts; produce resistant spores.

Haplotaxida Order of oligochaete worms comprising 3 suborders, Tubificina (freshwater), Haplotaxina (groundwater) and Lumbricina (terrestrial).

Haplotaxina Suborder of primitive groundwater oligochaete worms; clitellum comprising a single layer of cells; chaetae single or paired, simple or bifid; copulatory organs absent; comprises single family with about 20 species distributed throughout the Holarctic.

haptobenthos Those aquatic organisms that live closely applied to, or growing on, submerged surfaces; *cf.* benthos.

Haptomonadida Prymnesiophyceae *q.v.* treated as an order of the protozoan class Phytomastigophora.

haptonastic Used of growth movement of a plant in response to a touch or contact stimulus; **haptonasty.**

Haptophyta The Prymnesiophyceae *q.v.* treated as a phylum of Protoctista.

Haptorida Order of gymnostome ciliates containing large, active and voracious carnivores feeding mostly on other freshwater ciliates.

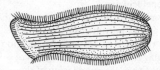

ciliate (Haptorida)

haptotropism An orientation response to a touch or contact stimulus; **haptotropic.**

hard pan A compacted layer in the B-horizon of a soil, typically rich in deposited salts, restricting drainage and root penetration.

hard tick Any of the tick family Ixodidae that have a rigid head plate and a forwardly directed head.

hardening The process of increasing cold resistance in plants.

hardiness The ability to survive exposure to sub-zero temperatures; cold resistance.

hare Leporidae *q.v.*

harem A group of females associated with a single male.

Harpacticoida Diverse order of typically free-living copepod crustaceans which are mainly epibenthic, burrowing or interstitial forms, occasionally planktonic or parasitic; characterized by the fusion of the inner branch (endopod) of the fifth leg to its base; feeding primarily on algae and a variety of microorganisms; contains about 3000 species from both marine and freshwater habitats.

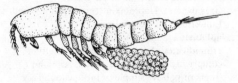

copepod (Harpacticoida)

harpactophagous Feeding by preying on other animals; **harpactophage, harpactophagy.**

Harpadontidae Bombay ducks; family of marine and brackish-water myctophiform teleost fishes comprising 5 species, widespread in the Indian Ocean; body elongate, cylindrical, head compressed with eyes forwardly directed; scales confined to posterior part of body; jaws very long bearing slender curved teeth; paired fins elongate.

Harpellales Order of trichomycete fungi occurring as obligate gut symbionts in freshwater insect larvae.

harrier Accipitridae *q.v.*

harvestman Opiliones *q.v.*

hatchetfish 1: Sternoptychidae *q.v.* 2: Gasteropelecidae *q.v.*

haustorium An organ formed by a parasite and used to absorb nutriments from the host.

Hawaiian region A zoogeographical region comprising the Hawaiian Archipelago; regarded either as a subdivision of the Polynesian subregion or as a separate region.

hawk Accipitridae *q.v.*

hawk moth Sphingidae *q.v.*

hawkfish Cirrhitidae *q.v.*

hawking Feeding in flight.

hawkweed Asterales *q.v.*

hawthorn Rosaceae *q.v.*

hazelnut Corylaceae *q.v.*

headstander Anostomidae *q.v.*

heart shell Veneroida *q.v.*

heart urchin Spatangoida *q.v.*

heat A period of sexual activity or receptiveness.

heath 1: Vegetation characteristic of low fertility, acidic, poorly drained soils, dominated by small leaved shrubs of Ericaceae (heathers and heaths) and Myrtaceae (myrtles). 2: Ericaceae *q.v.*

heather Ericaceae *q.v.*

heavy metal A metallic element of high specific gravity; including antimony, bismuth, cadmium, copper, gold, lead, mercury, nickel, silver, tin and zinc.

hebetic Pertaining to adolescence; juvenile.

hecto- (h) Prefix used to denote unit x 10^2; see metric prefixes.

hedgehog Erinaceidae *q.v.*

hedge sparrow Prunellidae *q.v.*

hekistotherm A plant with a requirement for a temperature of less than 10°C in the warmest month, typically occurring in regions with a mean annual temperature below 0°C; sometimes used to refer to those organisms living above the tree line in areas of heavy snow; **hekistothermic;** *cf.* megatherm, mesotherm, microtherm.

Heleophrynidae Small family containing 3 species of aquatic frogs (Anura) found in mountain streams of South Africa from the Cape to eastern Transvaal.

heleoplankton The planktonic organisms of small ponds and marshy habitats; **heleoplanktonic.**

helic Pertaining to marshes or marsh communities.

Heliconiaceae Lobster claw; family of Zingiberales containing over 100 species of large perennial herbs with leaves having a long petiole and expanded blade and prominent midrib; native mainly to tropical and subtropical America; flowers perfect, with 6 perianth segments, 5 stamens and an inferior ovary, arranged in alternate, compound inflorescences subtended by a large colourful bract shielding several small flowers; adapted for insect pollination.

heliconia (Heliconiaceae)

heliophilous Thriving under conditions of high light intensity; **heliophile, heliophily**; *cf.* heliophobous.

heliophobous Intolerant of high light intensity; shade-loving; **heliophobe**; *cf.* heliophilous.

heliophyllous Used of plants having leaves that can tolerate full sunlight; *cf.* sciophyllous.

heliophyte A plant that shows optimum growth under conditions of full sunlight; *cf.* sciophyte.

Helioporacea Order of octocorals in which the colony produces a rigid calcareous skeleton composed of aragonite crystals anchored basally or attached by stolons.

Heliornithidae Sungrebes; family containing 3 species of medium-sized aquatic gruiform birds, widespread in tropical streams, lakes and marshland; swim and dive well, feeding on a variety of fish and invertebrates; habits solitary, secretive, monogamous; nest in bushes.

heliosciophyte A plant which thrives in both sunlight and shade, but which grows best in sunny conditions.

heliotaxis A directed response of a motile organism towards (positive) or away from (negative) sunlight; **heliotactic.**

heliothermic Used of organisms that maintain a comparatively high body temperature by basking in sunlight; **heliotherm.**

heliotropism An orientation response to sunlight; **heliotropic.**

helioxerophilous Used of desert organisms thriving in both strong sunlight and drought conditions; **helioxerophile, helioxerophily.**

Heliozoa Class of actinopod protozoans in which the skeleton, when present, may be siliceous or chitinous and consists of independent scales, and which lack a capsular membrane; most species are freshwater, some marine; generally feed on small organisms caught with the axopods.

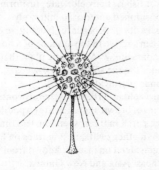

heliozoan (Heliozoa)

hellebore Ranunculaceae *q.v.*

helminthology The study of parasitic flatworms and roundworms.

Helminthomorpha Largest subclass of oniscomorphan diplopods (millipedes) comprising 11 orders in 7 superorders, Ommatomorpha, Monocheta, Anocheta, Diplocheta, Typhogena, Coelocheta and Merocheta; body typically elongate, cylindrical, each segment except first with 2 pairs of legs.

helobious Living in marshes.

Helodermatidae Gila monster; family containing 2 species of venomous lizards from Central America and southern United States; body length to 900 mm, limbs well developed; habits nocturnal, feeding on birds' eggs and small mammals; reproduction oviparous; producing a neurotoxin from a modified group of salivary glands.

gila monster (Helodermatidae)

helodric Pertaining to a swamp thicket.

Helogeneidae Family of small (to 100 mm) nocturnal Amazonian catfishes (Siluriformes) containing 3 species; dorsal fin lacking spine, anal fin elongate; 3 pairs of barbels present.

helohylophilous Thriving in wet or swampy forests; **helohylophile, helohylophily.**

helolochmophilous Thriving in meadow thicket habitats; **helolochmophile, helolochmophily.**

helolochmophyte A meadow thicket plant.

helophilous Thriving in marshes; **helophile, helophily.**

helophyte 1: A perennial plant with renewal buds, commonly on rhizomes, buried in soil or mud below water level; *cf.* Raunkiaerian life forms. 2: Any marsh or bog plant.

heloplankton The floating vegetation of a marsh.

helorgadophilous Thriving in swampy woodlands; **helorgadophile, helorgadophily.**

Helotiales Large order of discomycete fungi with a worldwide distribution, possessing club-shaped inoperculate asci; comprising many parasites of higher plants, some saprophytes, and soil- and dung-dwelling forms. (Picture overleaf).

helotism Symbiosis in which one symbiont enslaves the other.

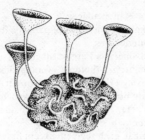

brown rot of peaches (Helotiales)

helper In an avian communal breeding system, a bird that assists in nest building, care of young or gives other aid, but is not related to either parent or offspring.

hemeranthous Flowering only during the day; **hemeranthy**; *cf.* nyctanthous.

hemerocology The study of the ecology of cultivated areas and culture communities.

hemerophilous Thriving in habitats influenced by the activities of man or under cultivation; **hemerophile**, **hemerophily**.

hemerophyte A cultivated plant.

Hemiascomycetes Class of ascomycotine fungi; characterized by the production of naked asci, not enclosed in ascocarp; includes saprophytic forms and plant and animal pathogens from terrestrial, freshwater and marine habitats; comprises 3 orders, Endomycetales, Protomycetales and Taphrinales.

hemiautophytic Used of a parasitic plant which is also capable of photosynthesis; **hemiautophyte**.

hemichimonophilous Used of plants that thrive under cold conditions and begin their growth even during frost; **hemichimonophile**, **hemichimonophily**.

Hemichordata Phylum of benthic marine invertebrates possessing pharyngeal gill slits for respiration, a body divided into proto-, meso- and metasome, paired coelomic cavities (except protosome), a mid-dorsal neurochord in the mesosome formed by invagination, an anterior evagination of the digestive tract termed a hemichord (not a homologue of chordate notochord), and a circulatory system comprising major dorsal and ventral vessels and a heart vesicle; development includes planktonic ciliated larva; 3 extant classes recognized, Enteropneusta, Pterobranchia and Planctosphaeroidea; includes acorn worms and the fossil graptolites; Stomochordata.

hemicleistogamic Used of a plant having flowers that only open partially before pollination.

hemicryptophyte A perennial plant with renewal buds at ground level or within the surface layer of soil; typically exhibiting degeneration of vegetative shoots to ground level at the onset of the unfavourable season; *cf.* Raunkiaerian life forms.

hemiepiphyte A plant that spends only part of its life cycle as an epiphyte, producing both aerial and subterranean roots at different times.

Hemigaleidae Weasel sharks; family containing 6 species of small to medium-sized (to 2.5 m), tropical, shallow water, carcharhiniform elasmobranch fishes; feed mainly on fishes; reproduction viviparous with yolk sac placenta.

hemigamotropic Used of flowers that alternate between open and partially closed; *cf.* agamotropic, gamotropic.

hemimetabolous Used of the pattern of development characterized by gradual changes, without metamorphosis or distinct separation into larval, pupal and adult stages; hemimetamorphic; **hemimetaboly**; *cf.* ametabolous, holometabolous.

Hemiodidae Hemiodontidae *q.v.*

Hemiodontidae Family containing 50 species of small to medium-sized tropical Central and South American freshwater characiform teleost fishes; body elongate, fusiform, dorsal fin positioned anterior to mid-body.

hemiparasite 1: A partial or facultative parasite that can survive in the absence of a host; meroparasite. 2: A parasitic plant that develops in the soil from free seeds.

hemipode Turnicidae *q.v.*

Hemiprocnidae Tree swifts; family containing 4 species of brown or grey swifts (Apodiformes) which perch in trees, spending less time in the air than other swifts; nest in trees on forest margins; feed on insects; known from India, southeast Asia and New Guinea.

Hemiptera Bugs; large and diverse order of

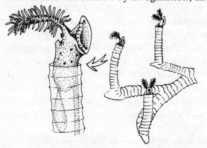

pterobranch (Hemichordata)

hemipterodean insects comprising about 35 000 species, including many of economic importance as crop pests, carriers of disease, and as predators; body often depressed, forewings typically leathery at base and membranous distally, but reduction or loss of wings common; mouthparts modified for piercing and sucking; feeding on plants, fungi and as predators; a few species are ectoparasites of vertebrates; most are terrestrial, many littoral on water surface, some aquatic; mimicry, aposematic coloration, procrypsis and sound production common within group.

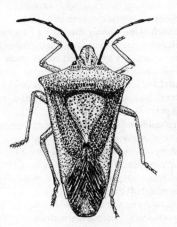

bug (Hemiptera)

Hemipterodea Superorder of mostly small neopteran insects comprising 6 orders, Psocoptera (book lice, bark lice), Anoplura (sucking lice), Mallophaga (chewing lice), Thysanoptera (thrips), Hemiptera (bugs) and Homoptera (aphids); wings and/or wing venation often reduced or absent; mouthparts frequently specialized for feeding on fluids; habits commonly gregarious but not fully social; Oligonephridia.

Hemiramphidae Half-beaks; family of surface-living marine and freshwater beloniform teleost fishes; body elongate, to 450 mm length, lower jaw prolonged well beyond upper; pectoral fins short; contains 60 species, many exploited as food-fishes and as baitfish.

hemisaprophyte A plant that may be autotrophic or saprotrophic at different times.

Hemiscyllidae Long-tailed carpet sharks; family containing 10 species of small (to 1 m) orectolobiform elasmobranch fishes found in shallow coastal waters of the tropical Indian Ocean and western Pacific.

hemitropic Used of flowers adapted for pollination by particular insect species; also used of the insects that pollinate specifically adapted flowers.

hemlock Apiaceae *q.v.*

henbane Solanaceae *q.v.*

Hepaticae Hepaticopsida *q.v.*

Hepaticopsida Liverworts; group of lower plants in which the dominant generation is the gametophyte; usually perennial, thallose or leafy in form with apical growth, and attached by rhizoids; sex organs develop from superficial cells; sperms biflagellate; sporophyte generation short-lived, determinate in growth and typically consisting of a dark brown spore-bearing capsule, a stalk (the seta) and an absorptive foot; includes 6 orders, Calobryales, Jungermanniales, Metzgeriales, Sphaerocarpales, Monocleales and Marchantiales; formerly known as the Hepaticae.

marchantialean liverwort (Hepaticopsida)

Hepialidae Ghost moths; widely distributed family of medium to large moths (Lepidoptera); wings generally rounded at apex; hindwings and forewings with similar venation.

Hepsetidae Monotypic family of small (to 350 mm), predatory, piscivorous African freshwater characiform teleost fishes widespread in tropical rivers; body elongate, pike-like; snout prolonged, median fins positioned posteriorly; eggs laid in floating bubble nests.

herb A plant having stems that are not secondarily thickened and lignified (non-woody) and which die down annually; herbaceous.

herbage Total vegetation available to a grazing animal; *cf.* forage.

herbarium 1: A collection of preserved (usually dried) plant specimens; *hortus siccus.* 2: The building in which such a collection is kept.

herbicide A chemical used to kill weeds or herbage.

herbicolous Living predominantly in
herbaceous habitats; **herbicole.**

herbivorous Feeding on plants; **herbivore,
herbivory.**

hercogamous Used of a flower having the
stamens and stigma positioned in such a way as
to prevent self-pollination; **hercogamy.**

hereditary Having a genetic basis; transmitted
from one generation to the next.

heredity The mechanism of transmission of
specific characters or traits from parent to
offspring.

heritability 1: The capacity of being inherited.
2: That part of the phenotypic variability that
is genetically based.

hermaphrodite Having both male and female
reproductive organs in the same individual
(animal) or the same flower (plant); bisexual;
the maturation of the male organs before the
female is protandrous hermaphroditism, the
female before the male is protogynous
hermaphroditism, and simultaneously is
synchronous hermaphroditism.

hermatypic Used of reef-forming corals that
contain symbiotic algae within the polyps;
hermatype.

hermit crab Anomura *q.v.*

Hernandiaceae Small family of woody shrubs,
trees and climbers (Laurales) widely
distributed in tropical regions; leaves
palmately veined; flowers typically with 6–10
perianth segments and 3–5 stamens, ovary
inferior.

heron Ardeidae *q.v.*

herpesian Pertaining to amphibians and
reptiles.

herpetogeny The history of colonization and
evolution in the establishment of the modern
amphibian and reptile faunas.

herpetology The study of amphibians and
reptiles; **herpetological.**

herpism Creeping locomotion; used of
protistans employing pseudopodia.

herpobenthos Those organisms growing or
moving through muddy sediments; *cf.*
benthos.

herpon Crawling organisms.

herring Clupeidae *q.v.*

herring smelt Argentinidae *q.v.*

hesmosis Colony fission in ants.

hesperidium A type of berry *q.v.*, such as an
orange, which has a leathery outer skin
(epicarp) and in which the locules contain fluid
filled trichomes.

Hesperiidae Skippers; large, widely distributed
family of butterflies (Lepidoptera) typically

skipper (Hesperiidae)

with short wings and large bodies; antennae
often club-shaped and widely separated;
larvae feed on grass.

heterauxesis The differential growth of body
parts resulting in a change of shape or
proportion with increase in size.

Heterenchelidae Family containing 7 species of
shallow marine anguilliform teleost fishes that
burrow into soft bottom sediments; body
cylindrical, naked, to 0.8 m in length; dorsal,
caudal and anal fins continuous, pectorals
absent; eyes vestigial; Atlantic distribution,
except for single Pacific species.

Heterobasidiomycetidae Subclass of saprobic
phragmobasidomycete fungi producing
fruiting bodies typically with a waxy or
gelatinous texture and basidiospores which are
usually discharged forcibly, then wind-
disseminated; comprises 2 orders,
Eutremellales and Septobasidiales.

Heterochlorida Heterochloridales *q.v.* treated
as an order of the protozoan class
Phytomastigophora.

Heterochloridales A class of simple
Xanthophyta; including both unicellular and
colonial forms; also classified as an order of
the protozoan class Phytomastigophora.

Heterococcales A class of Xanthophyta,
comprising coccoid cells organized into
various colonial forms.

Heterocorallia Small order of extinct corals
(Zoantharia) known only from the
Carboniferous.

Heterocyemida Order of rhombozoan
mesozoans endoparasitic in the renal glands of
squid and cuttlefish.

Heterodonta A subclass of mainly bivalve
molluscs characterized by a more-or-less fused
mantle forming inhalent and exhalent current
apertures frequently drawn out into siphons;
shell never with mother-of-pearl; contains 3
orders, Hippuritoida, Myoida and Veneroida.

Heterodontidae Heterodontiformes *q.v.*

Heterodontiformes Bullhead sharks; order of
inoffensive bottom-dwelling galeomorph
elasmobranch fishes; body length to 1.5 m;

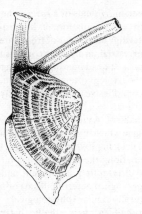

bivalve (Heterodonta)

mouth terminal, teeth differentiated, cusped
and non-cusped; dorsal fins with strong
anterior fin spine; feed on benthic
invertebrates and small fishes; reproduction
oviparous, egg cases of spiral shape; comprises
8 species in a single family found from the
littoral to 300 m.

Heterodontosaur An ornithopod dinosaur with
tusk-like canine teeth; known only from the
Upper Triassic.

heteroecious 1: Used of a parasite occupying
two or more different hosts at different stages
of the life cycle; *cf.* homoecious. 2: Used of a
non host-specific parasite. 3: Used of a
unisexual organism in which male and female
gametes are produced by different individuals.

heterogamete 1: A gamete produced by the
heterogametic sex *q.v.* 2: A gamete belonging
to one of two distinguishable types.

heterogametic Having two kinds of gametes,
one producing males and the other females;
heterogamy; *cf.* homogametic.

heterogametic sex The sex which is determined
by a pair of dissimilar sex chromosomes (X and
Y, or Z and W) or by a single unpaired sex
chromosome (X or Z); the sex which produces
two types of sex-determining gametes; *cf.*
homogametic sex.

heterogamy 1: The union of gametes
(heterogametes) of different shape or size. 2:
Alternation of two sexual generations, one
syngamic and the other parthenogenetic. 3:
The condition of a plant producing both male
and female gametes from one kind of flower;
heterogamic, heterogamous; *cf.* homogamy.

heterogeneous Having a non-uniform structure
or composition; **heterogeneity**; *cf.*
homogeneous.

heterogenesis 1: Spontaneous generation. 2:
Alternation of generations; *cf.* homogenesis.
3: The appearance of a mutant in a population;
heterogenetic.

Heterogloeales Order of non-motile algae
(Xanthophyceae) lacking both flagella and
pseudopodia; includes solitary and colonial
forms which may be free-living or attached in
freshwater or marine habitats.

heterogonic life cycle 1: A cycle of alternating
parthenogenetic and sexually reproducing
phases. 2: A cycle involving an alternation of a
parasitic and a free-living generation; *cf.*
homogonic life cycle.

heterogony 1: Allometric growth *q.v.*;
heterogonic. 2: Cyclic parthenogensis with an
alternation of one or more parthenogenetic
generations with a bisexual (amphigonic)
generation.

heterograft Tissue transplanted from one
organism to another of a different species.

heterometabolic Having a pattern of
development exhibiting an incomplete
metamorphosis.

heteromixis Sexual reproduction in fungi
involving the union of nuclei with different
genetic constitutions, each from a different
fungal thallus; *cf.* homomixis.

heteromorphic Having different forms at
different times or at different stages of the life
cycle; used of a plant having an alternation of
vegetatively dissimilar generations;
heteromorphous, heteromorphy; *cf.*
homomorphic.

heteromorphosis Regeneration of a part
following injury, with a structurally different
replacement part.

Heteromyidae Kangaroo rats, pocket mice;
family containing about 75 species of small
nocturnal terrestrial rodents found in arid
habitats of Nearctic and Neotropical regions;
cheek pouches present; tail elongate;
hindlimbs longer than forelimbs, adapted for
ricochetal locomotion; feed mainly on seeds.

Heteromyota Monotypic order of echiuran
marine worms in which the longitudinal
muscles of the trunk body wall are outside the
circular and oblique layers; up to 400 unpaired
nephridia present.

Heteronematales Order of colourless,
phagotrophic euglenoid flagellates, possessing
a complex ingestion apparatus.

Heteronemertea Order of anoplan nemerteans
with 3 or more layers of body wall muscles and
with the main components of the nervous
system located between the outer longitudinal

and middle circular muscle layers; typically found in marine sediments, beneath boulders or amongst algal holdfasts; Heteronemertini.

heterophagous Feeding on a wide variety of food items; used of a parasite utilizing a wide variety of hosts; **heterophage, heterophagy.**

heterophyte 1: A plant occurring in a wide range of habitats. 2: A dioecious *q.v.* plant. 3: A parasitic plant devoid of chlorophyll.

Heteropneustidae Family of freshwater catfishes (Siluriformes) comprising 2 species, found in stagnant pools and swamps from India to China; body elongate, compressed, to 700 mm in length, head depressed and armoured; dorsal and adipose fin reduced; 4 pairs of barbels present; accessory organ connected to gill chamber for aerial respiration.

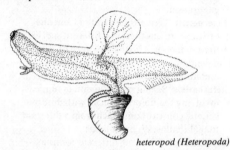
heteropod (Heteropoda)

Heteropoda Heteropods; a group of mainly planktonic marine snails (Mesogastropoda) which are free-swimming predators in near surface oceanic waters; characterized by a reduced or absent shell, a foot modified into a fin for swimming and a powerful proboscis for seizing prey.

Heteroptera The bugs (Insecta) classified as a suborder of the order Hemiptera; now treated as an order under the name Hemiptera *q.v.*

Heterosiphonales Class of Xanthophyta, including forms with a large multicellular thallus attached to the substrate by branched rhizoids; some producing resistant cysts.

heterosis Hybrid vigour; the increased vigour resulting from hybridization when measured against either parental stock.

heterosporous Having two kinds of spores, such as microspores and megaspores, that give rise to distinct male and female gametophyte generations, respectively; **heterosporic, heterospory**; *cf.* homosporous.

Heterostraci Alternative name for the Pteraspidimorphi *q.v.*

heterostyly A polymorphism of flowers in which the length of the style and stamens varies between individuals of a species, helping to prevent self-pollination; *cf.* homostyly.

Heterotardigrada Order of freshwater, marine and terrestrial tardigrades comprising about 175 species; head bearing appendages and lateral cirri; legs often digitate with simple claws or adhesive disks; feed mostly on algae or detritus.

heterothallism A mode of reproduction in fungi and algae which involves the interaction of two thalli, each haploid thallus being self-incompatible; the thalli may exhibit distinct sexual differentiation into male and female (morphological heterothallism) or may be structurally similar (physiological heterothallism); **heterothallic, heterothallous**; *cf.* homothallism.

heterothermic Cold-blooded; **heterothermal.**

heterotopic Occurring in a wide variety of habitats.

Heterotrichales A class of Xanthophyta, including many complex multicellular types, most of which are filamentous.

Heterotrichida Order of spirotrich ciliates containing large, often contractile forms, commonly with a dense, uniform covering of body cilia; found in free-living aquatic or soil habitats and occasionally in symbiotic associations.

heterotrophic 1: Obtaining nourishment from exogenous organic material; used of organisms unable to synthesize organic compounds from inorganic substrates; **heterotroph**; *cf.* autotrophic. 2: Used of plants occurring in a wide range of habitats on a wide variety of soil types, and of protistans utilizing a wide variety of food materials; **heterotrophy.**

heteroxenous Used of a parasite occupying more than one host during its life cycle; **heteroxeny**; *cf.* dixenous, monoxenous, oligoxenous, trixenous.

heterozygote A diploid individual formed by the fusion of gametes carrying different alleles at a given locus and which produces two kinds of gametes differing with respect to the allele at that locus; *cf.* homozygote.

heterozygous Having two different alleles at a given locus of a chromosome pair; **heterozygosis, heterozygosity**; *cf.* homozygous.

Hexacorallia Zoantharia *q.v.*

Hexactinellida Glass sponges; class of sponges with a skeleton composed of 6-rayed siliceous spicules which are sometimes fused into a reticulum; found chiefly in deeper oceans at depths of 200–2000 m but known to penetrate below 6000 m.

Hexactinosida Order of hexasterophoran
sponges characterized by a rigid skeleton
made up of 6-rayed spicules fused together by
secondary deposits of silica enveloping
adjacent rays; found in deep oceans down to a
depth in excess of 4000 m.

Hexagrammidae Greenlings, lingcod; family of
bottom-living North Pacific marine
scorpaeniform teleost fishes; body elongate,
to 1.5 m in length; contains 10 species, some
locally common and including important
commercial and sporting species.

Hexamerocerata Monofamilial suborder of
tropical pauropods comprising 7 species; body
with 11 pedigerous segments, antennae
strongly telescopic, mandibles modified for
grinding, tracheae present; feed on fungi and/
or as predators on other solid food.

Hexanchidae Cow sharks; family containing 6
species of hexanchiform sharks showing many
primitive features; mostly benthic in tropical
and subtropical waters to 2000 m depth,
feeding on fish and crustaceans but also taking
larger prey such as dolphins; body length 2–8
m, dorsal fin tall, mouth large, subterminal.

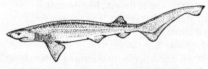

cow shark (Hexanchidae)

Hexanchiformes Small order of medium-sized
(2–8 m) benthic marine squalomorph sharks;
body stout to slender, moderately
compressed, with anal fin and single spineless
dorsal fin; rostrum trough-shaped; notochord
poorly constricted with only weakly calcified
vertebral centra.

Hexapoda An alternative name for the Insecta.

Hexasterophora Subclass of hexactinellid
sponges which are commonly firmly fixed to a
hard substratum, but occasionally rooted by
anchoring spicules; comprises 3 orders,
Hexactinosida, Lychniscosida and
Lyssacinosida; includes Venus' flower basket.

hibernaculum The domicile *q.v.* in which an
animal hibernates or overwinters.

hibernal Pertaining to the winter; *cf.* aestival,
hiemal, serotinal, vernal.

hibernation The act or condition of passing the
winter in a torpid or resting state; *cf.*
aestivation.

hibiscus Malvaceae *q.v.*

hickory Juglandaceae *q.v.*

hidroplankton Planktonic organisms that
achieve buoyancy by means of surface
secretions.

hiemal Pertaining to the winter; *cf.* aestival,
hibernal, serotinal, vernal.

high tide High water; the maximum height of
the incoming tide.

high water (HW) High tide; the highest surface
water level reached by the rising tide.

higher high water (HHW) The higher of two
high waters during any tidal day in areas where
there are marked inequalities of tidal height.

higher low water (HLW) The higher of two low
waters *q.v.* during any tidal day in areas where
there are marked inequalities of tidal height.

Himantandraceae A primitive family of
flowering plants (Magnoliales) consisting of a
single genus of large, aromatic trees native to
New Guinea, the Moluccas and northeast
Australia; flowers with 2 sepals, 4 petals and
numerous stamens and carpels.

Himantolophidae Footballfishes; family
containing 4 species of piscivorous
anglerfishes (Lophiiformes); body globose, to
600 mm length, spinose in male, armoured
with bony plates in female; males free-living
but much smaller than females, teeth absent;
lure very large and branched in female.

Hiodontidae Mooneyes; family of primitive
North American freshwater teleost fishes;
body compressed, herring-like, to about 0.5 m
length; comprises 2 species, popular locally as
game and food-fishes.

hip A type of false fruit (pseudocarp) typical of
roses, in which the receptacle of the flower is
cup-shaped and contains numerous achenes
q.v.

Hippocastanaceae Horse chestnut, buckeye;
family of Sapindales containing only 13 species
of shrubs and trees with palmately compound
leaves and large seeds with a hard coat; flowers
with 5 sepals, 4 or 5 petals, 5–8 stamens and a

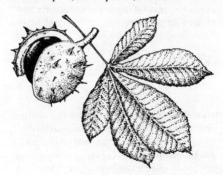

horse chestnut (Hippocastanaceae)

superior ovary, arranged in showy inflorescences.

Hippocrateaceae Family of Celastrales containing about 300 species of woody plants typically with latex canals; widespread in tropical regions.

Hippopotamidae Hippopotamuses; family containing 2 species of large herbivorous, amphibious ungulates (Artiodactyla) found in the Ethiopian region; incisors and canines tusk-like; feet bearing 4 digits, hooves vestigial; stomach 3-chambered; feed on land by night, spending daytime in water.

hippopotamus (Hippopotamidae)

Hipposideridae Old World leaf-nosed bats; family of tropical, insectivorous bats in which the nose has a prominent leaf and the nostrils open at the centre of a disk.

Hippuridaceae Mare's tail; family of Callitrichales containing a single, widely distributed, species of aquatic, perennial, rhizomatous herb producing small wind-pollinated flowers on erect stems.

Hippuritoida Rock oysters; order of marine heterodont bivalve molluscs found widely in tropical waters; typically filter-feeders characterized by unequal shell valves one of which is cemented to the substratum.

Hirudinea Leeches; subclass of clitellate worms (Hirudinoidea) comprising 2 orders, Rhynchobdellae and Arhynchobdellae; typically blood feeders on vertebrate hosts; body highly contractile, composed of 34 segments; several species of veterinary and medical importance.

leech (Hirudinea)

Hirudinidae Swallows; cosmopolitan family containing about 80 species of small swift-flying passerine birds that feed on insects caught on the wing; wings long and pointed, tail commonly forked; legs short; bill small and flattened with broad gape; habits gregarious,

often migratory; nest in crevices, burrows, or in cup-shaped mud nest on rockface.

swallow (Hirudinidae)

Hirudinoidea Class of clitellate worms (Annelida) comprising about 500 species in 3 subclasses, Branchiobdellida, Acanthobdellida and Hirudinea (leeches); body length 3–250 mm, with fixed number of segments each divided into rings (annuli); one or both ends bearing suckers; typically hermaphroditic employing cross fertilization; habits terrestrial, freshwater or marine, predacious or parasitic, mostly on vertebrates.

histology The branch of anatomy dealing with the structure of tissues; **histological.**

histophilous Inhabiting living host tissue; parasitic.

histophyte A parasitic plant.

H-layer The heavily transformed humus layer of a soil with no macroscopic plant remnants and having a strong mineral particle intermix; humified layer; see soil horizons.

hoatzin Opisthocomidae *q.v.*

Holacanthida Order of simple acantharians in which 10 diametral spines are simply crossed in the centre of the organism.

holandric Occurring only in males; used of a character carried on the Y chromosome in the heterogametic male sex; **holandry**; *cf.* hologynic.

Holarctic A zoogeographical region comprising the Palaearctic and Nearctic regions.

holard The total water content of the soil; *cf.* chresard, echard.

holaspis Late larval stage of trilobites in which trunk segmentation was distinct and the body had the basic adult configuration.

Holasteroida Order of irregular echinoids with an elongate apical system; first appearing as fossils in the Lower Cretaceous.

Holectypoida Order of shallow tropical echinoids (Gnathostomata) comprising only 3 extant species, but having rich fossil record; test circular but anus displaced from aboral pole to a position adjacent to the mouth; spines simple, erect and numerous.

holendophyte A parasitic plant which passes its entire life within its host.

holeuryhaline Used of organisms that freely inhabit fresh water, sea water and brackish water; *cf.* euryhaline, oligohaline, polystenohaline, stenohaline.

holly Aquifoliaceae *q.v.*

hollyhock Malvaceae *q.v.*

Holocene A geological epoch within the Quaternary period (*c.* 10 000 years B.P. to the present time); Recent; see geological time scale.

Holocentridae Squirrelfishes, soldierfishes; family containing 70 species of tropical marine beryciform teleosts, nocturnally active on rocky bottoms and reefs; body elongate, compressed, scales ctenoid; eyes large, teeth minute.

Holocephalii Subclass of cartilaginous fishes (Chondrichthyes) comprising the single order Chimaeriformes *q.v.*

holocyclic parthenogenesis Reproduction by a series of parthenogenetic generations alternating with a single sexually reproducing generation; *cf.* parthenogenesis.

hologamodeme A local interbreeding population comprising all those individuals which are able to interbreed with a high level of freedom; *cf.* deme.

hologamy A mode of reproduction in protistans involving gametes similar in size to the vegetative cell; *cf.* merogamy.

hologynic Occurring only in females; used of sex-linked characters in species with a heterogametic female sex; **hologyny**; *cf.* holandric.

Holometabola The largest superorder of neopteran insects comprising 9 orders, Neuroptera (alderflies, lacewings), Coleoptera (beetles), Strepsiptera, Hymenoptera (wasps, bees), Mecoptera (scorpion flies), Siphonaptera (fleas), Diptera (true flies), Trichoptera (caddis flies) and Lepidoptera (butterflies, moths); morphology exceptionally diverse; wings paired, or one pair specialized as balance organs, or wings absent; mouthparts mandibulate; metamorphosis complete, life cycle comprising distinct, larval and pupal stages; Neuropterodea; Endopterygota.

holometabolous Used of the pattern of development which is characterized by a distinct metamorphosis with well defined larval and adult stages; **holometaboly**; *cf.* ametabolous, hemimetabolous.

holomictic Used of a lake having complete free circulation throughout the water column at the time of winter cooling; *cf.* mictic.

holoparasite A parasite that cannot survive away from its host; an obligate parasite; *cf.* meroparasite.

holopelagic Used of aquatic organisms that remain pelagic throughout the life cycle; *cf.* meropelagic.

holophytic Synthesizing organic compounds from inorganic substrates in the manner of a green plant; photosynthetic; **holophyte**; *cf.* holozoic.

holoplankton Those organisms that are permanent members of the plankton; **holoplanktonic**; *cf.* meroplankton.

holosaprophyte A plant deriving nourishment entirely from decomposing organic matter; obligate saprophyte.

Holostei Loose assemblage of primitive bony fishes (Osteichthyes) comprising the Amiiformes (bowfin) and the Semionotiformes (gars).

holostylic The condition in which the hyomandibular plays no part in the suspension of the lower jaw in fishes, since the palatoquadrate arch is fused with the cranium; *cf.* amphistylic, autostylic, hyostylic.

Holothuroidea Sea cucumbers; class of echinozoan echinoderms in which the axis of radial pentamerous symmetry is horizontal; the test is typically soft with microscopic skeletal ossicles, usually sausage-shaped but occasionally spherical or flattened; mouth at oral pole surrounded by up to 30 tentacles, anus at opposite, aboral, end; sexes separate, fertilization external, hermaphroditism and viviparity rare; development may incorporate planktonic auricularia or vitellaria larvae; sea cucumbers are widespread, epifaunal to infaunal, littoral to hadal, and in the deep sea may comprise large part of benthic biomass; 3 subclasses recognized, Dendrochirotacea, Aspidochirotacea and Apodacea.

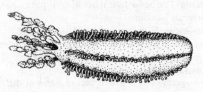

sea cucumber (Holothuroidea)

Holotrichia Diverse assemblage of ciliates, formerly classified as a subclass but not now

used; most are included in the
Kinetofragminophora *q.v.*

holotrophic 1: Used of a plant which synthesizes
all organic compounds from inorganic
substrates, typically by photosynthesis. 2:
Used of a predator that utilizes only one
species of prey organism.

holozoic Feeding entirely in the manner of an
animal by ingesting complex organic matter;
cf. holophytic.

Holsteinian interglacial An interglacial period
in the middle of the Quaternary Ice Age *q.v.*
in northern Germany and Poland.

Homalopteridae Hillstream loaches; family
containing 85 species of mostly small (to 130
mm) cypriniform teleost fishes found in fast
flowing hill streams of India and southeast
Asia; body anteriorly depressed, 3 or 4 pairs
of barbels present; paired fins may form
sucker; gas bladder partly enclosed in bony
capsule; includes Gastromyzontidae
sometimes treated as a distinct family.

Homalorhagida Order of kinorhynchs found
mostly in marine muddy sediments from the
subtidal down to several thousand metres;
characterized by a neck consisting of 6–8
plates, by the cavity of the pharynx which is
triangular in cross-section and by the lack of
articulated spines on the trunk.

Homalozoa Extinct subphylum of echinoderms
that exhibit no trace of radial symmetry;
known from the Cambrian to the Devonian;
Homoalozoa.

home That part of the habitat of an organism
that is utilized for resting and breeding; *cf.*
hotel.

home range The area, usually around the
domicile, over which an animal normally
travels in search of food; *cf.* territory.

home site The location of the domicile or resting
place regularly used by a particular animal.

homeostasis The maintenance of a relatively
steady state or equilibrium in a biological
system by intrinsic regulatory mechanisms;
homeostatic.

homing The behavioural act of returning to an
original location.

Hominidae Humans; family of omnivorous
terrestrial primates comprising a single Recent
species, *Homo sapiens*, and the extinct
australopithecines.

homiogamy The union of two gametes of the
same sex; **homiogamic.**

homobium The symbiosis between a fungus and
an alga in a lichen.

homochronous Occurring at the same time, or

age in successive generations; simultaneous.

homoecious 1: Used of a parasite occupying the
same host throughout its life cycle; *cf.*
heteroecious. 2: Host-specific.

homoeosis The modification of one segmental
appendage or structure to resemble a different
appendage or structure of the same
homologous series.

homogamete The gamete produced by the
homogametic sex *q.v.*

homogametic Having only one kind of gamete;
homogamy; *cf.* heterogametic.

homogametic sex The sex which is determined
by a pair of similar sex chromosomes (X); that
sex which produces only one type of gamete;
cf. heterogametic sex.

homogamy 1: The simultaneous maturation of
male and female reproductive organs in a
flower or hermaphrodite organism. 2: The
condition of a plant producing only one kind
of flower; *cf.* heterogamy.

homogeneous Similar throughout; of uniform
structure or composition; **homogeneity**; *cf.*
heterogeneous.

homogenesis Non-alternation of generations;
the succession of morphologically similar
generations; **homogenetic**; *q.v.* heterogenesis.

homogonic life cycle A life cycle in which all
generations are either parasitic or free-living;
cf. heterogonic life cycle.

homoiosmotic Used of organisms that maintain
a relatively constant internal osmotic pressure;
cf. poikilosmotic.

homoiothermic Used of animals that regulate
their body temperature independent of
ambient temperature fluctuations; warm-
blooded; **homoiothermy**; *cf.* poikilothermic.

homologous Used of structures or traits having
common ancestry but not necessarily retaining
similarity of structure, function or behaviour;
homology; *cf.* analogous.

homologous chromosomes Structurally similar
chromosomes having identical genetic loci in
the same sequence and which pair during
nuclear division.

homolotropism Growth orientation in a
horizontal direction; **homolotropic.**

homomixis Sexual reproduction in fungi
involving the union of genetically similar
nuclei from one thallus; *cf.* heteromixis.

homomorphic Used of a plant with a
dimorphous life cycle in which two different
types of individuals in the life cycle are
morphologically similar; isomorphic;
homomorphous, homomorphy; *cf.*
heteromorphic.

homoplasy Structural resemblance due to parallelism or convergent evolution rather than to common ancestry; **homoplasia.**

Homoptera Order of phytophagous hemipterodean insects comprising about 45 000 species including cicadas, plant/frog/leaf/tree hoppers, psyllids, whiteflies, aphids, coccoids, scale insects and mealybugs; often classified as a subgroup of Hemiptera; hindlegs commonly saltatory; mouthparts modified for piercing and sucking; many species are plant pests or disease vectors.

leafhopper (Homoptera)

Homoscleromorpha Primitive subclass of demosponges found in shallow marine waters from the intertidal zone down to 500 m; characterized by a skeleton of equal-rayed, 4-axonid siliceous spicules, and by viviparity as they produce an amphiblastula larva; comprises a single order, Homosclerophorida.

homosporous Having only one kind of spore; isosporous; **homospory;** *cf.* heterosporous.

homostyly The condition in which all the flowers of a species have styles of similar length; *cf.* heterostyly.

homothallism A mode of reproduction in fungi and algae in which each thallus produces both male and female sex cells, and is self-fertile; **homothallous;** *cf.* heterothallism.

homotropic Used of a flower fertilized by its own pollen.

homozygote An individual with identical alleles at the two homologous loci of a chromosome pair; *cf.* heterozygote.

homozygous Having identical alleles at a given locus of a chromosome pair; *cf.* heterozygous.

honesty Brassicaceae *q.v.*

honey bee Apidae *q.v.*

honey locust Caesalpiniaceae *q.v.*

honey possum Tarsipedidae *q.v.*

honeycreeper Drepanididae *q.v.*

honeyeater Meliphagidae *q.v.*

honeyguide Indicatoridae *q.v.*

honeysuckle Caprifoliaceae *q.v.*

honing The process of tooth sharpening.

hook order A social dominance hierarchy found in horned mammals established by the aggressive use of horns.

Hookeriales Order of mosses (Bryidae) occurring mostly in southern hemisphere and tropics; typically yellowish-green, often shiny, mat-forming perennial plants found in moist shady forests.

hoopoe Upupidae *q.v.*

hop Cannabaceae *q.v.*

Hoplestigmataceae Small family of Violales containing only 2 species of trees, native to tropical Africa.

Hoplichthyidae Ghost flathead; family of little-known Indo-Pacific deep-water scorpaeniform teleost fishes; body elongate, head broad and depressed, naked.

Hoplocarida Mantis shrimps; subclass of large, predatory marine malacostracan crustaceans comprising a single extant order, Stomatopoda.

Hoplonemertea Order of enoplan nemerteans in which the proboscis is regionally differentiated and armed with one or more stylets; typically found free-living in marine habitats but some freshwater and terrestrial representatives are known.

Horaichthyidae Monotypic family of tiny (to 30 mm) freshwater cyprinodontiform teleost fishes from India; body translucent, anal fin in male modified as gonopodium for sperm transfer; right pelvic fin absent in female; reproduction oviparous, fertilization internal.

horehound Lamiaceae *q.v.*

horizon Any horizontal stratum within a sediment, soil profile, water column or geological series.

hormone A chemical substance secreted by an endocrine gland, that produces a specific physiological response on some remote target tissue.

hormone weedkillers Synthetic organic herbicides that are similar in effect to natural plant hormones (auxins), promoting, retarding or modifying the growth and development of plants.

hornbeam Corylaceae *q.v.*

hornbill Bucerotidae *q.v.*

horned pondweed Zannichelliaceae *q.v.*

horntail Siricidae *q.v.* (Symphyta).

hornwort Anthocerotopsida *q.v.*

horse Equidae *q.v.*

horse botfly Gasterophilidae *q.v.*

horse chestnut Hippocastanaceae *q.v.*

horse radish Brassicaceae *q.v.*

horsefly Tabanidae *q.v.*

horsehair worm Nematomorpha *q.v.*

horseshoe bat Rhinolophidae *q.v.*

horseshoe crab Xiphosura *q.v.*

horseshoe worm Phoronida *q.v.*

horsetail Equisetophyta *q.v.*

hortal Used of an ornamental plant that has escaped from cultivation and can be found growing wild.

hospitating Offering refuge or acting as a host (hospitator) to another organism (the hospite).

host 1: Any organism that provides food or shelter for another organism; may be a definitive host (infected by the mature adult stage) or an intermediate host (infected by developmental stages). 2: An animal that is the recipient of a tissue graft.

host specificity The extent to which an adult parasite is restricted in the variety of host species utilized.

hot water plant Gesneriaceae *q.v.*

hotel That part of the habitat of an organism that is used for resting, breeding and feeding; *cf.* home.

hottentot fig Aizoaceae *q.v.*

hound shark Triakidae *q.v.*

house spider Agelenidae *q.v.*

housefly Muscidae *q.v.*

houseleek Crassulaceae *q.v.*

hover fly Syrphidae *q.v.*

howler monkey Cebidae *q.v.*

Hoxnian interglacial An interglacial period in the middle of the Quaternary Ice Age *q.v.* in the British Isles.

Huaceae Small family of Violales native to tropical Africa; comprises only 3 species of woody plants, each with a characteristic garlic-like odour.

Hugoniaceae Family of Linales containing about 60 species of woody plants widespread in tropical regions.

Humboldt Current A cold surface ocean current that flows northwards off the Pacific coast of South America, arising from the Antarctic Circumpolar Current and forming part of the South Pacific Gyre; see ocean currents.

Humic Gley soil An intrazonal soil with a thick peaty surface layer over an underlying gley horizon, formed in forest swamps and wet meadows.

humic lake A lake rich in organic matter mainly in the form of suspended plant colloids and larger plant fragments but having low nutrient content.

humicolous Living on or in the soil; **humicole.**

humidity The amount of water vapour in the atmosphere.

humification The transformation of dead plant material into humus.

Humiriaceae Small family of Linales containing about 50 species of evergreen trees and shrubs often with an aromatic juice; found mainly in tropical America.

hummingbird Trochilidae *q.v.*

humpback whale Balaenopteridae *q.v.*

humphead Kurtidae *q.v.*

humus A dark brown amorphous material formed by the partial decomposition of plant and animal remains; representing the organic constituent of soil although also found in colloidal suspension in water.

hunting spider Lycosidae *q.v.*

husbandry The cultivation, production and management of plants and animals; used especially with reference to domesticated species.

hutia Capromyidae *q.v.*

hyacinth Liliaceae *q.v.*

Hyaenidae Hyenas; small family of dog-like terrestrial mammals (Carnivora) comprising 4 species found in Ethiopian, Oriental and southern Palaearctic regions; forelimbs longer than hindlimbs; feed mainly on carrion using powerful jaws.

Hyaenodontidae Extinct family of carnivorous mammals known from most of the Tertiary period; slimly built, with elongated jaws.

hybrid Offspring of a cross between genetically dissimilar individuals; in taxonomy, often restricted to the offspring of interspecific crosses.

hybridization Any crossing of individuals of different genetic composition, typically belonging to separate species, resulting in hybrid offspring.

capybara (Hydrochoeridae)

Hydatellales Order of Commelinidae comprising a single family of only 5 species of small submerged or emergent aquatic annuals; with naked, unisexual flowers pollinated by water or self-pollinated; native to Australia and New Zealand.

Hydnoraceae Family of Rafflesiales containing only 10 species of leafless herbs lacking chlorophyll which parasitize the roots of other plants and produce large, fleshy and malodorous flowers at the soil surface.

Hydrangeaceae Hydrangea; family of Rosales containing about 170 species of mostly woody plants with large flowers, widespread in temperate and subtropical parts of the northern hemisphere; flowers have 4–10 sepals and petals, 4 or more stamens, a superior ovary and are arranged in cymose inflorescences.

hydrarch succession An ecological succession commencing in a habitat with abundant water; hydrosere; *cf.* mesarch succession, xerarch succession.

hydroanemophilous Having airborne spores that are discharged after wetting of the spore-producing structure; **hydroanemophily.**

Hydrobatidae Storm petrels; cosmopolitan family containing 21 species of oceanic sea birds (Procellariiformes); feed mainly on fishes and crustaceans; breed in colonies on islands, nesting in burrows or crevices; single egg per clutch.

hydrobiology The study of life in aquatic habitats.

hydrobios The sum total of all aquatic life; that part of the Earth's surface occupied by aquatic organisms; *cf.* geobios, halobios, limnobios.

hydrocarpic Used of aquatic plants that are pollinated above water but develop below the surface.

hydrochamaephyte An aquatic plant producing renewal buds up to 250 mm from the lake floor; an aquatic chamaephyte.

Hydrocharitales Elodea, turtle grass; order of Alismatidae comprising a single cosmopolitan family of about 100 species of submersed, partly emergent or free-floating perennial herbs found in freshwater or coastal marine habitats; lacking vessels; flowers with 3 sepals and petals.

hydrochimous Pertaining to wet winters.

Hydrochoeridae Capybaras; family containing 2 species of large semiaquatic Neotropical hystricomorph rodents; body robust, head large, tail vestigial; feet webbed with strong claws; habits diurnal, herbivorous; capybaras

are the largest living rodents, weighing up to 50 kg.

hydrochoric Dispersed by the agency of water; **hydrochory.**

hydrochthophyte A plant growing normally in water and producing emergent leaves or shoots.

hydrocleistogamy Self-pollination in a flower that remains closed because it is submerged; **hydrocleistogamic.**

hydrocolous Living in an aquatic habitat; **hydrocole.**

Hydrocorallina Diverse group comprising the milleporine and stylasterine corals; hydrocorals.

hydrogenic soil A soil formed under waterlogged conditions.

hydrogeophyte An aquatic plant producing renewal buds on a buried rhizome; an aquatic geophyte.

hydrohemicryptophyte An aquatic plant producing renewal buds at the water/sediment interface; an aquatic hemicryptophyte.

Hydroida Order of Hydrozoa in which the polyp generation is dominant; most polyps are colonial, connected by a branching stolon system which extends over the substrate; polyps usually have a non-living chitinous exoskeleton, mouth generally surrounded by tentacles; polyps may be polymorphic with feeding gastrozoids and reproductive gonozoids which bud off free medusae; includes some freshwater forms, such as *Hydra*, but most are marine.

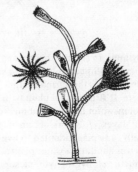

hydrozoan polyps (Hydroida)

hydrologic cycle Water cycle *q.v.*

hydromegathermic Used of organisms inhabiting warm wet environments, such as tropical rainforests; **hydromegatherm.**

Hydrometridae Water measurer; cosmopolitan family containing about 100 species of slender, stick-like, hemipteran insects found crawling among aquatic vegetation or walking on the surface film of water.

Hydromyxomycetes Alternative name for the Labyrinthulata (slime nets).

hydronasty Orientation movement in plants in response to changes in atmospheric humidity; **hydronastic.**

Hydrophiidae Sea snakes; family of venomous marine snakes (Serpentes) with laterally compressed tails, valvular nostrils and salt-secreting glands in the head.

hydrophilous 1: Thriving in wet or aquatic habitats; *cf.* hydrophobic. 2: Pollinated by waterborne pollen; **hydrophile, hydrophily.**

hydrophobic Intolerant of water or wet conditions; water-repellent; **hydrophobe, hydrophoby;** *cf.* hydrophilous.

Hydrophyllaceae Virginian waterleaf, Californian bluebell; family of Solanales containing about 250 species of mostly rough-hairy, herbs, common in dry habitats in western United States; flowers regular with 5 sepals, 5 blue or purple petals, normally 5 stamens and a superior ovary.

baby blue eyes (Hydrophyllaceae)

hydrophyllium In colonial hydrozoan coelenterates (Siphonophora), the shield-like zooid that covers the cormidium.

hydrophyte 1: A perennial plant with renewal buds below water and with submerged or floating leaves; *cf.* Raunkiaerian life forms. 2: Generally, any plant adapted to live in water or very wet habitats; a water plant; **hydrophytic;** *cf.* hygrophyte, mesophyte, xerophyte.

hydroponics A system of plant culture in which the growing plants have their roots immersed in a nutrient rich solution or in an inert substrate which is irrigated with nutrients.

hydrosere An ecological succession commencing in a habitat with abundant water, typically on the submerged sediments of a standing water body.

hydrosphere The global water mass, including atmospheric, surface and subsurface waters.

Hydrostachyaceae Family of about 20 species of submerged aquatic perennial herbs native to Africa and Madagascar.

hydrostatic pressure The pressure exerted by a column of water; pressure increases by about 1 atmosphere per 10 m of depth down a water column.

hydrotaxis The directed response of a motile organism towards (positive) or away from (negative) a water or moisture stimulus; **hydrotactic.**

hydrothermal Pertaining to hot water; used especially of hot water springs or ocean floor vents.

hydrotribophilous Thriving in badlands; **hydrotribophile, hydrotribophily.**

hydrotribophyte A badlands plant.

hydrotropic Exhibiting a trend towards greater water content in an ecological succession.

hydrotropism Orientation in response to water; **hydrotropic.**

Hydrozoa Class of marine and occasionally freshwater cnidarians in which the life cycle typically involves alternation between an asexually reproducing, attached polyp and a sexually reproducing, planktonic medusa; medusae usually small and transparent; polyps mostly colonial and arising from stolons, typically with a chitinous, or sometimes calcified, exoskeleton, often polymorphic with individual polyps specialized for different roles; comprises 7 orders.

hygric Pertaining to moisture, or to moist or humid conditions.

Hygrobiidae Screech beetles; family of primitive water beetles (Coleoptera) found in muddy pools; bodies highly convex, eyes prominent; swimming by alternate leg movements.

hygrochastic Used of a fruit in which dehiscence is induced by moisture; **hygrochasy;** *cf.* xerochastic.

hygrocolous Living in moist or damp habitats; **hygrocole.**

hygrokinesis A change in the rate of movement of an organism in response to a change in humidity; **hygrokinetic.**

hygroklinokinesis A change in the rate of random movement of an organism expressed as a change in the frequency of turning movements (rate of change of direction) in response to a humidity stimulus; **hygroklinokinetic.**

hygroorthokinesis A change in the rate of

random movement of an organism expressed as a change in the linear velocity in response to a humidity stimulus; **hygroorthokinetic.**

hygropetric Used of an organism living in the surface film of water on rocks.

hygrophilous Thriving in moist habitats; **hygrophile, hygrophily**; *cf.* hygrophobic.

hygrophobic Intolerant of moist conditions; **hygrophobe, hygrophoby**; *cf.* hygrophilous.

hygrophyte A plant living in a wet or moist habitat, typically lacking xeromorphic features; **hygrophytic**; *cf.* hydrophyte, mesophyte, xerophyte.

hygroscopic Readily absorbing and retaining moisture from the atmosphere; **hygroscopicity.**

hygrotaxis The directed response of a motile organism to moisture; **hygrotactic.**

hygrotropism Orientation in response to humidity or moisture; **hygrotropic.**

hylacolous Living among trees, or in woodland or forest; **hylacole.**

Hylidae Diverse family containing about 400 species of New World tree frogs (Anura); typically arboreal, few species terrestrial and fossorial, eggs and larvae usually aquatic; some species lay eggs on vegetation over water, others carry eggs in pits on back of female and have direct or attenuated development; widespread in tropical and temperate Americas and West Indies, with small group of species in tropical Eurasia.

gibbon (Hylobatidae)

hylophilous Living or thriving in forests; **hylophile, hylophily.**

hylotomous Used of wood-cutting insects.

Hymenogastrales Small order of gasteromycete fungi; often in mycorrhizal association with roots of green plants but sometimes found growing with algae on damp soil; produce underground fruiting bodies.

Hymenomycetes Class of basidiomycotine fungi in which the basidia are organized into a fertile layer (hymenium) that is exposed at maturity; comprises 2 orders, Exobasidiales and Agaricales.

craterelle (Hymenomycetes)

Hymenoptera Diverse order of holometabolous insects comprising 130 000 species in 2 suborders, Symphyta and Apocrita, including sawflies, horntails, wasps, bees, ants, and many other groups; adults mandibulate, typically with 2 pairs of membranous wings; hindwings small, but may be vestigial or absent; larval morphology and habits very diverse; sociality widely exhibited.

Hymenostomata Subclass of Oligohymenophora, most of which occur as free-living freshwater forms, but some are marine and a few endosymbiontic; possessing compound ciliary organelles in the buccal area.

Hymenostomatida Order of hymenostome ciliates occurring mainly in freshwater and soil

tree frog (Hylidae)

Hylobatidae Gibbons; family containing 6 species of arboreal primates (Catarrhini) found in southeast Asia and Indonesia; limbs elongate, forelimbs longer than hindlimbs, adapted for brachiation; tail absent; ischial callosities present; feed on variety of plant and animal material.

hylodophilous Thriving in dry open woodland; **hylodophile, hylodophily.**

hylodophyte A dry open woodland plant.

hylophagous Feeding on wood; **hylophage, hylophagy.**

habitats, typically free-living but some symbiotic mainly with invertebrate hosts; includes *Paramecium* and *Tetrahymena*.

Hynobiidae Family containing about 30 species of small primitive Asiatic salamanders; fore and hindlimbs present; metamorphosis complete; adults typically terrestrial, larvae aquatic with external gills and caudal fins; fertilization external, spermatophores absent.

Hyolithida Extinct order of molluscs (Calyptoptomatida) with pyramidal shells; known from the Cambrian to the Permian.

hyostylic The condition in which the hyomandibular is the principal suspensional element for the lower jaw in fishes; *cf.* amphistylic, autostylic, holostylic.

hyperbenthic Living above but close to the substratum; **hyperbenthos**; *cf.* endobenthic, epibenthic.

Hyperiidea Suborder of exclusively marine pelagic amphipod crustaceans commonly found in association with gelatinous zooplankton such as salps, medusae and siphonophores; body elongate or globular; head large, typically with massive compound eyes.

Hypermastigida Order of large zooflagellates with numerous flagella, often arranged in a characteristic pattern, and a single nucleus; all are anaerobic and inhabit the hindguts of cockroaches and termites where they play an important role in the digestion of wood.

hypermetamorphic Used of an organism having two or more distinct metamorphoses in the life cycle or having an extensive metamorphosis; **hypermetamorphosis.**

Hyperoliidae Family of mostly small (to 50 mm) frogs (Anura) from Africa, Madagascar and Seychelles Islands; eggs typically laid on vegetation over still water; tadpoles aquatic.

Hyperotreti Myxiniformes *q.v.*

hyperparasite An organism parasitic upon another parasite; superparasite; **hyperparasite, hyperparasitism.**

hyperplasia Excessive growth due to increase in cell number; **hyperplastic**; *cf.* hypertrophy.

hypersaline Having a high salinity, well in excess of normal sea water, typical of enclosed bodies of water with high evaporation rates.

hypertrophication Over-enrichment with nutrients; **hypertrophicated.**

hypertrophy Excessive growth due to increase in cell size; *cf.* hyperplasia.

Hyphochytridiomycetes Class of microscopic true fungi which parasitize algae and aquatic fungi or live as saprobes on plant debris; characterized by asexual reproduction involving the production of motile zoospores possessing an anterior tinsel-type flagellum; thallus unicellular or filamentous; hyphae sometimes with cross walls; also classified as the phylum Hyphochytridiomycota of the Protoctista.

Hyphochytridiomycota Hyphochytridiomycetes *q.v.*; treated as a phylum of Protoctista.

Hyphomycetales Large order of hyphomycete fungi containing over 6000 species of mostly saprophytic forms, sometimes parasitic on plants and animals; vegetative and fertile hyphae are septate and loosely arranged at the surface of the colony, not forming an organized fruiting body.

Hyphomycetes An artificial class of deuteromycotine fungi containing the asexual states of ascomycotine and basidiomycotine fungi in which the reproductive hyphae and the propagative spores (conidia) are not formed inside an enclosed structure; comprises 4 orders, Hyphomycetales, Stilbellales, Tuberculariales and Agonomycetales.

hyphydrogamous Having waterborne pollen transported below the water surface; *cf.* ephydrogamous.

Hypnidae Monotypic family of shallow-water torpediniform fishes from Australia; body disk broader than long containing very powerful electric organs, tail short bearing 2 small dorsal fins; pelvics enlarged and united ventrally; mouth large, teeth tricusped.

Hypnobryales Large order of mosses (Bryidae); typically large perennial, mat-forming plants growing on soil, logs, humus and rocks, commonly found in forested habitats in the northern hemisphere; stems prostrate to erect, variously branched, spore capsule typically asymmetrical and inclined on long stalk.

hypnody A prolonged resting period (diapause) during development; **hypnodic.**

hypnoplasy Arrested development resulting in failure to reach normal size.

hypnosis A state of dormancy in seeds that retain the capacity for normal development; **hypnotic.**

hypobiosis The condition in which only minimal outward signs of metabolic activity are present in a dormant organism.

hypobiotic Living in sheltered microhabitats.

hypocarpogean Producing subterranean fruit; **hypocarpogenous**; *cf.* amphicarpogean.

Hypocreales Large order of pyrenomycete fungi containing many pathogens, including apple canker and pea and bean wilt, some saprobic forms and some parasites of other fungi; typically with persistent asci and forcibly discharged ascospores; ascocarps usually brightly coloured, often fleshy.

hypogean Living or germinating underground; **hypogeous.**

hypogenous Living or growing underneath an object or on its lower surface.

hypogyny The typical arrangement of the parts of a flower in which the sepals, petals and stamens are inserted below the ovary (a superior ovary); **hypogynous**; *cf.* epigyny, perigyny.

hypolimnion The cold bottom water zone below the thermocline in a lake; **hypolimnetic**; *cf.* epilimnion.

hypolithic Living beneath rocks or stones.

hyponasty Upward growth; **hyponastic.**

hyponeuston Organisms living immediately below the surface film of a water body; a component of the neuston; **hyponeustonic**; *cf.* epineuston.

hypophloeodal Living or occurring immediately below the surface of bark; **hypophloeodic.**

Hypophthalmidae Monotypic family of large planktotrophic tropical South American freshwater catfishes (Siluriformes); body naked, dorsal fin with stout spine, anal fin elongate; 3 pairs of barbels present; gill rakers long and complex.

hypophyllous Growing on the lower surface of leaves.

Hypoptychidae Sand eel; monotypic family of small (to 80 mm) shallow marine gasterosteiform teleost fishes found in the western North Pacific; body elongate, upper jaw protractile, scales and pelvic fins absent.

hyporheic Pertaining to saturated sediments beneath or beside streams and rivers.

Hyposittidae Coral-billed nuthatch; family of passeriform birds comprising a single species from Madagascar; bill stout, hook-tipped; feeds on insects gleaned from trees.

Hypostomata Subclass of kinetofragminophoran ciliates characterized

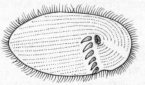

ciliate (Hypostomata)

by the basically ventral position of the cytostome and its associated cytopharyngeal basket; contains free-living, microphagous or algivorous forms as well as many commensals and parasites.

hypothermia The condition of a homoiothermic animal which has a lower than normal temperature; **hypothermic.**

Hypotremata Rays; an older name used either for the order Rajiformes or for the superorder Batoidea.

Hypotrichida Order of dorsoventrally flattened spirotrich ciliates which exhibit a unique type of creeping-darting locomotion; mostly free-living in various habitats but a few are commensals.

hypsilophodont A bipedal ornithischian dinosaur known from the Upper Triassic to the Upper Cretaceous.

Hypsithermal period A postglacial interval (*c.* 9000–2500 years B.P.) during which mean temperatures were above those existing at the present time; equivalent to Boreal, Atlantic and Subboreal periods combined; *cf.* Xerothermal period.

Hyracodontidae Extinct family of long-legged, browsing rhinoceroses known from the Eocene to the Miocene; teeth similar to modern rhinoceroses; 3 toes on each foot.

Hyracoidea Order of herbivorous mammals (Ferungulata) comprising a single extant family, Procaviidae (hyraxes).

hyrax Procaviidae *q.v.*

Hysteriales Order of loculoedaphomycetid fungi containing saprobic, epiphytic, parasitic or lichenized forms; typically apothecioid with bitunicate asci.

hysterophyte A plant obtaining nourishment from dead or decaying organic matter; saprophyte; **hysterophytic.**

Hystricidae Old World porcupines; family containing 15 species of small to large herbivorous terrestrial hystricomorph rodents in which the body is covered by stiff unbarbed spines; found primarily in the Oriental and Ethiopian regions; habits nocturnal, feeding on roots, bulbs and fruit.

Hystricomorpha Diverse suborder of rodents comprising 17 families and including porcupines, guinea-pigs, viscachas and chinchillas; widespread in Neotropical region but also found throughout the Old World.

i

ibis Threskiornithidae *q.v.*

Icacinaceae Family of Celastrales containing about 400 species of woody plants with a mainly tropical distribution; flowers typically unisexual, fruit commonly a drupe.

icefish Salangidae *q.v.*

Icelidae Family containing 5 species of small (to 100 mm) Arctic–boreal shallow marine sculpins (Scorpaeniformes); body slender; gill cover membrane united below the throat.

ichneumon fly Ichneumonidae *q.v.*

Ichneumonidae Ichneumon flies; diverse family containing about 15 000 species of usually slender parasitic wasps (Hymenoptera) in which the female often has a conspicuous elongate ovipositor; eggs typically deposited on the larvae or pupae of holometabolous insects, but also on spiders or their eggs, on which the ichneumon larvae feed as internal or external parasitoids.

ichneumon fly (Ichneumonidae)

ichnite A fossil track or footprint; ichnolite.

ichnocoenosis An assemblage of trace fossils; **ichnocoenose.**

ichnofauna The animal traces of an area.

ichnoflora The plant traces of an area.

ichnofossil A trace fossil.

ichnology The study of traces made by organisms, both fossil and recent.

ichthyofauna The fish fauna of a given region.

ichthyology The study of fishes; **ichthyological.**

ichthyoneuston The ichthyological component of the neuston, typically fish eggs or fry; **ichthyoneustont.**

ichthyophagous Feeding on fishes; piscivorous; **ichthyophage, ichthyophagy.**

Ichthyophiidae Family of terrestrial Asiatic caecilians (Gymnophiona); body length to 500 mm, scales reduced, may be absent anteriorly; small tail present; larvae aquatic with branched gills; feed at night on worms and arthropods; contains about 30 species found in southeast Asia, from India to Borneo.

Ichthyosauria Order of typically shark-like aquatic reptiles known from the Triassic to the Cretaceous.

ichthyosaur (Ichthyosauria)

Icosteidae Ragfish; monotypic family of bathypelagic perciform teleost fishes from the North Pacific; body slender, to 2 m in length, strongly compressed, naked; pelvic fins absent in adults; skeleton mostly cartilaginous.

Ictaluridae Family containing about 40 species of small to medium-sized (to 1.6 m) freshwater catfishes (Siluriformes) native to eastern North America; body smooth, dorsal fin short with stout spine; some have venom gland at base of pectoral fin; 4 barbels present around mouth; adipose fin large; some species important in pond aquaculture.

Icteridae Blackbirds, orioles, cowbirds; diverse New World family containing about 95 species of small to large passerines found in a variety of forest, grassland and open arid habitats; plumage black or colourful, patterns bold or cryptic; bill short to long, slender to robust, with broad gape; habits solitary to gregarious, arboreal to terrestrial, monogamous to polygamous; feed on invertebrates, seeds, fruit, nectar; nest colonial to solitary, on or off ground.

identical twins Monozygotic *q.v.* twins.

Idiacanthidae Black dragonfishes; family containing 3 species of luminescent deep-sea stomiiform teleost fishes; body eel-like, naked, dorsal and anal fins long, pectorals absent, pelvics absent in male only; sexual dimorphism marked, female body length 400 mm, male 70 mm; larvae transparent with eyes on long stalks.

idiobiology That aspect of biology concerned with study of individual organisms.

idiochromosome A sex chromosome.

idiogamy Self-fertilization; **idiogamous.**

Idiospermaceae Family of evergreen trees (Laurales) comprising a single species native to rainforests of northeast Australia; possesses a massive embryo with 3 or 4 cotyledons.

idiotype The total hereditary determinants of an individual, comprising both chromosomal and extrachromosomal factors.

idiotypic Sexual; pertaining to sex or sexual reproduction.

igneous rock Rock formed by solidification of molten magma; *cf.* metamorphic rock, sedimentary rock.

Iguanidae Iguanas; large and extremely diverse family of mainly New World arboreal and terrestrial lizards (Sauria); length to 2 m, body usually compressed; tail autotomic, non-prehensile; coloration commonly different between males and females; habits usually diurnal, insectivorous, occasionally herbivorous; reproduction oviparous; contains about 600 species found in Central and South America, Caribbean, Galapagos Islands, with a few representatives in other Pacific Islands and Madagascar.

iguana (Iguanidae)

Iguanodont A bipedal ornithischian dinosaur known from the Jurassic to the Cretaceous.

Illiciaceae Star anise; small family of glabrous, aromatic evergreen shrubs or trees which accumulate a toxic lactone; contains about 40 species native to southeast Asia and the Caribbean; usually with solitary, axillary flowers which are bisexual, with 7 or more perianth segments, 4 or more stamens and superior ovaries.

Illiciales Small order of woody plants (Magnoliidae) with mostly solitary flowers and poorly differentiated sepals and petals; pollen grains unusual in possessing either 3 or 6 germinal furrows.

Illinoian glaciation A glaciation of the Quaternary Ice Age *q.v.* in North America with an estimated duration of 100 000 years.

illuvial layer A zone of illuviation; the B-horizon of a soil profile.

illuviation The process of deposition and precipitation of material in the B-horizon, that was leached or eluviated from the A-horizon of a soil.

imagination The final stage of development that produces the imago (adult).

imago The adult stage of an insect; **imagine**; *cf.* instar.

imbibition The non-active absorption or uptake of water; **imbibe.**

imbibitional water That part of the soil water held within the lattice structure of colloidal matter.

Imbricata Extinct order of eocrinoids known from the Lower Cambrian in which the stalk is composed of overlapping plates.

immature Used of the developmental stages of an organism preceding the attainment of sexual maturity.

immature soil A recently formed soil in which profile development is incomplete for the prevailing climatic and biological conditions.

immersed Used of an aquatic plant growing entirely beneath the water surface.

immigration The movement of an individual or group into a new population or geographical region; **immigrant**; *cf.* emigration.

immobilization The conversion of an element from an inorganic to an organic state by an organism; *cf.* biological mineralization.

immunity The state in which a host is more or less resistant to an infective agent; includes inherited natural immunity and acquired resistance arising from exposure to the agent and consequent stimulation of an immune response.

immunology The study of immunity and immunological reactions.

impatiens Balsaminaceae *q.v.*

Impennes Superorder of birds comprising a single order, Sphenisciformes and family, Spheniscidae (penguins).

imperfect flower A unisexual flower; a flower lacking either pistils or stamens.

impervious Not permeable; usually with reference to the passage of water or air; **impermeable.**

implantation 1: In mammals, the attachment of an embryo to the uterus wall. 2: The insertion of a graft into the host or stock.

impoundment An artificially enclosed body of water; typically with fluctuating water levels and high turbidity.

imprinting The imposition of a stable behaviour pattern in social species by exposure, during a particular period in early development, to one of a restricted set of stimuli.

Inadunata Extinct subclass of sea lilies (Crinoidea) in which the arms were completely free; known from the Ordovician to the Permian.

Inarticulata Class of lamp shells (brachiopods) characterized by the absence of hinge teeth and sockets, the valves held together by muscles alone; pedicle absent to elongate; lophophore without skeletal support; gut terminates in a functional anus; habits epifaunal to infaunal, intertidal to 7500 m; contains about 45 species in 2 extant orders, Lingulida, and Acrotretida.

inbreeding Mating or crossing of individuals more closely related than average pairs in the population; endogamy; *cf.* outbreeding.

inch worm Geometridae *q.v.*

incidence of infection The number of infected hosts as a proportion of the host population; *cf.* intensity of infection.

incipient About to become or occur; used of the initial stage in the development of a structure or event.

inclination Aspect; the direction of a slope face.

incompatibility 1: The inability to unite, fuse or form any homogeneous or viable association; **incompatible.** 2: Failure of either cross fertilization or self-fertilization as a result of structural, physiological or genetic factors. 3: Failure of a graft to unite with the host or stock tissue.

incomplete dominance The partial expression of both alleles at a given locus so that the phenotype of the heterozygote is intermediate between both homozygotes.

incomplete flower A flower that lacks one or more of the four basic parts (pistil, stamen, petal and sepal); *cf.* complete flower.

incross Breeding or mating between homozygotes; *cf.* intercross.

incubation 1: The maintenance of eggs under conditions favourable for hatching; **incubate.** 2: The period between infection by a pathogen and the appearance of symptoms.

indeciduous Persistent; everlasting; evergreen; **indeciduate.**

indehiscent Used of fruits that do not open spontaneously when ripe; **indehiscence.**

independent assortment, law of Mendel's second law; that the random distribution of alleles to the gametes results from the random orientation of the chromosomes during meiosis.

indeterminate growth Growth that continues

throughout the life span of an individual such that body size and age are correlated; *cf.* determinate growth.

index fossil A fossil apparently restricted to a narrow stratigraphic unit or time unit.

index species Indicator species *q.v.*

Indian almond Combretaceae *q.v.*

Indian Equatorial Countercurrent A warm surface ocean current that flows eastwards in the equatorial Indian Ocean and forms the southern winter countercurrent to the North Equatorial Current; see ocean currents.

Indian North Equatorial Current A warm surface ocean current that flows westwards in the equatorial Indian Ocean, generated largely by the northeasterly winds of the winter monsoon; see ocean currents.

Indian paintbrush Scrophulariaceae *q.v.*

Indian pipe Monotropaceae *q.v.*

Indian South Equatorial Current A warm surface ocean current that flows westwards in the central Indian Ocean and forms the northern limb of the South Indian Gyre and the southern limb of the North Indian Ocean circulation; see ocean currents.

indicator An organism, species or community characteristic of a particular habitat, or indicative of a particular set of environmental conditions.

indicator species A species, the presence or absence of which is indicative of a particular habitat, community or set of environmental conditions; index species.

Indicatoridae Honeyguides; family containing 15 species of small Old World arboreal birds (Piciformes) distributed from Africa to Borneo; feed on insects, honey and beeswax; lay eggs in nests of other hole-nesting birds; habits solitary, non-migratory, monogamous; newly hatched young use sharp hooked bill to kill other young in nest.

indifferent species A species exhibiting no obvious affinity for any particular community; a species of low fidelity *q.v.*; companion species.

indigenous Native to a particular area; used of an organism or species occurring naturally in an environment or region.

individual distance The minimal distance that an animal endeavours to maintain between itself and other individuals of the species.

individualism A symbiosis in which the two symbionts are intimately associated, appearing to form a single organism not resembling either of the partners; **individuation.**

indole acetic acid The most important auxin *q.v.* of higher plants.

Indo-Malaysian subkingdom A subdivision of the Palaeotropical kingdom *q.v.*

Indostomidae Monotypic family of tiny (to 30 mm) freshwater teleost fishes inhabiting shallow weedy lakes or streams of northern Burma and Thailand; body slender with armour of bony plates; dorsal fin with 5 isolated fin spines; formerly placed with sticklebacks (Gasterosteiformes) but now assigned to a separate order, Indostomiformes.

indri Indriidae *q.v.*

Indriidae Indriid lemurs; family containing 4 species of mainly arboreal prosimian lemurs found only in Madagascar; hindlimbs longer than forelimbs with feet partly webbed; tail long or short; habits diurnal to nocturnal, herbivorous; solitary or in small groups.

indri (Indriidae)

induction 1: The determination of the developmental fate of one cell mass by another. 2: A hereditary transformation mediated by foreign DNA in a multicellular organism.

inductive method A scientific method involving the formulation of universal statements, such as hypotheses or theories, from singular statements, such as empirical observations or experimental results; *cf.* deductive method.

induration The process of hardening, as of sediments or rocks rendered hard by compaction, heat or pressure; **indurate.**

industrial melanism An increase in the frequency of melanistic (dark pigmented) morphs as a response to changes in selection, in habitats blackened by industrial pollution.

inert Inactive; physiologically quiescent.

infauna The total animal life within a sediment; endobenthos; *cf.* epifauna.

infection The process of invasion of a host by a parasite or microorganism; the pathological

condition resulting from the growth of infectious organisms within a host; **infect.**

infectious disease Any disease transmitted without physical contact.

inferior ovary Epigyny *q.v.*

infertile Non-reproductive; not fertile; **infertility.**

infestation Invasion by parasites or pests.

infiltration The process by which water seeps into a soil, influenced by soil texture, aspect and vegetation cover.

inflorescence The arrangement and sequence of development of flowers on a flowering shoot; includes capitulum (head), catkin, corymb, cyme, panicle, raceme, spadix, spike.

infradian rhythm A biological rhythm with a period less than 24 hours.

infrahaline Used of fresh water having a salinity of less than 0.5 parts per thousand.

infralittoral 1: The depth zone of a lake permanently covered with rooted or adnate macroscopic vegetation; often divided into upper (with emergent vegetation), middle (with floating vegetation) and lower (with submerged vegetation) zones. 2: The upper subdivision of the marine littoral zone, typically dominated by algae having a lower limit at the depth at which illumination is at about 1% of the surface level; sometimes used for the depth zone between low tide and 100 m; see marine depth zones.

infralittoral fringe The transitional marine zone immediately below the eulittoral, between the intertidal and sublittoral zones; see marine depth zones.

infrared radiation That part of the solar radiation spectrum *q.v.* above the wavelength 780 nm.

Infusoria Ciliophora *q.v.*

ingesta The total intake of substances into the body; **ingest.**

ingestion The action of taking in food material; feeding.

Ingolfiellidea Small suborder of primitive amphipod crustaceans that live interstitially in marine bottom sediments or subterranean ground waters; typically having an elongate vermiform body lacking side plates.

inheritance The transmission of genetic information from ancestors or parents to descendants or offspring.

inhibition Any process that acts to restrain reactions or behaviour.

ink cap Coprinaceae *q.v.*

innate Inherited; present at birth; used of behaviour that is instinctive and not learned.

inoculation The introduction or insertion of a pathogen into a host organism by a vector; or of an organism into an experimental culture.

inoculum The individual or group of individuals comprising the founders of a colony or newly established population.

inorganic Pertaining to, or derived from, non-biological material; used of non-carbon-chain compounds; *cf.* organic.

Inozoida The sole order of Pharetronidia *q.v.*

inquilinism A symbiosis in which one organism (the inquiline) lives within another without causing harm to its host; also used for the relationship between a species living within the burrow, nest or other domicile of another species; **inquilinous.**

Insecta Insects; a class of uniramian arthropods, by far the largest and most diverse of all classes of living organisms with about 1 million recognized species representing perhaps only 10–20 per cent of the world fauna; body of adult typically divided into head, thorax and abdomen; body length about 0.2–280 mm; head bearing pre-oral unjointed mandibles, one pair of antennae and 2 pairs of maxillae (the second fused to form labium); thorax with 3 pairs of legs on pro-, meso- and metathoracic segments; the mesothorax and metathorax may also bear wings; respiration takes place through tracheae; insects exhibit an immense variety of life history strategies and are of great ecomonic importance as pests, pollinators, parasites and vectors of disease; divided into 2 subclasses, Apterygota (wingless insects) and Pterygota (winged insects), and 28 extant orders; Hexapoda.

caddis fly (Insecta)

insecticide A chemical used to kill insects; used generally for any chemical agent used to destroy invertebrate pests.

Insectivora Diverse order of small eutherian mammals comprising 8 families, Erinaceidae (hedgehogs), Talpidae (moles), Tenrecidae (tenrecs), Chrysochloridae (golden moles), Solenodontidae, Soricidae (shrews), Macroscelididae (elephant shrews) and Tupaiidae (tree shrews); widespread distribution except Australian region, southern South America and oceanic islands.

insectivorous Feeding on insects; **insectivore, insectivory.**

insemination The introduction of sperms into the genital tract of a female; impregnation.

insessorial Adapted for perching.

insolation Exposure to solar radiation.

inspiration The process by which oxygen is absorbed by plants and animals.

instar Any intermoult stage in the development of an arthropod; *cf.* imago.

instinctive behaviour Relatively complex, highly stereotyped behaviour exhibited in response to an environmental stimulus and directed towards a predictable end product.

insular Pertaining to islands; used of an organism that has a restricted or limited habitat or range.

insulosity The percentage of the area of a lake, within the shore line, occupied by islands.

intensity of infection The number of individuals of a particular parasite species in each infected host.

interbreeding Mating or hybridization between different individuals, population, varieties, races or species.

intercalated Inserted between adjacent structures or strata.

interception That part of total precipitation retained on the surface of vegetation before reaching the ground and returned to the atmosphere by evaporation.

intercross The breeding or mating of heterozygotes; *cf.* incross.

interdemic Between local interbreeding populations.

interface The contact surface between two contiguous substances or zones.

interferon A low-molecular-weight protein formed by cells in response to a virus.

interfertile Capable of interbreeding.

interglacial A warm period between two Ice Ages or glaciations; see Quaternary Ice Age.

interlittoral The shallow subtidal marine zone to a depth of about 20 m.

intermediate host The host occupied by juvenile stages of a parasite prior to the definitive host *q.v.* and in which asexual reproduction frequently occurs.

intersex An individual of a bisexual species possessing reproductive organs and secondary sexual characters partly of one sex and partly of the other, but which is of one sex only genetically.

interspecies A cross between two distinct species; interspecific hybrid; **interspecific.**

interstade A climatic episode within a glacial stage during which a secondary retreat or standstill occurred; **interstadial**; *cf.* stade.

intersterile Unable to interbreed; **intersterility.**

interstitial Pertaining to, or occurring within, the pore spaces (interstices) between sediment particles.

intertidal zone The shore zone between the highest and lowest tides; littoral; see marine depth zones.

intraspecific Within a species; between individuals or populations of the same species.

intrazonal soil An immature soil showing more or less well developed characteristics that reflect the influence of some local factors, over the basic effects of climate and vegetation as soil-forming factors; transitional soil; *cf.* azonal soil, zonal soil.

intrinsic Originating or occurring within an individual, group or system.

intrinsic rate of increase The potential rate of growth of a population in an infinite environment.

introgression The spread of genes of one species into the gene pool of another by hybridization and backcrossing.

introrse dehiscence Spontaneous opening of a ripe fruit from outside inwards; *cf.* extrorse dehiscence.

inundatal Liable to flooding; used of plants that occupy sites susceptible to inundation during wet weather.

invasion The mass movement or encroachment of organisms from one area into another; **invader.**

inversion A structural change of a chromosome resulting from the reversal of a segment such that the genes occur in a reversed sequence.

invertebrate 1: Used of those animals that lack a vertebral column. 2: Formerly one of the primary divisions in the classification of the Animal Kingdom, containing over 97 per cent of all species.

inviability A measure of the number of individuals failing to survive in one cohort or group, relative to another.

ionosphere The outer layer of the atmosphere more than 80 km above the Earth's surface, characterized by free electrically charged particles.

Iowan glaciation The penultimate glaciation of the Quaternary Ice Age *q.v.* in North America, with an estimated duration, together with the Wisconsin glaciation, of about 70 000 years.

Ipnopidae Family of little-known abyssobenthic myctophiform teleost fishes; body elongate, to 450 mm in length, subcylindrical with snout depressed; photophores absent; eyes variable, large and lensless or degenerate and covered with skin.

Ipswichian interglacial The interglacial period preceding the Devensian glaciation in the British Isles; see Quaternary Ice Age.

Irenidae Leafbirds; family of colourful passerine birds found in forests of southern Asia and Indonesia; feed mainly on fruits and insects; nest in trees.

Iridaceae Iris, crocus, gladiolus, freesia, montbretia; cosmopolitan family of Liliales comprising about 1500 species of perennial, mostly geophytic herbs, arising from bulbs, corms or rhizomes and dying back each year; leaves narrow and parallel-veined; seeds with fleshy endosperm containing reserves of hemicellulose, protein and oil; inflorescences usually spike-like and terminal, each flower bisexual, with 6 perianth segments united at base to form a tube, 3 stamens and a typically inferior ovary.

iris (Iridaceae)

iris Iridaceae *q.v.*

Irminger Current A cold surface ocean current that flows westwards and northwards around the southern coast of Greenland, derived in part from a deflection of the North Atlantic Drift; see ocean currents.

Iron Age An archaeological period dating from about 3000 years B.P. in Europe; a period of human history characterized by the smelting of iron for industrial purposes.

Irregulares Anthocyathea *q.v.*

irritability Sensitivity; the ability to perceive and respond to stimuli; **irritable.**

irruption An irregular abrupt increase in population size or density typically associated with favourable changes in the environment and often resulting in the mass movement of the population.

Ischnocera Suborder of chewing lice (Phthiraptera) containing about 1800 species, parasitic on placental mammals and birds.

Ischnochitonida Large order of chitons distributed worldwide but especially abundant in tropical and subtropical waters; characterized by the shell valves, the first has an anterior area only, valves 2 to 7 have median and lateral areas, valve 8 has median and posterior areas; surrounding girdle well developed, often with embedded scales or bristles.

Ischyrinioida Extinct order of rostroconch molluscs known only from the Ordovician.

isidium An outgrowth on a lichen which may break off and develop into a lichen thallus as it contains both algal and fungal cells.

isoalleles Alleles which are so similar in their phenotypic effects that special techniques are required to distinguish them.

isobar A line on a chart or map connecting points of equal atmospheric or hydrostatic pressure.

isobath A line on a chart or map connecting points of equal depth; bathymetric contour.

Isobryales Order of mosses (Bryidae) largely confined to the southern hemisphere; typically large perennial mat-forming plants growing on tree trunks, branches and rocks in forested habitats; exhibit extensive development of primary and secondary stems and the sporophyte is borne laterally.

isochronal line A line on a chart or map connecting points (localities) at which a given event occurs simultaneously, as in the mean date of arrival of certain migratory species.

Isochrysidales Order of scale-covered flagellate algae (Prymnesiophyceae), most of which are marine; scales occasionally calcareous, forming coccoliths.

Isocrinida Order of stalked articulate crinoids in which true cirri radiate at intervals from the ossicles of the stalk; contains about 30 extant species in 2 families.

isodeme A line on a map connecting points of equal population density; isodemic line.

Isoetales Quillworts; aquatic lower plants (Isoetopsida) with grass-like leaves containing air spaces spirally arranged on typically short stubby stem; heterosporous (producing megaspores and microspores).

Isoetopsida Subdivision of Lycopodiophyta, comprising the quillworts (Isoetales) and spike mosses (Selaginellales).

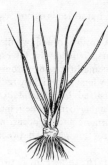

quillwort (Isoetales)

Isofilobranchia A superorder of bivalve molluscs comprising a single order, Mytiloida *q.v.*

isogametic Producing gametes that are similar in size, shape and behaviour; having gametes (isogametes) not differentiated into male and female; *cf.* anisogametic.

isogamy The fusion of gametes of similar size, shape and behaviour; *cf.* anisogamy.

isogenetic Of the same origin.

isogenic Homozygous *q.v.*; having the same set of genes; syngenic; *cf.* allogenic.

isohaline A line on a chart or map connecting points of equal salinity.

isohyet A line on a chart or map connecting points of equal rainfall.

Isolaimida Small order of free-living enoplian nematodes found in sandy soils which are infrequently cultivated; characterized by the apparent absence of cephalic chemosensory organs (amphids) and by the presence round the mouth of 6 hollow tubes and 2 whorls each of 6 sensory papillae.

isolating mechanism Any intrinsic or extrinsic mechanism or barrier to the free exchange of genes between populations.

isolation The separation of two populations so that they are prevented from interbreeding, whether by extrinsic (premating) or intrinsic (postmating) mechanisms.

isolume A line on a chart or map connecting points of equal light intensity.

isomerism 1: Repetition of similar parts. 2: The possession of equal numbers of serially homologous structures.

isometric growth Growth in which relative proportions of the body parts remain constant with change in total body size; *cf.* allometric growth.

isomorphic Used of a plant having an alternation between diploid and haploid generations which are morphologically similar in appearance; homomorphic; *cf.* heteromorphic.

isomorphism Superficial similarity between individuals of different species or races; **isomorphous.**

Isonidae Family containing 6 species of surf-dwelling atheriniform teleost fishes found primarily in the Indian Ocean and western Pacific; body deep, robust, transparent with silvery lateral stripe.

isophagous Used of predators or parasites which are selective with regard to food or hosts but which are not restricted to a single food type or host species.

isophenous Sharing or producing similar phenotypic effects; **isophenic.**

isophot A line on a chart or map connecting points of equal incident radiation.

isopiptesis A line on a chart or map connecting points at which a given migratory species arrives on the same date.

Isopoda Cosmopolitan order of small to medium-sized peracaridan crustaceans found in marine, freshwater and terrestrial habitats and as parasites; body usually depressed; thoracic legs uniramous and with short coxae; young leaving brood pouch at manca stage; comprises about 4000 species in 9 suborders, including the Oniscoidea (woodlice), Asellota, Flabellifera, Epicaridea, Anthuridea, Gnathiidea, Microcerberidea, Valvifera and Phreatoicoidea.

Isoptera Termites, white ants; order of small social insects that construct nests on or in hollow trees, rotten wood, or the ground, which vary in size from a few centimetres with less than a hundred individuals to huge earth mounds measuring several metres with millions of colony members; caste structure includes morphologically distinct reproductives, soldiers and workers; adults (imagos) possess 2 pairs of long wings,

compound eyes and a robust pigmented cuticle; secondary reproductives (neotenics) are typically blind, wingless and unpigmented; soldiers are usually blind, sterile and wingless with large heads and strong jaws; workers are mostly sterile and blind, with soft white cuticle (white ants); about 2000 species recognized, ubiquitous in tropics except for arid zones, rare in temperate regions; termites are of immense economic importance, damaging timber buildings, trees and crops; they are primary energy transformers of wet tropical ecosystems and form the principal diet of many predators.

termites (Isoptera)

isoseismal A line on a chart or map connecting points of equal earthquake intensity.

Isospondyli Diverse assemblage of herring-like fishes, comprising the Clupeiformes, Elopiformes, Ostariophysi and Osteoglossidae.

isosporous Having asexually produced spores of only one kind; homosporous; **isospory;** *cf.* heterosporous.

isotherm A line on a chart or map connecting points of equal temperature.

Istiophoridae Billfishes, marlin, sailfish; family containing 10 species of large (to 4.5m) surface-living oceanic scombroid teleost fishes (Perciformes); head produced into pointed bill, jaw teeth present; dorsal and pelvic fins elongate; feed on other fishes using bill to stun prey.

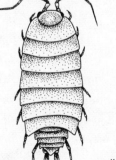

woodlouse (Isopoda)

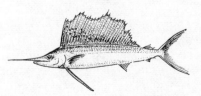

sailfish (Istiophoridae)

iteration The procedure used in computation, in which the operations are repeated until a good fit is obtained.

iteroparous Used of organisms that have repeated reproductive cycles; **iteroparity**; *cf.* semelparous.

ivy Araliaceae *q.v.*

Ixodida Arachnid ticks; subgroup of parasitiform mites (Acari) specialized as blood-feeding ectoparasites of terrestrial vertebrates – mammals, birds and reptiles; chelicerae modified for cutting skin; comprises about 800 recognized species.

Ixonanthaceae Family of Linales containing about 28 species of trees and shrubs of tropical distribution.

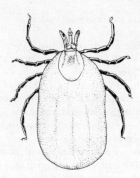

tick (Ixodida)

j

jacamar Galbulidae *q.v.*
jacana Jacanidae *q.v.*
Jacanidae Jacanas; family containing 8 species
of tropical freshwater charadriiform shore
birds having long legs with extremely long toes
for walking on floating vegetation such as lily
pads; habits gregarious, dive and swim well,
but flight slow; feed mostly on aquatic
invertebrates and small fish; nest on clump of
floating plant material.

jacana (Jacanidae)

jack Carangidae *q.v.*
jackal Canidae *q.v.*
jack-in-the-pulpit Araceae *q.v.*
jactitation A method of seed dispersal in which
seeds are tossed from the fruit by jerking
movements of the plant.
jaguar Felidae *q.v.*
jasmine Oleaceae *q.v.*
javelinfish Anotopteridae *q.v.*
jaw worm Gnathostomulida *q.v.*
jawfish Opisthognathidae *q.v.*
jay Corvidae *q.v.*
jelly fungi Eutremellales *q.v.*
jellyfish Scyphozoa *q.v.*
Jenynsiidae Family of viviparous
cyprinodontiform teleost fishes containing 3
species found in fresh waters of Argentina;
anal fin of male modified as gonopodium for
sperm transfer.
jerboa Dipodidae *q.v.*
jewel beetle Buprestidae *q.v.*
jigger Tungidae *q.v.*
jimsonweed Solanaceae *q.v.*
jingle shell Ostreoida *q.v.*
Joinvilleaceae Small family of Restionales
comprising 2 species of coarse, erect herbs,
with unbranched stems which are hollow
except at the nodes and with alternate, rather

grass-like leaves; native in the Pacific Islands;
sometimes included in the Flagellariaceae.
jojoba Simmondsiaceae *q.v.*
jollytail Galaxiidae *q.v.*
Juglandaceae Walnut, hickory; family of
aromatic trees containing about 60 species
widely distributed in the temperate and
subtropical northern hemisphere; flowers
inconspicuous and wind-pollinated, ovary
inferior; fruit is a nut.

walnut (Juglandaceae)

Juglandales Order of Hamamelidae containing
two families of aromatic trees with small wind-
pollinated flowers borne in catkins.
Julianiaceae Family of Sapindales containing
only 5 species of trees and shrubs, with well
developed resin ducts; found in tropical
America.
Julida Order comprising about 600 species of
helminthomorphan diplopods (millipedes)
found in Eurasia and North America; body
length to 50 mm, a few species to 150 mm;
typically inhabiting deciduous forest litter,
sometimes caves and open ground.
jumping mouse Zapodidae *q.v.*
jumping plant louse Psyllidae *q.v.*
jumping spider Salticidae *q.v.*
Juncaceae Rushes; family of herbs or, rarely,
shrubs with basal alternate leaves, each with a
slender parallel-veined blade; flowers small,
inconspicuous and typically wind-pollinated;
widespread in distribution, particularly in the
northern hemisphere.
Juncaginaceae Family of Najadales comprising
about 20 species of cyanogenic, commonly
rhizomatous, herbs of bogs and other wet,
often saline habitats; with emergent or floating
inflorescences; widespread in temperate and
cold regions.
Juncales Order of Commelinidae comprising 2
families, Juncaceae (rushes) and Thurniaceae.
Jungermanniales Large and diverse order of
liverworts (Hepaticopsida) containing 43
families; plants typically leafy in organization,

having 2 lateral rows of leaves and a ventral
row of smaller underleaves; growth occurring
by means of apical cell; smooth rhizoids
usually present; female sex organs terminal,
incorporating the apical cell; stalked
sporophyte capsule apical and uniformly 4-
valved.

jungle Dense seral vegetation especially
characteristic of tropical regions with a high
level of precipitation.

juniper Cupressaceae *q.v.*

Jurassic A geological period of the Mesozoic
era (*c.* 210–140 million years B.P.); see
geological time scale.

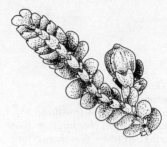

liverwort (Jungermanniales)

k

K selection Selection producing superior competitive ability in stable predictable environments so that the population is maintained at or near the carrying capacity of the habitat; *cf. r* selection.

Kainogaea Caenogaea *q.v.*

Kainozoic Cenozoic *q.v.*

Kamchatka Current Oyashio Current *q.v.*; see ocean currents.

Kamptozoa Entoprocta *q.v.*

kangaroo Macropodidae *q.v.*

kangaroo-paw Haemodoraceae *q.v.*

kangaroo rat Heteromyidae *q.v.*

Kansan glaciation A glaciation of the Quaternary Ice Age *q.v.* in North America with an estimated duration of 90 000 years.

kapok tree Bombacaceae *q.v.*

karstic Pertaining to irregular limestone strata permeated by streams, typically with sinks, caves and other subterranean passages.

karyogamy The fusion of gametic nuclei; **karyogamic.**

karyogenetic Heritable; not subjected to direct environmental influences.

karyokinesis Nuclear division; *cf.* cytokinesis.

karyology The branch of cytology dealing with the study of nuclei, especially the structure of chromosomes.

Karyorelictida Order of gymnostome ciliates characterized by the primitive nature of their dual nuclear system, in which the macronucleus is diploid and non-dividing rather than polyploid; all are interstitial, usually in marine sediments, and most are carnivorous.

ciliate (Karyorelictida)

karyotype 1: The chromosome complement of a cell, individual, or group. 2: The structural characteristics of the chromosome set. 3: Those individuals having an identical complement of chromosomes.

kasidoron Juvenile stage of fishes in the family Gibberichthyidae *q.v.*

katabatic wind A dense mass of cold air flowing down an incline in response to gravity.

katatropism An orientation response downwards; **katatropic.**

katharobic Inhabiting pure water, or water of very low organic content; **katharobe.**

katydid Tettigoniidae *q.v.*

kelp Laminariales *q.v.*

kelp fly Coelopidae *q.v.*

kelpfish Chironemidae *q.v.*

Kelvin scale (K) A temperature scale that takes absolute zero (−273.16°C) as its starting point, zero K; each degree on the Kelvin scale is equal to one degree Celsius.

kenapophyte A plant colonizing cleared land.

Kentucky coffee tree Caesalpiniaceae *q.v.*

keratinophilic Thriving on horny (keratin-rich) substrates; **keratinophile, keratinophily.**

keystone predator The dominant predator, often the top predator in a given food web, having a major influence upon community structure.

Kickxellales Order of zygomycete fungi containing mostly soil- and dung-inhabiting species.

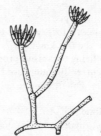

mould (Kickxellales)

killer whale Delphinidae *q.v.*

killifish Cyprinodontidae *q.v.*

kilo- (k) Prefix used to denote unit x 10^3; see metric prefixes.

kin selection Selection acting on one or more individuals and favouring or disfavouring the survival and reproduction of relatives (other than offspring) that possess the same genes by common descent, as in the case of selection for altruistic behaviour between genetically related individuals.

kinesis A change in rate of random movement of an organism as a result of a stimulus; the linear or angular velocity changes in response to an alteration in the intensity of the stimulus, but no orientation of movement occurs with respect to the direction of the stimulus; **kinetic;** *cf.* orthokinesis, klinokinesis.

Kinetofragminophora Class of ciliates either with a complete cover of cilia or with cilia limited to the ventral surface, characterized by the specialized cilia around the cytostome which are derived from the ends of the cilia rows on the general body surface; includes 4 subclasses which utilize a variety of habitats from free-living in marine sands to endoparasitic on other animals; most are holozoic, feeding on other ciliates, but many feed on bacteria.

Kinetoplastida Order of small free-living and parasitic zooflagellates possessing 1 or 2 flagella arising from a pit or pocket and a DNA-containing nucleoid (the kinetoplastid) near the basal bodies of the flagella; reproduce by binary or multiple fission; includes *Trypanosoma* which causes Chagas' disease and sleeping sickness in man, and nagana in cattle.

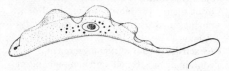

trypanosome (Kinetoplastida)

king crab Xiphosura *q.v.*

kingdom The highest category in the hierarchy of classification; the five kingdom classification of living organisms proposed by Whittaker and currently widely used is Monera, Protista, Fungi, Plantae and Animalia.

kingfisher Alcedinidae *q.v.*

Kinorhyncha Kinorhynchs, Echinodera; phylum of small, free-living marine invertebrates found in subtidal muds worldwide and characterized by a segmented body, the possession of a pseudocoelom and a chitinous cuticle; body divided into an eversible head, a neck of 1 segment and a trunk

kinorhynch (Kinorhyncha)

of 11 segments; the neck usually consists of a series of plates which retract over the head when it is withdrawn into the trunk; sexes separate; gut possessing mouth and anus; locomotion by repeated eversion of the head; comprises 2 orders, Cyclorhagida and Homalorhagida.

Kinosternidae Musk turtles, mud turtles; family containing about 20 species of aquatic freshwater turtles (Testudines: Cryptodira) widespread in North, Central and South America; body length to 300 mm, shell domed with horny scutes; neck elongate, feet webbed; may produce foul-smelling secretions from cloacal glands.

kinship Possession of a recent common ancestor; the condition of being related.

kipuka An island of vegetation isolated by lava flows.

kissing bug Reduviidae *q.v.*

kite Accipitridae *q.v.*

kiwi Apterygidae *q.v.*

Klebsormidiales Small order of predominantly freshwater green algae in which the thallus comprises unbranched filaments lacking holdfast cells.

kleptobiosis An interspecific association found in some social organisms in which one species steals food from the stores of another species but does not live or nest in close proximity to it.

kleptoparasitism A form of parasitism found in some social organisms in which a member of one species steals the prey or food stores of another species to feed its own progeny; **kleptoparasite.**

klinogeotropism The drooping tendency displayed by the tip of a climbing plant during nutation *q.v.*; **klinogeotropic.**

klinokinesis A change in rate of random movement of an organism (kinesis) in which the rate of change of direction (frequency of turning movements) varies with the intensity of the stimulus; *cf.* orthokinesis.

klinophyte A terrestrial plant that is dependent upon other plants for physical support.

klinotaxis An orientation response of a motile organism (taxis) in which the direction of movement results from a comparison of the intensity of the stimulus on each side of the body by successive deviations from side to side; **klinotactic;** *cf.* tropotaxis, telotaxis.

knephopelagic Pertaining to the middle pelagic zone of the sea extending from about 30 m to the lower limit of the photic zone.

knephoplankton Planktonic organisms of the

twilight zone between about 30 and 500 m depth; *cf.* phaoplankton, skotoplankton.

Kneriidae Family containing 12 species of African freshwater teleost fishes (Gonorhynchiformes) found mostly in fast-running waters; body slender, subcylindrical, mouth ventral, protractile, teeth absent; gas bladder may function as accessory respiratory organ.

knifefish Gymnotidae *q.v.*

knifejaw Oplegnathidae *q.v.*

knotweed Polygonales *q.v.*

koala Phascolarctidae *q.v.*

kollaplankton Planktonic organisms rendered buoyant by encasement in gelatinous envelopes.

Komodo dragon The largest living lizard, reaching a length of up to 3 m; feeds on deer, pigs and other vertebrates, often as carrion; known from several Indonesian islands; Varanidae.

Komodo dragon

krait Elapidae *q.v.*

Krameriaceae Family of Polygalales containing about 15 species of hemiparasitic shrubs and perennial herbs native to the New World.

kremastoplankton Planktonic organisms that possess modified appendages or surface structures to reduce rates of sinking.

krill Euphausiacea *q.v.*; commonly used for *Euphausia superba* which is an abundant food species of baleen whales, found in the Southern Ocean.

krotovina A former animal burrow in one soil horizon that has been filled with organic material or material from another horizon.

krummholz A discontinuous belt of stunted forest or scrub typical of windswept alpine regions close to the tree line; *cf.* elfin forest.

K-selected species A species characteristic of a relatively constant or predictable environment, typically with slow development, relatively high competitive ability, late reproduction, large body size and iteroparity; *cf. r*-selected species.

Kuroshio Current A warm surface ocean current that flows northwards and eastwards in the North Pacific, fed from the North Equatorial Current and forming the western limb of the North Pacific Gyre; see ocean currents.

Kuroshio Extension A warm surface ocean current that flows eastwards in the North Pacific forming the northern limb of the North Pacific Gyre.

Kurtidae Humphead; family comprising one brackish and one freshwater species of perciform teleost fishes found in India, Australia, New Guinea and China; body deep, to 600 mm length; single dorsal fin present, caudal deeply forked; eggs carried by male on supraoccipital crest.

Kutorginida Extinct order of lamp shells (Brachiopoda) with biconvex, calcareous shells; known from the Lower Cambrian.

Ky Abbreviation denoting 1000 years.

kymatology The study of waves and wave motion.

Kyphosidae Family containing over 30 species of marine teleost fishes (Pericformes) with relatively short heads and a small mouth; many feed on plants; widespread in warm to temperate waters.

l

Labiatae Lamiaceae *q.v.*

labile Plastic; readily modified.

Laboulbeniales Large order of parasitic pyrenomycete fungi found on insects, mites and diplopods; transmitted from one generation to the next by spores passed on by physical contact.

Labrador Current A cold surface ocean current that flows southwards off the east coast of Canada; see ocean currents.

Labridae Wrasses; diverse family containing about 400 species of tropical and temperate marine teleost fishes (Perciformes); body form variable, often elongate, colourful, length to 300 mm, few larger forms to 3 m; mouth protractile; single dorsal fin consisting of spinose and soft-rayed sections, commonly use pectoral fins instead of caudal fin for swimming; sexual dimorphism common.

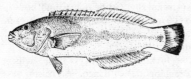

wrasse (Labridae)

Labroidei Suborder of mainly tropical and warm temperate coastal marine teleost fishes (Perciformes) comprising about 500 species in 3 families Labridae (wrasses), Scaridae (parrotfishes) and Odacidae.

laburnum Fabaceae *q.v.*

Labyrinthodontia Extinct subclass of primitive amphibians known from the Palaeozoic and Triassic; attaining a maximum body length of several metres.

Labyrinthulamycota Slime nets; a phylum of

Protoctista also classified as a phylum of Protozoa, the Labyrinthulata *q.v.*

Labyrinthulata Slime nets; phylum of protozoans in which the trophic stage typically consists of mobile cells contained within an anastomosing network of ectoplasmic extensions produced at the cell surface; nutrients are absorbed through these extensions which occur on algae, angiosperms and organic detritus in marine and estuarine waters; reproduction involves the production of biflagellate zoospores.

Lacertidae Family of Old World terrestrial lizards (Sauria); typically agile forms, with well developed legs and tails capable of autotomy; habits mainly diurnal, insectivorous, some forms frugivorous; reproductive oviparous, one species viviparous; containing about 210 species; distributed through Palaearctic, Ethiopian and Oriental regions.

lacewing Neuroptera *q.v.*

Lacistemataceae Small family of Violales containing about 20 species of tanniferous shrubs and small trees with unbranched multicellular hairs; native to tropical America.

Lactoridaceae Primitive family of flowering plants (Magnoliales) containing a single species of shrubs, *Lactoris fernandeziana*, endemic to one of the Juan Fernandez Islands off the coast of Chile.

lacustrine Pertaining to, or living in, lakes or ponds; **lacustral.**

ladybird Coccinellidae *q.v.*; ladybug.

Laemodipodea Caprellidea *q.v.*

Lagenidiales Order of aquatic oomycete fungi comprising about 65 species, most of which are obligate parasites of algae or invertebrates.

Lagerberg–Raunkiaer cover scale A scale for estimating cover of a plant species, comprising 4 categories of percentage ground cover, 1 (0–10%), 2 (11–30%), 3 (31–50%), 4 (51–100%).

Lagomorpha Cosmopolitan order of small terrestrial herbivorous mammals comprising

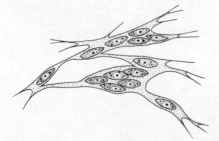

slime net (Labyrinthulamycota)

rabbit (Lagomorpha)

about 60 species in 2 familes, Ochotonidae (pikas) and Leporidae (hares, rabbits); incisor teeth chisel-like, separated from cheek teeth by wide space (diastema); ears typically elongate, tail short or absent, hindlimbs strong for hopping locomotion; habits often nocturnal or crepuscular; may inhabit simple to complex burrows; formerly known as the Duplicidentata.

lair-flora The plants inhabiting ground manured by animals.

lake A large body of fresh or saline standing water with negligible current, having a narrow peripheral beach largely devoid of vegetation as a result of wave action.

Lamarckism Inheritance of acquired characters; the theory that changes in use and disuse of an organ result in changes in size and functional capacity and that these modified characters acquired by organisms in response to environmental factors are transmitted to the offspring.

Lamellibranchia Bivalvia *q.v.*

Lamiaceae Mint, thyme, catnip, lavender, horehound, balm, deadnettle, rosemary, sage; large cosmopolitan family containing about 3200 species of aromatic herbs or shrubs commonly with epidermal glands containing ethereal oils; stems often square and leaves frequently hairy and arranged in opposite pairs; flowers bisexual, typically with a 5-toothed, cleft calyx, a 2-lipped, 5-lobed corolla, 2 or 4 stamens, and a superior ovary; fruit composed of 4 single-seeded nutlets; formerly known as the Labiatae.

sage (Lamiaceae)

Lamiales Order of Asteridae containing 4 families of plants producing iridoid compounds, alkaloids or aromatic oils.

lamina A thin layer; the smallest recognizable unit in a bedded sediment.

laminar flow The streamlined movement of water, in which water particles appear to move

in smooth paths, and one layer of water slides over another; *cf.* turbulence.

Laminariales Kelp; order of often large brown algae, common in the lower intertidal and sublittoral zones of temperate and polar seas; exhibiting alternation between a filamentous gametophyte and a sporophyte differentiated into holdfast, stipe and blade.

kelp (Laminariales)

Lamnidae Mackerel sharks; cosmopolitan family of large (6–9 m) pelagic lamniform elasmobranch fishes; coastal and oceanic, very active and offensive, feeding on fishes, marine mammals, birds and squid; contains 5 species including great white shark and mako, frequently recorded in attacks on people and boats; also porbeagle and salmon sharks.

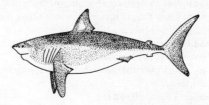

mackerel shark (Lamnidae)

Lamniformes Order of pelagic marine, coastal to oceanic, galeomorph elasmobranch fishes (sharks); nasoral groove and barbels absent; mouth large, teeth small; feed on a variety of fishes, marine mammals and reptiles, and invertebrates; reproduction ovoviviparous; contains about 15 species in 7 families; includes the mako, great white, thresher, mackerel and megamouth sharks.

lamp shell Brachiopoda *q.v.*

lamprey Petromyzoniformes *q.v.*

Lamprididae Opah or moonfish; family containing a single species of mesopelagic marine teleost fishes widespread in tropical and temperate waters feeding on squid, small fishes and crustaceans; body oval,

opah (Lamprididae)

compressed, to 1.8 m in length; teeth absent; dorsal and anal fins elongate, retractile; pectorals long and curved.

Lampridiformes Order of pelagic marine teleosts having uniquely protrusible upper jaw; contains 35 species in 10 families, including opah, ribbonfishes, oarfishes, tapetails and tube-eye.

Lampyridae Fireflies, glow worms; family of mostly nocturnal small beetles (Coleoptera) typically with luminous organs near the tip of the abdomen which are used to produce mating signals; larvae found in soil and litter, predatory on various invertebrates; contains about 2000 species.

lancefly Lonchaeidae *q.v.*

lancelet Cephalochordata *q.v.*

lancetfish Alepisauridae *q.v.*

land breeze A light wind blowing off the land at night when the sea surface is warmer than the land; *cf.* sea breeze.

land-bridge A more or less continuous connection between adjacent landmasses, forming a potential route for migration and dispersal.

langur Cercopithecidae *q.v.*

Laniidae Shrikes; family containing about 65 species of small to medium-sized passerine birds found in variety of brush, grassland and open habitats worldwide; bill stout, compressed, hooked; wings and tail long; habits solitary, aggressive, some species migratory; feed on insects and other invertebrates that are frequently impaled on thorns; cup-shaped nest located in tree.

lantern tree Elaeocarpaceae *q.v.*

lanterneye Anomalopidae *q.v.*

lanternfish Myctophidae *q.v.*

Lanthanotidae Earless monitor; family containing a single species of lizard (Squamata) found in Borneo; can burrow and swim, feeding on various invertebrates and small fishes; body elongate with short limbs, tail not autotomous.

lapidicolous Living under or among stones; **lapidicole.**

Lapidognatha Araneomorph spiders; largest of the 3 suborders of spiders (Araneae) comprising about 75 families and some 33 000 species; abdomen unsegmented, spinnerets located posteriorly; chelicerae move at right angles to the body axis.

Lardizabalaceae Small family of Ranunculales containing about 30 species of mostly twining woody vines native principally to southeast Asia but also to Chile.

Laridae Gulls, terns; cosmopolitan family containing about 85 species of small to large charadriiform seabirds; coloration mostly black and/or white; habits gregarious, monogamous, migratory, nesting on ground, on cliffs or in trees; feed on variety of aquatic animals and carrion.

gull (Laridae)

lark Alaudidae *q.v.*

larkspur Ranunculaceae *q.v.*

larva A stage in development between hatching and attainment of adult form or maturity.

Larvacea Appendicularia *q.v.*

larvicidal Lethal to larvae.

larviparous Producing eggs that are hatched internally with release of free-living larvae; **larvipary**; *cf.* oviparous, ovoviviparous, viviparous.

larvivorous Feeding on larvae; **larvivore, larvivory.**

lasion Fouling; a dense growth of aquatic organisms, particularly on submerged objects projecting into free water.

latebricolous Living in holes; **latebricole.**

Laterite soil A zonal soil, highly weathered, having a reddish brown clay subsoil, low in minerals and soluble constituents, formed in forested warm temperate to tropical regions; Latosol.

laterization A soil-forming process occurring in tropical regions in which the minerals are converted to clay and pronounced leaching results in the accumulation of hydroxides of aluminium and iron in the deeper layers; **laterite.**

Latimeriidae Coelacanth; monotypic family of marine tassel-finned fishes found off the Comoro Islands in the Indian Ocean at a depth of 150–400 m; length to 2.75 m, body bearing cosmoid scales; swim bladder fat-filled.

coelacanth (Latimeriidae)

Latosol Laterite soil *q.v.*

Latridae Trumpeters; family containing 10 species of marine perciform teleost fishes found in cool temperate waters of the southern hemisphere; body oval, compressed, to 1 m length, dorsal fin elongate with many short fin spines.

Lauraceae Laurels, bay laurel, avocado pear; large family of aromatic evergreen trees or shrubs (Laurales) widespread in subtropical and tropical regions; flowers generally small with a pistil consisting of a simple carpel with a single ovule; fruit typically fleshy.

avocado pear (Lauraceae)

Laurales One of the more archaic orders of flowering plants (Magnoliidae) comprising about 2500 species of mostly woody plants, commonly found in warm moist regions; flowers perigynous or epigynous, typically containing a single functional ovule; pollen grains mostly inaperturate or biaperturate.

Laurasia The northern supercontinent formed by the break up of Pangaea *q.v.* in the Mesozoic (*c.* 150 million years B.P.), and comprising North America, Greenland, Europe and Asia, excluding India; Eurasia; *cf.* Gondwana.

laurel Lauraceae *q.v.*

laurophilous Thriving in sewers and drains; **laurophile, laurophily.**

laurophyte A sewer or drain plant.

lavender Lamiaceae *q.v.*

law An empirical generalization; a statement of a biological principle that appears to be without exception at the time it is made, and has become consolidated by repeated successful testing.

layer Stratum; a horizontal zone of vegetation; four such zones are recognized by plant ecologists – tree layer, shrub (bush) layer, field (herb) layer, and ground (moss) layer.

LD_{50} Lethal dose, 50%; the dose level required to kill half the organisms in a test population per unit time.

leaching The removal of readily soluble components, such as chlorides, sulphates and carbonates, from soil by percolating water; *cf.* flushing.

leaf beetle Chrysomelidae *q.v.*

leaf insect Phyllidae *q.v.*

leaf spot fungus Diaporthales *q.v.*

leafbird Irenidae *q.v.*

leaffish Nandidae *q.v.*

leafhopper Cicadellidae *q.v.* (Homoptera).

leaf-nosed bat Phyllostomatidae *q.v.*, Hipposideridae *q.v.*

learning The production and accumulation of adaptive changes in individual behaviour as a result of experience.

leatherback turtle Dermochelyidae *q.v.*

leatherjacket Tipulidae *q.v.*

Lebiasinidae Pencil fishes; family containing 50 species of mostly small colourful tropical South American freshwater characiform teleost fishes; body spindle-shaped, dorsal and anal fins small; complex courtship behaviour exhibited by some forms; typically surface-living in weedy habitats with still or slow moving water; extremely popular amongst aquarists.

Lecanicephalidea Order of tapeworms parasitizing elasmobranch fishes; characterized by a scolex divided into an anterior portion which is either cushion-like or armed with retractable tentacles and a posterior portion which usually has 4 simple suckers.

Lecithoepitheliata Order of freshwater, terrestrial and marine free-living turbellarians found throughout the world; characterized by a complex pharynx and ovovitellaria, organs that produce both oocytes and yolk cells.

lecithotrophic Pertaining to developmental stages that feed upon yolk, and to eggs rich in yolk.

Lecythidales Brazil nut tree; small order of Dilleniidae comprising about 400 species of woody plants commonly with rather large flowers; best developed in tropical rainforests especially in South America; comprises the family Lecythidaceae.

Leeaceae Family of Rhamnales containing about 70 species of erect herbs or shrubs found in the Palaeotropical region; flowers usually small, perfect and borne in large cymose inflorescences.

leech Hirudinea *q.v.*

leeward Pertaining to the side facing away from a wind, ice, or water current.

legume A pod; a dry dehiscent fruit developed from a single carpel and containing one or more seeds; splits along ventral and dorsal sutures on ripening, to form 2 valves; typical of members of the Fabales *q.v.*

Leguminosae Older name for the Fabaceae, the only family of the Fabales *q.v.*

leimocolous Living in moist grassland or meadowland; **leimocole.**

leimonapophyte A plant introduced into grassland or meadowland.

leimophyte A wet-meadow plant.

Leiognathidae Ponyfishes; family containing 20 species of Indo-Pacific, bottom-living, coastal marine, brackish and freshwater teleost fishes (Perciformes); body compressed, to 300 mm in length, skin covered with copious mucus; head naked, eyes large, mouth highly protrusible.

Leiopelmatidae Family of primitive frogs (Anura) comprising 4 species from New Zealand and western North America; eggs laid onto damp ground, developing into froglets before hatching.

Leitneriales Order of Hamamelidae containing a single North American species of tanniferous shrubs with resin canals and small wind-pollinated flowers.

lek 1: An assembly area for communal courtship display. 2: A communal prenuptial display, specifically those displays involving ritualised contests between competitors, often for the use of symbolic sites; **lekking.**

lemming Cricetidae *q.v.*

Lemnaceae Duckweed; small family of Arales containing about 30 species of small free-floating, thalloid plants without roots or with short unbranched roots, widely distributed in still-water habitats; reproduce mainly by vegetative budding, rarely producing inflorescences.

lemon Rutaceae *q.v.*

lemon shark Carcharhinidae *q.v.*

Lemurian realm The zoogeographical region comprising the island of Madagascar; Malagasy region.

Lemuridae Lemurs; family containing 15 species of mostly arboreal prosimians found only in Madagascar and Comoro Islands; limbs long and slender, tail well developed; habits diurnal or nocturnal, feeding on variety of plant material and insects.

Lemuroidea A superfamily of the mammalian suborder Strepsirhini, comprising the lemurs (Lemuridae) and lorises (Lorisidae).

lemur (Lemuridae)

Lennoaceae Family of Lamiales containing only 5 species of fleshy root-parasites lacking chlorophyll and commonly provided with stalked, glandular hairs; native to the New World.

Lentibulariaceae Bladderwort, butterwort; cosmopolitan family of Scrophulariales containing over 200 species of insectivorous herbs of aquatic or wet habitats, with submerged photosynthetic organs and traps formed from modified stems or leaves; flowers irregular and bisexual, usually with a 2-lipped corolla, 2 stamens and superior ovaries.

lentic Pertaining to static, calm or slow-moving aquatic habitats; *cf.* lotic.

lentil Fabaceae *q.v.*

leopard Felidae *q.v.*

lepidophagous Feeding on scales; used of fishes that feed on the external scales of other fishes; **lepidophage, lepidophagy.**

Lepidopleurida Order of primitive chitons (Mollusca) distributed worldwide but containing only about 50 species; characterized by the shell valves; the first has a single area, valves 2 to 8 have central and lateral areas and valve 8 has an additional medial projection.

Lepidoptera Butterflies, moths; large order of holometabolous insects comprising about 138 000 species in 5 suborders, Zeugloptera, Dacnonympha, Exoporia, Monotrysia and Ditrysia; adults typically with slender coiled sucking proboscis (maxillae); mandibles reduced, non-functional; body and wings covered with scales; the membranous fore- and hindwings coupled together as single functional unit, wings rarely reduced or absent; larvae (caterpillars) usually phytophagous, occasionally scavengers, rarely predatory or parasitic.

lepidopterid Used of flowers adapted for pollination by butterflies and moths.

lepidopterophilous Pollinated by butterflies and moths; **lepidopterophily.**

Lepidosauria Subclass of Reptilia comprising about 5600 extant species in 2 orders, Rhynchocephalia (tuatara) and Squamata (lizards, snakes, amphisbaenians).

Lepidosirenidae South American lungfish; monotypic family of freshwater fishes found in stagnant weedy backwaters of the Amazon basin; body elongate, cylindrical, length to 1.25 m; pectoral and pelvic fins short and filamentous; swim bladders paired functioning as lungs; larvae with 4 pairs of feathery external gills; adults aestivate during the dry season.

Lepisosteidae Gars; family containing 7 species of medium-sized (to 1.5 m) fishes found in rivers, lakes and occasionally coastal marine waters of Central America, Cuba, and eastern North America; body very elongate bearing heavy ganoid scales; jaws prolonged with needle-like teeth; swim bladder vascularized aiding aerial respiration; larvae with adhesive organ on ventral surface of snout.

gar (Lepisosteidae)

Leporidae Hares, rabbit; Lagomorpha *q.v.*

leporin Anostomidae *q.v.*

Lepospondyli Extinct group of amphibians known from the Palaeozoic.

leptocephalus A type of larval stage found in a variety of marine bony fishes; that of the common eel is transparent, with a leaf-like body and tiny eyes.

Leptochariidae Monotypic family of small (to 0.75 m) shallow-water carcharhiniform sharks of the tropical and temperate eastern Atlantic Ocean; feed on fishes, crustaceans and cephalopods; reproduction viviparous.

Leptodactylidae Diverse family of mainly terrestrial Neotropical frogs (Anura), few species aquatic or arboreal; reproductive strategies variable, ranging from fully aquatic eggs and larvae to terrestrial eggs and direct development; contains about 650 species distributed mostly in South America but extending into West Indies, Central America and southern United States.

leptodactylid frog (Leptodactylidae)

Leptolinida Alternative name for the Hydroida *q.v.*

leptology The study of minute structures or particles.

Leptomedusae Suborder of predominantly marine cnidarians (Hydroida) in which the polyps are typically colonial with a horny rather than calcareous exoskeleton; free medusae tall and bell-shaped with gonads usually on the stomach.

Leptomitales Water moulds; order of oomycete fungi comprising about 20 species of saprophytes usually found on plant remains in stagnant water or waterlogged soils.

Leptomyxida Order of Acarpomyxa containing multinucleate or similar uninucleate amoeboid organisms of diverse form, including reticulate sheets, non-reticulate fans, branched amoebae and unbranched slug-like amoebae; found in freshwater and soil habitats.

leptopel Large organic molecules or aggregates of colloidal proportions suspended in water.

leptophyll A Raunkiaerian leaf size class *q.v.* for leaves having a surface area less than 25 mm^2.

Leptoscopidae Sandfishes; family containing 3 species of small (to 90 mm) slender, sand-burrowing teleost fishes (Perciformes) found in cool coastal waters around Australia and New Zealand.

Leptosomatidae Coural; family containing a single species of coraciiform bird found in scrub and forest habitats of Madagascar and the Comoro Islands; bill stout, feeds on insects and lizards.

Leptostraca Order containing about 25 species of typically benthic marine phyllocaridan crustaceans; head and most of thorax enclosed by large bivalve carapace hinged along the mid-dorsal line; eyes stalked; long setae on thoracic legs form a ventral brood chamber.

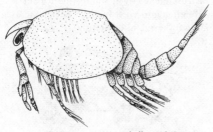

phyllocarid (Leptostraca)

Leptotyphlopidae Family containing about 65 species of small (to 300 mm) worm-like burrowing snakes widespread in tropical and subtropical regions; feed mainly on insects and worms; reproduction oviparous.

Lessepsian migration Migration between the Red Sea and the Mediterranean by way of the Suez Canal.

lestobiosis A symbiosis between social insects in which colonies of a small species inhabit the nests of a larger species and prey on its brood or rob the food stores.

lethal Pertaining to, or causing, death by direct action.

lethal trait Any inherited character that causes the death of an organism before it can reproduce.

Lethrinidae Emperors; family containing 20 species of Indo-Pacific and west African coastal marine perciform teleost fishes; body perch-like, to 1 m in length; mouth weakly protractile, lips fleshy; single dorsal fin present.

lettuce Asterales *q.v.*

Leucettida Order of calcinian sponges varying in form from encrusting to massive, lobed or branching; choanocyte chambers tubular to spherical; widely distributed to a maximum depth of about 500 m.

leucism Paleness, as of coloration.

Leucosoleniida Order of calcaronian sponges typically consisting of a cluster of upright tubes, lined with choanocytes and with

terminal oscula, united by a basal system of stolon-like tubes; found from the intertidal to over 2400 m.

sponge (Leucosoleniida)

levée A sediment bank.

liana 1: Gnetidae *q.v.* 2: A woody, free-hanging, climbing plant.

Lias A geological epoch of the early Jurassic period (*c.* 210–175 million years B.P.); see geological time scale.

Liceales Diverse order of Myxogastromycetidae (plasmodial slime moulds) characterized by the absence of lime from the fruiting body which typically lacks a true capillitium and columella (stalk).

slime mould (Liceales)

Liceida The slime mould order Liceales *q.v.*, treated as an order of the protozoan group Myxogastria.

lichen A type of thalloid lower plant formed as an association between a fungus (the mycobiont) and an alga or blue-green bacterium (the phycobiont); thallus may be encrusting, leafy or even shrubby according to species.

lichen

lichenicolous Living in or on a lichen;
lichenicole.

lichenism The symbiotic relationship between
a fungus and an alga that comprise a lichen.

lichenology The study of lichens.

lichenometry A method for dating rocks based
on a knowledge of the growth rates of
encrusting lichens; **lichenometric.**

lichenophilous Thriving on, or having an
affinity for, lichens or lichen-rich habitats;
lichenophile, **lichenophily.**

life cycle 1: The sequence of events from the
origin as a zygote, to the death of an
individual. 2: Those stages through which an
organism passes between the production of
gametes by one generation and the production
of gametes by the next.

life form The characteristic structural features
and method of perennation of a plant species;
cf. Raunkiaerian life forms.

life history The significant features of the life
cycle through which an organism passes, with
particular reference to strategies influencing
survival and reproduction.

life span Longevity; the maximum or mean
duration of life of an individual or group.

lightfish Gonostomatidae *q.v.*

lignicolous Growing on or in wood; **lignicole.**

ligniperdous Used of organisms that destroy
wood.

lignivorous Feeding on wood; **lignivore,
lignivory.**

lignophilic Thriving on or in wood; **lignophile,
lignophily.**

lignum vitae Zygophyllaceae *q.v.*

lilac Oleaceae *q.v.*

Liliaceae Lilies; asparagus, aspidistra, bluebell,
daffodil, hyacinth, day lily, lily of the valley,
merry bells, narcissus, spider plant, tulip;
large family of perennial herbs generally
arising from a rhizome, bulb or corm and dying
back to the ground each year, usually with
basal leaves which are typically narrow and

lily (Liliaceae)

parallel-veined; flowers mostly insect-
pollinated and usually trimerous with a
superior ovary and many seeds; comprising
nearly 4000 species, widely distributed
particularly in dry temperate to subtropical
regions.

Liliales Cosmopolitan order of
monocotyledons (Liliidae) comprising about
8000 species in 15 families; mostly autotrophic
herbs often with rhizomes, corms or bulbs, and
commonly with trimerous flowers having 6
stamens and adapted for insect pollination.

Liliidae Subclass of monocotyledons
(Liliopsida) comprising only 2 orders of
terrestrial, epiphytic or aquatic herbs often
with rhizomes, corms and bulbs; commonly
mycotrophic and sometimes lacking
chlorophyll; also includes some woody forms;
leaves usually alternate, simple and entire,
mostly without a petiole; flowers
characteristically showy, with septal nectaries
and adapted for pollination by insects.

Liliopsida Monocotyledons; class of
Magnoliophyta; herbaceous or, sometimes,
woody plants lacking typical secondary
growth, vascular bundles scattered in stem or
in 2 or more rings; vessels often confined to
roots or absent; root system adventitious;
leaves usually parallel-veined; embryo
typically with a single cotyledon; consists of 65
families in 5 subclasses, Alismatidae,
Arecidae, Commelinidae, Liliidae and
Zingiberidae.

lily of the valley Liliaceae *q.v.*

Limacomorpha Superorder of pentazonian
diplopods (millipedes) containing a single
order, Glomeridesmida.

lime 1: Rutaceae *q.v.* 2: Tiliaceae *q.v.*

limicolous Inhabiting mud; **limicole.**

limiting factor Any environmental factor, or
group of related factors, which exist at
suboptimal level and thereby prevent an
organism from reaching its full biotic
potential.

limivorous Feeding on mud; **limivore, limivory.**

Limnanthaceae Family of Geraniales
containing 11 species of small annual herbs
producing mustard oils, native to North
America.

limnetic Pertaining to lakes or to other bodies
of standing fresh water; often used with
reference only to the open water of a lake away
from the bottom; **limnic.**

limnetic zone The surface zone of a lake above
the compensation depth.

Limnichthyidae Family containing 6 species of

small (to 70 mm) Indo-Pacific tropical marine teleost fishes (Perciformes) found in shallow coastal waters burrowing in sandy sediments; body slender with pointed snout overlapping lower jaw; dorsal and anal fins elongate, lacking fin spines.

limnicolous Living in lakes; **limnicole.**

limnion All lakes, ponds, pools and other bodies of standing fresh water.

limnobiology The study of organisms living in lakes, ponds and other standing freshwater bodies; limnology.

limnobiont A freshwater organism; **limnobiontic.**

limnobios The total life of fresh waters; that part of the Earth's surface occupied by freshwater organisms; *cf*. geobios, halobios, hydrobios.

Limnocharitaceae Water poppy; family of Alismatales containing about 12 species of perennial aquatic herbs, either free-floating or rooted and emergent; native to tropical and subtropical regions.

limnocrene A pool of spring-water with no outflow.

limnocryptophyte A marsh or pond plant with perennating organs below the surface.

limnodic Pertaining to salt marshes.

limnodophilous Thriving in salt marshes; **limnodophile limnodophily.**

limnodophyte A salt-marsh plant.

limnology The study of lakes, ponds, and other standing waters, and their associated biota; **limnological.**

Limnomedusae Suborder of predominantly freshwater cnidarians (Hydroida) in which the polyps may be solitary or colonial and the exoskeleton thin or lacking; medusae with hollow tentacles and the gonads on the stomach or radial canals.

limnophilous Thriving in lakes or ponds; **limnophile, limnophily.**

limnophyte A marsh or pond plant.

limnoplankton Planktonic organisms of freshwater lakes and ponds.

Limoida File shells; order of about 125 species of widely distributed, marine, bottom-living, pteriomorphian bivalves characterized by shell and hinge structure, and by single posterior shell-closing muscle.

limophagous Feeding on mud; **limophage, limophagy.**

limpet 1: Basommatophora *q.v.*; freshwater limpets. 2: Archaeogastropoda *q.v.*

Linaceae Common flax; family of about 220 species of often cyanogenic herbs and shrubs

common flax (Linaceae)

widely distributed in temperate and subtropical regions; flowers regular, bisexual and usually with 5 sepals and petals, 5, 10 or more stamens and staminodes, and superior ovaries.

Linales Order of Rosidae containing five families of woody or herbaceous plants, with most species belonging to the Erythroxylaceae and Linaceae.

linden Tiliaceae *q.v.*

lineage A line of common descent; line.

lingcod Hexagrammmidae *q.v.*

Linguatulida Alternative name for the Pentastomida *q.v.*

Lingulida Small order of inarticulate brachiopods comprising about 20 species , one of which, *Lingula*, has existed unchanged for 400 million years; pedicle elongate, foramen absent; intertidal to 130 m in tropical and warm temperate regions inhabiting U-shaped burrows in sandy sediments.

Liniphiidae Money spiders; large family of dark-coloured spiders (Araneae) which construct sheet webs on vegetation; adults hang to the underside of the web, running out over it to catch prey.

linkage The presence of specific genes on the same chromosome such that the traits are not independently assorted; the greater the proximity of these genes the smaller the chance of their separation by crossing over, and the stronger the linkage.

Linnaean classification The system of hierarchical classification and binomial nomenclature established by Linnaeus.

Linophrynidae Family of tiny (to 80 mm) deep-sea anglerfishes (Lophiiformes) found over a depth range from 100 to 3000 m; female body rounded, mouth large, teeth long and fang-like; male parasitic on female.

lion Felidae *q.v.*

lipid A diverse class of hydrocarbon compounds including sterols, carotenes, xanthophylls, waxes, phosphoglycerides and acylglycerols; serving a variety of structural and storage functions.

Lipogenyidae Monotypic family of small (to 400 mm) eel-like notacanthiform teleost fishes found in western North Atlantic to about 1500 m depth; body compressed with blunt snout and small mouth devoid of teeth, dorsal and anal fins spinose.

anglerfish (Linophrynidae)

Liposcelidae Cosmopolitan family of psocopteran insects, comprising about 145 species many of which cause damage to a variety of stored products; body characteristically flattened; includes the familiar book louse which causes damage to old papers and dried natural history material.

Lipotyphla Suborder of placental mammals comprising hedgehogs and moon rat (Erinaceidae), shrews (Soricidae), moles (Talpidae), solenodon (Solenodontidae), tenrecs (Tenrecidae) and golden mole (Chrysochloridae).

lipoxenous Used of a parasite that leaves its host after feeding; **lipoxeny.**

lipstick tree Bixaceae *q.v.*

liptocoenosis An assemblage of dead organisms and their biogenic products.

Lissamphibia Subclass of amphibians comprising all living forms; 3 orders are recognized, Caudata (salamanders), Anura (frogs, toads) and Gymnophiona (caecilians); contains about 3140 living species in 36 families; cosmopolitan distribution, but frequently absent from oceanic islands.

Lissocarpaceae Small family of Ebenales containing only 2 species of small trees, native to tropical South America.

litchi Sapindaceae *q.v.*

lithic Pertaining to rock.

lithification The physiochemical process that produces rock from sedimentary deposits.

Lithistida Order of tetractinomorph sponges found mostly in tropical and warm temperate seas at depths of 30 to 2000 m; possessing spicules (desmas) which interlock or articulate with one another to form a rigid skeleton.

Lithobiida Diverse order of mostly temperate anamorphan chilopods comprising about 1450 species in 4 families; eyes composed of small clusters of ocelli, antennae filiform; trunk 18-segmented, bearing 15 pairs of legs; tergal plates alternating large and small.

Lithobiomorpha Lithobiida *q.v.*

lithocarp A fossil fruit.

lithodomous Used of an organism that lives in holes in rock, or that bores into rock.

lithology The study of rocks and rock forming processes; petrology.

lithophagic 1: Used of organisms that erode or bore into rock; **lithophage, lithophagy.** 2: Eating small stones, as in some birds.

lithophilous Thriving in stony or rocky habitats; **lithophile, lithophily.**

lithophyl A fossil leaf.

lithophyte A plant growing on rocks or stones.

lithosere An ecological succession originating on an exposed rock surface; *cf.* sere.

Lithosol A very stony shallow azonal soil without pronounced horizons, and with incomplete surface layers comprising imperfectly weathered rock fragments; skeletal soil.

lithosphere The rigid crustal plates of the Earth.

lithostratigraphy The organization and classification of rock strata according to their lithological character; **lithostratigraphic.**

lithotomous Used of an organism that burrows into rock.

lithoxyle Fossilized wood.

Litobothridea Small order comprising 4 species of tapeworms parasitic in thresher sharks; characterized by a scolex comprising a single apical sucker, and by a segmented body with each proglottis containing a single set of reproductive organs.

Litopterna Extinct order of hoofed ungulates known from the Eocene to the Pleistocene of South America.

litter 1: Recently fallen plant material which is only partially decomposed and in which the organs of the plant are still discernible, forming a surface layer on some soils. 2: Those animals produced at a multiple birth.

Little Ice Age An interval characterized by expanding mountain glaciers (*c.* 4000–2000 years B.P.); Medithermal period.

littoral 1: Pertaining to the shore. 2: The shore

of a lake to a depth of about 10 m. 3: The intertidal zone of the seashore; sometimes used to refer to both the intertidal zone on the seashore and the adjacent continental shelf to a depth of about 200 m; subdivided into supralittoral, eulittoral (intertidal zone), infralittoral and circalittoral; see marine depth zones.

live-bearer Poeciliidae *q.v.*

liverwort Hepaticopsida *q.v.*

living fossil A species that has persisted to the present time with little or no change over a long period of geological time; a species with characters shared only with an otherwise extinct group.

lizard Sauria *q.v.*

lizard fish Synodontidae *q.v.*

llama Camelidae *q.v.*

L-layer The surface litter layer of a soil profile comprising loose fragmented material in which the original plant structures are still readily discernible; see soil horizon.

loach 1: Cobitidae *q.v.* 2: Homalopteridae *q.v.*

loam A friable soil comprising a mixture of clay, silt, sand and organic matter.

Loasaceae Family of Violales containing about 200 species of mostly herbs commonly producing mineralized and sometimes stinging hairs; occurring mostly in the New World.

Lobata Order of pelagic ctenophores found at various depths from polar to tropical seas; characterized by a body that is compressed in the tentacular plane but expanded each side of the mouth into a pair of oral lobes, and by the lack of tentacular sheaths.

Lobeliaceae Lobelia; family of herbs and small trees often treated as part of the family Campanulaceae *q.v.*

Lobosa Class of amoeboid protozoans with lobose pseudopodia comprising 2 subclasses, Gymnamoeba and Testacealobosa.

Lobotidae Tripletail; family containing 4 species of marine, brackish and freshwater perciform teleost fishes widespread in warm coastal waters; body deep, perch-like, with rounded dorsal, anal and caudal fins; length to 1 m.

lobster 1: Astacidea *q.v.* 2: Palinura *q.v.* (spiny lobsters).

lobster claw Heliconiaceae *q.v.*

lochmocolous Inhabiting thickets; **lochmocole.**

lochmodophilous Thriving in dry thickets; **lochmodophile, lochmodophily.**

lochmodophyte A plant inhabiting dry thickets.

lochmophilous Thriving in thickets; **lochmophile, lochmophily.**

lochmophyte A plant inhabiting thickets.

loculicidal dehiscence Spontaneous opening of a ripe fruit along the centre line.

Loculoanoteromycetidae Subclass of loculoascomycete fungi in which the ascocarp is determinate in growth and flask shaped with the asci forming a compact layer; contains 2 orders, Chaetothyriales and Verrucariales.

Loculoascomycetes Class of ascomycotine fungi characterized by a fruiting body (ascocarp) produced by cell division or by interweaving of vegetative hyphae; sexual fusion of nuclei occurs within cells of inner portion (locule) of the ascocarp; contains 7 orders grouped into 4 subclasses, Loculoedaphomycetidae, Loculoanoteromycetidae, Loculoparenchemycetidae and Loculoplectascomycetidae.

Loculoedaphomycetidae Subclass of loculoascomycete fungi in which the ascocarp is determinate in growth, the asci form a layer and are separated by sterile hyphae or cells (pseudoparaphyses); comprises 3 orders, Hysteriales, Melanommatales and Pleosporales.

Loculoparenchemycetidae Subclass of loculoascomycete fungi in which the ascocarp is determinate in growth, the asci form a compact layer and are typically separated by the cells of the locule; contains 2 orders, Asterinales and Dothideales.

Loculoplectascomycetidae Subclass of loculoascomycete fungi containing a single order, Myriangiales *q.v.*

locus The position of a given gene on a chromosome; **loci.**

locust Caelifera *q.v.* (Orthoptera).

loess A fine unconsolidated wind-blown sediment.

Loganiaceae Family of Gentianales containing about 500 species of mainly tropical or subtropical woody or herbaceous plants commonly accumulating bitter substances; flowers typically with 4 or 5 sepals, petals and stamens, and a superior ovary; strychnine is obtained from seeds of one species of the largest genus, *Strychnos*.

lomentum A dry dehiscent fruit developed from a single carpel; contains one or more seeds released by fracture of valves on maturity.

Lonchaeidae Lanceflies; cosmopolitan family containing about 500 species of small blue-black flies (Diptera) in which body has a metallic sheen; widespread in forested habitats; larvae feed mostly on decaying plant material.

longevity The average life span of the individuals of a population under a given set of conditions.

long-horned beetle Cerambycidae *q.v.*

longneck eel Derichthyidae *q.v.*

longshore Used of currents or movement parallel to the coastline.

long-tailed tit Aegithalidae *q.v.*

loon Gaviidae *q.v.*

looper The caterpillar larva of geometrid moths typically showing a looping locomotion; includes the inch worms.

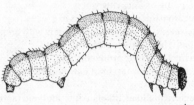

looper

loosejaw Malacosteidae *q.v.*

loosestrife Primulaceae *q.v.*

Lophiidae Goosefishes; family containing about 25 species of bottom-living anglerfishes (Lophiiformes) widespread on continental slopes; body naked, length to 1.2 m, head and anterior body broad, depressed, bearing spines and ridges; mouth large, teeth long and pointed.

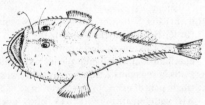

goosefish (Lophiidae)

Lophiiformes Anglerfishes; order of mostly deep-water teleost fishes comprising about 215 species in 15 families, including also goosefishes, frogfishes, batfishes and sea devils; dorsal fin ray (illicium) often greatly extended bearing a lure (esca) to attract prey; sexual dimorphism may be pronounced with males much smaller than females; in some families males ectoparasitic on females.

lophophilous Thriving on hill tops; **lophophile**, **lophophily**.

Lophophorata Lophophorates; a diverse assemblage of coelomate invertebrates characterized by two longitudinal coelomic compartments divided by a septum; includes 3 extant phyla, Brachiopoda, Bryozoa and Phoronida.

lophophyte A hill-top plant.

Lophotidae Crestfishes; family containing 3 species of little-known mesopelagic lampridiform teleost fishes; body ribbon-like, to 1.3 m length, bearing tiny deciduous scales; anal fin absent, dorsal elongate, pelvics present or absent; a dark brown ink can be discharged from a sac opening into the cloaca.

Loranthaceae Showy mistletoe; large family of Santalales containing about 900 species of brittle, evergreen shrublets hemiparasitic on branches of trees; found in tropical and subtropical regions; flowers usually with inferior ovary, producing a berry as the fruit.

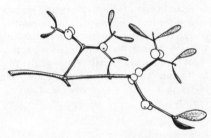

mistletoe (Loranthaceae)

Loricariidae Family of herbivorous South American fresh and brackish water catfishes (Siluriformes); body heavily armoured with bony plates; mouth ventral, suctorial; dorsal and adipose fins often with stout spine; comprises 400 species; popular amongst aquarists, used to control algae in aquarium tanks.

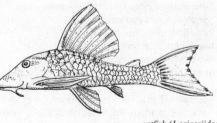

catfish (Loricariidae)

Loricifera Recently discovered phylum of microscopic (up to 400 μm) bilaterally symmetrical animals found in the interstices of marine sediments; body comprising an eversible head, neck and thorax all of which are retractible into a loricate abdominal region. (Picture overleaf).

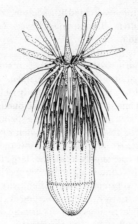

loriciferan (Loricifera)

Lorisidae Lorises; family containing 5 species of slow-moving arboreal prosimians found in central Africa, India and southeast Asia; body typically slender, limbs subequal length, tail short to rudimentary; habits nocturnal, feeding on variety of small animals and plant material; often hang suspended beneath branch of tree.

loris (Lorisidae)

lotic Pertaining to fast running-water habitats, such as rivers and streams; *cf.* lentic.

lotus lilies Nelumbonaceae *q.v.*

louse Phthiraptera *q.v.*

lousewort Scrophulariaceae *q.v.*

low tide Low water; the minimum height of the falling tide.

low water (LW) The lowest surface water level reached by the falling tide; low tide.

lower high water (LHW) The lower of two high waters during any tidal day where there are marked inequalities of tidal height.

lower low water (LLW) The lower of two low waters during any tidal day where there are marked inequalities of tidal height.

lowest astronomical tide The lowest low water produced only by the gravitational effects of the sun and moon.

Lowestoft glaciation Anglian glaciation *q.v.*

Lowiaceae Family of Zingiberales comprising only 6 species of perennial herbs with silica cells next to the vascular bundles; with malodorous flowers presumably pollinated by flies; native to southern China, Malaysia and the Pacific islands.

Lucanidae Stag beetles; family containing 1200 species of often large beetles (Coleoptera) with enlarged, modified mandibles in the males which may be used for fighting; adults feeding on nectar or non-feeding; larvae feeding on decaying wood.

Lucernariidae Stauromedusae *q.v.*

lucicolous Living in open habitats with ample light; **lucicole**; *cf.* umbraticolous.

luciferous Light-producing; bioluminescent.

lucifugous Intolerant of light; **lucifugal**; *cf.* luciphilous.

Luciocephalidae Pikehead; monotypic family of predatory freshwater perciform teleost fishes from the Malay Archipelago; length to 180 mm, dorsal and anal fins positioned posteriorly, mouth large and protractile; utilize atmospheric oxygen through accessory respiratory structure, the suprabranchial organ.

luciphilous Thriving in open, well-lit habitats; **luciphile**, **luciphily**; *cf.* lucifugous.

luffa Cucurbitaceae *q.v.*

lugworm Arenicolidae *q.v.* (Capitellida).

lumachelle An accumulation of shells in a sediment.

Lumbricina Suborder of oligochaete worms (order Haplotaxida) containing the earthworms and related freshwater forms; clitellum comprising multiple layers of cells; testis in segments X and/or XI usually with single pair of male pores; ovaries in segment XIII with female pores on XIV; body length may exceed 4 m; contains about 3000 species in 15 families.

Lumbriculida Order of freshwater oligochaete worms comprising a single family; body robust with 4 pairs of simple chaetae per segment; pharynx eversible; distribution mainly northern hemisphere, many species restricted to Lake Baikal.

luminescence The emission of light by biochemical processes without the production of heat; *cf.* phosphorescence.

lumpfishes Cyclopteridae *q.v.*; lumpsuckers.

lunar day The time interval between consecutive moonrises; 24.8 hour; *cf.* solar day.

lungfish 1: Ceratodontidae *q.v.* (Australian

lungfish). 2: Lepidosirenidae *q.v.* (South American lungfish). 3: Protopteridae *q.v.* (African lungfish).

lungwort Boraginaceae *q.v.*

lupin Fabaceae *q.v.*

Lusitanian Having affinities with the Iberian peninsula.

luticolous Inhabiting mud; **luticole.**

Lutjanidae Snappers; family containing 225 species of demersal and mid-water, tropical marine teleost fishes (Perciformes) found in coastal and shelf waters; body elongate to moderately deep, compressed, to 1 m in length; upper jaw protrusible; single dorsal fin present.

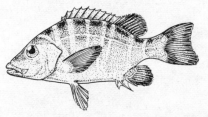

snapper (Lutjanidae)

Lycaenidae Blues, coppers, hair-streaks; large family of often brilliant metallic-coloured, small to medium-sized butterflies (Lepidoptera) found mainly in the tropics and subtropics; caterpillars frequently associated with ants.

Lychniscosida Small order of hexasterophoran sponges which are typically stalked and cup-shaped, or formed as an ovoid mass of anastomosing tubes.

Lycoperdales Puff balls, earthstars; widely distributed order of gasteromycete fungi producing epigeal, more or less globose, fruiting bodies; spores released by dehiscence or by rupturing.

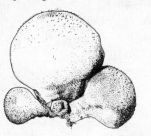

puffballs (Lycoperdales)

lycophore First larval stage of a cestodarian tapeworm, possessing 10 hooks.

Lycopodiophyta Division of small to medium-sized plants possessing true roots, stems and leaves; stem solid and lacking long internodes; needle or scale-like leaves each with single vein; sporangia embedded in leaf bases or in axils; contains aquatic terrestrial and epiphytic species; comprises 2 subdivisions, Lycopodiopsida and Isoetopsida.

Lycopodiopsida Clubmoss; terrestrial or epiphytic plants with needle or scale-like leaves arranged spirally on stem; roots borne adventitiously on stem; sporangium bearing leaves may resemble vegetative leaves or may be aggregated to form cones.

clubmoss (Lycopodiopsida)

Lycosidae Wolf spiders, hunting spiders; family of medium-sized to large spiders (Araneae) which run over the ground to capture prey; some burrow or make funnel-shaped webs.

wolf spider (Lycosidae)

lygophilous Thriving in dark or shaded habitats; **lygophile, lygophily.**

Lymantriidae Tussock moths; widely distributed family containing about 2000 species of medium-sized moths; typically lacking a proboscis; antennae usually bipectinate in both sexes; larvae often hairy and brightly coloured; includes the gypsy moth and other species that cause serious damage to foliage of trees.

lynx Felidae *q.v.*

lyrebird Menuridae *q.v.*

lysocline The oceanic depth zone between the

gypsy moth (Lymantridae)

carbonate dissolution depth (about 4000 m) and the carbonate compensation depth (about 5000 m), representing the major facies change between well preserved and poorly preserved calcareous elements of the ocean floor sediment.

Lyssacinosida Order of hexasterophoran sponges which are typically cup-shaped, vase-shaped or tubular with a stalk that is either fixed to a hard substratum or anchored in loose sediment; found in all oceans to depths greater than 6000 m.

Lythraceae Purple loosestrife, crape myrtle, water willow; family of Myrtales containing about 500 species of mostly herbs widespread in tropical regions; characterized by flowers with usually 4–8 sepals, petals and stamens.

m

Macaronesia A biogeographical area encompassing the islands off the coast of N.W. Africa and Europe, including the Azores, Canaries, Cape Verde Islands and Madeira.

macaw Psittacidae *q.v.*

maceration The fragmentation and separation of parts of a specimen for microscopic examination, often by chemical treatment with strong acids.

Machaeridia Extinct class of echinoderms known from the Ordovician to the Devonian; body enclosed in an elongate, bilaterally symmetrical shell composed of calcite plates.

mackerel Scombridae *q.v.*

mackerel shark Lamnidae *q.v.*

macrobenthos The larger organisms of the benthos, exceeding 1 mm in length; *cf.* meiobenthos, microbenthos.

macrobiota Large soil organisms, exceeding about 40–50 mm in length; *cf.* mesobiota, microbiota.

Macrocephenchelyidae Monotypic family of little-known Indo-Pacific marine anguilliform teleost fishes (eels); body smooth, strongly compressed posteriorly, to 500 mm length; dorsal, caudal and anal fins continuous; snout short and blunt.

macroclimate 1: The climate of a major geographical region. 2: The conditions of temperature, precipitation, relative humidity, sunshine and other meteorological factors, recorded about 1.5 m above ground level to avoid topological, vegetational and soil inflences; *cf.* microclimate.

Macrodasyida Order of gastrotrichs found in sandy sediments of marine and brackish water habitats; characterized by pharyngeal pores, and a typically strap-shaped body bearing several adhesive tubes anteriorly.

macroevolution Major evolutionary events or trends such as the origin of higher groups, above species level; *cf.* microevolution.

macrogamete The larger of two anisogametes, usually regarded as the ovum or female gamete; *cf.* microgamete.

macronutrients Those nutrients required in relatively large amounts for optimal growth; *cf.* micronutrients.

macrophagous Feeding on relatively large food particles or prey; **macrophage, macrophagy**; *cf.* microphagous.

macrophanerophyte Megaphanerophyte *q.v.*

macrophyll A Raunkiaerian leaf size class *q.v.* for leaves having a surface area between 18 225 and 164 025 mm^2.

macrophyte A large plant, used especially of aquatic forms such as kelp; **macrophytic.**

macrophytophagous Feeding on higher plant material only; **macrophytophage, macrophytophagy.**

macroplankton Large planktonic organisms 20–200 mm in diameter.

Macropodidae Kangaroos, wallabies; family containing 50 species of small to large (over 2.5 m) terrestrial or arboreal diprotodont marsupials found in Australia, Tasmania and New Guinea; hindlimbs much larger than forelimbs for hopping (ricochetal) locomotion; tail often long and stout; marsupium opening anteriorly; mostly herbivorous.

kangaroo (Macropodidae)

Macrorhamphosidae Snipefishes; family containing 11 species of tropical and warm

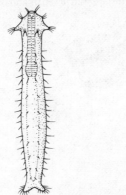

gastrotrich (Macrodasyida)

temperate marine gasterosteiform teleost fishes; body deep and compressed, to 300 m in length, bearing bony lateral plates; snout elongate with tiny apical mouth; first dorsal fin with 4–8 spines.

Macroscelididae Elephant shrews; family containing 15 species of small (length to 200 mm) African mammals (Insectivora); tail elongate, hindlimbs much longer than forelimbs, locomotion saltatorial.

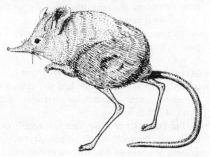

elephant shrew (Macroscelididae)

macroscopic Of relatively large size; used of a structure or occurrence visible to the naked eye, or with the aid of a hand lens; *cf.* microscopic.

macrosmatic Used of an organism possessing a highly developed sense of smell; *cf.* microsmatic.

macrospore The larger of two kinds of haploid spores produced by vascular plants, regarded as the female spore; *cf.* microspore.

Macrostomida Order of free-living turbellarians found interstitially or on the surface of plants and rocks in marine and freshwater habitats; characterized by a simple pharynx, straight gut, paired excretory ducts, and by the absence of yolk glands and of flagella on the sperm.

macrothermophilous Thriving in the tropics; thriving in warm habitats; **macrothermophile, macrothermophily.**

Macrouridae Rattails, grenadiers; family containing about 250 species of primarily bottom-living deep-sea gadiform teleost fishes characterized by a long tapering tail lacking caudal fin; pelvics thoracic; barbel usually present, gas bladder lacking connection to

inner ear; photophores often present; body length to 900 mm.

Macrura The long-tailed decapod crustaceans, including the shrimp-like and lobster-like forms; formerly used as a suborder but now replaced by the suborder Dendrobranchiata and part of the Pleocyemata.

Macrurocyttidae Family of small (to 150 mm) marine zeiform teleost fishes from South America and the Philippines; body strongly compressed, mouth and eyes large; scales small and thin, dorsal fin tall and spinose.

madder Rubiaceae *q.v.*

Madeira vine Basellaceae *q.v.*

madescent Becoming moist.

madid Wet or moist.

Madreporaria Scleractinia *q.v.*

madrone Ericaceae *q.v.*

mafic Pertaining to rocks rich in magnesium and iron.

Magelonida Order of detritus-feeding polychaete worms that inhabit fragile mucus-lined burrows in soft sediments; body thread-like, divided into 2 distinct regions; prostomium fused to peristomium, bearing long tentacular palps; pharynx eversible, unarmed; contains about 45 species in a single family, sometimes included in the order Spionida.

maggot Grub-like larval stage of dipterous insects.

magnetic anomaly A reversal in the magnetic polarity in rocks or sediments indicating a reversal of the Earth's magnetic field.

magnetotropism Orientation in response to a magnetic field; **magnetotropic.**

Magnoliaceae Magnolia, tulip tree; family of archaic flowering plants (Magnoliales) containing about 210 species found mostly in warm regions; flowers often large with poorly differentiated sepals and petals; stamens

tulip tree (Magnoliaceae)

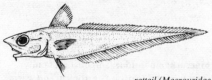

rattail (Macrouridae)

typically ribbon-shaped blades with embedded pollen sacs.

Magnoliales Order of rather archaic flowering plants (Magnoliidae), mostly occurring in tropical or moist warm temperate regions; plants woody, with simple leaves, commonly producing isoquinoline alkaloids; flowers typically without clearly differentiated sepals and petals, often pollinated by beetles.

Magnoliidae A subclass of dicotyledons (Magnoliopsida) comprising those forms that have retained one or more of a suite of primitive characters, the most important of which is the possession of pollen with a single germinal aperture, or a simple pistil and numerous stamens; contains 8 orders, Aristolochiales, Illiciales, Laurales, Magnoliales, Nymphaeales, Papaverales, Piperales and Ranunculales.

Magnoliophyta Flowering plants, angiosperms; the dominant land vegetation of the Earth with over 220 000 species; characterized by aggregation of sexual reproductive structures with specialized shoots (flowers) which typically comprise 4 kinds of modified leaves; sepals, petals, stamens (male organs) and carpels (female organs); ovules are enclosed within an ovary and mature into seeds in a fruit; pollen grains are produced by the stamens and germinate on the receptive stigma of the carpel, producing a pollen tube that delivers 2 sperm nuclei into the embryo sac of the ovule; flowering plants extend through the fossil record from the Lower Cretaceous to the present.

Magnoliopsida The dicotyledons; the largest class of flowering plants containing about 170 000 woody or herbaceous species; leaves usually net-veined and differentiated with a stalk (petiole) and an expanded blade; embryo typically has 2 cotyledons, although there are exceptions.

magpie Corvidae *q.v.*

mahogany Meliaceae *q.v.*

maidenhair tree Ginkgoatae *q.v.*

maize Poaceae *q.v.*

mako shark Lamnidae *q.v.*

Malacobothrii Alternative name for the Digenea *q.v.*

malacology The study of molluscs.

malacophilous Pollinated by snails; **malacophily.**

Malacopterygii Little-used grouping of lower fishes; characterized by soft-rayed fins lacking spines, pectoral fins set low on body and pelvics set far back.

Malacosteidae Loosejaws; family containing 10 species of luminescent deep-sea stomiiform teleost fishes found from surface to depth of 4000 m; body elongate, naked, to 300 mm in length; jaws very long with massive gape; membrane absent between the sides of the lower jaw; photophores minute, distributed over head and body.

Malacostraca The higher crustaceans; diverse class of crustaceans with a segmented body divided into 3 functional regions, head (5 segments), thorax (8 segments) and abdomen (6 or 7 segments); contains about 21 000 species in 3 subclasses, Eumalacostraca, Hoplocarida and Phyllocarida; includes crabs, lobsters, shrimps, hoppers, slaters, woodlice and many other groups.

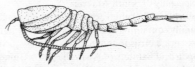

cumacean (Malacostraca)

Malacotylea Alternative name for Digenea *q.v.*

Malapteruridae Electric catfishes; family of nocturnal Old World freshwater teleost fishes (Siluriformes) comprising only 2 species, found in quiet backwaters of central and western Africa; body robust, to 1.2 m in length, naked, eyes small; anal and adipose fins small; 3 pairs of barbels present; produce powerful electric shocks to stun prey.

malaxation The act of chewing; mastication.

male The sperm-producing form of a bisexual or dioecious organism; symbolized ♂; *cf.* female.

Malesherbiaceae Small family of Violales containing about 25 species of herbs or half shrubs native to dry habitats in the Andes.

Mallophaga Chewing lice; order of small wingless hemipterodean insects that live

chewing louse (Mallophaga)

exclusively as obligate ectoparasites of birds and mammals; body depressed, mouthparts specialized for chewing; eggs attached to hair or feathers; some may be of economic importance in domestic animals.

mallow Malvaceae *q.v.*

Malm A geological epoch of the late Jurassic period (*c.* 160–140 million years B.P.).

malnutrition A deficiency condition in which one or more necessary nutrients is available in an insufficient amount for normal growth and maintenance.

Malpighiaceae Barbados cherry; family of Polygalales containing about 1200 species of woody plants commonly provided with pick-shaped unicellular hairs and with pentamerous flowers; found mostly in warm regions, especially of South America.

Maluridae Thornbills, chats, wrens; diverse family containing about 100 species of small passerine birds found in forest, grassland and open arid habitats of Australia, New Guinea and New Zealand; coloration cryptic or bold; habits solitary to gregarious, arboreal to terrestrial, non-migratory; feed largely on insects, seeds and fruit; domed nest of grass, on or off the ground.

Malvaceae Hollyhock, hibiscus, cotton, mallow; large, cosmopolitan family of herbs or soft shrubs generally covered with stellate hairs; cotton is the seed hairs of *Gossypium* species; flowers usually with 5 sepals, 5 free petals and numerous stamens and carpels, joined into rings.

hibiscus (Malvaceae)

Malvales Order of Dilleniidae probably derived from the Theales, containing 5 families of mostly trees and shrubs producing mucilage in special cells or canals.

mamba Elapidae *q.v.*

Mammalia Mammals; class of warm-blooded vertebrates characterized by mammary glands, epidermal hair, enucleate red blood cells, muscular diaphragm, 3 middle ear bones, a single pair of bones (dentary) forming lower jaw articulating directly with cranium, left aortic arch only present in adults; habits primarily terrestrial but also includes aquatic, fossorial, arboreal and aerial forms; comprises the Prototheria (egg-laying mammals) and Theria (viviparous mammals).

mammalology The study of mammals.

mammoth Extinct Pleistocene elephants (Elephantidae *q.v.*) found in steppe and tundra habitats; tusks elongate and strongly curved; often with long hair.

Mammutidae Mastodontidae *q.v.*

manakin Pipridae *q.v.*

manatee Trichechidae *q.v.*

manca Juvenile stage found in some peracaridan crustaceans in which the last pair of thoracic limbs is lacking.

Mandibulata A former subphylum of arthropods comprising 6 classes, Crustacea, Insecta, Chilopoda, Diplopoda, Pauropoda and Symphyla; the Crustacea is now regarded as an independent subphylum or phylum, the 5 remaining classes (plus the Onychophora) comprising the Uniramia.

mandrill Cercopithecidae *q.v.*

mango Anacardiaceae *q.v.*

mangrove 1: A tidal salt-marsh community dominated by trees and shrubs, particularly of the genus *Rhizophora*, many of which produce adventitious aerial roots. 2: Combretaceae *q.v.* 3: Rhizophorales *q.v.*

Manidae Pangolins, scaly anteaters; family containing 7 species of nocturnal, terrestrial or arboreal mammals (Pholidota) found in Ethiopian and Oriental regions; body covered with armour of large overlapping scales; mouth tubular with long protrusible sticky tongue, teeth absent; limbs bearing strong claws; feed mainly on ants and termites.

pangolin (Manidae)

mannikin Estrildidae *q.v.*

manometabolous Used of a pattern of development in some insects in which a minor or very gradual metamorphosis occurs without a resting stage; **manometaboly.**

manta ray Mobulidae *q.v.*

mantis Mantodea *q.v.*

mantis shrimp Stomatopoda *q.v.* (Hoplocarida).

Mantodea Mantises, praying mantis; order comprising about 1800 species of medium to large (10–140 mm) predatory orthopterodean insects in which body shape is highly adapted for camouflage; head very mobile, eyes large; forelegs large, spinose, raptorial for grasping prey; forewings slender and leathery, hindwings membranous with broad fan; wings often reduced in female; most feed on insects and spiders seized by rapid strike of forelimbs, the mantis waiting motionless for the prey to arrive; in some species female eats male headfirst during copulation; distribution mainly tropical and subtropical.

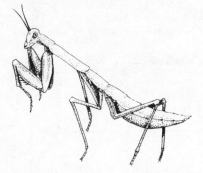

preying mantis (Mantodea)

maple Aceraceae *q.v.*

maquis Scrub woodland; stunted woodland typically found on poor soil in deforested, semi-arid regions.

Marantaceae Prayer plant, peacock plant, never never plant; family of Zingiberales comprising about 400 species of perennial herbs with few leaves, and with flowers modified into a complex mechanism for insect pollination; found throughout the tropics; stamens all petaloid, one bearing a single marginal pollen sac.

Marattiopsida Subdivision of tropical terrestrial ferns (Filicophyta) in which the massive, thick-walled sporangia within a cluster (sorus) are fused; fronds often fleshy and undivided, unroll circinately; fronds fall off leaving swellings on rhizome; plants with mucilage ducts.

Marcgraviaceae Family of Theales containing about 100 species of lianas or glabrous epiphytic shrubs with clinging roots; flowers often pollinated by hummingbirds; native to tropical America.

Marchantiales Order of liverworts (Hepaticopsida); typically large plants organized as a dorsoventrally flattened, linear thallus which commonly bears ventral scales and possesses both smooth and tuberculate rhizoids; sporophyte capsule releases spores by dehiscence or disintegration.

mare's tail Hippuridaceae *q.v.*

margin shell Neogastropoda *q.v.*

mariculture Cultivation, management and harvesting of marine organisms in their natural habitat or in specially constructed channels or tanks.

marigold Asterales *q.v.*

marine Pertaining to the sea.

marine humus The products of organic decomposition that accumulate in solution or suspension in the sea.

marine depth zones See diagram on page 236 for zonation of the marine province.

maritime Living in the sea; having a special affinity for the sea.

marlin Istiophoridae *q.v.*

marmoset Callitrichidae *q.v.*

marmot Sciuridae *q.v.*

marsh An ecosystem of more or less continuously waterlogged soil dominated by emersed herbaceous plants, but without a surface accumulation of peat.

Marsileales Water ferns; small order of ferns (Filicopsida) growing underwater, on muddy shores or in ditches; creeping rhizomes rooted in mud, producing long stalked fronds; heterosporous, producing micro and megasporangia inside hard nut-like structures attached to leaf stalks.

marsupial Metatheria *q.v.*; Marsupialia.

marsupial mole Notoryctidae *q.v.*

marsupial mouse Dasyuridae *q.v.*

Marsupicarnivora Order of carnivorous and omnivorous marsupials comprising 5 families, Didelphidae (opossums), Thylacinidae (Tasmanian wolf), Dasyuridae (native cats, marsupial mice), Myrmecobiidae (numbat) and Notoryctidae (marsupial mole); includes terrestrial, arboreal, semiaquatic and fossorial forms.

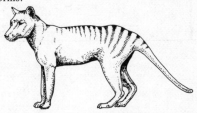

Tasmanian wolf (Marsupicarnivora)

226

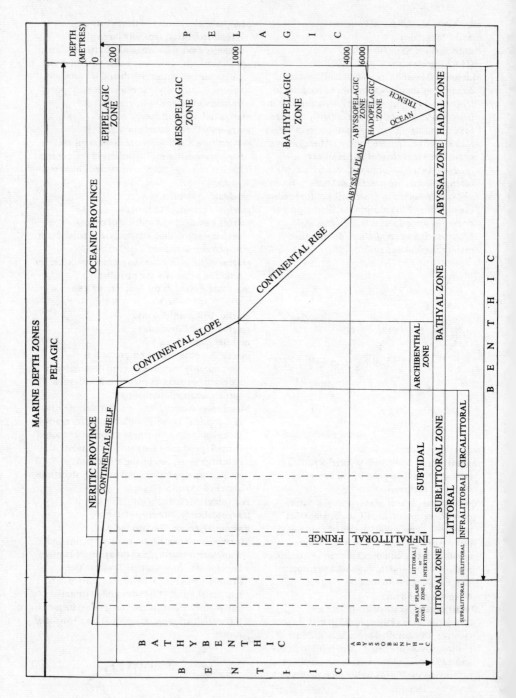

marine depth zones

marten Mustelidae *q.v.*

Martyniaceae Unicorn plants; small family of herbs typically clothed in sticky hairs; showy flowers arranged in terminal inflorescence; fruit a capsule with the persistent style forming a horn or hooked process; known mainly from tropical and subtropical America.

mason wasp Eumenidae *q.v.*

mast year A year in which seed production is exceptionally high, usually followed by a period of several years during which relatively few seeds are produced.

Mastacembelidae Spiny eels; family of nocturnal burrowing, fresh and brackish water teleost fishes (Perciformes) found in tropical Africa and southern Asia; body eel-like with many free spines in front of long dorsal fin; snout bearing fleshy tentacle.

mastication The act of chewing.

Mastigomycotina Subdivision of fungi characterized by the production of motile flagellate zoospores during the life cycle; thallus unicellular to mycelial; cosmopolitan in aquatic habitats and soil; also found as parasites; comprises 3 classes distinguished by flagellar type, Chytridiomycetes, Hyphochytridiomycetes and Oomycetes, each of which is sometimes treated as a distinct phylum of the Protoctista.

Mastigophora True protozoan flagellates; subphylum of Sarcomastigophora characterized by 1 to many flagella although 2 or 4 are the commonest numbers; about one third of the species are symbionts or parasites of other animal hosts, free-living forms are abundant in both aquatic and terrestrial habitats; this subphylum comprises 3 classes, the zooflagellates (Zoomastigophora), the phytoflagellates (Phytomastigophora) and the dinoflagellates (Dinoflagellata), the last 2 groups typically possess chloroplasts and are now commonly regarded as algae.

mastigopus Late (megalopal) larval stage of sergestoid shrimps (Decapoda).

Mastodontidae Mastodon; extinct family of elephant-like animals (Hyracoidea) known from the Lower Miocene at least to the end of the Pleistocene; bodies shorter and heavier than modern elephants; tusks usually present in both upper and lower jaws.

maternal Pertaining to, or derived from, the female parent; *cf.* paternal.

maternal inheritance Inheritance in which characters or traits in the offspring are determined by cytoplasmic factors, such as mitochondria or chloroplasts, that are carried by the female gamete.

maternal sex determination The condition in which the sex of the offspring is determined by the idiotype of the female gamete.

mating Pairing of unisexual individuals for purposes of reproduction; **mate.**

matromorphic Resembling the female parent; *cf.* patromorphic.

maturation The attainment of sexual maturity; differentiation of the gametes; the increasing complexity or precision of behaviour patterns during growth to sexual maturity, not learned from prior experience.

matutinal Pertaining to the morning; used of plants that flower in the morning.

Maxillopoda Group of crustaceans containing the Mystacocarida, Cirripedia, Copepoda, Branchiura and Ostracoda; sometimes regarded as a level of organization.

copepod (Maxillopoda)

Mayacaceae Small family of Commelinales comprising 3 species of freshwater herbs, either free-floating or submerged and rooted in the substratum; flowers aerial each with 3 sepals, petals, stamens and carpels; found in tropical and warm temperate America and Africa.

mayfly Ephemeroptera *q.v.*

meadow An area of closed herbaceous vegetation dominated by grasses.

meadow beauty Melastomataceae *q.v.*

meadow sweet Rosaceae *q.v.*

mealybug Coccoidea *q.v.* (Homoptera).

mean Average; equal to the sum of the observations divided by the number of observations.

mean high water (MHW) The average height of all high waters recorded at a given place.

mean high water neap (MHWN) The average height of all high waters recorded at a given place during quadrature *q.v.*

mean high water spring (MHWS) The average height of all high waters recorded at a given place during syzygy *q.v.*

mean higher high water (MHHW) The average height of all higher high waters *q.v.* recorded at a given place.

mean low water (MLW) The average height of all low waters recorded at a given place.

mean low water neap (MLWN) The average height of all low waters recorded at a given place during quadrature *q.v.*

mean low water spring (MLWS) The average height of all low waters recorded at a given place during syzygy *q.v.*

mean lower low water (MLLW) The average height of all lower low waters *q.v.* recorded at a given place.

mean range The difference in height between mean low water and mean high water.

mean sea level (MSL) The average height of the surface of the sea determined from all stages of the tide.

mean tide level (MTL) The average of the observed heights of high water and low water.

mean water level (MWL) The mean surface level determined by averaging the height of the water at a given place at hourly intervals over a prolonged period of time.

mechanical weathering The breakdown of parent rock material without chemical changes, as part of the soil-forming process.

mechanotropism Orientation in response to a mechanical stimulus; **mechanotropic.**

Mecoptera Scorpion flies, snow fleas; order of mostly diurnal holometabolous insects that inhabit moist forests feeding on nectar or preying on other insects; head prolonged into rostrum, mouthparts mandibulate; wings narrow, membranous, subequal, occasionally reduced or absent; legs long and slender; larvae caterpillar-like; about 450 species in 3 suborders, Protomecoptera, Eumecoptera and Neomecoptera.

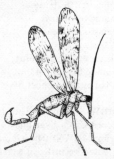

scorpion fly (Mecoptera)

Medeolariales Monotypic order of discomycete fungi containing a single species parasitic on the stems of the North American plant *Mededa virginiana*; characterized by its unorganized ascocarp.

Mediterranean climate A climate characterized by hot dry summers and mild wet winters.

Medithermal period An interval characterized by expanding mountain glaciers (*c.* 4000–2000 years B.P.); Little Ice Age; *cf.* Altithermal period, Anathermal period.

medusa fish Centrolophidae *q.v.*

Medusagynaceae Family of Theales containing a single species of glabrous trees native to the Seychelles.

Medusandraceae Family of Santalales containing a single tree species native to rainforests of tropical Africa.

mega- (M) Prefix meaning large, great, greater than usual; used to denote unit x 10^6.

Megachiroptera Suborder of Old World fruit bats (Chiroptera) comprising a single family Pteropodidae; these bats lack the echolocation mechanism of Microchiroptera.

flying fox (Megachiroptera)

Megadermatidae False vampires; family containing 5 species of carnivorous microchiropteran bats found in Australian, Oriental and Ethiopian regions; typically feeding on small vertebrates including fishes, amphibians, reptiles and mammals, sometimes on insects.

megafauna Large animals visible to the naked eye.

megaflora Large plants visible to the naked eye.

Megagaea A zoogeographical area comprising the Palaearctic, Nearctic, Ethiopian and Oriental Regions; *cf.* Arctogaea, Neogaea, Notogaea, Palaeogaea.

megagamete The larger of two anisogametes, usually regarded as the ovum or female gamete; macrogamete; *cf.* microgamete.

Megalomycteridae Family of deep-sea lampridiform teleost fishes containing 5 species characterized by extremely large olfactory organs giving rise to the common name, large-nose fishes.

Megalonychoidea Ground sloths; extinct superfamily of ground-dwelling edentates known from the Oligocene to the Pleistocene; 5 digits on each limb, often with very long claws; teeth simple and reduced.

megalopa Late larval stage of many malacostracan crustaceans in which at least one pair of swimming pleopods is present.

Megalopidae Tarpon; family containing 2 species of large (to 2.5 m) marine teleost fishes (Elopiiformes), highly prized as game fish by anglers.

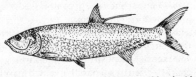

tarpon (Megalopidae)

megaloplankton Largest of the planktonic organisms, typically greater than 10 mm in diameter.

Megaloptera Dobsonflies, alderflies; suborder containing about 300 species of primitive neuropteran insects in which the adults have 2 pairs of large wings but perform slow, clumsy flight; adults probably non-feeding; larvae aquatic, active predators with abdominal gills; sometimes treated as an order distinct from the Neuroptera.

megaphanerophyte A tall phanerophyte *q.v.* with renewal buds more than 30 m above ground level; macrophanerophyte.

megaphyll A Raunkiaerian leaf size class *q.v.* for leaves having a surface area greater than 164 025 mm².

megaplankton Planktonic organisms of the largest size group, 200 mm to 2 m.

Megapodiidae Family containing 12 species of medium-sized pheasant-like birds (Galliformes), with very large strong feet; found in wooded areas of the Indo-Pacific from Australia to Polynesia; habits gregarious, monogamous; feed on seeds, fruit and small invertebrates; eggs laid in mounds of rotting vegetation or in sand, incubated by the heat of the nest.

megapode (Megapodiidae)

megascopic Used of a structure or occurrence observable with the naked eye, or with the aid of a hand lens; *cf.* microscopic.

megaspore The larger of two kinds of haploid spores produced by vascular plants; regarded as the female spore; *cf.* microspore.

megasporogenesis The production of megaspores in vascular plants.

megatherm A plant favouring warm habitats, and with a requirement for a minimum temperature of 18°C in the coldest month; **megathermic**; *cf.* hekistotherm, mesotherm, microtherm.

meiobenthos Small benthic organisms that pass through a 1 mm mesh sieve but are retained by a 0.1 mm mesh; *cf.* macrobenthos, microbenthos.

meiofauna The small interstitial animals that pass through a 1 mm mesh sieve but are retained by a 0.1 mm mesh.

meiosis Reduction division; two successive divisions of a diploid nucleus preceding the formation of the haploid gametes or meiospores, each of which contains one of each pair of the homologous chromosomes of the parent cell; **meiotic.**

meiospore A haploid cell produced by meiosis that forms the gametophyte by mitotic division.

meiotherm A plant thriving in cool temperate habitats; **meiothermous.**

meiotic parthenogenesis Parthenogenesis *q.v.* in which meiosis is preserved and diploidy is reinstated either by fusion of haploid nuclei within a single gamete or by the formation of a restitution nucleus.

Melamphaidae Circumtropical family containing 30 species of small (to 150 mm) bathypelagic beryciform teleost fishes; body moderately deep, rounded, bearing large cycloid scales; head cavernous; teeth minute, setiform; caudal fin with 3–4 procurrent spines.

Melanconiales Order of coelomycete fungi containing about 1000 species in a single family most of which are saprobic on plant material or parasitic; produce fruiting bodies bearing conidia on upper surface with sterile stromatic tissue restricted to basal part.

Melanesia A geographical area and an ethnological unit, sometimes considered as a distinct biogeographical unit separate from Polynesia; comprising the continental islands of the West Pacific south of the Equator, and some islands of volcanic origin; includes New Guinea, New Caledonia, Fiji, Vanuatu,

Bismarck archipelago and the Solomon Islands.

melangeophilous Thriving in or on black loam; **melangeophile**, **melangeophily**.

melanism An increase in the amount of black or dark pigment in an organism, population or group; melanic; *cf.* industrial melanism.

Melanocetidae Family containing 10 species of small (to 130 mm) bathypelagic anglerfishes (Lophiiformes); body rounded, naked, mouth large bearing sharp curved teeth in female; male free-living, smaller than female with small mouth lacking teeth; lure absent in male.

anglerfish (Melanocetidae)

Melanogastrales Small order of gasteromycete fungi found in terrestrial or occasionally marine habitats; usually saprophytic or sometimes mycorrhizal; fruiting bodies typically produced underground.

Melanommatales Large order of loculoedaphomycetid fungi containing saprobic and lichenized forms as well as parasites of other plants and fungi; includes forms in which the sterile pseudoparaphyses (which separate the asci) are trabeculate and the ascospores are symmetrical.

Melanonidae Family containing 3 species of little-known pelagic gadiform teleost fishes; 2 dorsal fins and single long anal fin present; photophores absent.

Melanosporales Order of pyrenomycete fungi typically found in soil, often in association with other fungi; characterized by globose ascocarp with semitransparent walls and deliquescent asci.

Melanostomiatidae Black dragonfishes; family containing 90 species of luminescent pelagic marine teleost fishes found from surface to depth of 4500 m; body elongate, black, naked, to 350 mm in length; photophores minute,

dragonfish (Melanostomiatidae)

numerous on head and body; barbel present below mouth.

Melanostomiidae Melanostomiatidae *q.v.*

Melanotaeniidae Rainbowfishes; family containing 20 species of small (to 100 mm) colourful fresh and brackish water atheriniform teleost fishes from Australia and New Guinea; body deep, compressed, anterior dorsal fin spinose.

Melastomataceae Meadow beauty, purple glory tree; large family of Myrtales containing about 4500 species of mostly herbs and shrubs characterized by subparallel-veined leaves and specialized stamens; widespread in tropical and subtropical regions, especially of South America; flowers normally with 4–5 sepals, 4–5 petals, and 8–10 stamens.

Meleagridae Turkeys; family containing 2 species of large terrestrial gallinaceous birds found in forests of eastern United States and Mexico; habits gregarious, polygamous; feeding on insects, variety of seeds and other plant material; nest solitary, in hollow in ground; important as domesticated fowl.

turkey (Meleagridae)

Meliaceae Mahogany; family of Sapindales containing about 550 species of mostly trees and shrubs with bitter bark and spirally pinnate leaves; found mainly in warm regions; flowers usually with stamens forming a tube, and a superior ovary.

Melianthaceae Family of Sapindales containing 8 species of shrubs and trees native to Africa.

Meliolales Black mildews; order of pyrenomycete fungi widely distributed throughout moist tropical regions; forming dark-coloured mycelium over surface of leaves and stems of higher plants and feeding by means of specialized hyphae which form haustoria penetrating the host cells.

Meliphagidae Honeyeaters; family containing about 170 species of small colourful passerine birds found in forest to open arid habitats of

Australia, New Zealand, New Guinea, southern Africa and many oceanic islands; bill variable, short to long, straight or curved; tongue divided distally and frilled; habits gregarious, arboreal, aggressive; feed on insects, nectar and fruit; cup or domed nest placed in tree.

meliphagous Feeding on honey; **meliphage, meliphagy.**

Melittidae Diverse family of solitary, non-parasitic bees (Hymenoptera) which dig nests in soil or build nests in pre-existing cavities, especially in wood; adults and larvae feed on nectar and pollen; some larvae spin cocoons.

melittology The study of bees.

melittophilous 1: Thriving in association with bees; used of an organism that spends part of its life cycle in association with bees. 2: Pollinated by bees; **melittophile, melittophily.**

melliferous Honey-producing.

mellisugous Feeding on honey; honey sucking.

mellivorous Feeding on honey; **mellivore, mellivory.**

Meloidae Blister beetles; family of often brightly coloured beetles (Coleoptera) usually found on flowers; adults producing a chemical (cantharidin) which causes skin blisters; larvae feeding on eggs of other insects and on food stores in bees' nests; contains about 3000 species, most common in warm dry regions.

Membracidae Treehoppers; family of mostly host-specific insects (Homoptera) feeding on the sap of trees and shrubs; nymphs gregarious and often attended by ants as they produce honeydew; contains about 2400 species abundant in subtropical and tropical regions.

Mendelian character Any character that is inherited in accordance with Mendel's second law of independent assortment *q.v.*

Mendelian inheritance The inheritance of characters through nuclear chromosomes.

Mendelism Inheritance in accordance with the chromosome theory of heredity.

Mendel's laws First law: that in sexual organisms the two members of an allele pair or pair of homologous chromosomes separate during gamete formation and that each gamete receives only one member of the pair; law of segregation. Second law: that the random distribution of alleles to the gametes results from the random orientation of the chromosomes during meiosis; law of independent assortment.

Mendonciaceae Family of Scrophulariales containing about 60 species of twining shrubs with jointed twigs; native to South America

and tropical Africa.

menhaden Clupeidae *q.v.*

Menidae Moonfish; family of small (to 200 mm) Indo-Pacific marine teleosts fishes (Perciformes) found mostly in deep water off coastal reefs; body deep and strongly compressed, upper profile almost straight, ventral profile deeply convex; fin spines absent; first pelvic fin ray elongate.

Meniscotheriidae Extinct family of ungulates known from the Palaeocene to the Eocene.

Menispermaceae Moonseed; family of vines or climbing shrubs commonly containing poisonous terpenoids; includes about 400 species mostly from the subtropics and tropics; flowers usually with 2 or 3 petals, and numerous stamens and carpels; fruit a drupe.

menotaxis A directed response of a motile organism at a constant angle to the stimulus source; **menotactic.**

mensuration The act of measuring; quantitative estimation.

Menuridae Lyrebirds; family containing 2 species of ground-living passerine birds found in dense forests of southeastern Australia; wings short and rounded, tail long and very ornate with lyre-shaped outer feathers, legs long and stout; habits solitary, secretive, monogamous, run well but fly poorly; feed on ground invertebrates.

lyrebird (Menuridae)

Menyanthaceae Buckbean; cosmopolitan family of Solanales containing about 35 species of aquatic or semiaquatic herbs that produce iridoid compounds; flowers bisexual, with 5 sepals, petals and stamens and a superior ovary.

meraspis Second larval stage of trilobites in which the pygidium was located behind the cephalon but trunk segmentation was absent.

merdicolous Living on or in dung; **merdicole.**

merdivorous Feeding on dung and faecal matter; **merdivore, merdivory.**

meridional Southern; pertaining to, or situated in, the south.

meristic variation Variation in the number of structures or parts.

Merliida Order of sclerosponges containing mostly encrusting forms known from depths of 5 to 160 m in the Atlantic and Indo-Pacific Oceans; when present the basal skeleton comprises mostly calcite.

Merlucciidae Hakes; family containing 12 species of moderately deep water gadiform teleost fishes having widespread distribution, but most abundant in Atlantic and eastern Pacific Oceans; tail elongate, tapering; 1–2 dorsal fins and single anal fin present; photophores and barbels absent.

mermaid's wine glass Dasycladales *q.v.*

Mermithida Small order of enoplian nematodes found only as parasites of insects, slugs and snails; characterized by an exceptionally long and thin body and by the arrangement of oesophageal glands in 2 rows.

Merocheta Superorder of helminthomorphan diplopods comprising a single order Polydesmida, the largest order of millipedes.

merogamy A mode of reproduction in protistans involving gametes that are smaller than the vegetative cell and are produced by multiple fission; **merogamete**; *cf.* hologamy.

meromictic Used of a lake that is permanently stratified due to the presence of a density gradient (pycnocline) resulting from chemical stratification; *cf.* mictic.

meroparasite A partial or facultative parasite that can survive in the absence of the host; *cf.* holoparasite.

meropelagic Used of aquatic organisms that are only temporary members of the pelagic community; *cf.* holopelagic.

Meropidae Bee eaters; family containing 24 species of small colourful birds (Coraciiformes) that feed on insects caught on the wing; possessing slender pointed bills,

bee eater (Meropidae)

often decurved; wings long and pointed; nest in burrows in banks; often migratory and widely distributed in tropical and warm temperate parts of the Old World from Europe to Australia.

meroplankton Temporary members of the planktonic community; **meroplanktonic**; *cf.* holoplankton.

Merostomata Class of large aquatic arthropods (Chelicerata) comprising 2 orders, Xiphosura (king crabs) and the extinct Eurypterida (water scorpions); prosoma covered by a carapace, abdomen with gills and a tail spine.

merry bells Liliaceae *q.v.*

Mertensian mimicry Mimicry in which a moderately offensive form is the model and a fatally offensive form is the mimic; *cf.* mimicry.

mesarch succession An ecological succession beginning in a habitat with a moderate amount of water; *cf.* hydrarch succession, xerarch succession.

Mesaxonia Superorder of ungulates in which the body weight is carried on the third digit; comprising a single order, Perissodactyla.

mesic Pertaining to conditions of moderate moisture or water supply; used of organisms occupying moist habitats.

mesic-supralittoral The wettest part of the supralittoral *q.v.* zone of the seashore, occasionally washed by splashing waves; splash zone.

Mesitornithidae Family containing 3 species of small terrestrial forest-dwelling birds from Madagascar tentatively assigned to the order Gruiformes; legs short and strong, wings small; efficient runners; flight weak, probably flightless.

mesobenthos Those organisms inhabiting the sea-bed in the archibenthal zone, between 200 and 1000 m depth.

mesobiota Soil organisms of intermediate size, from about 40–50 mm in length to a size just visible with the aid of a hand lens; *cf.* macrobiota, microbiota.

mesocosm A medium-scale enclosed experimental facility, used especially for assessing the impact of pollutants on natural ecosystems.

Mesogastropoda An order of prosobranch molluscs containing about 10 000 species of marine, freshwater and terrestrial snails including conches, cowries, fig shells, frog shells, heteropods, periwinkles, triton shells, tun shells, violet snails, wentletraps and winkles; characterized by a primitively

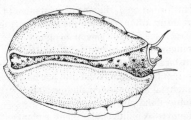

cowrie (Mesogastropoda)

conical, spiral shell usually with right-handed coiling and typically lacking mother-of-pearl (nacre), and by an asymmetrical mantle cavity in which the water current enters from the anterior left side and a single gill is typically present on the left side; usually with internal fertilization.

mesohaline Pertaining to brackish water having a salinity between 3 and 10 parts per thousand or sea water having a salinity between 30 and 34 parts per thousand; *cf.* oligohaline, polyhaline.

mesohalobous Pertaining to planktonic organisms living in brackish water with a salinity between 5 and 20 parts per thousand.

mesohydrophyte A plant thriving under damper conditions than a true mesophyte; **mesohydrophytic.**

mesohylile Pertaining to moist forest habitats.

Mesolithic An archaeological period *c.* 10 000–4000 years B.P.; Middle Stone Age.

mesolithion Organisms inhabiting cavities in rock.

mesomorphic Possessing intermediate characters or traits.

Mesonychidae Extinct family of condylarth mammals known from the Palaeocene to the Oligocene; upper cheek teeth triangular, adapted for shearing; may have fed on carrion.

mesopelagic The pelagic zone of intermediate depth, 200–1000 m; see marine depth zones.

mesophanerophyte A medium-sized phanerophyte *q.v.* with renewal buds 8–30 m above ground level.

mesophilic 1: Thriving under intermediate or moderate environmental conditions; sometimes restricted to conditions of moderate moisture, or to moderate temperature; **mesophile, mesophily.** 2: Used of microorganisms having an optimum for growth between 20 and 45°C.

mesophyll A Raunkiaerian leaf size class *q.v.* for leaves having a surface area between 2025 and 18 225 mm²; sometimes subdivided into notophyll (2025–4500 mm²) and mesophyll (4500–18 225 mm²).

mesophyte A plant thriving under intermediate environmental conditions of moderate moisture and temperature, without major seasonal fluctuations; **mesophytic;** *cf.* hydrophyte, hygrophyte, xerophyte.

mesophythmile Pertaining to the floor of a lake at depths between 6 and 25 m.

Mesophytic The period of geological time during the first appearance of the angiosperms; *cf.* Aphytic, Archaeophytic, Caenophytic, Eophytic, Palaeophytic.

mesoplankton Planktonic organisms of intermediate body size, 0.2–20 mm.

mesopleustophyte Any large plant floating freely between the surface and the floor of a lake.

mesopsammon Those organisms living in the interstitial spaces of a sandy sediment.

mesosaprobic Pertaining to a polluted aquatic habitat having reduced oxygen concentration and a moderately high level of organic decomposition; **mesosaprobe.**

Mesosauria Extinct order of lightly built, anapsid reptiles adapted to life in fresh water; known from Late Carboniferous and Early Permian.

Mesotardigrada Order of tardigrades (water bears) found in Japanese thermal springs (65°C); head lacking cephalic appendages but lateral cirri present; legs bearing 6–10 simple claws.

Mesothelae Suborder of spiders (Araneae) having a segmented abdomen bearing mid-ventral spinnerets; chelicerae move parallel to body axis; contains about 12 species typically inhabiting trapdoor burrows in the ground.

mesotherm A plant favouring intermediate temperature conditions, with a minimum of 22°C in the warmest month and a range of 6–18°C in the coldest month; **mesothermic;** *cf.* hekistotherm, megatherm, microtherm.

mesothermophilous Thriving in temperate regions; **mesothermophile, mesothermophily.**

mesotraphent Used of an aquatic plant characteristic of water bodies with intermediate nutrient concentrations; *cf.* eutraphent, oligotraphent.

mesotrophic 1: Having intermediate levels of primary productivity; pertaining to waters having intermediate levels of the minerals required by green plants; *cf.* dystrophic, eutrophic, oligotrophic. 2: Sometimes used of organisms that are incompletely autotrophic; **mesotrophy.**

Mesozoa Small phylum of ciliated, multicellular animals found as endoparasites in various

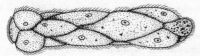

mesozoan (Mesozoa)

marine invertebrates; characterized by a solid, two-layered body lacking skeletal, muscular, nervous, digestive and excretory elements; all are obligate parasites and their simple organization may be secondary, a result of parasitic adaptation; found on both sides of the Atlantic, and on the west coast of America.

Mesozoic A geological era (*c.* 245–65 million years B.P.) comprising Cretaceous, Jurassic and Triassic periods; see geological time scale.

mesquite Mimosaceae *q.v.*

messenger RNA RNA that carries the code for a protein from the DNA in the nucleus to the ribosomes, and acts as a template for the synthesis of that protein; mRNA.

Metabasidiomycetidae Small subclass of tropical to temperate phragmobasidiomycete fungi in which the fruiting bodies are mainly gelatinous to waxy; comprises about 110 species.

metabolism 1: The totality of the synthetic and degradative biochemical processes of living organisms. 2: In ecological energetics, sometimes used as equivalent to respiration *q.v.*

metabolous Used of a pattern of development that includes a metamorphosis.

metachrosis The ability to change colour.

metagynic Used of a hermaphroditic organism that matures first as a male, followed by the transition to a functional female condition; **metagyny.**

metalimnion The zone of steep temperature gradient (thermocline) between the epilimnion and the hypolimnion in a lake; **metalimnetic.**

metallophyte A plant confined to substrates with very high levels of heavy metals.

Metameria A grouping of segmented animals, including the annelids and arthropods, exhibiting metamerism *q.v.*

metamerism The division of the body along an anteroposterior axis into a series of segments, each segment containing representatives of all the organ systems of the body; **metameric.**

metamorphic rock Rock formed by the restructuring of pre-existing rocks under the effects of high temperature and pressure; *cf.* igneous rock, sedimentary rock.

metamorphosis A marked structural transformation during the development of an organism, often representing a change from larval stage to adult; **metamorphic.**

metanauplius Late nauplius larval stage of crustaceans, with more than 3 pairs of limbs present but no functional thoracic limbs.

Metaphyta A kingdom including all the multicellular plants.

metaphyte A multicellular plant; a plant with completely differentiated tissues.

metaplasia 1: An evolutionary state characterized by maximum vigour and diversification of organisms; *cf.* anaplasia, cataplasia. 2: The change of one type of tissue into another; **metaplastic.**

metastasis 1: A change of state, form, position, or function. 2: The transportation of pathogenic organisms or cancerous cells around the host body.

Metatheria Marsupials; infraclass of viviparous mammals (Theria) comprising the single superorder Marsupialia with 4 extant orders, Marsupicarnivora, Peramelina, Paucituberculata and Diprotodontia; young typically develop within abdominal pouch (marsupium); uteri separate, vaginae paired, yolk sac placenta usually present, penis bifurcate; found in Australian and Neotropical regions, with single Nearctic species; Marsupialia.

metathetely The retention of some juvenile characters by an insect adult that has developed through the normal number of moults; **metathetelic.**

metatrophic Used of organisms that utilize organic nutrients as the source of both carbon and nitrogen.

Metazoa A kingdom including all the multicellular animals; Eumetazoa *q.v.*

metazoan A multicellular animal; Eumetazoa *q.v.*

meteorological tide The change in water level resulting from meteorological factors, largely barometric pressure and wind; **meteorologic tide.**

meteorology The study of weather or local atmosphere conditions; **meteorological**; *cf.* climatology.

methanogenic Methane-producing, as of certain autotrophic and chemolithotrophic bacteria.

metochy A neutral symbiosis *q.v.* in which a guest organism freely inhabits the nest of a host species without demands upon the resources of the colony.

metoecious Used of a parasite that is not host-specific; *cf.* ametoecious.

metoxenous Used of a parasite that occupies different hosts at different stages of the life cycle.

metric prefixes Standard prefixes used in conjunction with the metric units of the International System; see page 236.

liverwort (Metzgeriales)

Metzgeriales Small order of liverworts (Hepaticopsida); plants usually thallose with a flattened axis and lateral expansions, typically with smooth rhizoids; sex organs found dorsally on leading axes or on specialized reduced branches but not involving apical cells.

mezereon Thymelaeaceae *q.v.*

Miacidae Extinct family of early carnivores (Fissipedia) known from the Palaeocene and Eocene; typically small, with long bodies and short legs; probably arboreal forest-dwellers.

micelle The individual particles of the dispersed phase of a colloid in which the dispersion medium is a liquid.

Michurinism The theory that the genetic constitution of the scion can be modified by grafting.

micro- (μ) 1: Prefix meaning small, short. 2: Used to denote unit x 10^{-6}; see metric prefixes.

microaerophilic 1: Thriving in a free oxygen concentration significantly less than that of the atmosphere; **microaerophile, microaerophily.** 2: Used of an environment in which the partial pressure of oxygen is significantly below normal atmospheric levels but which is not fully anaerobic *q.v.*

Microascales Small order of plectomycete fungi containing a single family of typically saprophytic forms found in soil, dung and organic debris; characterized by dark, hairy or appendaged ascocarps which contain evanescent asci arranged singly or in short chains in the inner tissues.

microbenthos Microscopic benthic organisms less than 0.1 mm in length; *cf.* macrobenthos, meiobenthos.

microbiology The study of microorganisms.

microbiota Microscopic soil organisms not visible with the aid of a hand lens; *cf.* macrobiota, mesobiota.

microbivorous Feeding on microorganisms, in particular bacteria; **microbivore, microbivory.**

Microcerberidea Suborder of minute interstitial isopod crustaceans having an elongate slender body form, found in subterranean waters of Central America, the Mediterranean region, and from Africa to India.

Microchiroptera Cosmopolitan suborder of mainly small bats that use echolocation for navigation; many are insectivorous, but other diets include fruit, nectar, vertebrates and blood.

bat (Microchiroptera)

microclimate The climate of the immediate surroundings or habitat, differing from the macroclimate *q.v.* as a result of the influences of local topography, vegetation and soil; bioclimate.

Microdesmidae Wormfishes; family containing 30 species of eel-like tropical or subtropical marine gobioid teleost fishes found burrowing in muddy estuarine habitats or on coral reefs; body length to 300 mm, single dorsal fin present; median fins may be continuous with caudal.

microevolution Minor evolutionary events usually viewed over a short period of time, consisting of changes in gene frequencies, chromosome structure or number within a population over a few generations; *cf.* macroevolution.

microfauna 1: Small animals not visible to the naked eye. 2: A localized group of animals. 3: The animals of a microhabitat.

microflora 1: Small plants not visible to the naked eye. 2: A localized flora. 3: The plants of a microhabitat.

microfossil A microscopic fossil.

microgamete The smaller of two anisogametes, usually regarded as the sperm or male gamete; *cf.* macrogamete.

Metric prefixes and multiplication factors

Multiplication factor						Prefix	Symbol
1 000 000 000 000 000 000					10^{18}	exa	E
1 000 000 000 000 000					10^{15}	peta	P
1 000 000 000 000					10^{12}	tera	T
1 000 000 000					10^{9}	giga	G
1 000 000					10^{6}	mega	M
10 000					10^{4}	myria	
1 100					10^{3}	kilo	k
100					10^{2}	hecto	h
10					10^{1}	deca	da
0.1					10^{-1}	deci	d
0.01					10^{-2}	centi	c
0.001					10^{-3}	milli	m
0.000 001					10^{-6}	micro	μ
0.000 000 001					10^{-9}	nano	n
0.000 000 000 001					10^{-12}	pico	p
0.000 000 000 000 001					10^{-15}	femto	f
0.000 000 000 000 000 001					10^{-18}	atto	a

metric prefixes

microgametogenesis The development of microgametes (spermatozoa).

microhabitat A small specialized habitat.

Microhylidae Diverse family containing about 230 species of narrow-mouthed frogs (Anura) with characteristic larval morphology; tadpoles have single median spiracle; beak, barbels and denticles absent; most forms have aquatic eggs and larvae, some have attenuated development or direct development from terrestrial eggs; distributed worldwide, except for Australian and Palaearctic regions.

microhylid frog (Microhylidae)

micromelittophilous Pollinated by small bees; **micromelittophily.**

micromyiophilous Pollinated by small flies; **micromyiophily.**

micron Micrometre; a derived metric unit of length, equal to 10^{-6} m; (μm).

Micronesia An assemblage of many small oceanic islands within the warmer part of the western Pacific Ocean; a recognized geographical and ethnological area which is sometimes considered to be a distinct biogeographical unit, separate from Polynesia.

micronutrients Trace elements; nutrients required in minute quantities for optimal growth; *cf.* macronutrients.

microorganisms Organisms of microscopic or ultramicroscopic size; commonly includes bacteria, blue-green algae, yeasts, some lichens and fungi, protistans, viroids and viruses.

micropalaeontology The study of microscopic fossils – those fossils ranging in size from 1 or 2 cm to a few microns, that must be studied by light or electron-microscopy.

microphagous Feeding on relatively minute particles or on very small prey; **microphage**, **microphagy**; *cf.* macrophagous.

microphanerophyte A small phanerophyte *q.v.* with renewal buds 2–8 m above ground level.

microphilic Thriving only within a narrow range of temperature; **microphile, microphily.**

microphyll A Raunkiaerian leaf size class *q.v.* for leaves having a surface area between 225 and 2025 mm^2.

microphyte A microscopic plant; **microphytic.**

microphytic Used of a plant community comprising only lichens or algae.

microplankton Small planktonic organisms, 20–200 μm in diameter.

micropredator An organism that feeds on a prey organism which is larger than itself and to which it attaches temporarily.

Micropterigidae Family of primitive, small, metallic-coloured moths (Zeugloptera) with functional mandibles; adults feed on pollen.

microscopic Used of an organism or occurrence that cannot be observed without the use of a microscope; *cf.* macroscopic.

microsere An ecological succession within a microhabitat, often failing to reach a stable climax.

microsmatic Used of an organism possessing a poorly developed sense of smell; *cf.* macrosmatic.

Microspora Microsporidians; phylum of intracellular parasitic protozoans occurring in nearly every major animal group, especially arthropods; characterized by the lack of mitochondria and by the production of tiny unicellular spores equipped with an internal, coiled polar filament which is extruded as part of the infection mechanism; includes 2 subclasses, Microsporea and Rudimicrosporea; also treated as a class, Microsporida, of the protoctistan phylum Cnidosporida.

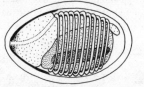

microsporidian (Microspora)

Microsporales Small order of freshwater green algae in which the thallus comprises an unbranched uniseriate filament and vegetative reproduction occurs by fragmentation.

microspore The smaller of two kinds of haploid spores produced by vascular plants, regarded as the male spore; *cf.* megaspore.

Microsporea Subclass of sporulating microsporidian protozoans that produce spores consisting of the sporoplasm, spore wall and extrusion apparatus; includes 2 orders based on the degree of development of the extrusion apparatus, Microsporidia and Minisporidia.

Microsporida Microspora *q.v.*; treated as a class of the protoctistan phylum Cnidosporida.

Microsporidia Order of Microsporea *q.v.* typically with marked development of the extrusion apparatus and a variety of spore sizes and shapes.

microsporogenesis The production of microspores in vascular plants; the production of the male gametophyte (pollen).

microtherm A plant favouring relatively cold habitats with a minimum temperature of 6°C in the coldest month and a range of 10–22°C in the warmest month; **microthermic**; *cf.* hekistotherm, megatherm, mesotherm.

microthermic Used of organisms requiring only a minimum of heat, or favouring relatively low temperatures; **microtherm**.

microthermophilous Thriving in boreal regions; **microthermophile, microthermophily.**

microwhipscorpion Palpigradi *q.v.*

microzoophilous Pollinated by small animals; **microzoophily.**

Mictacea Order of peracaridan crustaceans lacking an epipod on the thoracic legs and lacking a carapace and functional eyes; known from one deep-sea species and one cavernicolous species from marine caves on Bermuda.

mictic Pertaining to patterns of water circulation in a lake; *cf.* amictic, dimictic, holomictic, meromictic, monomictic, polymictic.

midden A refuse heap; used especially in archaeology.

midge Chironomidae *q.v.* (Nematocera *q.v.*).

mid-ocean ridge A topographical feature of the ocean floor comprising rifts and mountain ridges representing the sites of ocean floor upwelling and spreading.

mignonette Resedaceae *q.v.*

migration 1: Periodic or seasonal movement, typically of relatively long distance, from one habitat or climate to another; any general movement that affects the range of distribution of a population or individual; **migrate**; *cf.* dispersion. 2: Movement of a pathogen within the host body. 3: Gene flow; exchange of genetic information between populations.

migrule A unit step or agent of migration.

mildew Any fungal disease in which the fungal mycelium is visible as pale patches on the leaves of the host plant.

milkfish Chanidae *q.v.*

milkweed Asclepiadaceae *q.v.*

milkwort Polygalaceae *q.v.*

millennial Pertaining to periods of thousands of years; *cf.* decennial, secular.

millennium A unit of time equal to 1000 years or 1 millicron.

Milleporina Fire coral; order of Hydrozoa in which the polyps are colonial and form a massive calcareous skeleton; medusae are free-swimming and formed in special cavities in the colony; found in coral reefs in tropical seas; known as fossils from the Cretaceous to Recent times.

Millericrinida Order of deep-sea stalked articulate crinoids in which the cirri are absent, the stalk elongate and slender, and the attachment is by a terminal calcareous disk; 5 arms are present and may be branched; the group is largely extinct with only 8 living species.

millet Poaceae *q.v.*

milli (m) Prefix meaning thousand, thousandth; used to denote unit x 10^{-3}; see metric prefixes.

millipede Diplopoda *q.v.*

mimesis Mimicry *q.v.*; **mimetic.**

mimetic polymorphism Polymorphism in which the various morphs resemble other unpalatable or venomous species in order to deceive the operator (predator).

mimic To imitate; any organism that imitates another.

mimicry The close resemblance of one organism (the mimic) to another (the model) to deceive a third (the operator); mimesis; *cf.* aggressive mimicry, Batesian mimicry, Mertensian mimicry, Müllerian mimicry.

Mimidae Mockingbirds; family containing about 30 species of medium-sized passerine birds found in a variety of forest, woodland and arid habitats of the New World from Canada to Argentina; habits solitary, arboreal to terrestrial, feeding mostly on insects and fruit; cup-shaped nest of grass and twigs, on or off the ground.

mockingbird (Mimidae)

Mimosaceae Acacia, mimosa, mesquite; large family of Fabales containing about 2000 species of sometimes spiny, mostly woody leguminous plants with mostly bipinnately compound leaves; widespread in tropical to subtropical regions especially in dry areas;

flowers regular, with 4–10, or many stamens; formerly known as the subfamily Mimosoideae of the family Leguminosae.

Mindel glaciation A glaciation of the Quaternary Ice Age *q.v.* in the Alpine area, with an estimated duration of 90 000 years.

Mindel–Riss interglacial An interglacial period in the middle of the Quaternary Ice Age *q.v.* in the Alpine area.

mineral replacement A mode of fossilization in which organic material is removed and replaced by minerals.

minerotrophic Used of organisms nourished by minerals.

Minisporidia Order of Microsporea *q.v.* typically with tiny rounded spores with only a minimal development of the extrusion apparatus.

mink Mustelidae *q.v.*

minke whale Balaenopteridae *q.v.*

minnow Cyprinidae *q.v.*

mint Lamiaceae *q.v.*

Miocene A geological epoch within the Tertiary period (*c.* 26–5 million years B.P.); see geological time scale.

miracidium Free-swimming ciliated larval stage of a digenean trematode parasite that hatches from the egg and infects the molluscan intermediate host and develops asexually into the redia q.v.

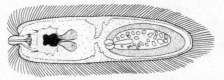

miracidium

Mirapinnidae Hairyfish; family of bizarre lampridiform teleost fishes known from a single immature specimen collected in the North Atlantic; body length to 55 mm, with dense hair-like covering; pelvic fins large, fan-like; pectorals small; caudal fin with overlapping lobes.

mire An ecosystem in which the vegetation is rooted in wet peat; a collective term for various bogs and fens.

Mischococcales Order of mostly freshwater algae (Xanthophyceae) comprising unicellular forms with a distinct cell wall but lacking contractile vacuoles and an eyespot.

miscible Used of liquids that can be mixed together in any ratio without separation; **miscibility.**

Misodendraceae Family of Santalales

containing 10 species of dioecious shrubs which are hemiparasitic on the branches of *Nothofagus* trees in temperate forested regions of South America.

Misophrioida Order of primitive marine copepod crustaceans containing about 16 species of near-bottom living planktonic predators or scavengers restricted mainly to the deep sea but also known from shallow water and marine caves; characterized by a carapace covering the first leg-bearing thoracic segment.

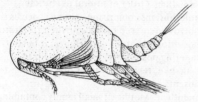

copepod (Misophrioida)

mistletoe Viscaceae *q.v.*

mite Acari *q.v.*

mite harvestmen Suborder comprising about 50 species of small (up to 3 mm) litter-dwelling harvestmen (Opiliones) that superficially resemble true mites (Acari) except that the abdomen is segmented.

mitosis The process of chromosome division and separation that takes place in a dividing cell, producing daughter cells of equivalent chromosomal composition to the parent cell; typically 4 main stages are recognized, prophase, metaphase, anaphase and telophase; **mitotic.**

mitospore A diploid spore produced by mitosis.

Mitrastemonaceae Family of Rafflesiales containing only 2 species of chlorophyll-less endoparasites the vegetative body of which resembles a fungal mycelium penetrating the roots of its host.

mitre shell Neogastropoda *q.v.*

Mitsukurinidae Goblin shark; monotypic family of rare deep-water bottom-dwelling lamniform elasmobranch fishes; body length to 2.5 m, head flattened, snout elongate and blade-like, eyes very small.

mixed layer A layer of water well mixed by wave action or thermocline convection.

mixis The fusion of gametes; karyogamy and karyomixis.

mixolimnion The upper low-density region of periodic free circulation above the monimolimnion *q.v.* in the meromictic lake.

mixosaur One of the earliest of the

ichthyosaurs, without the well developed caudal fin of later forms.

mixotrophic Used of organisms that are both autotrophic and heterotrophic; **mixotroph, mixotrophy.**

mnemotaxis A directed response of a motile organism to a memory stimulus; **mnemotactic.**

moa Dinornithidae *q.v.*

mobbing A collective attack by a group of animals, on a predator that is too large or aggressive to be repelled by individual effort.

Mobulidae Manta rays, devil rays; family of large, tropical and subtropical marine, myliobatiform elasmobranch fishes containing 10 species; body disk much broader than long, distinctly demarkated from head; pectoral fins forming pair of vertical lobes in front of head; tail filamentous distally, caudal fin absent; mouth broad with many minute teeth; feed on small pelagic fishes and crustaceans filtered by gill plates.

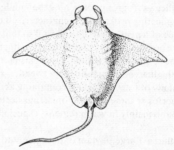

manta ray (Mobulidae)

moccasin Viperidae *q.v.*

Mochokidae Family of crepuscular African freshwater catfishes (Siluriformes) found in slow-moving rivers and swamps; some swim actively in an inverted posture beneath vegetation; body naked, dorsal fin short with stout spine; 3 pairs of barbels present; contains about 150 species; popular in aquarium trade; also known as the Synodidae.

mockingbird Mimidae *q.v.*

Moeritherioidea Extinct suborder of small primitive elephant-like mammals (Proboscidea) known from the Upper Eocene to the Oligocene; exhibited early stage of tusk evolution.

mojarra Gerreidae *q.v.*

mole 1: Talpidae *q.v.*, Chrysochloridae *q.v.* (placental moles). 2: Notoryctidae *q.v.* (marsupial mole).

mole cricket Gryllotalpidae *q.v.*

mole rat 1: Spalacidae *q.v.* 2: Rhizomyidae *q.v.* 3: Bathyergidae *q.v.*

Molidae Sunfishes, trunkfishes; family containing 3 species of large (to 4 m) surface-living oceanic tetraodontiform teleost fishes; body deep, compressed; mouth small with jaw teeth fused to form beak; dorsal and anal fins tall.

sunfish (Molidae)

Mollicutes A group containing the smallest self-replicating prokaryotes, characterized by the lack of a peptidoglycan cell wall so that they are bound only by a plasma membrane; all but one species are parasitic; mycoplasms.

Molluginaceae Common carpetweed; small family of Caryophyllales containing about 100 species of sometimes saponiferous herbs found mainly in warm regions, especially in Africa.

Mollusca Large phylum of unsegmented animals with a ventral gliding surface or foot and a dorsal mantle bearing more-or-less imbricating calcareous scales primitively, or a solid calcareous shell in more advanced forms; posterior mantle cavity containing a pair of respiratory gills (ctenidia) and the excretory and genital openings; mouth typically possessing a radula; head commonly with sensory tentacles and eyes; most internal organs contained within dorsal visceral mass, protected by shell secreted by the mantle; molluscs are widespread in marine, freshwater

whelk (Mollusca)

and terrestrial habitats, mostly free-living but occasionally parasitic; comprising nearly 50 000 species in 8 classes, Bivalvia, Caudofoveata, Cephalopoda, Gastropoda, Monoplacophora, Polyplacophora, Scaphopoda and Solenogastres.

molluscicide A chemical used to kill snails, slugs and other molluscs.

Molossidae Free-tailed bats; family containing about 80 species of insectivorous microchiropteran bats found in both Old and New World.

Molpadiida Order of littoral to abyssal holothurians comprising about 85 species that burrow in soft sediments with their respiratory posterior end at the surface; test robust, tapering posteriorly, tentacles digitate or claw-like, respiratory trees present, tube-feet absent.

Momotidae Motmots; small family containing 9 species of colourful Neotropical forest birds (Coraciiformes) that feed on a variety of invertebrates and small vertebrates caught mainly on the wing; typically nest in burrrows in banks.

motmot (Momotidae)

monacmic Exhibiting one abundance peak per year; *cf.* diacmic, polyacmic.

monandrous Used of a female that mates with a single male; **monandry**; *cf.* polyandrous.

monaxenic Used of a mixed culture of an organism with one prey species; *cf.* axenic, dixenic.

Monera The major kingdom of prokaryotes *q.v.* in which the genetic material comprises a single long molecule of DNA arranged as the chromonema, and the main site of metabolic activity is the plasma membrane; reproduction essentially asexual, neither meiosis nor mitosis is exhibited; comprises 3 divisions, Bacteria, Cyanophycota and Prochlorophycota.

moneran Pertaining to the prokaryotic organisms of the kingdom Monera; *cf.* protistan.

money spider Liniphiidae *q.v.*

mongoose Viverridae *q.v.*

Moniliformida Order of archiacanthocephalan thorny-headed worms found worldwide as parasites of mammals and occasionally birds; characterized by a cylindrical proboscis with long, straight rows of hooks, each with a simple root.

Moniligastrida Order of primitive oligochaete worms typically found in damp primary forests of southeast Asia, from Japan to southern India; body may exceed 1 m length; comprises a single family with about 115 species.

Monimiaceae Large family of woody plants (Laurales) containing about 450 species native to subtropical and tropical regions, particularly in the southern hemisphere; characterized by having flowers in cymes, with sepals and petals frequently not differentiated.

monimolimnion The perennially stagnant high-density, usually saline, deep-water layer in a meromictic lake *q.v.*; *cf.* mixolimnion.

monitor lizard Varanidae *q.v.*

monkey hopper Caelifera *q.v.*

monkey puzzle tree Araucariaceae *q.v.*

monkfish Squatinidae *q.v.*

Monoblepharidales Order of chytridiomycete fungi comprising about 20 species of microscopic saprobes living in temperate freshwater habitats or in tropical and subtropical soils; characterized by a mycelium-like thallus and sexual reproduction by fusion of a non-motile oosphere and motile male planogametes; also treated as a class of the protoctistan phylum Chytridiomycota, under the name Monoblepharida.

monocarpic Producing a single fruit, or having only one fruiting period, during the life cycle; *cf.* polycarpic.

Monocentridae Pine-cone fishes; family containing 2 species of small (to 230 mm) pelagic Indo-Pacific beryciform teleost fishes; body compressed, bearing armour of heavy scales; eyes large, teeth setiform; ventral surface of lower jaw with luminous organ containing symbiotic bacteria.

Monocheta Superorder of helminthomorphan diplopods (millipedes) comprising a single order, Stemmiulida.

monochronic Occurring only once.

Monocleales Small order of liverworts (Hepaticopsida) found in South and Central America, the Caribbean and New Zealand; organized as a dichotomously branching thallus with mostly smooth rhizoids on its ventral surface, some rhizoids having rudimentary pegs; sporophyte capsule produced on long, massive stalk releasing spores by dehiscence.

liverwort (Monocleales)

monoclinous Used of a flower having both male and female organs; hermaphrodite; perfect; **monocliny**; *cf.* diclinous.

Monocotyledoneae Monocotyledons; one of the great divisions of flowering plants in which the embryo typically has one cotyledon; now known as the Liliopsida *q.v.*

monoculture The cultivation or culture of a single crop or species to the exclusion of others.

Monocyathea Extinct class of usually solitary, marine animals (Archaeocyatha) known from the Lower and Middle Cambrian.

Monodactylidae Moonfish; family containing 5 species of small (to 200 mm) coastal marine and brackish-water teleost fishes (Perciformes) found in western Africa and the Indo-Pacific; body very deep and compressed, dorsal and anal fins elongate, pelvics vestigial.

monodomic Used of colonies of social insects that occupy only one nest; *cf.* polydomic.

Monodontidae Family of odontocete mammals comprising the narwhal and the beluga or white whale, found in cold waters of north polar seas; in the male narwhal the left incisor of the upper jaw develops into a long (up to 2.5 m) spiral tusk; feed on large invertebrates and fishes.

monoecious 1: Used of individuals having both male and female reproductive organs; hermaphrodite. 2: Used of a bisexual plant species having separate male and female flowers (unisexual) on the same individual; ambisexual; monecious; **monoecism**, **monoecy**; *cf.* dioecious, trimonoecious, trioecious.

monoestrous Having a single breeding period in a sexual season; *cf.* anestrus, dioestrus, polyoestrous.

monogamous Pertaining to the condition in which a single male and female form a prolonged and more or less exclusive breeding relationship; **monogamic, monogamy**; *cf.* polygamous.

Monogenea Subclass of trematodes comprising about 1100 species, mostly ectoparasitic on the gills and skin of fishes but also found on amphibians, reptiles and crustaceans; characterized by a direct life cycle involving only one host and without intermediate multiplication by larval stages, and by an anterior holdfast (prohaptor) of one or more suckers and a well developed posterior holdfast (opisthaptor) with suckers, clamps or hooks; most species are oviparous, producing typically operculate eggs containing a ciliated oncomiracidium larva.

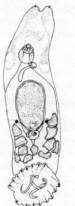

monogenetic fluke (Monogenea)

monogenetic 1: Pertaining to a symbiont having only one host throughout its life cycle; *cf.* digenetic, trigenetic. 2: Used of asexually reproducing organisms, or of asexual reproduction.

monogenic Producing only male or only female offspring; **monogeny**; *cf.* allelogenic, amphogenic, arrhenogenic, thelygenic.

monogenous Pertaining to asexual reproduction or asexually reproducing organisms.

Monognathidae Family containing 6 species of small (to 100 mm) bathypelagic gulper-eels (Anguilliformes) found in tropical and subtropical waters; upper jaws absent, tail long and tapering; pectoral fins small or absent.

Monogonata Large class of rotifers characterized by a single germovitellarium, by a maximum of 2 toes on the foot and by a small

male with a single testis; comprises 3 orders, Collothecaceae, Flosculariaceae and Ploima.

monogoneutic Producing only one brood per year or season; **monogoneutism**; *cf.* digoneutic, trigoneutic, polygoneutic.

monogony Asexual reproduction; monogenesis.

monogyny 1: The mating of a male with only one female. 2: The presence of a single queen in a colony of social insects; **monogynous**; *cf.* oligogyny, polygyny.

Monohysterida Order of chromadorian nematodes found in marine, fresh- and brackish-water habitats, and in soil; loosely characterized by simple and spiral to circular amphids; typically possessing neck setae; generally the cephalic sensory organs are combined so that the second and third circlets form a single whorl of 10 setae.

Monomastigales Order of Prasinophyceae *q.v.* known from both freshwater and marine habitats; characterized by a covering of organic scales over the cell surface and flagella.

monometrosis The founding of a colony of social organisms by a single fertile female, as from a queen in some social insects; **monometrotic**; *cf.* pleometrosis.

monomictic Used of a lake having a single period of free circulation or overturn per year, with consequent disruption of the thermocline; may be either cold monomictic or warm monomictic; *cf.* mictic.

monomorphic 1: Pertaining to a population or taxon showing no genetically fixed discontinuous variation, therefore comprising a single discrete morph; *cf.* dimorphic, polymorphic. 2: Having only one type of flower on an individual plant; *cf.* dimorphic.

Mononchida Order of enoplian nematodes found in soil and freshwater habitats throughout the world; mostly predatory on soil microorganisms; characterized by small cup-like chemosensory organs (amphids) positioned on the head just behind the lateral lips and by the arrangement of head sensilla in one whorl of 6 and another of 10.

monophagous Utilizing only one kind of food; feeding upon a single species or food plant; host-specific; **monophage, monophagy**; *cf.* oligophagous, polyphagous.

monophyletic Derived from the same ancestral taxon; used of a group sharing the same common ancestor; **monophyly**; *cf.* polyphyletic.

Monoplacophora Small class of primitive,

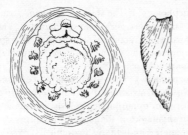

Neopilina (Monoplacophora)

shelled molluscs found in the sea from 200 m to abyssal depths, characterized by a single, typically conical shell formed by the dorsal mantle and up to 8 paired foot retractor muscles and up to 6 pairs of gills around the foot; includes *Neopilina*.

monoplanetic Having a single motile stage during a life cycle; **monoplanetism**; *cf.* diplanetic, polyplanetic.

monoploid Having a single set of chromosomes with a basic chromosome number; equivalent to the lowest haploid number of a polyploid series; **monoploidy**; *cf.* ploidy.

Monoraphideae Order of pennate diatoms (Pennales) possessing a fully developed raphe on one of the valves; also treated as a suborder, Monoraphidineae.

Monorhina A group of jawless fishes possessing a single nostril between the eyes; includes the lampreys.

monospermy The penetration of an ovum by a single sperm during normal fertilization; **monospermic**; *cf.* dispermy, polyspermy.

monotokous Having only one offspring per brood; fruiting only once during a life cycle; *cf.* ditokous, oligotokous, polytokous.

monotopic Occurring or arising in a single locality or geographical area; **monotopy**; *cf.* polytopic.

Monotremata The egg-laying mammals; Prototheria *q.v.*; including the modern duck-billed platypus (Ornithorhynchidae) and spiny anteater (Tachyglossidae).

Monotropaceae Indian pipe; small family of Ericales containing about 12 species of strongly mycotrophic herbs which have reduced leaves, lack chlorophyll and depend on their associated fungus for food, water and minerals.

monotropism Orientation movement in one direction only; **monotropic**.

Monotrysia Suborder of small to minute lepidopteran insects comprising about 950 species, in which the adult proboscis is rudimentary; larvae mainly leaf or seed-pod miners.

monoxenous Used of a parasite utilizing a single host species during its life cycle; **monoxeny**; *cf.* dixenous, heteroxenous, oligoxenous, trixenous.

monozygotic Used of twins derived from a single fertilized egg; such twins are thus genetically identical and of the same sex; identical (twins); *cf.* dizygotic.

monsoon A seasonal climatic pattern in which a cool dry period alternates with a hot wet period.

Monsoon Current A warm surface ocean current that flows eastwards in the northern part of the Indian Ocean produced by the southwesterly winds of the summer monsoon; North East Monsoon Drift; see ocean currents.

Monstrilloida Order of bizarre marine copepods (Crustacea) which live as parasites inside polychaetes and echinoderms during early development, then emerge as non-feeding, planktonic adults, lacking mouthparts and a gut.

monstrosity Any abnormal, malformed, or markedly aberrant individual.

montane Pertaining to mountainous regions; the cool moist upland habitat below the tree line, dominated by evergreen trees.

montbretia Iridaceae *q.v.*

monticolous Living in mountainous habitats; **monticole**.

moon rat Erinaceidae *q.v.*

mooneye Hiodontidae *q.v.*

moonfish 1: Lampridae *q.v.* 2: Menidae *q.v.* 3: Monodactylidae *q.v.*

moonseed Menispermaceae *q.v.*

moor An open elevated region of wet acidic peat typically dominated by heathers, sedges and some grasses.

moose Cervidae *q.v.*

mor A forest soil type characterized by a discrete humus layer separated from the underlying mineral soil, typically highly acidic; a raw humus produced under cool moist conditions; *cf.* mull.

Moraceae Fig, breadfruit, mulberry; large family of Urticales widespread in subtropical and tropical regions; mostly woody plants producing milky latex and often with mineralized cell walls; fruits typically produced in infructescences with a fleshy receptacle, often edible. (Picture overleaf).

moray eel Muraenidae *q.v.*

morel Pezizales *q.v.*

fig (Moraceae)

mores Groups of organisms having similar ecological requirements and behavioural attributes.

moribund Dying; close to death.

Moringaceae Small family of Capparales containing about 10 species of xerophytic, deciduous trees found from Africa to India; fruit a long pod splitting into to 3 valves to release large, 3-winged seeds.

Moringuidae Spaghetti eels; family containing 10 species of marine anguilliform teleost fishes of the western Atlantic and Indo-Pacific; body very slender, smooth, cylindrical; dorsal and anal fins small, pectorals reduced or absent; sexual dimorphism pronounced.

Mormonilloida Small order containing 2 widely distributed species of marine planktonic copepod crustaceans; characterized by a median ventral genital aperture in the female and by the lack of the fifth swimming legs.

Mormoopidae Family containing 8 species of insectivorous bats (Microchiroptera) in which there is no nose leaf, but the nostrils form part of a fleshy pad with the lips; found in Central and South America.

Mormyridae Elephant fishes; family containing about 100 species of mostly small (to 0.5 m) primitive teleost fishes found in African freshwater habitats; body elongate, eyes reduced, caudal musculature electrogenic; habits benthic or pelagic, commonly crepuscular.

morning glory Convolvulaceae *q.v.*

morph A form; any of the individuals of a polymorphic group; any phenotypic or genetic variant; any local population of a polymorphic species exhibiting distinctive morphology or behaviour.

morphocline A graded series of character states of a homologous character.

morphogenesis 1: The totality of the process of embryological development and growth. 2: The evolution of morphological structures; **morphogenetic.**

morphology 1: The study of form and structure of organisms, rocks and sediments. 2: The form and structure of an organism, with special emphasis on external features; **morphologic, morphological.**

morphometry The measurement of external form and structure; the characterization of form for quantitative analysis; **morphometric.**

morphoplankton Planktonic organisms rendered buoyant by anatomical specializations such as oil droplets or gas vesicles, or in which the rate of sinking is reduced by structural features or diminutive body size.

morphosis Variation in morphogenesis of an individual induced by environmental changes.

mortality Death rate as a proportion of the population expressed as a percentage or as a fraction; often used in a general sense as equivalent to death; *cf.* natality.

morwong Cheilodactylidae *q.v.*

mos A community or assemblage of species living together but without mutual interdependence.

mosaic An organism comprising tissues of two or more genetic types; usually used with reference to plants.

mosquito Culicidae *q.v.* (Nematocera).

moss Bryopsida *q.v.*

moss animal Bryozoa *q.v.*

moss campion Caryophyllaceae *q.v.*

moss rose Portulaceae *q.v.*

Motacillidae Wagtails, pipits; widely distributed family containing about 50 species of passerine birds found in open grassland; body and tail elongate, bill slender; feed mainly on insects; nest on the ground.

wagtail (Motacillidae)

moth Lepidoptera *q.v.*

motile With the capacity for movement; exhibiting movement; **motility.**

motmot Momotidae *q.v.*

mould 1: A fungus which produces a furry or velvet-like mycelium or spore mass over the surface of its host, as in the bread mould (Mucorales *q.v.*). 2: A cavity or space in a sediment remaining after the organic material of an organism has been removed by the action of soil water.

mouldering Decomposition of organic matter under conditions of low oxygen resulting in the formation of carbon-rich residues; *cf.* decay, putrefaction.

moultinism A polymorphism in which different strains have a different number of larval stages.

mountain ash Rosaceae *q.v.*

mountain beaver Aplodontidae *q.v.*

mountain laurel Ericaceae *q.v.*

mouse 1: Muridae *q.v.* (Old World mice). 2: Cricetidae *q.v.* (New World mice).

mousebird Coliidae *q.v.*

mousefish Gonorhynchidae *q.v.*

mouse-tailed bat Rhinopomatidae *q.v.*

Mozambique Current A warm surface ocean current that flows southwards through the Mozambique Channel; see ocean currents.

mRNA Messenger RNA *q.v.*

mucilaginous Comprising or resembling mucilage.

mucivorous Feeding on plant juices or mucilage; **mucivore, mucivory.**

muck Highly decomposed plant material typically darker and with higher mineral content than peat.

Mucorales Bread mould; large order of zygomycete fungi containing forms saprobic on dung or organic debris, and parasites of invertebrates, other fungi and plants; also treated as a class of the phylum Zygomycota.

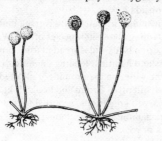
bread mould (Mucorales)

mud Fine sediment particles of diameter less than 0.002 mm; fine-grained, slimy, marine or lacustrine detrital sediment; see sediment particle size.

mud lobster Thalassinidea *q.v.*

mud plantain Pontederiaceae *q.v.*

mud puppy Proteidae *q.v.*

mud salamander Ambystomatidae *q.v.*

mud turtle Kinosternidae *q.v.*

mudminnow Umbridae *q.v.*

mudsnail Neogastropoda *q.v.*

Mugilidae Mullets, grey mullets; family containing 70 species of coastal marine, brackish and freshwater teleost fishes (Perciformes) widespread in tropical to temperate waters; body stout, to 0.9 m in length, with blunt-snouted head; anterior dorsal fin composed of 4 strong spines; lateral line usually absent; feed by sucking up mud containing algae, invertebrates and detritus.

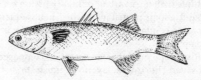
mullet (Mugilidae)

Mugiloidei Suborder of perciform teleosts comprising a single family Mugilidae (mullets).

Mugiloididae Sandperches; family containing 25 species of Indo-Pacific tropical marine teleost fishes (Perciformes) found in shallow water, burrowing in sand or concealed amongst stones or coral; body elongate, to 300 mm in length, head and mouth large, jaws protractile; 2 dorsal fins contiguous; Parapercidae.

mulberry Moraceae *q.v.*

mull A forest soil type in which the organic matter is intermixed with mineral soil without a discrete humus layer; *cf.* mor.

mullein Scrophulariaceae *q.v.*

Müllerian mimicry Imitative similarity, typically based on warning coloration amongst a number of mimic species all of which are unpalatable or otherwise offensive to a predator (the operator); *cf.* mimicry.

Müller's larva Free-swimming larval stage of some polyclad turbellarians (Platyhelminthes); possessing 8 ciliated lobes or lappets.

mullet 1: Mugilidae *q.v.* (grey mullets). 2: Mullidae *q.v.* (red mullets).

Mullidae Goatfishes, red mullets; family containing 60 species of bottom-dwelling tropical and warm-temperate, shallow marine

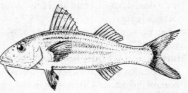
red mullet (Mullidae)

and estuarine teleost fishes (Perciformes); body elongate, strongly convex dorsally, flattened ventrally, length to 600 mm; 2 large chin barbels present; 2 short dorsal fins.

multiparous 1: Producing 2 or more offspring in a single brood; pluriparous; *cf.* biparous, uniparous. 2: Having produced more than one previous brood.

multiple alleles Two or more different forms of a gene occupying a specific locus.

multiple fission A variant of binary fission *q.v.* found in some protistans in which a subsequent division commences before the earlier one is complete.

multistratal Pertaining to vegetation composed of several horizontal layers; *cf.* unistratal.

Multituberculata Extinct order of probably herbivorous, rodent-like mammals (Prototheria) known from the late Jurassic to the Eocene; regarded as a side branch from the main lineage of mammalian evolution.

Multivalvulida Order of Myxosporea *q.v.* characterized by having spores with more than 2 shell valves.

multivoltine Having several broods or generations per year; *cf.* voltine.

mune A group of organisms having similar ecological requirements and a characteristic behavioural response.

Muraenesocidae Pike congers; family containing 15 species of tropical and warm temperate anguilliform teleost fishes found mainly in shallow sandy marine habitats; body smooth, to 2 m length; jaws bearing strong canine-like teeth; eyes, pectoral fins and gill openings well developed; dorsal, caudal and anal fins continuous.

Muraenidae Moray eels; family containing 100 species of nocturnal predatory or scavenging anguilliform teleost fishes widespread on reefs and rocky bottoms of shallow tropical and temperate seas; body smooth, to 3 m length, compressed; dorsal and anal fins continuous with caudal; pelvic and pectoral fins absent; jaws and teeth well developed.

Muraenolepididae Family of little-known Southern Ocean gadiform teleost fishes containing only 3 species; body length to 200 mm; dorsal, caudal and anal fins continuous; pelvics jugular.

mural Pertaining to, or growing on, walls; rupestral.

Muridae Old World rats, mice; large cosmopolitan family of myomorph rodents comprising about 450 species in 100 genera; habits diverse, may be terrestrial, arboreal,

fossorial or semiaquatic; herbivorous or insectivorous.

Musaceae Banana, plantain; small family of Zingiberales comprising about 35 species of coarse, tree-like perennial herbs dying back to the ground after flowering; leaves with a petiole and expanded simple blade; flowers strongly nectariferous and adapted for pollination by birds and bats, producing a fleshy fruit; confined to tropical and subtropical regions of the Old World; flowers finger-shaped, arranged in racemes with brightly coloured bracts, each with 6 petals and stamens and an inferior ovary.

banana (Musacaceae)

muscarian Used of flowers that attract flies by a putrid odour.

Musci True mosses; large class of bryophytes characterized by a typically filamentous protonema from which arise buds that develop into the gametophyte; spore capsule opening by means of a lid; usually known as the Bryopsida *q.v.*

Muscicapidae Flycatchers; diverse family containing about 300 species of Old World passerine birds widespread in forest and grassland habitats; feeding on insects caught on the wing, on the ground, or gleaned from trees; bill often strong and hooked; habits solitary to gregarious, arboreal to terrestrial, monogamous or polygamous; nest in tree, bank or rock crevice.

muscicolous Living on or in mosses or moss-rich communities; **muscicole.**

Muscidae Houseflies, stable fly; cosmopolitan family of small to large flies (Diptera) comprising over 3000 species many of which are pests of livestock and may be vectors of disease; the larvae feed mostly on decaying plant and animal material or dung, and some are parasitic.

muscology The study of mosses; bryology.

mushroom Agaricales *q.v.*

musk Scrophulariaceae *q.v.*

housefly (Muscidae)

musk turtle Kinosternidae *q.v.*

muskeg A grassy bog habitat, with scattered, stunted conifers.

muskmelon Cucurbitaceae *q.v.*

Musophagidae Touracos; family containing 18 species of arboreal birds found in African forests; feed mainly on seeds and fruit; regarded by some authorities as suborder of Cuculiformes (cuckoos).

Muspiceida Small order of enoplian nematodes found as parasites in vertebrates; typically lacking amphids and cephalic sensory papillae; digestive tube reduced.

mussel Mytiloida *q.v.*

mussel, pearly freshwater Unionoida *q.v.*

mussel shrimp Ostracoda *q.v.*

mustard tree Salvadoraceae *q.v.*

Mustelidae Weasels, stoats, otters, badgers, mink; family of mostly small terrestrial, arboreal or semiaquatic carnivorous mammals (Carnivora) widespread in Holarctic, Neotropical, Ethiopian and Oriental regions; body typically elongate with short limbs and long tail; contains about 65 species including also martens, sable, skunk, ermine, wolverine and polecat.

badger (Mustelidae)

mutagen Any agent that produces a mutation or enhances the rate of mutation; **mutagenic**, **mutagenesis.**

mutant Any organism, gene, or character that has undergone a mutational change.

mutation 1: A sudden heritable change in the genetic material, most often an alteration of a single gene by duplication, replacement or deletion of a number of DNA base pairs. 2: An individual that has undergone such a mutational change; mutant.

Mutica A mammalian cohort comprising the single order Cetacea (whales).

Mutillidae Cow killers, velvet ants; family of wasps (Hymenoptera) in which the female is brightly coloured and wingless, and the male dark with wings; females parasitize bees, wasps, beetles or other insects, typically laying eggs on relatively immobile stages of the host enclosed in a cell, cocoon or puparium.

mutualism A symbiosis in which both organisms benefit, frequently a relationship of complete dependence; interdependent association.

mycelium The mass of filamentous hyphae that comprises the vegetative stage of many fungi.

mycetogenic Produced by fungi.

mycetology The study of fungi; mycology.

mycetophagous Feeding on fungi; **mycetophage**, **mycetophagy.**

Mycetophilidae Fungus gnats; family containing about 2000 species of small fragile flies (Diptera); larvae feed mostly on fungi or decaying vegetation and are economically important as pests of mushroom crops.

mycetozoan A slime mould; Myxomycota *q.v.*

mycobiont The fungal partner of a lichen (algal/fungal symbiosis); *cf.* phycobiont.

mycobiota The fungal flora of an area or habitat.

mycocriny The decomposition of organic material by the action of fungi.

mycology The study of fungi.

mycophagous Feeding on fungi; **mycophage**, **mycophagy.**

mycoplasma Any organism in the class Mollicutes *q.v.*

mycorrhiza The close physical association between a fungus and the root system of a plant, which enables the roots to take up nutrients more efficiently than if uninfected.

mycotic Produced by the agency of fungi; **mycosis.**

mycotrophic Used of plants that live in a symbiosis with a fungus and are nutritionally dependent upon it, as in those plants with mycorrhizal associations; **mycotrophy.**

Myctophidae Lanternfishes; cosmopolitan family of small to medium-sized (to 150 mm) pelagic deep-sea myctophiform teleost fishes; body compressed, eyes and mouth large;

photophores numerous on ventral surface of head and body; adipose fin present; contains about 240 species, some locally abundant, many exhibiting diurnal migration between surface and 500–1000 m.

Myctophiformes Order of pelagic and benthic teleost fishes found in littoral to abyssal marine habitats, comprising about 400 species in 16 families, including lanternfishes, lizard fishes, barracudinas, lancetfishes, daggertooths, pearleyes; body commonly bearing photophores, adipose fin present, pelvics abdominal; many species hermaphroditic.

mygalomorph spider Orthognatha *q.v.*

myiophilous Pollinated by flies; **myiophily.**

Myliobatidae Eagle rays; family containing 20 species of medium to large, tropical and subtropical, marine and estuarine elasmobranch fishes; body disk much broader than long and distinctly demarkated from head; pectoral fins produced anteriorly as subrostral lobe; tail filamentous distally, caudal fin absent; mouth bearing series of grinding plates; feed on benthic molluscs and crustaceans dug up using pectoral fins.

Myliobatiformes Stingrays, eagle rays, devil rays; order of small to large (to 6 m) benthic or epipelagic, marine, brackish or freshwater, ray-like elasmobranch fishes; body disk strongly depressed, may be much broader than long, tail slender to whip-like; single dorsal fin present or absent; mouth small; reproduction viviparous; comprises about 150 species in 7 families.

Mymaridae Fairy flies; family of extremely tiny hymenopterous insects (Apocrita) with narrow, hair-fringed wings exhibiting reduced venation; all are egg parasites of other insects.

Myobatrachidae Family containing about 95 species of mainly terrestrial frogs (Anura) from Australia, a few species arboreal or aquatic; eggs and larvae typically aquatic, but other reproductive strategies include gastric brooding, retention of tadpoles in lateral pouches, and direct development.

Myocastoridae Coypu; a monotypic family of large semiaquatic hystricomorph rodents native to the southern Neotropical region, but recently established in the Holarctic; body robust, limbs short, feet webbed; excavate burrows in earth banks, feeding mainly on aquatic vegetation.

Myodocopa Class of benthic, epibenthic and planktonic marine ostracod crustaceans found from the surface to abyssal depths; contains

600 species in 2 orders, Myodocopida and Halocyprida.

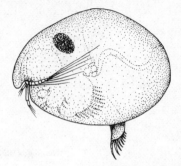

ostracod (Myodocopa)

Myodocopida Order of mostly marine myodocopan ostracods, typically benthic and mesoplanktonic, occasionally abyssopelagic; contains about 380 species in 5 families.

Myoida Order of marine and freshwater heterodont bivalves which are generally burrowing, filter feeders possessing well developed siphons; shell morphology variable, 2 shell-closing muscles present, hinge often degenerate; includes the soft-shelled clams, piddocks and shipworms.

Myomorpha Large, cosmopolitan suborder of small terrestrial and arboreal mammals (Rodentia) comprising 9 families and including rats, mice, dormice, voles, jerboas, gerbils, hamsters, lemmings, and many others.

Myoporaceae Emu bushes; family of Scrophulariales containing about 125 species of shrubs and small trees producing iridoid compounds and with scattered small secretory cavities; widely distributed in southern hemisphere; flowers with 5 fused sepals, a 5-lobed tubular corolla, 4 stamens and a superior ovary.

myria- 1: Prefix used to denote unit x 10^4. 2: Prefix meaning many, innumerable.

Myriangiales Order of mostly tropical loculoascomycete fungi in which the asci are scattered through the poorly differentiated inner tissue of the ascocarp; most are external saprobes on plants, some parasitic on plants and animals.

Myriapoda Heterogeneous assemblage of arthropods comprising the centipedes (Chilopoda), millipedes (Diplopoda), Pauropoda and Symphyla; trunk segments not differentiated into thorax and abdomen; sometimes treated as a subphylum or class of Uniramia *q.v.*

Myricales Wax myrtle, sweet fern; small order

of Hamamelidae comprising a single family (Myricaceae) of about 50 species of aromatic shrubs or small trees bearing small flowers in catkins.

Myristicaceae Family of primitive plants (Magnoliales) containing mostly trees producing aromatic oils and hallucinogenic compounds, widespread in tropical regions; flowers usually tiny, unisexual, with stamens united as a column; includes the nutmeg tree.

Myrmecobiidae Numbat; monogeneric family of small (to 400 mm) terrestrial marsupial anteaters; tongue long, sticky, and protrusible; marsupium absent.

myrmecochorous Dispersed by the agency of ants; **myrmecochore, myrmecochory.**

myrmecoclepty A symbiosis between ant species in which the guest species steals food directly from the host species.

myrmecodomatia Plant structures inhabited by ants or termites.

myrmecodomus Used of a plant affording shelter to ants.

myrmecology The study of ants.

myrmecolous Living in ant or termite nests; **myrmecocole.**

Myrmecophagidae Anteaters; family of Neotropical insectivorous mammals (Edentata) containing a single terrestrial, and two mainly arboreal, species; head elongate, oral region tubular; teeth absent, tongue elongate, protrusible and sticky; forelimbs powerful with large claws for breaking open nests of ants and termites.

giant anteater (Myrmecophagidae)

myrmecophagous Feeding on ants or termites; **myrmecophage, myrmecophagy.**

myrmecophilous 1: Thriving in association with ants; used of organisms that spend part of their life cycle in ant or termite nests; **myrmecophile, myrmecophily;** *cf.* myrmecophobous. 2: Pollinated by ants or termites.

myrmecophobous Used of organisms that repel ants or termites; **myrmecophobe, myrmecophoby;** *cf.* myrmecophilous.

myrmecophyte A plant having specialized structures for sheltering ants or termites, or

having a mutual interdependence with ants or termites.

myrmecosymbiosis A symbiosis between an ant and its host plant; **myrmecosymbiotic.**

myrmecotrophic Pertaining to plants and animals that provide food for ants.

myrmecoxenous Used of plants that provide both food and shelter for ants and termites; **myrmecoxeny.**

Myrmeleontidae Ant lions; family containing about 600 species of generally nocturnal flying, neuropteran insects often with a long abdomen and highly patterned wings; larvae typically lie in wait for prey insects, some constructing conical pitfall traps.

Myrothamnaceae Family containing 2 species of glabrous, aromatic shrubs found in Africa and Madagascar having leaves that fold up and turn blackish during the dry season but expand and turn green again after rain.

Myrsinaceae Large family of Primulales containing about 1000 species of woody plants containing yellow to brown resinous material in ducts or cavities; widespread in tropical and subtropical regions; flowers typically arranged in umbels or panicles.

Myrtaceae Myrtle, eucalyptus, gum tree; large family containing about 3000 species of shrubs or trees producing ethereal oils in scattered cavities; found in tropics, subtropics and in temperate Australia; flowers typically perfect, borne in cymes and with 4–5 sepals and petals, large numbers of stamens and an inferior ovary.

myrtle (Myrtaceae)

Myrtales Large order of Rosidae containing over 9000 species of trees, shrubs and herbs; largest families are the Melastomataceae and Myrtaceae.

myrtle Myrtaceae *q.v.*

Mysidacea Opossum shrimps; order of predominantly marine shrimp-like peracaridan crustaceans containing about 780 species, epibenthic or pelagic, littoral to abyssal; includes a few freshwater and cavernicolous forms; typically with a pair of conspicuous statocysts in the tail fan.

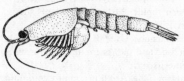

opossum shrimp (Mysidacea)

mysis The late zoeal phase of larval development found in many shrimp-like malacostracan crustaceans.

Mystacinidae Family containing a single species of omnivorous bat (Microchiroptera) from New Zealand; body with thick fur; wings tucked into pockets when at rest.

Mystacocarida Class of minute crustaceans inhabiting the interstitial spaces in shallow marine sediments; characterized by larval-type cephalic appendages and by a slender trunk bearing 4 pairs of limb buds; contains 9 species in a single order, Mystacocaridida.

Mystacocaridida Sole order of the Mystacocarida *q.v.*

Mysticeta Baleen whales; suborder of large marine mammals (Cetacea) in which teeth are absent and upper jaw carries a filter curtain of long baleen plates; skull bilaterally symmetrical; blowhole paired; found in all oceans but more common at high latitudes; most are migratory; feed mainly on crustaceans; comprises 3 families, Balaenidae (right whales), Eschrichtiidae (grey whale), Balaenopteridae (rorquals).

Mytiloida Mussels; an order of typically marine or estuarine pteriomorphian bivalves which are commonly sedentary as adults, attached by byssus threads to the substratum; characterized by their gill structure and by stiff ciliated disks uniting the gill filaments.

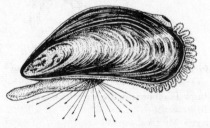

mussel (Mytiloida)

Myxiniformes Hagfishes; order of agnathan vertebrates (Cephalaspidomorphi); body eel-like, naked, with copious mucous secretion;

paired fins and dorsal fin absent; eyes reduced; up to 16 pairs of gill openings; feed on dead or moribund fish and invertebrates; comprises about 30 species in a single family, Myxinidae; also known as the Hyperotreti.

Myxogastria Subclass of plasmodial slime moulds (Eumycetozoa), more commonly treated as the Myxogastromycetidae *q.v.* in the class Myxomycetes.

Myxogastromycetidae Subclass of plasmodial slime moulds *q.v.* (Myxomycetes) exhibiting two plasmodial types, a large phaneroplasmodium which has well developed margins, polarity and grows better in drier conditions producing either many fruiting bodies or a single massive one, and a microscopic protoplasmodium which gives rise to a single small sporangium; fruiting body arises from elongate column of protoplasm which swells apically.

Myxomycetes Plasmodial slime moulds *q.v.*, treated as a class of the Myxomycota and classified with the fungi; comprises 3 subclasses, Ceratiomyxomycetidae, Myxogastromycetidae and Stemonitomycetidae.

Myxomycota 1: Slime moulds; group of primitive organisms with an amoeba-like colonial stage in the life cycle; treated as fungi, comprising cellular slime moulds *q.v.* (Acrasiomycetes) and plasmodial slime moulds *q.v.* (Myxomycetes, Plasmodiophoromycetes and Protosteliomycetes). 2: Plasmodial slime moulds *q.v.*, treated as a phylum of Protoctista.

Myxophaga Small suborder of minute (0.5–2.5 mm) beetles (Coleoptera) comprising about 60 species in 4 families.

Myxosporea Class of Myxozoa *q.v.* containing mostly parasites of marine and freshwater fishes; possessing 1–6 polar capsules and usually 2, but up to 6, valves; comprises 2 orders, Bivalvulida and Multivalvulida.

Myxosporida Myxozoa *q.v.*; treated as a class of the protoctistan phylum Cnidosporida.

myxotrophic Used of an organism obtaining nutrients through the ingestion of partices; **myxotrophy.**

Myxozoa Phylum of sporulating parasitic organisms commonly classified in the Protozoa but representing primitive multicellular organisms as they have spores composed of several specialized cells; spores comprise generative cells (sporoplasms), capsule cells containing coiled polar filaments and valve cells that develop into shell valves; known as parasites of invertebrates and lower vertebrates; also treated as a class, Myxosporida, of the protoctistan phylum Cnidosporida.

Myzopodidae Family containing a single species of bat (Microchiroptera) from Madagascar bearing adhesive suckers on all 4 limbs.

Myzostomata Class of small aberrant annelid worms comprising about 120 species in 7 families all of which are parasitic on or in echinoderms, mostly crinoids, but also ophiuroids and asteroids; body oval or flattened, lacking external segmentation, head indistinct, pharynx muscular and unarmed, parapodia uniramous.

n

naiad Unionoida *q.v.*

Najadaceae Water nymph; cosmopolitan family comprising about 35 species of submerged annual herbs found in fresh or brackish water with small, unisexual flowers borne singly in axils.

Najadales Order of Alismatidae containing 10 families of aquatic or semiaquatic herbs; leaves commonly basal and sheathing; flowers inconspicuous, without differentiated sepals and petals, pollinated by wind or water.

namatophilous Thriving in brooks and streams; **namatophile, namatophily.**

namatophyte A plant inhabiting a stream or brook.

nanander A dwarf male.

Nandidae Leaffishes; family containing 10 species of small (to 200 mm) predatory freshwater teleost fishes (Perciformes) that feed voraciously on other small fishes by rapid extension of the jaws whilst floating leaf-like at the water surface; body deep, mouth large and highly protrusible.

nanism The condition of being stunted or smaller than normal, or of having restricted growth; dwarfism; *cf.* gigantism.

nano- (n) Prefix denoting unit x 10^{-9}; nanno-.

nanofossils Very small marine fossils, usually algae, typically close to the resolution limit of a light microscope and studied by electron microscopy.

nanophanerophyte A dwarf phanerophyte *q.v.* with renewal buds less than 2 m above ground level.

nanophyll A Raunkiaerian leaf size class *q.v.* for leaves having a surface area between 25 and 225 mm^2.

nanoplankton Minute planktonic organisms with a body diameter between 2 and 20 μm; nannoplankton; formerly used for those organisms that passed through a 0.03–0.04 mm mesh silk bolting cloth.

Narcinidae Family of small (to 0.75 m) torpediniform fishes; body disk circular to wedge-shaped, tail stout, dorsal fins large; eyes small to vestigial; mouth highly protractile; contains about 15 species found from the littoral to depths of 100 m, rarely to 1000 m.

narcissus Liliaceae *q.v.*

Narcomedusae Suborder of marine cnidarians (Trachylina) in which the polyp generation is reduced or absent; medusae with scalloped margin, solid tentacles and gonads on the stomach wall; mouth opening directly into stomach.

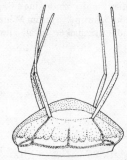

narcomedusa (Narcomedusae)

narcotropism Orientation movements resulting from the effects of narcotics; **narcotropic.**

Narkidae Family of small (less than 0.5 m) torpediniform fishes; body disk ovoid to circular; tail bearing lateral folds; dorsal fins present or absent; mouth small, protractile; eyes small to vestigial; contains 10 species found from the littoral zone to 200 m.

narwhal Monodontidae *q.v.*

Nassellarida Order of radiolaria (Polycystina) in which the capsular membrane is perforated at one pole and the skeleton has a bilateral symmetry.

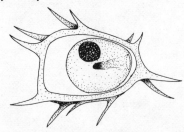

radiolarian (Nasellarida)

Nassulida Order of hypostomatan ciliates with reduced circumoral ciliature (the hypostomial frange); common in aquatic and soil habitats.

nastic Pertaining to plant movement in response to a diffuse stimulus, and to structural curvature resulting from differential growth of opposite surfaces; **nasty.**

nasturtium Tropaeolaceae *q.v.*

Natalidae Family of agile, long-legged, insectivorous bats (Microchiroptera) containing 8 species found mostly in Central and South America; nose simple, without a leaf.

natality Birth rate; the number of offspring produced per female or head of the population per unit time; the production of new individuals by birth, germination or fission; *cf.* mortality.

Natantia The shrimp-like decapod crustaceans; formerly used as a suborder but now divided between the suborders Dendrobranchiata (Penaeidea) and Pleocyemata (Caridea and Stenopodidea).

natatorial Adapted for swimming; **natant.**

native Indigenous; living naturally within a given area; used of a plant species that occurs at least partly in natural habitats and is consistently associated with certain other species in these habitats.

native cat Dasyuridae *q.v.*

natural classification A hierarchical classification based on hypothetical phylogenetic relationships such that the members of each category in the classification share a single common ancestor.

natural history The study of nature, natural objects and natural phenomena.

natural selection The non-random and differential reproduction of different genotypes acting to preserve favourable variants and to eliminate less-favourable variants; viewed as the creative force that directs the course of evolution by preserving those variants or traits best adapted in the face of natural competition.

naturalized Used of an alien or introduced species that has become successfully established.

nauplius Earliest larval stage of crustaceans; typically with 3 pairs of functional limbs, lacking compound eyes but usually with a simple median eye.

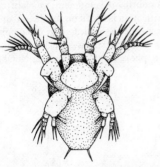

nauplius

Nautiloidea Pearly nautilus; a subclass of cephalopod molluscs characterized by a chambered, external, spiral shell containing gas for buoyancy and by the numerous adhesive cirri or tentacles arranged around the mouth; also possessing 4 gills in the mantle cavity and pinhole camera-like eyes without lenses; contains a few living species but has an extensive fossil record dating from the Upper Cambrian.

nealogy The study of young and immature organisms.

neandertals Group of homonids that lived in the mid-Palaeolithic about 100 000–35 000 years B.P., before and during the last Ice Age, known from fossils found in Europe, North Africa and western Asia; they share many features with modern man and are placed in the same species, but as a separate subspecies, *Homo sapiens neandertalensis*.

neanic 1: Adolescent; pertaining to the larval phase that precedes the adult condition. 2: Used of characters which first appeared, in evolution, in early ontogenetic stages; *cf.* ephebic.

neap tide The tide of minimum range occurring at the time of first and third quarters of the moon, when the gravitational attraction of the sun and the moon act at right angles to each other during quadrature *q.v.*; *cf.* spring tide.

Nearctic region A zoogeographical region *q.v.* comprising North America, Greenland and northern Mexico; subdivided into Alleghany, Californian, Canadian and Rocky Mountain subregions.

Nebraskan glaciation A glaciation of the Quaternary Ice Age *q.v.* in North America with an estimated duration of 100 000 years.

necrocoenosis An assemblage of dead organisms.

necrocoleopterophilous Pollinated by carrion beetles; **necrocoleopterophily.**

necrogenous 1: Growing on, or inhabiting, dead bodies. 2: Used of organisms or factors that premote decay.

necrology The study of decomposition, fossilization and other processes affecting plant and animal remains after death.

necrophagous Feeding on dead material; saprophytic; **necrophage, necrophagy.**

necrophoresis The transport of dead individuals away from a colony, as in some social insects.

necrophytophagous Feeding on dead plant material; **necrophytophage, necrophytophagy.**

necrotrophic symbiosis A symbiosis established between two living organisms in which one symbiont continues to use the other as a food source even after complete or partial death has occurred.

nectariferous Nectar-producing.

Nectariniidae Sunbirds; family containing about 115 species of small active Old World passerines found in moist to arid wooded habitats, from Africa to Australia; male plumage often colourful; bill slender, curved, serrated, tongue tubular; habits solitary to gregarious, arboreal, non-migratory; feed on insects, nectar and fruit; nest pouch-like, suspended in tree.

nectarivorous Feeding on nectar; **nectarivore**, **nectarivory**.

nectism Swimming by means of cilia.

nectobenthic Swimming off the sea bed.

nectochaete A free-swimming post-trochophore larval stage in some polychaetes in which body segmentation is apparent and chaetae-bearing parapodia are present.

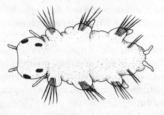

nectochaete

nectocalyx Medusa-shaped zooid lacking a manubrium in colonial hydrozoan coelenterates (Siphonophora).

Nectonematoida Class of exclusively marine horsehair worms found as parasites in the body cavities of a variety of decapod crustaceans; characterized by a double row of swimming hairs along each side of the body.

needlefish Belonidae *q.v.*

neidioplankton Planktonic organisms possessing some form of swimming apparatus.

nekrophytophagous Feeding on dead plant material; **nekrophytophage**, **nekrophytophagy**.

nektobenthic Swimming off the sea bed.

nektobenthos Organisms typically associated with the benthos that swim actively in the water column at certain periods; **nektobenthic**.

nekton Those actively swimming pelagic organisms able to move independently of water currents; typically within the size range 20 mm to 20 m; **nektonic**; *cf.* plankton.

Nelumbonaceae Lotus lilies; small family of aquatic plants (Nymphaeales) with floating or emergent leaves and a large solitary flower raised above water on long peduncle.

Nemaliales Large diverse order of algae containing erect filamentous or pseudoparenchymatous forms; found in both marine and freshwater habitats.

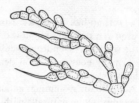

red seaweed (Nemaliales)

nematicide A chemical used to kill nematode worms.

Nemathelminthes Phylum comprising the Nematoda *q.v.* and Nematophora *q.v.*; often treated as separate phyla

Nematocera Suborder of mainly slender-bodied delicate flies (Diptera) comprising about 32 000 species and including crane flies, sand flies, mosquitoes, blackflies, gnats and midges.

Nematoda Nematodes, roundworms, eelworms; a phylum of about 12 000 species of unsegmented, bilaterally symmetrical worm-like organisms which are extremely abundant in marine and freshwater habitats, in soil, and as parasites of plants and animals; characterized by the possession of a cuticle secreted by hypodermal cells, a pseudocoelom, a terminal mouth surrounded by lips and 3 rings of sensilla (6 + 6 + 4), and a subterminal anus; sexes usually separate, female system opening through a ventromedial vulva, male system opening through anus, male often provided with cuticular sex organs (spicules); 3 juvenile stages during development; much of classification is based on the degree of cephalization as indicated by the numbers of sense organs; comprises 2 classes, Adenophorea and Secernentea; Nemata.

nematode (Nematoda)

nematology The study of nematodes; **nematodology.**

Nematomorpha Gordian worms, horsehair worms; phylum of elongate, pseudocoelomate worms found as parasites of arthropods when juvenile but emerging into aquatic habitats as ephemeral free-living adults; the adults lack a functional digestive tract and are probably dependent on food reserves built up during parasitic phase; characterized by a single layer of longitudinal body wall muscles and by the opening of the genital ducts through a cloaca, shared with the anus; comprises a marine class, the Nectonematoida, and a freshwater class, the Gordioida.

horsehair worm (Nematomorpha)

Nemertea Ribbon worms; phylum of unsegmented, bilaterally symmetrical, acoelomate worms found mainly in intertidal or subtidal marine habitats living either on the substrate or burrowing in soft sediments, occasionally found in freshwater or damp terrestrial situations; characterized by an eversible muscular proboscis, ciliated epidermis and gut epithelium, and an alimentary system with mouth and anus; typically predatory carnivores or scavengers, some can reach lengths of several metres; mostly dioecious and oviparous with rare exceptions; Nemertini.

Nemertodermatida Order of turbellarian flatworms comprising only 5 species found free-living in subtidal sediments and as parasites in holothurians; characterized by the possession of sperm with a single flagellum and of a statocyst with 2 statoliths.

Nemichthyidae Snipe eel; family containing 10 species of bathypelagic anguilliform teleost fishes; body slender, naked, compressed, dorsal fin very long and continuous with anal, pectorals present; jaws markedly prolonged, often divergent.

nemoricolous Living in open woodland; **nemoricole.**

nemorose Living in open woodland.

Neobatrachia Suborder of aquatic, terrestrial and arboreal anurans (Amphibia) containing the majority of modern frogs; distribution worldwide, especially diverse in the tropics; reproductive strategies very varied; comprises about 2700 species.

Neoceratiidae Monotypic family of tiny (to 60 mm) deep-sea anglerfishes (Lophiiformes); female with very large mouth bearing long slender, articulating teeth; lure absent; male parasitic on female, with vestigial eyes and small teeth.

neo-Darwinism The modern theory of evolution that combines both natural selection and population genetics, in which the Darwinian concept of spontaneous variation is explained in terms of mutation and genetic recombination.

Neoechinorhynchida Order of eoacanthocephalan thorny-headed worms found as parasites of marine and freshwater fishes, amphibians and turtles, and utilizing crustaceans, molluscs and insects as intermediate hosts; characterized by the lack of spines on the trunk.

Neogaea A zoogeographical area originally comprising both Nearctic and Neotropical regions; now generally used to refer to the Neotropical region only; Neogea; *cf.* Arctogaea, Megaea, Notogaea, Palaeogaea.

Neogastropoda Order of advanced prosobranch snails containing about 5000 carnivorous and scavenging species, most of which are marine, and including augurs, basket shells, cone shells, crown conches, dogwhelks, dove shells, false conches, margin shells, mitre shells, mudsnails, olive shells, pagoda shells, tulip shells, volutes and whelks; characterized by a highly variable shell with a pronounced canal for the anterior incurrent siphon; shell lacking mother-of-pearl layer, operculum chitinous; mantle cavity with a single gill.

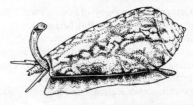

cone shell (Neogastropode)

Neogene A division of the Tertiary period comprising the Pliocene and Miocene epochs (*c.* 26–1.6 million years B.P.); see geological time scale.

Neognathae Superorder of neornithine birds containing about 23 Recent orders and comprising all living groups except Impennes (penguins) and Palaeognathae (ratites, tinamous).

Neogregarinida Order of gregarines which reproduce by multiple fission (merogony) within their insect hosts as well as by gamete and sporozoite formation.

neoichnology The study of tracks, burrows and other trace structures made by living organisms; *cf.* palaeoichnology.

neo-Lamarckism Inheritance of acquired characters; the theory that characters acquired by organisms as a response to environmental factors are assimilated into the genome and transmitted to the offspring.

Neolampadoida Small order of irregular echinoids with non-petaloid ambulacra found in tropical seas.

Neolithic An archaeological period *c.* 4000–2500 years B.P.; a period of human history characterized by the use of polished stone tools; the late Stone Age; New Stone Age.

Neoloricata Subclass of Polyplacophora *q.v.* containing all living chitons, and some fossils.

Neomeniamorpha Order of solenogastres characterized by a typically blunt body shape and by a variable mantle covering; Neomeniida.

neonatal Newborn; recently hatched; **neonate.**

neontology The study of living organisms; *cf.* palaeontology.

Neoophora Subclass of turbellarian flatworms found mainly in marine and freshwater habitats; with divided ovary.

Neopilina Living representative of the primitive class of molluscs, Monoplacophora *q.v.*

Neoptera Infraclass of pterygote insects in which the wings can be folded onto the body at rest; contains the majority of extant insects in 3 superorders, Orthopterodea, Hemipterodea and Holometabola.

Neopterygii Large group of bony fishes including the Holostei and Teleostei.

Neorhabdocoela Order of free-living and parasitic turbellarians found in freshwater and marine habitats; characterized by the bulbous, complex pharynx, a straight gut without branchings or diverticula and separate yolk glands and ovaries.

Neornithes Subclass of Aves comprising all 9000 species of Recent birds; 4 superorders recognized, Odontognathae (fossil), Palaeognathae (ratites, tinamous), Impennes (penguins) and Neognathae.

Neoscopelidae Family of pelagic myctophiform teleost fishes containing 5 species widespread in tropical and subtropical seas from surface to 1000 m or more; body compressed, eyes small to large, photophores usually present.

Neoselachii Sharks and rays; group of elasmobranch fishes containing all living species in 4 superorders, Squalomorphii, Batoidea, Squatinimorphii and Galeomorphii.

Neostethidae Family containing 15 species of tiny (to 35 mm) fresh- and brackish-water cyprinodontiform teleost fishes from southeastern Asia; body slender, translucent; pelvic fins reduced or absent in female, specialized as copulatory organ in male; reproduction oviparous, fertilization internal.

neoteny Paedomorphosis *q.v.* produced by retardation of somatic development, such that sexual maturity is attained in an organism retaining juvenile characters; **neotenic.**

Neotropical kingdom One of six major phytogeographical areas *q.v.* characterized by floristic composition; comprises Andean, Caribbean, Juan Fernandez, Venezuela, and Guianan and Pampas regions.

Neotropical region A zoogeographical region *q.v.* comprising South America, West Indies and Central America south of the Mexican plateau; subdivided into Antillean, Brazilian, Chilean, and Mexican subregions.

Neozoic The period of geological time from the end of the Mesozoic (*c.* 65 million years B.P.) to the present.

Nepenthaceae East Indian pitcher plants; family of insectivorous, tanniferous shrubs, often climbing or epiphytic, with leaf blade carrying an open pitcher at the end of a distal tendril.

East Indian pitcher plant (Nepenthaceae)

Nepenthales Order of Dilleniidae containing about 200 species of insectivorous perennial herbs or shrubs with leaves modified as traps or pitchers; comprises Droseraceae, Nepenthaceae and Sarraceniaceae.

nepheloid layer The turbid layer of bottom ocean water carrying very fine suspended particulate matter.

Nepidae Water scorpions; family of aquatic heteropteran bugs which live submerged in still waters, breathing through a tubular tail extending to the surface; front legs raptorial and used for grasping prey.

nepotism Showing favouritism to a relative.

Nerillida Order of minute (to 1.5 mm) polychaete worms containing 30 species in a single family, found in marine, brackish and freshwater sediments; body transparent, lacking distinct segmentation; pharynx eversible and armed; parapodia uniramous.

neritic Pertaining to the shallow waters overlying the continental shelf; *cf.* oceanic.

neritopelagic Inhabiting shallow coastal waters over the continental shelf.

neritoplankton Plankton of shallow continental shelf waters.

nervicolous Living on or in the veins of leaves; **nervicole.**

nest epiphyte An epiphyte with a twisted growth form which accumulates humus within the tangle.

nest odour The distinctive odour of a nest by which its inhabitants are able to distinguish the nest from others, and from the surrounding environment.

nest parasitism Symbiosis between two termite species in which colonies of one species live in, and feed on, the walls of the nests of a second (host) species.

net photosynthesis Apparent photosynthesis measured as the net uptake of carbon dioxide into the leaf; equal to gross photosynthesis less respiration.

net precipitation That part of total precipitation that actually reaches the ground; calculated as total precipitation less interception (that part of precipitated water intercepted by plant surfaces).

net primary production The total assimilation of organic matter by an autotrophic individual or population less that consumed by the catabolic processes of respiration; *cf.* gross primary production.

net production The total assimilation of organic matter by an individual, population or trophic unit less that consumed by respiration; often

referred to simply as production *q.v. cf.* gross production.

net secondary production The total assimilation of organic matter by a primary consumer *q.v.* individual or population less that consumed by the catabolic processes of respiration; gross secondary production.

Nettastomatidae Duckbill eels; family containing 20 species of tropical and temperate deep-sea anguilliform teleost fishes; body smooth, soft, skeleton weakly developed, tail filiform, gill openings small, jaws slender and prolonged; pectoral fins absent.

nettle Urticaceae *q.v.*

Neuradaceae Small family of Rosales containing about 10 species of densely hairy, prostrate annual plants native to deserts in Africa and the Middle East.

Neuroptera Order of primitive holometabolous insects typically having 2 pairs of subsimilar membranous wings and chewing mouthparts; comprises about 4500 species in 3 suborders, Megaloptera (alderflies, dobsonflies), Raphidioidea (snakeflies) and Plannipennia (lacewings, dusty wings, spongilla flies, owl flies, ant lions).

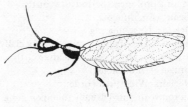

snake fly (Neuroptera)

Neuropterodea Holometabola *q.v.*

neuston Small to medium-size organisms that live on (epineuston) or under (hyponeuston) the surface film of water bodies; **neustont, neustonic;** *cf.* pleuston.

neustonology The study of neuston – those organisms living at the air/water interface of water bodies.

neustophagous Feeding on organisms living at the surface film of a water body (neuston); **neustophage, neustophagy.**

neuter Of neither sex; with imperfectly developed or non-functional reproductive organs.

never never plant Marantaceae *q.v.*

New World monkey Cebidae *q.v.*

New World pitcher plant Sarraceniaceae *q.v.*

New Zealand lemonwood Pittosporaceae *q.v.*

newt Salamandridae *q.v.*

niche The ecological role of a species in a community; conceptualized as the multidimensional space, of which the coordinates are the various parameters representing the condition of existence of the species, to which it is restricted by the presence of competitor species; sometimes used loosely as an equivalent of microhabitat in the sense of the physical space occupied by a species; *cf.* fundamental niche, realized niche.

nidus 1: A nest, domicile, breeding place, or point of origin. 2: The focus or primary site of an infection.

night lizard Xantusiidae *q.v.*

nightingale Turdidae *q.v.*

nightjar Caprimulgidae *q.v.*

niphic Pertaining to snow.

Nippotaeniidea Small order of tapeworms comprising 3 species parasitic in freshwater teleost fishes; characterized by a rounded scolex with an apical sucker, a small subcylindrical body, a single set of reproductive organs in each proglottis and a covering of minute spines over the entire body surface.

nidicolous Living in a nest; used also of young animals, especially birds, that remain in the nest for a prolonged period after birth; **nidicole**; *cf.* nidifugous.

nidification Nest building.

nidifugous Used of young animals, especially birds, that leave the nest soon after birth; **nidifuge**; *cf.* nidicolous.

Nidulariales Widespread order of gasteromycete fungi usually found on dung and plant debris; producing small, sessile fruiting bodies which are typically nest-shaped; includes cannonball fungi (*Sphaerobolus*) and bird's-nest fungus (*Cyathus*).

bird's-nest fungus (Nidulariales)

nit An egg belonging to a phthirapteran louse, especially the human head and body lice.

nitration The oxidation of nitrite to nitrate, especially by aerobic soil microorganisms.

nitrification The oxidation of ammonia to nitrite, and nitrite to nitrate, by aerobic microorganisms.

nitrogen cycle The biogeochemical cycle of nitrogen involving fixation, nitrification, decomposition by putrefaction and denitrification.

nitrogen fixation The reduction of gaseous nitrogen to ammonia or other inorganic or organic compounds by microorganisms or lightning.

nitrophilous Thriving in soil rich in nitrogenous compounds; **nitrophile, nitrophily.**

nitrophyte A plant thriving in soil rich in nitrogenous compounds.

nitrozation The oxidation of ammonia to nitrite by aerobic soil microorganisms.

nival Pertaining to snow.

nivation Erosion of the land surface by the action of snow.

niveoglacial Pertaining to the combined action of snow and ice.

nivicolous Living in snow or snow-covered habitats; **nivicole.**

Noctilionidae Fish-eating bats; family containing 2 species of large bats (Microchiroptera) with long legs and large feet equipped with well developed claws; includes an insectivorous and a piscivorous species, both from tropical America.

Noctilucales Order of typically large, motile, marine phagotrophic dinoflagellates, including the bioluminescent *Noctiluca*.

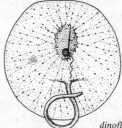

dinoflagellate (Noctilucales)

noctilucent Used of bioluminescent organisms emitting light during darkness.

Noctuidae Cutworms, armyworms; family of small to very large mostly nocturnal moths (Lepidoptera) containing about 25 000 species; larvae feeding on leaves, flowers, buds or boring into plant stems and fruits, and many causing serious damage to crops.

nocturnae Pertaining to night-flying insects; *cf.* diurnae.

nocturnal 1: Active during the hours of darkness. 2: Lasting only one night; *cf.* crepuscular, diurnal.

Nolanaceae Family of Solanales containing
about 65 species of herbs or small shrubs with
internal phloem in small strands; native to
western South America and the Galapagos;
flowers typically with 5 fused sepals, a 5-lobed
bell-shaped corolla and a superior ovary; fruit
usually a nutlet.

nomad 1: A wandering organism. 2: A pasture
plant.

nomadic Pertaining to the habit of wandering
from place to place, usually within a well
defined territory; nomadism.

Nomeidae Driftfishes; family containing 15
species of meso- and bathypelagic stromateoid
teleost fishes (Perciformes); length to 1 m,
mouth small; juveniles live in association with
jellyfish and siphonophores.

nomenclature The system of scientific names
applied to taxa, or the application of these
names; nomenclatural.

nomocolous Living in pasture; nomocole.

nomophilous Thriving in pastures; nomophile,
nomophily.

non-available water The amount of water still
remaining in the soil when a plant reaches the
permanent wilting point.

Non-calcic Brown soil A zonal soil with a light
red or reddish brown acidic A-horizon over a
light brown to red B-horizon, formed in
subhumid climates under a mixture of forest
prairie transitional vegetation; Shantung soil.

noosphere That part of the biosphere altered or
influenced by the activities of man.

normoxic Used of a habitat having the normal
atmospheric level of oxygen; normoxicity; cf.
anoxic, oligoxic.

North Atlantic Drift A north-easterly
continuation of the Gulf Stream current into
the eastern North Atlantic; see ocean
currents.

North Atlantic Gyre The major clockwise
circulation of surface ocean water in the North
Atlantic Ocean; see ocean currents.

North Equatorial Countercurrent A warm
surface ocean current that flows eastwards in
the tropical Pacific Ocean; see ocean currents.

North Equatorial Current A major warm
surface ocean current that flows westwards in
the North Pacific and forms the southern limb
of the North Pacific Gyre; see ocean currents.

North Pacific Gyre The major clockwise
circulation of surface ocean water in the North
Pacific; see ocean currents.

Northern coniferous forest biome The
circumboreal forest belt across North America
and Eurasia bordering to the north with the

Tundra biome q.v., also found on high
mountains at lower latitudes; characterized by
long cold winters with deep snow, short warm
summers, podzol soils; dominated by
evergreen needle-leaved trees and dense mats
of lichens, mosses and low shrubs, deer,
rodents and seasonal swarms of insects; taiga.

Norway lobster Astacidea q.v.

nosogenic Pathogenic; disease-producing.

nosology The study and classification of
disease.

nosophyte A pathogenic plant.

Nostocales Order of blue-green algae
(Cyanophycota) found in freshwater, soil and
marine habitats and as symbionts in lichens;
characterized by a filamentous organization
lacking true branching.

Notacanthidae Spiny eels; family containing 9
species of notacanthiform teleost fishes with
elongate cylindrical body, length to about 1.2
m, blunt snout and small mouth; dorsal and
anal fins as series of spines lacking external fin
membrane; gill openings well developed.

Notacanthiformes Small order of eel-like deep-
sea teleost fishes comprising about 25 species;
body tapering posteriorly with weak skeleton,
caudal fin reduced or absent, pelvics united,
snout elongate; development includes
leptocephalus larval stage.

Notaspidea Umbrella shell; order of
opisthobranch molluscs comprising about 150
species that feed on sponges, ascidians and
other invertebrates; characterized by the
radula teeth formula, by their discarding
outgrown teeth and by the production of an
acidic secretion when disturbed.

noterophilous Thriving in mesic habitats;
noterophile, noterophily.

Nothosauria Extinct suborder of marine
reptiles (Sauropterygia) with long necks and
limbs adapted for swimming, known from the
Triassic.

Notodontidae Prominents; family of large
robust moths (Lepidoptera) comprising over
2000 species widespread in temperate and
tropical regions; larvae of some species are
economically important pests causing damage
to the foliage of a variety of fruit and forest
trees.

Notogaea A zoogeographical area originally
comprising both Australian and Neotropical
regions, now generally used to refer to the
Australian region only; Notogea; cf.
Arctogaea, Megagaea, Neogaea, Palaeogaea.

Notonectidae Water boatmen; family of aquatic
hemipteran insects, comprising about 200

species, that characteristically swim upside down.

notophyll A leaf size category added to the Raunkiaerian leaf size class *q.v.* for leaves having a surface area between 2025 and 4500 mm^2.

Notopteridae Featherbacks; family containing 6 species of primitive nocturnal teleost fishes found mainly in weedy lentic freshwater habitats of the Old World tropics; body compressed, to 0.8 m length, anal fin continuous with caudal; dorsal and pelvic fins reduced or absent.

Notoryctidae Marsupial mole; family of small (to 200 mm) blind fossorial mammals comprising a single genus and 2 species found in arid sandy habitats of Australia; marsupium opens posteriorly; burrow actively close to sand surface, feeding on insects and earthworms.

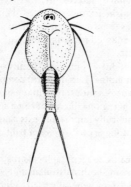

marsupial mole (Notoryctidae)

Notostigmata Opilioacariformes *q.v.*

Notostraca Tadpole shrimps; sole order of branchiopodan crustaceans inhabiting temporary freshwater bodies in arid areas; typically with a shield-shaped carapace covering head and thorax, and a slender abdomen; feed on detritus or as predators; contains only 11 extant species within a single family, Triopidae.

tadpole shrimp (Notostraca)

Nototheniidae Cod icefishes, Antarctic cods; family containing 50 species of Southern Ocean, bottom-dwelling teleost fishes

(Perciformes) abundant in Antarctic waters; body elongate, robust, to 1 m length; mouth protractile; 2 dorsal fins and 2–3 lateral lines present; blood may contain a glycoprotein antifreeze; red blood cells absent in some forms.

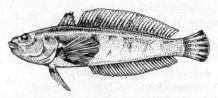

antarctic cod (Nototheniidae)

Notothenioidei Suborder of southern hemisphere marine teleost fishes (Perciformes) found primarily in cold Antarctic waters, comprising about 100 species in 4 families; many have blood antifreeze as protection against subzero temperatures.

Notoungulata Extinct order of ungulates known mainly from the Oligocene of South America; characterized by their unusual ear structure.

nucivorous Feeding on nuts; **nucivore**, **nucivory**.

nucleotide A subunit of the DNA and RNA molecules, comprising phosphoric acid, a purine or pyrimidine base and a sugar.

Nuculoida Nut shells; order of primitive protobranch bivalves found in marine sediments; characterized by 2 more or less equal shell-closing muscles.

Nuda Class of ctenophores lacking tentacles, comprising a single order, the Beroida.

nudation The production of an area of bare ground by natural or artificial means.

Nudibranchia Sea slugs; large cosmopolitan order of exclusively marine slug-like opisthobranch molluscs which lack a shell and are usually bilaterally symmetrical; most are carnivores with specialized diets, many are brightly coloured and have an elaborate body form covered with papillae, tufts or finger-like processes.

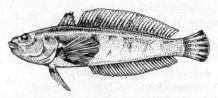

sea slug (Nudibranchia)

numbat Myrmecobiidae *q.v.*

numerical taxonomy Classification based on the numerical comparison of large numbers of equally weighted characters, scored consistently for all the groups under consideration, and in which individuals are grouped solely on the basis of observable similarities.

Numididae Guinea fowl; family containing 7 species of medium-sized gallinaceous birds (Galliformes) found in forests of Africa and Madagascar; habits solitary, monogamous, feeding on insects, seeds, and plant material; nest in hollow on the ground.

nunatak hypothesis The theory that apparent anomalies of modern plant distribution patterns can be explained in terms of survival through periods of glaciation in ice-free areas (nunataks) surrounded by ice.

nuptial flight The mating flight of social insects.

nurse shark Ginglymostomatidae *q.v.*

nut A dry indehiscent fruit derived from more than one carpel but containing only a single seed; the wall of the fruit (pericarp) is usually woody; includes acorn and hazelnut.

nut shell Nuculoida *q.v.*

nutation A rhythmic or rotational movement of the growing tip of a plant.

nuthatch 1: Sittidae *q.v.* 2: Hyposittidae *q.v.* (coral-billed nuthatch).

nutmeg Myristicaceae *q.v.*

nutrition Ingestion, digestion and/or assimilation of food by plants and animals; **nutritive.**

Nyctaginaceae Garden four-o'clock, bougainvillea; family of Caryophyllales containing about 300 species of herbs or woody plants, widespread in tropical and subtropical regions; characterized by individual flowers typically surrounded by bracts which may be large and showy, but lacking petals.

nyctanthous Flowering only during the night; **nyctanthic, nyctanthy**; *cf.* hemeranthous.

Nycteribiidae Bat flies; family of small fragile wingless flies (Diptera) that are blood-feeding ectoparasites of bats.

Nyctibiidae Pottoos; family of cryptically coloured, nocturnal insectivorous birds (Caprimulgiformes) with small bills, large gapes and no rictal bristles; nest in tree crevices.

nyctigamous Used of flowers that open at night and close during the day; **nyctigamy.**

nyctinasty Orientation movements of plants during the night; **nyctinastism, nyctinastic.**

nyctipelagic Pertaining to organisms that migrate into surface waters at night.

nyctiperiod A period of darkness; a dark phase in a light/dark cycle; *cf.* photoperiod.

nyctitropism An orientation response occurring at night; **nyctitropic.**

nymph A stage in the development of exopterygote insects, between hatching and the reorganization involved in attaining the adult stage.

Nymphaeaceae Water lilies; family of aquatic plants containing about 50 species having a cosmopolitan distribution; leaves floating, flowers large and aerial, borne singly on long peduncles.

water lily (Nymphaeaceae)

Nymphaeales Water lilies; an ancient order of aquatic herbs found in still waters; tissues with conspicuous air passages; flowers solitary; comprises 5 families, Barclayaceae, Cabombaceae, Ceratophyllaceae, Nelumbonaceae and Nymphaeaceae.

Nymphalidae Large family of small to large, colourful butterflies (Lepidoptera) with reduced, non-functional forelegs and long, hair-like scales; eggs ribbed, caterpillars with spines; contains about 5000 species including the admirals, emperors, fritillaries and tortoiseshell butterflies.

red admiral (Nymphalidae)

nymphiparous Giving birth to offspring at the nymphal stage of development.

Nyssaceae Sour gum, dove tree; small family of Cornales containing 8 species of trees and shrubs found in eastern North America and eastern Asia.

O

oak Fagaceae *q.v.*

oarfish Regalecidae *q.v.*

oats Poaceae *q.v.*

Obik Sea An epicontinental sea joining the Arctic Sea and the Tethys Sea during the Eocene epoch (*c.* 54–38 million years B.P.).

obligate Essential; necessary; unable to exist in any other state or relationship; *cf.* facultative.

obligate anaerobe An organism surviving only in the absence of molecular oxygen.

obliterative shading A graded coloration of a body which neutralizes relief giving the appearance of a flat surface.

Obolellida Extinct order of inarticulate lamp shells (Brachiopoda) known only from the Cambrian.

occidental Western; westerly.

ocean currents See chart on page 263 for major ocean surface currents.

Oceania The islands of the Pacific and surrounding seas.

oceanic Pertaining to the open ocean waters beyond the edge of the continental shelf; *cf.* neritic.

oceanic climate A climate characteristic of continental margins and islands in which the annual temperature range is less than the average for that latitude because of the proximity of an ocean or sea.

oceanic eddy A body of water rotating within the main current system, or deflected from the main current into adjacent areas.

oceanic island A volcanic island formed independently of continental land masses; *cf.* continental island.

oceanodromous Used of organisms that migrate only within the oceanic province; *cf.* anadromous, catadromous, potamodromous.

oceanography The study of the oceans.

oceanophilous Thriving in oceanic habitats; **oceanophile, oceanophily.**

oceanophyte An oceanic plant.

ocelot Felidae *q.v.*

Ochnaceae Small family of tropical, mostly glabrous, evergreen trees or shrubs, especially common in Brazil; typically with regular, bisexual flowers having 5 sepals, 5 petals, 5, 10 or many stamens and a superior ovary.

Ochotonidae Conies, pika; family containing 14 species of lagomorph mammals found in mountainous regions of the northern hemisphere; ears short; forelimbs with 5 digits, hindlimbs with 4; tail not visible externally.

ochthophilous Thriving on banks; **ochthophile, ochthophily.**

ochthophyte A plant living on banks.

ocotillo Fouquieriaceae *q.v.*

Octocorallia Alcyonaria *q.v.*; octocorals.

Octodontidae Bush rats; family containing 7 species of rat-like ground-dwelling or burrowing rodents (Hystricomorpha) known only from South America; tail hairy, typically with apical tuft.

Octopoda Octopuses; order of carnivorous cephalopod molluscs with a short sac-like body, and 8 arms more or less connected by webbing; shell usually absent; funnel lacking a valve.

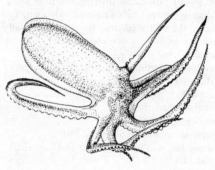

octopus (Octopoda)

octopus Octopoda *q.v.*

Odacidae Family of colourful marine teleost fishes resembling wrasses in which the jaws form a parrot-like beak; contains 8 species found in cool coastal waters of Australia and New Zealand.

Odobenidae Walrus; now treated as part of the family Otariidae.

Odonata Dragonflies, damselflies; order of large paleopterous insects comprising 3

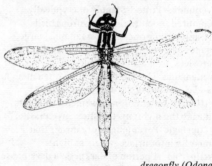

dragonfly (Odonata)

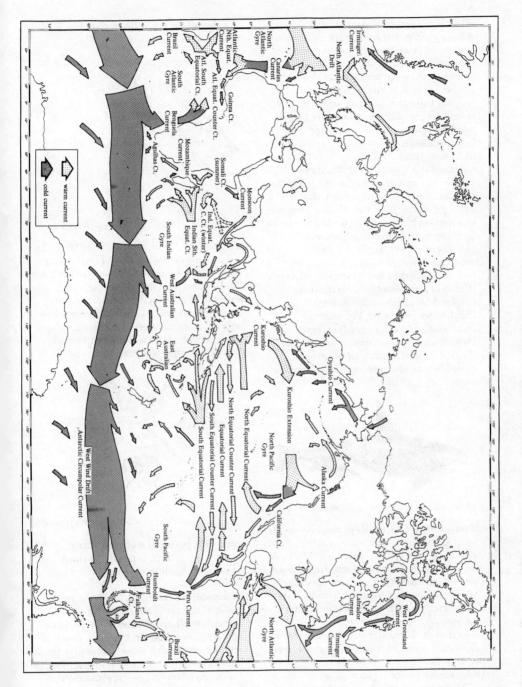

Ocean currents

suborders, Zygoptera (damselflies), Anisozygoptera and Anisoptera (dragonflies); adults with powerful predatory mouthparts, large compound eyes, and short slender antennae; the wings are long and slender, of similar size and distally rounded; larvae typically aquatic with strong hinged labrum (mask) for grasping prey; larvae feed on aquatic larvae, tadpoles, worms and small fishes; adults predatory on flying insects taken in flight or at rest.

Odontaspididae Sand tiger sharks; family containing 4 species of inoffensive pelagic or benthic lamniform elasmobranch fishes, cosmopolitan in tropical and warm temperate seas from surface to 1000 m, feeding mostly on bony fishes, but also crustaceans and cephalopods; body length to 3.5 m.

Odontoceta Toothed whales; suborder of carnivorous marine and freshwater mammals (Cetacea) comprising 6 extant families, Platanistidae (freshwater dolphins), Delphinidae (dolphins), Phocoenidae (porpoises), Monodontidae (belugas), Physeteridae (sperm whales) and Ziphiidae (beaked whales); teeth typically present, baleen absent, skull asymmetrical, blowhole single.

killer whale (Odontoceta)

Odontognathae Superorder of neornithine birds known as fossils from the Cretaceaous period; probably possessed teeth in both jaws; includes the large aquatic form, *Hesperornis*.

Odontostomatida Order of spirotrich ciliates with a laterally compressed, typically wedge-shaped body and reduced ciliature; found in polysaprobic freshwater habitats.

odour trail A chemical trace (trail pheromone) deposited by one animal for another to follow.

oecology Ecology *q.v.*; **oecological.**

Oedogoniales Order of green algae (Chlorophycota) found attached to various substrata in freshwater habitats; characterized by a thallus comprising filaments of uninucleate cells, each with a reticulate chloroplast, and by the production of zoospores with a ring of flagella; also treated

as a class of the protoctistan phylum, Chlorophyta.

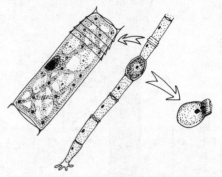

green alga (Oedogoniales)

Oegophiurida Order of primitive ophiuroidean echinoderms represented by a single living species from Indonesia, but with an extensive fossil record from the Palaeozoic.

Oestridae Warble flies, botflies; family of robust flies (Diptera) the larvae of which are subcutaneous parasites of mammals, especially ungulates and rodents; several species are important pests of livestock.

botfly (Oestridae)

oestrus The period of maximum sexual receptivity, or heat, in female mammals, usually also the time of release of the eggs; estrus.

Ogcocephalidae Batfishes; family containing about 55 species of bizarre bottom-living anglerfishes (Lophiiformes) widespread on soft offshore sediments; body depressed, to 400 mm length, disk-shaped anteriorly, dorsal and lateral surfaces spinose or rugose; pectoral fins set on elbow-like bases; glandular lure located beneath rostrum; mouth small.

O-horizon The upper soil horizon having more than 20% organic matter; subdivided into litter layer (L- or Ol-layer), fermentation layer (F- or Of-layer), and humified layer (H- or Oh-layer); see soil horizons.

oil A triacylglycerol that is liquid at room temperature and which typically functions as an energy storage compound.

oilfish Comephoridae *q.v.*

okapi Giraffidae *q.v.*

Olacaceae Family of Santalales containing about 250 species of evergreen woody plants, widespread in tropical and subtropical regions; many are hemiparasitic, attaching to roots of other plants; flowers usually green or white, with sepals reduced to narrow lobes, 4–6 petals and up to twice as many stamens.

Old World monkey Cercopithecidae *q.v.*

Oleaceae Olive, lilac, ash, forsythia, jasmine, privet, syringa, fringe tree; family of Scrophulariales containing about 600 species of woody plants nearly cosmopolitan in distribution but best developed in Asia and Malaysia; flowers usually perfect and regular, with 4 sepals and petals, 2 stamens and a superior ovary.

olive (Oleaceae)

oleaginous Producing or containing oil.

oleander Apocynaceae *q.v.*

oleaster Elaeagnaceae *q.v.*

Oligacanthorhynchida Order of archiacanthocephalan thorny-headed worms found worldwide as parasites of mammals and, rarely, birds, and utilizing insects or occasionally snakes as intermediate hosts; characterized by a large trunk which is not pseudosegmented, a subspherical proboscis and the presence of protonephridial glands.

oligoaerobic Pertaining to organisms that grow in conditions of low oxygen concentration.

Oligocene A geological epoch within the Tertiary period (*c.* 38–26 million years B.P.); see geological time scale.

Oligochaeta Earthworms, pot worms, freshwater ringed worms; class of clitellate annelids comprising 3 orders, Lumbriculida, Moniligastrida and Haplotaxida; body length 1 mm to 4 m, parapodia absent, chaetae sparse; head simple without sensory appendages; typically hermaphroditic but ensuring cross fertilization by copulation.

earthworm (Oligochaeta)

oligogyny 1: The mating of a single male with a few females. 2: The presence of a few queens within a single colony of social insects; **oligogynous**; *cf.* monogyny, polygyny.

oligohalabous Used of planktonic organisms living in sea water of less than 5 parts per thousand salinity.

oligohaline 1: Used of organisms that are tolerant of only a moderate range of salinities; *cf.* euryhaline, holeuryhaline, polystenohaline, stenohaline. 2: Pertaining to brackish water having a salinity between 0.5 and 3.0 parts per thousand, or sea water having a salinity between 17 and 30 parts per thousand; *cf.* mesohaline, polyhaline.

Oligohymenophora Class of ciliates with cilia on general body surface arranged in uniform rows when present; the oral ciliary apparatus is distinct from the somatic cilia and consists, in part, of a well defined undulating membrane; widely distributed, free-living and symbiotic.

trichodinid ciliate (Oligohymenophora)

oligolectic Used of insects that collect pollen from only a few different types of flowers.

Oligonephridia Hemipterodea *q.v.*

oligonitrophilous Thriving in habitats having a low nitrogen content; **oligonitrophile**, **oligonitrophily**.

oligopelic Used of an area, substrate or habitat containing only small amounts of clay.

oligophagous Utilizing only a small variety of food species; **oligophage**, **oligophagy**; *cf.* monophagous, polyphagous.

Oligopygoida Extinct order of irregular echinoids known only from the Eocene; genital plates fused, ambulacra petaloid.

oligosaprobic Pertaining to polluted habitats having a high oxygen concentration, low levels of dissolved organic matter, and a low level of organic decomposition; **oligosaprobe.**

oligothermic Tolerating relatively low temperatures; *cf.* polythermic.

oligotokous Having only a few offspring per brood; *cf.* ditokous, monotokous, polytokous.

oligotraphent Used of an aquatic plant characteristic of water bodies having low nutrient concentration; *cf.* eutraphent, mesotraphent.

Oligotrichida Large order of free-living spirotrich ciliates with reduced body ciliature and often with a lorica; contains the tintinnids which are abundant in the marine pelagic environment.

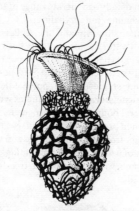

tintinnid (Oligotrichida)

oligotrophic 1: Having low primary productivity; used of substrates or water bodies low in nutrients. 2: Used of any organism requiring only a small nutrient supply, or restricted to a narrow range of nutrients. 3: Used of a lake in which the hypolimnion does not become depleted of oxygen during the summer; *cf.* dystrophic, eutrophic, mesotrophic. 4: Pertaining to insects that visit only a small variety of plant species.

oligoxenous Used of a parasite utilizing a few host species during the life cycle; **oligoxeny;** *cf.* dixenous, heteroxenous, monoxenous, trixenous.

oligoxic Used of a habitat having reduced levels of molecular oxygen; **oligoxicity;** *cf.* anoxic, normoxic.

Oliniaceae Small family of Myrtales containing 8 species of tanniferous shrubs or trees confined to tropical and southern Africa and to St Helena.

olive Oleaceae *q.v.*

olive shell Neogastropoda *q.v.*

olm Proteidae *q.v.*

olynthus Hollow vase-like stage in the development of some calcareous sponges.

Olyridae Family of small (to 100 mm) Old World freshwater catfishes (Siluriformes) found in streams of India and Burma, comprising a single genus and 3 species; body elongate, naked, eyes reduced; 4 pairs of barbels present; dorsal fin lacking spine, caudal forked with long upper lobe; gas bladder partially enclosed in bony capsule.

ombratropism An orientation response to rain; **ombratropic.**

ombrocleistogamic Used of a flower that remains closed because of rain and within which self-pollination occurs; **ombrocleistogamy.**

ombrophilous Thriving in habitats having abundant rain; **ombrophile, ombrophily;** *cf.* ombrophobic.

ombrophobic Intolerant of prolonged rain; **ombrophobe, ombrophoby;** *cf.* ombrophilous.

ombrophyte 1: A plant growing in situations exposed to direct rainfall. 2: An epiphyte that absorbs rainwater by aerial assimilation through specialized structures.

ombrotiphic Pertaining to temporary pools of water formed from melting snow or rain.

ombrotrophic Pertaining to organisms that obtain nutrients largely from rainwater; **ombrotrophy.**

Ommatomorpha Superorder of helminthomorphan diplopods (millipedes) comprising a single order, Polyzoniida.

omnicolous Living on a wide variety of substrata; **omnicole.**

omnivorous Feeding on a mixed diet of plant and animal material; **omnivore, omnivory.**

Omosudidae Hammerjaw; monotypic family of predatory pelagic myctophiform teleost fishes found in warm seas from surface to 1300 m; body compressed, naked, length to 250 mm; mouth very large bearing strong sharp teeth; eyes well developed; photophores absent.

Onagraceae Fuchsia, evening primrose, sundrop, water purslane, willow herb; family of Myrtales containing about 650 species of herbs or shrubs found mainly in temperate and subtropical regions; flowers bisexual, typically with 4 sepals, 0–4 petals, 8 stamens in 2 whorls and an inferior ovary.

Oncoceratida Extinct order of thick-shelled cephalopod molluscs (Nautiloidea) known from the Ordovician to the Carboniferous.

fuchsia (Onagraceae)

invertebrates; contains about 70 species; also treated as subphylum of Uniramia *q.v.*

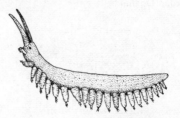

onychophoran (Onychophora)

oncosphere First larval stage of tapeworms, possessing 6 hooks.

Oncothecaceae Family of Theales containing a single glabrous, tanniferous evergreen tree species; known from New Caledonia.

Oneirodidae Family containing 35 species of small (to 200 mm) deep-sea anglerfishes (Lophiiformes); found at depths of 300–3000 m; body typically naked, or with sparse spines in female; males free-living, lacking teeth; feed mainly on crustaceans.

Oniscoidea Pill bugs, slaters, sow bugs, woodlice; suborder of isopod crustaceans adapted to terrestrial habitats, containing about 3500 species distributed worldwide; eyes sessile, thoracic legs ambulatory; abdominal limbs (pleopods) modified for reproduction and aerial respiration; many able to roll into a ball (conglobation).

Oniscomorpha Superorder of pentazonian diplopods (millipedes) in which the body comprises only 12 or 13 segments and is modified for enrolment into a compact ball; posterior legs of male specialized for clasping female; gut coiled or sinuous; comprises 2 orders, the largely northern hemisphere Glomerida and southern hemisphere Sphaerotheriida.

ontogeny The course of growth and development of an individual to maturity; **ontogenetic.**

Onychophora Phylum of primitive terrestrial tracheate arthropods typically found in humid forest litter; body cylindrical, 15–150 mm in length, metamerically segmented; head bearing single pair of antennae, eyes at base of antennae; mandibles present; legs sac-like, unsegmented, annulate (lobopods); sexes separate, oviparous to viviparous; habits largely nocturnal, feeding on small

oocyte The cell that produces eggs (ova) by meiotic division; *cf.* spermatocyte.

oogamy The fertilization of large non-motile eggs by motile sperms; usually applied to algae.

oogenesis Female gametogenesis; the formation, development, and maturation of female gametes (ova); ovogenesis.

oogonium The typically unicellular, female reproductive organ in which oospheres are produced in some algae and fungi.

oology The study of eggs.

Oomycetes Class of fungi (Mastigomycotina) containing parasitic or saprobic forms characterized by motile zoospores with a posteriorly directed whiplash flagellum and an anteriorly directed tinsel flagellum; usually without cross walls; cell walls composed mostly of cellulose; comprises 4 orders, Saprolegniales, Leptomitales, Lagenidiales and Peronosporales; sometimes classified as a phylum, Oomycota, of the Protoctista.

Oomycota Oomycetes *q.v.*, treated as a phylum of the Protoctista.

oophagous Feeding on eggs; **oophage, oophagy.**

oosphere The female gamete prior to fertilization.

oospore The fertilized female gamete.

ooze A fine-grained pelagic deposit comprising at least 30% undissolved sand or silt-sized skeletal remains of marine organisms, the remainder being amorphous clay-sized material; *cf.* coccolith ooze, diatomaceous ooze, foraminiferal ooze, globigerina ooze, pteropod ooze, radiolarian ooze, siliceous ooze.

opah Lampridae *q.v.*

Opalinata Subphylum of Sarcomastigophora; protozoans with a uniform covering of short flagella arranged in oblique rows over the body and with 2 to many nuclei, reproducing by binary or multiple fission and by the production of flagellated sexual gametes; all species are found in the intestine or rectum of cold-blooded vertebrates; comprises a single order, Opalinida.

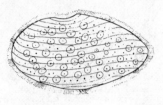

ciliate (Opalinata)

Opheliida Order of deposit-feeding polychaete worms found in soft sediments from littoral to abyssal depths, comprising about 140 species in 2 families; prostomium and peristomium naked, pharynx eversible, unarmed; parapodia biramous.

Ophichthidae Snake eels, worm eels; large family of littoral to bathyal anguilliform teleost fishes; body smooth, to 2 m length, slender, tail pointed (snake eels) or fringed with fin (worm eel); some snake eels totally devoid of fins; contains about 270 species; many burrow tail first in sandy or muddy sediments.

Ophidia Alternative name for the Serpentes *q.v.*; snakes.

Ophidiidae Cusk eels, brotulas; family of coastal to deep-water marine and freshwater gadiform teleost fishes; body elongate, tail tapering, length to 1.5 m; some species ovoviviparous with internal fertilization; contains 35 species, including the brotulas which are sometimes treated as a separate family, Brotulidae.

Ophiocistioidea Extinct class of free-living echinozoans with a dome-shaped test lacking arms; known from the Ordovician to the Devonian.

Ophioglossopsida Subdivision of small to medium-sized, mostly terrestrial ferns

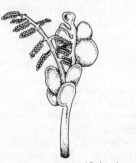

moonwort (Ophioglossopsida)

(Filicophyta) with a distinct fertile portion of frond held erect; stems fleshy, subterranean and mycorrhizal, fronds fleshy to membranous and do not unroll circinately; sporangia massive with 2000–5000 spores.

ophiology The study of snakes.

ophiophagous Feeding on snakes; **ophiophage, ophiophagy.**

ophiopluteus The pluteus *q.v.* larval stage of a brittle star.

ophiotoxicology The study of snake venoms.

Ophiurida Largest of 3 extant orders of ophiuroidean echinoderms comprising about 2000 species found at all depths from intertidal to deep sea; arms unbranched with vertebral articulation that permits lateral but limited vertical movement, spines typically simple; contains the majority of living ophiuroids.

Ophiuroidea Brittle stars, snake stars, basket stars; subclass of asterozoan echinoderms comprising about 2000 extant species in 3 orders, Oegophiurida, Phrynophiurida and Ophiurida; arms slender, tapering only gradually and distinctly offset from central disk; typically 5-radiate, although arms may be branched in some species; tube-feet without suckers, anus absent; sexes usually separate, fertilization external, development involving vitellaria and ophiopluteus larvae, but may be direct; asexual reproduction by fission may occur in some groups; most species feed as scavengers or utilize suspended matter; brittle stars range from intertidal to deep-sea and may be extremely abundant, forming a major component of the benthos.

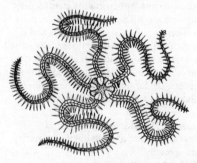

brittle star (Ophiuroidea)

Opiliaceae Family of Santalales containing about 50 species of mostly evergreen woody plants which are hemiparasitic on roots of other plants; widespread in tropical and subtropical regions.

Opilioacariformes Monofamilial suborder of primitive mites (Acari); Notostigmata.

Opiliones Harvestmen; order of typically small-bodied long-legged arthropods (Arachnida) that are frequently predators of other small arachnids, molluscs and worms, or scavenge dead plant and animal material; cephalothorax and abdomen compact, joined by broad waist; legs mostly elongate but may be short in some species; single pair of tracheae present; female with long ovipositor; comprises about 4500 species in 30 families.

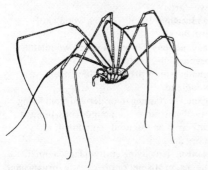

harvestman (Opiliones)

Opisthobranchia Subclass of gastropod molluscs containing about 2000 species of mainly marine, bottom-living or littoral forms; characterized by detorsion of the body, which may exhibit secondary bilateral symmetry, by reduction or loss of the shell and the mantle cavity with its associated structures, by a typically flat and creeping foot and by a head primitively covered with a cephalic shield and bearing paired tentacles and rhinophores; mostly hermaphrodites, with internal fertilization; comprises 9 orders, the largest of which is the Nudibranchia (sea slugs).

Opisthocomidae Hoatzin; monotypic family of medium-sized tree-dwelling birds found in forests of nothern South America; flight weak; feet strong for climbing; habits gregarious, monogamous, feeding on leaves and fruit; nest solitary, typically in tree over water; young with claws on wings.

Opisthocomiformes Order of neognathous birds comprising a single family, Opisthocomidae (hoatzin).

Opisthognathidae Jawfishes; family containing 30 species of shallow-water, tropical marine perciform teleost fishes; body elongate, to 500 mm in length; mouth large bearing strong jaw teeth; dorsal and anal fins long; some forms inhabit burrows in sandy sediments; males sometimes incubate eggs orally.

Opisthoproctidae Barreleyes; family containing 11 species of small (to 150 mm) bizarre deep-sea salmoniform teleost fishes characterized by projecting tubular eyes providing binocular vision; photophores often present within the orbit.

barreleye (Opisthoproctidae)

Opisthorchiida Order of digenetic trematodes with 3 hosts in the life cycles; eggs develop into miracidia, sporocysts and rediae in prosobranch snails, cercariae infect fishes and form encysted metacercariae and the adult fluke is found in another vertebrate; characterized by a single pair of flame cells in the miracidium larva.

Oplegnathidae Knifejaws, parrotfishes; family of Indo-Pacific shallow marine perciform teleost fishes comprising 4 species; teeth fused into parrot-like beak used to scrape food from hard surfaces; body deep, compressed, to 900 mm in length; single dorsal fin present.

opophilous Thriving or feeding on sap; **opophile**, **opophily.**

opophyte A parasitic plant.

opossum Didelphidae *q.v.*

opossum shrimp Mysidacea *q.v.*

opportunistic species A species adapted for utilizing variable, unpredictable or transient environments, typically with a high dispersal ability and a rapid rate of population growth; *r*-selected species.

optimal Most favourable; used of the levels of environmental factors best suited for growth and reproduction of organisms; *cf.* pessimal.

optimal foraging theory That selection favours a strategy in which a predator utilizes prey in a manner that optimizes net energy gain per unit feeding time.

optimal yield The maximum sustainable rate of increase of a population under a given set of environmental conditions.

orange Rutaceae *q.v.*

orang-utan Pongidae *q.v.*

Orbiniida Order of errant marine polychaete worms comprising about 200 species in 2 families; typically burrowing deposit feeders found at all depths from littoral to abyssal; prostomium conical, without appendages; pharynx usually eversible and unarmed; parapodia biramous lacking acicula and cirri.

orb-web spider Araneidae *q.v.*

orchid bee Apidae *q.v.*
Orchidaceae Orchids; well defined family of
monocotyledons comprising 15 000 to 20 000
species of strongly mycotrophic terrestrial or
often epiphytic herbs; cosmopolitan in
distribution but most abundant and diverse in
tropical forests; characterized by numerous,
often bizarre specializations for pollination by
particular species of insects; usually green and
commonly with crassulacean acid metabolism;
producing from a thousand to several millions
of tiny seeds with a minute undifferentiated
embryo which require association with an
appropriate fungus for successful germination.

orchid (Orchidaceae)

Orchidales Order of Liliidae comprising 4
families of strongly mycotrophic herbs,
sometimes without chlorophyll; leaves
parallel-veined or sometimes reduced to
scales, flowers basically trimerous and
adapted for pollination by insects; producing
abundant tiny seeds with a minute
undifferentiated embryo and no endosperm.
Ordovician A geological period within the
Palaeozoic (*c.* 504–441 million years B.P.); see
geological time scale.
Orectolobidae Wobbegongs; family containing
7 species of large (to 3.5 m) orectolobiform
elasmobranch fishes found on coral and rocky
reefs of the western Pacific Ocean; body and
head depressed, nostrils with long barbels;
mouth terminal, teeth strongly differentiated;
feed on fishes, crustaceans and cephalopods,
occasionally offensive to man; reproduction
ovoviviparous.
Orectolobiformes Carpet sharks;
circumtropical order of marine or brackish-
water galeomorph elasmobranch fishes,
primarily benthic, occasionally pelagic; mouth
terminal or subterminal, teeth

undifferentiated; dorsal fins without spines;
feed on bottom invertebrates and small fishes;
reproduction oviparous or ovoviviparous, egg
cases without spiral flanges; comprising about
30 species in 7 families; mostly inoffensive, a
few species dangerous if provoked.
Oreosomatidae Oreos; cosmopolitan family
containing 10 species of deep-water zeiform
teleost fishes; body compressed, to 400 mm
length, scales cycloid or ctenoid, mouth and
eyes large.
orgadocolous Living in open woodland;
orgadocole.
orgadophilous Thriving in open woodlands;
orgadophile, orgadophily.
orgadophyte A plant inhabiting open
woodland.
organic Pertaining to or derived from living
organisms, or to compounds containing
carbon as an essential component; *cf.*
inorganic.
organism Any living entity; plant, animal,
fungus, protistan, or prokaryote; **organismal,
organismic.**
organogenesis The formation of organs during
development; **organogeny.**
orgasm Intense sexual excitement or climax.
Oribatei Diverse subgroup of acariform mites
(Acari) containing about 145 families;
commonly part of the soil fauna these mites are
also widespread amongst lichens, mosses, and
algae; most are fungivorous, none is parasitic;
Cryptostigmata.
Oribatida Oribatei *q.v.*
oriental Eastern; easterly.
Oriental region A zoogeographical region *q.v.*
comprising India, Sri Lanka, Malay peninsula,
Sumatra, Philippines, Borneo, Djawa (Java)
and Bali; subdivided into Indian, Indo-
Chinese, Indo-Malayan and Sri Lankan
subregions.
oriole 1: Oriolidae *q.v.* 2: Icteridae *q.v.*
Oriolidae Orioles; family containing about 25
species of colourful Old World passerine birds
found in tropical forests from Africa to
Australia; habits solitary, arboreal, feeding on
fruit and insects; cup-shaped nest of twigs
placed in tree.
ornithichite A fossilized track or footprint of a
bird.
Ornithischia Order of Mesozoic dinosaurs
possessing a bird-like pelvis; included both
tetrapodal and bipedal forms; all were
herbivorous.
ornithocoprophilous Thriving in habitats rich in
bird droppings; **ornithocoprophile,**

ornithocoprophily.

Ornithogaea The zoogeographical region comprising New Zealand and Polynesia.

ornithogenic Pertaining to sediments rich in bird droppings.

ornithology The study of birds; **ornithological.**

ornithophilous Pollinated by birds; **ornithophily.**

Ornithopoda A suborder of ornithischian dinosaurs characterized by their bipedal gait; included the hadrosaurs and iguanodonts.

Ornithorhynchidae Platypus; monotypic family of egg-laying mammals (Monotremata) found in streams and lakes of Tasmania and eastern Australia; body covered with soft fur; feet webbed bearing strong claws; jaws toothless in adult, prolonged and strongly depressed to form large duck-bill; excavate burrows in banks for protection and as brood chamber; feed mainly on aquatic invertebrates.

duck-billed platypus (Ornithorhynchidae)

Orobanchaceae Family of Scrophulariales containing about 150 species of herbaceous root parasites that lack chlorophyll; commonly producing iridoid compounds and orobanchin; widespread in northern hemisphere.

orogeny The process of mountain formation; **orogenic, orogenesis.**

orographic Pertaining to relief factors such as hills, mountains, plateaux, valleys and slopes.

orographic desert A rain shadow desert on the leeward side of a mountain range.

orographic rain Rainfall produced by supersaturation of an air mass forced upwards by a mountain range; *cf.* convective rain, frontal-cyclonic rain.

orohylile Pertaining to an alpine or subalpine forest community or habitat.

orophilous Thriving in subalpine, or in mountainous regions; **orophile, orophily.**

orophyte A subalpine plant; **orophytic.**

orphan Of, or relating to, a virus that has no known association with disease.

ortet The original organism from which a clone was derived; *cf.* ramet, genet.

Orthida Extinct order of articulate lamp shells (Brachiopoda) known from the Cambrian to the Permian.

Orthognatha Mygalomorph spiders; diverse suborder of mainly tropical spiders (Araneae) comprising about 1800 species and including

various trapdoor, funnel web and purse web groups, and a family of large tarantulas; abdomen unsegmented, spinnerets located posteriorly; chelicerae move parallel to body axis.

orthoheliotropism An orientation response towards sunlight; **orthoheliotropic.**

orthokinesis A change in the rate of random movement of an organism (kinesis) in which the rate of locomotion (linear velocity) varies with the intensity of the stimulus; *cf.* klinokinesis.

Orthonectida Class of mesozoans found as endoparasites in marine brittle stars, bivalve molluscs and polychaete worms; characterized by the alternation of an asexually reproducing stage and a free-swimming sexual stage.

Orthonychidae Rail babblers; family containing about 20 species of ground-dwelling passerine birds found in forest and open habits of Australia and New Guinea; wings short, legs and feet large and robust; habits solitary, secretive, non-migratory, feeding on insects; nest on ground or in burrow.

orthophototropism An orientation response towards light; **orthophototropic.**

Orthopsida Extinct order or regular echinoids with simple ambulacral plates, known from the Jurassic and Cretaceous.

Orthoptera Grasshoppers, crickets, locusts, katydids; order of orthopterodean insects

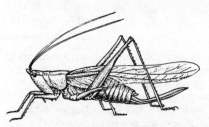

longhorned grasshopper (Orthoptera)

comprising about 20 000 species found in almost all terrestrial habitats from subterranean burrows to tree canopies, and forests to deserts; body length 5–120 mm, hindlimbs usually specialized for jumping; forewings leathery or parchment-like, hindwings forming membranous fan, or may be reduced; many species produce sounds (stridulate) by rubbing together forewings, hindlimbs or hindlimbs against forewings; most are phytophagous, some omnivorous, a few predatory; as crop pests the group is of immense economic importance.

Orthopterodea Superorder of neopteran insects that retain primitive mandibulate mouthparts; wings variously modified or absent; contains 10 extant orders, Blattaria (cockroaches), Mantodea (mantises), Isoptera (termites), Grylloblattaria, Orthoptera (grasshoppers, locusts), Phasmoptera (stick insects), Dermaptera (earwings), Embiidina, Plecoptera (stoneflies) and Zoraptera.

Orthotrichales Order of dull tuft- or mat-forming mosses (Bryidae) that grow on rocks or tree trunks; stems with terminal sporophytes; leaves mostly papillae.

orthotropism Orientation or growth in a straight line; **orthotropic.**

Orycteropodidae Aardvark; monotypic family of insectivorous mammals (Tubulidentata) found in open grassland and brushland of Ethiopian region; body sparsely covered with bristly hairs, limbs stout with strong nails for digging, tongue long and protrusible; habits nocturnal, solitary, semifossorial, feeding mainly on ants and termites.

aardvark (Orycteropodidae)

oryctology The study of fossils; palaeontology.

Oryziatidae Ricefishes; family containing 7 species of fresh- and brackish-water cyprinodontiform teleost fishes found in southeastern Asia; body elongate, exhibiting weak sexual dimorphism; reproduction oviparous, fertilization external.

Osmeridae Smelts; family of marine and freshwater salmoniform teleost fishes found in the Arctic, and northern Atlantic and Pacific Oceans; body elongate, silvery, compressed and rather fragile, to 400 mm in length, cycloid scales thin; adipose fin present; contains 10 species which are anadromous or migrate upstream for spawning; locally abundant, smelts are important commercially as food-fish.

osmoconformer An organism having a body fluid of the same osmotic concentration as the surrounding medium; *cf.* osmoregulator.

osmophilic Thriving in a medium of high osmotic concentration; **osmophile, osmophily.**

osmoregulation Control of the volume and composition of body fluids.

osmoregulator An organism that maintains the osmotic concentration of its body fluid at a level independent of the surrounding medium; *cf.* osmoconformer.

osmosis Diffusion of a solvent through a differentially permeable membrane from a region of low solute concentration to one of high solute concentration; **osmotic.**

osmotaxis A directed reaction of a motile organism to a change or gradient of osmotic pressure; **osmotactic.**

osmotic water Water held in close contact by clay particles that is less available to soil organisms than capillary water *q.v.*

osmotrophic Used of organisms that are capable of absorbing organic nutrients directly from the external medium; **osmotrophy.**

osmotropism An orientation response to an osmotic stimulus; **osmotropic.**

osphresiology Study of the sense of smell.

osprey Pandionidae *q.v.*

Ostariophysi Group of bony fishes possessing Weberian ossicles; includes the Cypriniformes, Gonorhynchiformes and Siluriformes.

Osteichthyes Bony fishes; group comprising all true bony fishes, about 18 000 species in 450 families; skull with sutures, teeth fused to bone; swim bladder frequently present functioning as lung or hydrostatic organ; internal fertilization rare.

Osteoglossidae Bonytongues; family containing 6 species of primitive freshwater teleost fishes from tropical South America, Africa,

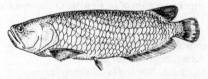

bonytongue (Osteoglossidae)

southeast Asia and Australia; body elongate, to 4.5 m in length; anal and dorsal fins long, pelvics posterior to pectorals; gas bladder richly vascularized and serving as accessssory respiratory organ; parental care may involve nest building or oral incubation.

osteology Study of the structure and development of bones.

Osteostraci Extinct order of jawless vertebrates with a fish-like body and a flattened head covered by a bony plate with a series of gill openings; known from the Silurian to the Devonian.

Ostraciontidae Boxfishes; family containing about 25 species of tropical shallow marine tetraodontiform teleost fishes commonly found on sea-grass beds; body enclosed within body carapace, length to 600 mm, fins small and paddle-like; may release toxin when alarmed.

Ostracoda Mussel shrimps; class of small aquatic crustaceans containing about 5700 species inhabiting marine, brackish and freshwater habitats from littoral to abyssal zones, and occasionally terrestrial; body entirely enclosed in hinged bivalved carapace ranging in length from 0.1 to 30 mm; comprises 3 subclasses, Myodocopa, Podocopa and Palaeocopa; contains at least 10 000 fossil species stretching through the fossil record from the Cambrian to Recent times.

Ostreoida Oysters, scallops, jingle shells, saddle oysters, window-pane shells; order of pteriomorphian bivalve molluscs with variable, often unequal, shell valves; characterized by 1 or 2 unequal shell-closing muscles, and by the lack of ventral fusion of the mantle, which has tentaculate margins.

ostrich Struthionidae *q.v.*

Ostropales Order of predominantly saprophytic discomycete fungi found on decaying wood in warm climates; characterized by sessile, flask-shaped fruiting bodies and filiform ascospores.

Otariidae Sealions, fur seals, walruses; family of aquatic mammals (Carnivora) widespread in cold marine waters, comprising about 15 species; fore- and hindlimbs modified as flippers; ear pinnae present; feeding mostly on fishes but diet may include squid, other molluscs and other large invertebrates.

Otididae Bustards; family containing 24 species of medium-sized to large terrestrial gruiform birds found in open grassland habitats of Africa through Asia to Australia; neck and legs long, wings broad; habits cursorial, gregarious, feeding on variety of animals and plants, nesting in hollow on ground.

otter Mustelidae *q.v.*

otter shrew Tenrecidae *q.v.*

outbreeding Mating or crossing of individuals that are either less closely related than average pairs in the population, or from different populations; *cf.* inbreeding.

ovariicolous Living in ovaries; **ovariicole.**

ovary 1: The egg-producing reproductive organ of animals. 2: The swollen basal part of the carpel in flowering plants, containing one or more ovules.

ovenbird Furnariidae *q.v.*

overgrazing Grazing which exceeds the recovery capacity of the community and thus reduces the available forage crop or causes undesirable changes in the community composition.

overstorey The topmost layer of a forest community, formed by the tallest trees of the canopy.

overturn Thorough water circulation in the sea or a lake, often occurring seasonally and caused by density differentials resulting from changes of temperature.

ovigerous Egg-bearing; used of a female carrying eggs; berried.

oviparous Egg-laying; producing eggs that are laid and hatch externally; **oviparity**; *cf.* larviparous, ovoviviparous, viviparous.

oviposition The act or process of depositing eggs.

ovoviviparous Producing fully formed eggs that are retained and hatched inside the maternal body with the release of live offspring; **ovoviviparity**; *cf.* larviparous, oviparous, viviparous.

ovum An unfertilized egg cell; a female gamete; **ova.**

Oweniida Order of tubicolous, suspension- or deposit-feeding, polychaete worms comprising a single family of about 30 species; body elongate, prostomium and peristomium rudimentary, fused; pharynx unarmed, non-eversible.

owl 1: Tytonidae *q.v.* 2: Strigidae *q.v.*

owl-fly Ascalaphidae *q.v.* (Neuroptera).

Oxalidaceae Carambola, oxalis, wood sorrel; widespread family of Geraniales containing about 900 species of herbs or shrubs, often with tubers or bulbs, commonly accumulating oxalates; flowers typically regular, with 5

oxalis (Oxalidaceae)

sepals and petals, 10 stamens and a superior 5-celled ovary.

oxodic Pertaining to a peat bog community.

Oxyaenidae Extinct family of mustelid-like carnivores (Creodonta) which flourished during the Eocene.

oxygenated With adequate available oxygen; aerobic; *cf.* deoxygenated.

oxygenotaxis A directed response of a motile organism towards (positive) or away from (negative) an oxygen stimulus; **oxygenotactic.**

oxygenotropism An orientation movement induced by an oxygen gradient stimulus; **oxygenotropic.**

oxygeophilous Thriving in humus-rich habitats; **oxygeophile, oxygeophily.**

oxygeophyte A plant growing in humus.

oxygeophytic Pertaining to plant communities in humus-rich habitats.

oxylophilous Thriving in humus or humus-rich habitats; **oxylophile, oxylophily.**

oxylophyte A plant growing in a humus-rich habitat; **oxylophytic.**

Oxymonadida Order of zooflagellates characterized by the arrangement of flagella in 2 pairs in the motile stages; all species are commensal or symbiotic usually in insect digestive systems, rarely in vertebrate digestive systems.

oxymonad (Oxymonadida)

oxyphilous Thriving in acidic habitats; **oxyphile, oxyphily**; *cf.* oxyphobous.

oxyphobous Thriving in alkaline habitats; intolerant of acidic conditions; **oxyphobe, oxyphoby**; *cf.* oxyphilous.

oxyphyte A plant growing under acidic conditions.

oxytaxis A directed response of a motile organism to an oxygen stimulus; **oxytactic.**

oxytropism 1: An orientation response to an oxygen gradient stimulus; **oxytropic.** 2: An orientation response to an acid stimulus.

Oyashio Current A cold surface ocean current that flows southwards in the northwest Pacific; Oya Shio; Kamchatka Current; see ocean currents.

oyster Ostreoida *q.v.*

oystercatcher Haematopodidae *q.v.*

P

paca Dasyproctidae *q.v.*

Pachycephalidae Whistlers; family containing about 45 species of arboreal passerine birds found in a variety of forest and open arid habitats of the Indo-Pacific region; bill typically compressed and hooked, wings short and rounded; habits solitary to gregarious, non-migratory; feed largely on insects, commonly caught on the wing; sometimes classified in the Muscicapidae.

paddlefish Polyodontidae *q.v.*

paddleworm Phyllodocida *q.v.*

paedogenesis Precocious sexual maturation in an organism that is still at a morphologically juvenile stage.

paedomorphosis The retention of juvenile characters of ancestral forms by adults, or later ontogenetic stages, of their descendants; superlarvation; **paedomorphic**; *cf.* neoteny, progenesis.

paedoparthenogenesis Parthenogenetic reproduction occurring in larvae.

paedophagous Feeding on embryos and the young stages of other species; **paedophage**, **paedophagy.**

Paenungulata Superorder of herbivorous mammals (Ferungulata) comprising the hyraxes (Hyracoidea), elephant-like forms (Proboscidea), Pantodonta, Dinocerata, Embrithopoda, sea cows (Sirenia) and Pyrotheria.

Paeoniaceae Peony; small isolated family in the Dilleniales containing about 30 species of mainly Eurasian herbs or soft shrubs which commonly have large showy flowers with 5 sepals, 5 petals, many stamens and a superior ovary.

peony (Paeoniaceae)

pagoda flower Verbenaceae *q.v.*

pagoda shell Neogastropoda *q.v.*

pagophilous Thriving in foothills; **pagophile**, **pagophily.**

pagophyte A plant inhabiting foothills.

pair bonding The formation of a close and lasting association between a male and female animal, used particularly with reference to the cooperative rearing of young.

Palaeacanthocephala Large, diverse class of thorny-headed worms found as parasites of fishes, amphibians, reptiles, birds and mammals, often utilizing crustaceans as intermediate hosts; characterized by a retractable proboscis with hooks arranged in alternating rows, and a double-walled proboscis receptacle; comprises 2 orders, Echinorhynchida and Polymorphida.

Palaeanodonta Extinct suborder of primitive edentates found in North America from the Palaeocene to the Oligocene; cheek teeth peg-like, without enamel but canines well developed.

Palaearctic region A zoogeographical region *q.v.* comprising Europe, North Africa, western Asia, Siberia, northern China, Japan; Palearctic region; subdivided into European, Mediterranean, Siberian and Manchurian subregions.

palaeoagrostology The study of fossil grasses.

palaeoalgology The study of fossil algae.

palaeobiology Study of the biology of extinct organisms.

palaeobotany Study of the plant life of the geological past.

Palaeocene A geological epoch within the Tertiary period (*c.* 65–54 million years B.P.); see geological time scale.

palaeoclimatology Study of the climate during periods of past geological time; **palaeoclimate.**

Palaeocopa Subclass of marine ostracod crustaceans comprising a single order, Palaeocopida.

Palaeocopida Sole order of marine palaeocopan ostracods, comprising a single family, Punciidae, with only 2 species.

palaeodendrology The study of fossil trees.

Palaeodonta Extinct suborder of small artiodactyls found in North America during the Eocene and Oligocene.

palaeoecology Study of the ecology of fossil communities.

Palaeogaea A zoogeographical area originally comprising the Palaearctic, Ethiopian, Oriental and Australian regions; *cf.* Arctogaea, Neogaea, Megagaea, Notogaea.

Palaeogene A division of the Tertiary period (*c.* 65–26 million years B.P.) comprising the Palaeocene, Eocene and Oligocene epochs; see geological time scale.

Palaeognathae Superorder of mainly large, flightless, terrestrial birds comprising 5 extant orders, Struthioniformes (ostrich), Rheiformes (rheas), Casuariformes (emus, cassowaries), Apterygiformes (kiwis) and Tinamiformes (tinamous); also includes the subfossil Dinornithiformes (moas), Aepyornithidae (elephant birds) and Dromornithidae; distribution confined to southern hemisphere; the flightless orders are commonly referred to as ratites.

palaeoichnology The study of trace fossils; *cf.* neoichnology.

palaeolimnology Study of the geological history and development of inland waters.

Palaeolithic An archaeological period from about 10 000 to 3.5 million years B.P.; a period of human history characterized by the use of stone tools; the early, older or chipped Stone Age.

Palaeoloricata Extinct order of primitive chitons (Polyplacophora) found from the Cambrian to the Cretaceous.

Palaeonemertea Order of anoplan nemerteans with either 2 or 3 layers of musculature in the body wall and with the main components of the nervous system located in the inner longitudinal muscle layer or external to the muscle layers; Palaeonemertini.

ribbon worm (Palaeonemertea)

palaeontology The study of fossils; paleontology; *cf.* neontology.

palaeopalynology The study of fossil spores and pollen.

Palaeophytic The period of geological time between the development of the algae and the first appearance of gymnosperms, during which pteridophytes were abundant; Pteridophytic; *cf.* Aphytic, Archaeophytic, Caenophytic, Eophytic, Mesophytic.

palaeosol A soil formed under past environmental conditions.

Palaeotheriidae Extinct family of horse-like mammals (Perissodactyla) known from the Eocene and Oligocene.

Palaeotropical kingdom One of 6 major phytogeographical areas *q.v.*, characterized by its floristic composition; comprises African, Indo-Malaysian and Polynesian subkingdoms.

Palaeotropical Region Ethiopian region *q.v.*

Palaeozoic A geological era (*c.* 570–245 million years B.P.) comprising the Cambrian, Ordovician, Silurian, Devonian, Carboniferous and Permian periods; see geological time scale.

palaeozoology Study of the animal life of the geological past.

paleo- See palaeo-.

Paleoheterodonta Subclass of freshwater and marine bivalve molluscs characterized by a generally equal-valved shell, a dentate hinge, 2 shell-closing muscles and a broad ventral mantle opening which forms inhalent and exhalent current apertures only rarely drawn out into siphons; comprises the orders Trigonioida and Unionoida.

Paleoptera Infraclass of primitive winged insects comprising 2 extant orders, Ephemeroptera (mayflies) and Odonata (dragonflies, damselflies); eyes large, antennae inconspicuous; the large wings have a rich venation and frequently exhibit surface corrugation or fluting; these insects are unable to fold the wings flat against the body; developmental stages are usually aquatic.

Palinura Spiny lobsters, crawfish; infraorder of pleocyematan decapod crustaceans; third maxilliped leg-like, not chelate; abdomen well developed; includes many species of commercial importance.

palm Arecales *q.v.*

Palmae Palms; former name for the Arecaceae, sole family of the Arecales *q.v.*

Palmariales Small order of red algae including the commercially important *Palmaria palmata*, which is used for human consumption and as cattle feed.

palmigrade Plantigrade *q.v.*

Palpigradi Microwhipscorpions; order of small (to 3 mm) agile terrestrial arthropods (Arachnida) comprising a single family of about 60 species typically found in the soil beneath stones; chelicerae long and slender, chelate; abdomen bearing terminal flagellum.

paludal Pertaining to marshes.

paludicolous Inhabiting marshy habitats; paludicole.

paludification The process of bog expansion

resulting from the rising water table produced as drainage is impeded by the accumulation of peat.

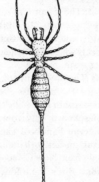

microwhipscorpion (Palpigradi)

palustrine Pertaining to wet or marshy habitats; **palustral**.

palynology The study of pollen and spores.

pan A soil horizon of highly compacted material, or one of high clay content.

PAN Peroxiacetyl nitrate, an atmospheric pollutant produced from automotive and industrial fumes under strong radiation.

Pancarida Superorder of eumalacostracan crustaceans comprising a single order, Thermosbaenacea *q.v.*

Pandaceae Family of Euphorbiales containing 26 species of dioecious trees or shrubs, native to tropical Africa, Asia and New Guinea; flowers usually unisexual, regular and pentamerous; fruit commonly a drupe.

Pandanales Screw pine; order of Arecidae comprising a single family, Pandanaceae, of about 500 species of trees, shrubs and woody climbers confined to the Old World and best developed in the tropics; stem often appears dichotomously branched and may have prop roots; leaves usually with an elongate and spiny blade; flowers small and unisexual.

Pandionidae Osprey; monotypic family of fish-eating raptors (Falconiformes) cosmopolitan in freshwater and coastal marine habitats; capture fish with powerful feet, diving from 20–50 m; habits solitary, monogamous, nesting in trees, rocks or on the ground.

Pangaea The single supercontinent formed about 240 million years B.P. comprising the present continental landmasses joined together, which began to break up about 150 million years B.P. (some authorities date the break up closer to 200 million years B.P.); *cf.* Gondwana, Laurasia, Tethys.

Pangasiidae Family containing 25 species of large (to about 3 m) southeast Asian freshwater catfishes (Siluriformes); body elongate, compressed; dorsal fin short with 1–2 spines, adipose fin small; 2 pairs of barbels present.

pangolin Manidae *q.v.*

panicle An inflorescence comprising several racemose parts; a compound raceme commonly found in grasses.

panicle

panmixis Unrestricted random mating; the free interchange of genes within an interbreeding population; **panmictic**.

panspermia The theory that life did not originate on Earth but was introduced from elsewhere in the Universe.

pansy Violaceae *q.v.*

Panthalassa The single large ocean surrounding Pangaea *q.v.*

Pantodonta Extinct order of ungulates (Ferungulata) known from the Palaeocene through to the Oligocene.

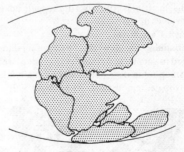

Pangaea

Pantodontidae Butterfly fish; monotypic family of small (to 100 mm) primitive teleost fishes inhabiting tropical west African fresh waters; dorsal and anal fins located posteriorly, pectorals large, pelvics with long fin rays; typically surface-dwelling, insectivorous, capable of jumping short distances.

Pantopoda Pycnogonida *q.v.*

Pantotheria Extinct infraclass of primitive, egg-laying mammals (Theria) including the ancestors of all marsupial and placental mammals; mostly insectivorous with shrew-like appearance; known from the Jurassic.

pantropical Extending or occurring throughout the tropics and subtropics.

Papaveraceae Poppies; family of herbs or soft-wooded shrubs with milky or coloured latex; flowers large typically with 2 sepals, 4 petals, many stamens and a superior ovary producing a nut or capsule containing many oily seeds; includes the opium poppy, *Papaver somniferum.*

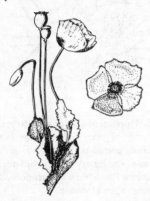

poppy (Papaveraceae)

Papaverales Order of dicotyledons (Magnoliidae) comprising 2 families of herbs or shrubs which produce isoquinoline alkaloids in specialized glands (laticifers) or cells; the fruit is typically a dry capsule opening by valves or pores.

papaya Caricaceae *q.v.*

Papilionidae Swallowtails; family of medium-sized to very large butterflies (Lepidoptera) of cosmopolitan distribution but most abundant in tropical regions; comprises about 500

swallowtail butterfly (Papillionidae)

species and includes the largest known butterflies, the birdwings.

papyrus Cyperaceae *q.v.*

parabiosis 1: The temporary suspension of physiological activity. 2: Utilization of the same nest by colonies of different species of social insects which nevertheless keep their broods separate; **parabiotic.**

parachrosis The process of colour change.

Paracrinoidea Extinct class of echinoderms (Pelmatozoa) with a box-like test and uniserial branchioles; known only from the Ordovician.

Paracryphiaceae Family of Theales containing a single tree species with unicellular hairs, known only from New Caledonia.

Paradisaeidae Birds of paradise; family containing 40 species of colourful and ornate passerine birds found in forests of New Guinea; often with elaborate plumes, crests, and other modified long feathers; habits solitary, arboreal, monogamous to polygamous; males with elaborate courtship behaviour; feed on insects, small vertebrates and fruits; cup-shaped nest placed in tree.

bird of paradise (Paradisaeidae)

paraheliotropism Movement of leaves to avoid or minimize exposure to sunlight; **paraheliotropic.**

Paralepididae Barracudinas; small family of myctophiform teleost fishes distributed worldwide in the deep sea; body slender, pointed anteriorly, with single mid-dorsal fin; contains about 50 species.

paralimnion The littoral zone of a lake from the water margin to the deepest level of rooted vegetation; **paralimnetic.**

parallel evolution The independent acquisition in two or more related descendant species of similar derived character states evolved from a common ancestral condition; parallelism; *cf.* convergent evolution.

parallelogeotropism An orientation response that brings the main axis of the body to the vertical; **parallelogeotropic.**

parallelotropism Orientation or growth in a straight line; **parallelotropic.**

parameter Any measurable factor.

Paramyxea Class of Ascetospora containing a single species parasitic in the intestinal cells of larval polychaetes; spore production is similar to that in the myxosporidia although no polar capsules are developed.

parapatric Used of populations whose geographical ranges are contiguous but not overlapping, so that gene flow between them is possible; **parapatry;** *cf.* allopatric, dichopatric, sympatric.

Parapercidae Mugiloididae *q.v.*

Parapithecidae Extinct family of anthropoid primates (Cercopithecoidea) known from the Eocene and Oligocene in Asia and Africa; short-faced, with jaws like those of tarsiers.

Parascyllidae Collared carpet sharks; family of small (to 1 m) bottom-dwelling orectolobiform elasmobranch fishes found in the western Pacific Ocean; body cylindrical, mouth subterminal, nostrils with short barbels; caudal fin lacking ventral lobe.

parasematic Pertaining to markings, structures or behaviour intended to deter, distract or mislead a predator.

parasexual Pertaining to organisms that achieve genetic recombination by means other than the regular alternation of meiosis and karyogamy *q.v.*; *cf.* eusexual.

parasite Any organism that is intimately associated with, and metabolically dependent upon, another living organism (the host) for completion of its life cycle, and which is typically detrimental to the host to a greater or lesser extent.

parasitic castration Reproductive death of a host organism resulting from a parasitic infection.

parasitic male A dwarf male that lives a parasitic existence attached to the body of its female, typically having well developed reproductive organs but an otherwise degenerate body form.

parasitic slime moulds A group of organisms with a plasmodial (multinucleate and lacking cell walls) trophic stage capable of amoeboid movement between host cells; comprising about 35 species all of which are intracellular parasites of algae, fungi or vascular plants; life cycle includes biflagellate zoospores; several species cause diseases of economically important plants, such as clubroot of cabbage and powdery scab of potatoes; classified as a phylum of Protoctista (Plasmodiophoromycota), as a class of fungal slime moulds (Plasmodiophoromycetes), or as a class of rhizopod protozoans (Plasmodiophorea).

Parasitica Parasitic wasps, gall wasps, fig wasps, chalcid wasps; diverse division of hymenopteran insects (suborder Apocrita) comprising over 100 000 species that are parasites or parasitoids of other arthropods, mainly insects and spiders, or plant tissues; female typically inserts egg into host with ovipositor; none of the parasitic wasps is eusocial; Terebrantia.

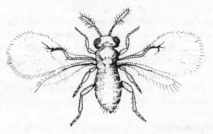

chalcidoid wasp (Parasitica)

Parasitiformes Suborder of medium-sized to large mites (Acari) comprising 3 subgroups, Mesostigmata, Holothyrina and Ixodida (arachnid ticks); ecologically very diverse, found in forest litter, moss and wood, feeding as predators, saprophages, fungivores or as parasites.

parasitism An obligatory relationship between individuals of two different species, in which the parasite is metabolically dependent on the host, and in which the host is typically adversely affected but rarely killed; **parasitic.**

parasitoid An organism with a mode of life intermediate between parasitism and predation; usually a species of hymenopteran in which the larva feeds within the living body of another organism eventually causing the death of the host.

parasitology The study of parasites and parasitism.

paratenic host A host that is not essential for the completion of a parasite's life cycle but is used as a temporary habitat or as a means of reaching the definitive host *q.v.*; *cf.* intermediate host.

parathermotropism Movement of leaves to avoid or minimize the heating effects resulting from exposure to sunlight; **parathermotropic.**

paratrophic Feeding in the manner of a parasite; parasitic; **paratrophism, paratrophy.**

Parazenidae Monotypic family of small (to 140 mm) marine zeiform teleost fishes reported from Japan and Cuba; body moderately compressed, mouth and eyes large; scales ctenoid.

Pareiasauridae Extinct group of primitive herbivorous reptiles (Cotylosauria) with limbs rotated in towards body; known from the Permian.

parenchymella larva Flagellated larval stage of demosponges.

parent material The inorganic deposit which has been modified by weathering, translocation or other pedogenic process, to give rise to different soil types.

Paridae Chickadees, titmice; family containing about 60 species of small acrobatic passerine birds found in forest and open woodland habitats of the northern hemisphere, Africa and Indonesia; bill typically short and stout; habits gregarious, arboreal, feeding mostly on insects, seeds and fruit; nest in tree hole or crevice.

Paromomyidae Extinct family of early primates that flourished during the Palaeocene; probably herbivorous, ground-dwelling, and about the size of a shrew.

parrot Psittacidae *q.v.*

parrotfish 1: Oplegnathidae *q.v.* 2: Scaridae *q.v.*

parrot's feather Haloragaceae *q.v.*

parsimonious The condition of being economic or frugal in use or application; **parsimony.**

parsley Apiaceae *q.v.*

parsnip Apiaceae *q.v.*

parthenocarpy The development of fruit without seeds, typically resulting from lack of fertilization; **parthenocarpic.**

parthenogamete A gamete capable of parthenogenetic development.

parthenogenesis The development of an individual from a female gamete without fertilization by a male gamete; *cf.* acyclic parthenogenesis, ameiotic parthenogenesis, androcyclic parthenogenesis, anholocyclic parthenogenesis, arrhenotoky, automictic parthenogenesis, cyclic parthenogenesis, haploid parthenogenesis, holocyclic parthenogenesis, meiotic parthenogenesis,

thelytoky.

parthenospore A spore produced from an unfertilized female gamete.

parturition The act of giving birth.

Parulidae Wood warblers; family containing about 125 species of small New World passerine birds found in variety of forest, woodland and open habitats; plumage often colourful; bill short, slender to robust; habits arboreal or terrestrial, some species migratory; feed on insects, fruit, nectar; nest on or off the ground.

pascual Pertaining to, or inhabiting, pasture.

pasculomorphosis Change in the structure of plants as a result of grazing by animals.

Passeridae Sparrows; family containing 27 species of small passerine birds found in variety of woodland and open habitats of the Old World, but also introduced and now widespread in many parts of New World and Australia; bill short, conical; habits gregarious, terrestrial, non-migratory; feed on seeds and insects; nest usually a loose mass of twigs and grass in a crevice or burrow.

sparrow (Passeridae)

Passeriformes Perching birds; the largest order of neognathous birds comprising about 70 families and over 5000 species; body form and habits extremely diverse; group includes great variety of flycatchers, swallows, larks, babblers, warblers, thrushes, wrens, shrikes, creepers, finches, sparrows, buntings, starlings and crows.

Passifloraceae Passion flower; large family of

passion flower (Passifloraceae)

Violales containing about 650 species of vines climbing by axillary tendrils, best developed in tropical America and Africa; flowers often large and typically with 5 sepals and petals, many thread-like appendages forming a central corona, 5 stamens, 3 styles and a superior ovary.

passion flower Passifloraceae *q.v.*

passive chamaephyte A type of chamaephyte *q.v.* in which the erect vegetation shoots persist through unfavourable seasons in a procumbent position.

passive uptake The absorption of ions by diffusion or other physical process not involving the expenditure of metabolic energy.

Pastonian interglacial An interglacial period of the Quaternary Ice Age *q.v.* in the British Isles.

Pataecidae Prowfishes; family containing 5 species of small (to 200 mm) Australian marine scorpaeniform teleost fishes; body strongly compressed, naked, sometimes papillose; dorsal fin very long, may be contiguous with caudal.

Paterinida Extinct order of inarticulate lamp shells (Brachiopoda) with phosphatic shells; known from the Cambrian to the Ordovician.

paternal Pertaining to, or derived from, the male parent; *cf.* maternal.

paternal sex determination The condition in which the sex of the offspring is determined by the idiotype of the male parent.

pathogenic Producing or capable of producing disease; **pathogen, pathogenicity.**

pathology The study of disease.

patobiontic Used of an organism that spends its entire life cycle in the forest litter; **patobiont.**

patocolous Used of an organism that inhabits the forest litter for only part of its normal life cycle; **patocole.**

patoxenous Used of an organism that occurs in the forest litter only by accident; **patoxene.**

patromorphic Resembling the male parent; *cf.* matromorphic.

Paucituberculata Order of small shrew-like South American marsupials comprising a single extant family, Caenolestidae; rat opossums.

paurometabolous Used of a pattern of development displaying a gradual metamorphosis in which changes are inconspicuous.

Pauropoda Pauropods; class of minute (0.5–1.5 mm) blind arthropods comprising about 500 species in 2 suborders, Hexamerocerata and Tetramerocerata; body with 9–11 pedigerous segments, tergites bearing 5 pairs of long setae; antennae branched, mouthparts reduced, blood vascular system absent; typically found in soil or litter habitats, feeding on fungi or decaying plant material.

pauropod (Pauropoda)

Pavlovales Small order of mostly marine flagellate algae (Prymnesiophyceae) lacking scales.

pawpaw Annonaceae *q.v.*

Paxillosida Order of primitive, intertidal to abyssal, sediment-dwelling asteroidean echinoderms characterized by a series of columnar ossicles on the aboral surface (paxillae) that form a barrier to exclude sediment from the upper surface; comprises about 400 extant species.

PBBs Polybrominated biphenyls; class of compounds related to PCBs.

PCBs Polychlorinated biphenyls; chlorinated hydrocarbons used in industry which occur as persistent pollutants, particularly in aquatic ecosystems.

pea Fabaceae *q.v.*

pea weevil Bruchidae *q.v.*

pea wilt Hypocreales *q.v.*

peach Rosaceae *q.v.*

peacock plant Marantaceae *q.v.*

peafowl Phasianidae *q.v.*

peanut Fabaceae *q.v.*

peanut worm Sipuncula *q.v.*

pear Rosaceae *q.v.*

pearl oyster Pterioida *q.v.*

pearl perch Glaucosomidae *q.v.*

pearleye Scopelarchidae *q.v.*

pearlfish Carapidae *q.v.*

pearly nautilus Nautiloidea *q.v.*

peat An accumulation of unconsolidated partially decomposed plant material found in more or less waterlogged habitats of fen (alkaline peat), bog (acid peat) and swamp.

peat moss Sphagnopsida *q.v.*

pebble A sediment particle between 4 and 64 mm in diameter; medium gravel; see sediment particle size.

peccary Tayassuidae *q.v.*

peck order A dominance hierarchy, used especially of birds.

Pecora Infraorder of advanced ruminants (Artiodactyla) containing the chevrotains

(Tragulidae), deer and giraffe (Cervoidea) and sheep, cattle and antelopes (Bovoidea); stomach 4-chambered; 2 functional digits in each limb; mostly with horns.

Pectobothrii Alternative name for the Monogenea *q.v.*

ped A unit of soil structure formed by natural processes, such as an aggregate, block or crumb.

pedalfer An acid soil type found in humid regions rich in aluminium and iron compounds, but low in carbonates; *cf.* pedocal.

Pedaliaceae Sesame; family of Scrophulariales containing about 80 species of terrestrial or sometimes aquatic herbs and shrubs, mostly tropical in distribution especially in arid areas or along sea coasts.

Pedetidae Springhaa; monotypic family of burrowing nocturnal rodents (Sciuromorpha); rabbit-like body, with long ears and a long tail; hindlimbs long for hopping locomotion, forelimbs short and clawed for digging; found in central and southern Africa.

Pediculidae Family of sucking lice (Anoplura) comprising 2 species parasitic on New World monkeys, gibbons, great apes and man; includes the human head and body lice.

Pedinomonadales Small order of uniflagellate Prasinophyceae *q.v.* lacking organic scales over cell and flagellar surfaces.

pediophilous Thriving in uplands; **pediophile**, **pediophily.**

pediophyte A plant of an upland community.

Pedipalpi A group of arachnids comprising the tailless whip scorpions (Amblypygi) and whip scorpions (Uropygi); sometimes treated as an order.

pedocal An alkaline soil type found in arid regions characterized by a high calcium carbonate content; *cf.* pedalfer.

pedology Study of the structure and formation of soils; soil science.

pedon 1: Those organisms that live on or in the substratum of an aquatic habitat. 2: The smallest vertical column of soil containing all the soil horizons at a given location.

pedosphere That component of the biosphere comprising the soil and soil organisms.

Pegasidae Sea moths; family of tropical Indo-Pacific shallow marine gasterosteiform teleost fishes; body broad, depressed, armoured with bony rings; length to 140 mm; snout elongate; pectoral fins wing-like.

pelagic Pertaining to the water column of the sea or lake; used of organisms inhabiting open

waters; **pelagial** *cf.* benthic.

pelagic sediment A deep-ocean sediment containing no significant terrigenous component.

human louse (Pediculidae)

pelagophilous Thriving in the open surface waters of the sea; **pelagophile**, **pelagophily.**

pelagophyte A plant living at the sea surface.

pelargonium Geraniaceae *q.v.*

Pelecanidae Pelicans; small family containing 8 species of large water birds, cosmopolitan in tropical and temperate coastal habitats and large lakes; feed on fish caught in large pouch-like bill, by plunging from flight; nest in colonies in trees or on ground.

Pelecaniformes Order of medium to large neognathous water birds, found in marine and freshwater coastal habitats worldwide; comprises 6 families including pelicans, gannets, cormorants and frigate birds.

pelican (Pelecanidae)

Pelecanoididae Diving petrels; small family containing 4 species of oceanic sea birds (Procellariiformes) found throughout the cold southern oceans; resembling the auks of northern seas; highly efficient divers, using wings underwater and feeding mainly on fish and crustaceans; nest in burrows or crevices.

Pelecypoda Bivalvia *q.v.*

pelican Pelecanidae *q.v.*

Pellicieraceae Family of Theales comprising a single species of glabrous mangrove trees native to the shores of the Pacific Ocean from Costa Rica to Colombia.

Pelmatozoa Subphylum of echinoderms typically exhibiting radial symmetry; attached by stalk (pelma) and with a globular body protected by a plated test; ambulacra extending onto arms; includes the Homalozoa and Crinozoa.

Pelobatidae Spadefoot toads; family of toad-like terrestrial or fossorial anurans containing about 50 species widely distributed from North America, Europe, North Africa to the Oriental region; eggs and larvae aquatic in ponds and occasionally in streams; adults lacking ribs; possess digging tubercles on hindfeet.

spadefoot toad (Pelobatidae)

Pelobiontida Order of elongate, multinucleate naked amoebae (Gymnamoeba) which feed by ingesting plant material at the posterior end; lack a contractile vacuole, mitochondria and a flagellate phase; found in black bottom muds of ponds and ditches.

pelochthophilous Thriving on mud-banks; **pelochthophile, pelochthophily.**

pelochthophyte A mud-bank plant.

Pelodryadidae Family of Australo-New Guinean tree frogs (Anura); most species arboreal, a few terrestrial; body length to about 140 mm, tips of digits bearing expanded disks for adhesion; eggs and larvae aquatic.

Pelodytidae Family containing 2 species of European toads (Anura) with long, leaping hindlimbs; related to the spadefoot toads (Pelobatidae).

pelogenous Mud-producing or clay-producing.

Pelomedusidae Small family containing 14 species of side-necked turtles (Pleurodira) inhabiting lakes and rivers of Africa,

Madagascar and South America; head fully retractile; omnivorous, feeding mostly in shallow water.

pelophilous Thriving in habitats rich in clay; **pelophile, pelophily.**

pelophyte A plant living in clay or muddy soil.

pelopsammic Pertaining to or comprising both clay and sand.

Pelycosauria Extinct order of synapsid reptiles containing both carnivorous and herbivorous species; many carried large, sail-like dorsal fins; known from the Upper Carboniferous and Lower Permian.

pelycosaur (Pelycosauria)

Penaeaceae Small family of Myrtales containing about 20 species of tanniferous evergreen shrubs native to Cape Province of South Africa.

Penaeidea Shrimps, prawns; infraorder of dendrobranchiate decapod crustaceans typically with a long abdomen, the first segment of which is not overlapped by the pleura of the second; includes many commercially important species.

pencil fish Lebiasinidae *q.v.*

pencil urchin Cidaroida *q.v.*

penguin Spheniscidae *q.v.*

Penicillata Subclass of diplopods (millipedes) comprising a single order, Polyxenida.

penicillium Eurotiales *q.v.*

Peninoida Order of rare echinoids (Diadematacea) containing only about 15 species, found on the continental slope and outer shelf.

Pennales Pennate diatoms; order of primarily freshwater algae found in the plankton and in damp terrestrial situations; characterized by a basically bilateral symmetry and by a slit between the valves, called a raphe, through which slime is exuded; may exhibit a gliding locomotion; sexual reproduction involves nonflagellate male gametes; also treated as a class of the phylum Bacillariophyta.

Pennatulacea Sea pens, sea pansies; order of complex octocorals adapted for living on soft substrata in both warm and cold seas; the colony comprises an elongate axial polyp (oozoid) giving rise to dimorphic secondary polyps, some of which are specialized as siphonozooids pumping water into the colony to maintain its turgor pressure.

sea pen (Pennatulacea)

pentacrinoid A stalked, larval stage of a sea lily (Crinoidea), attached at the base and with a distal crown showing the rudiments of developing arms and the 5-radial symmetry.

Pentamerida Extinct order of thick-shelled, articulate brachiopods known from the Cambrian to the Devonian.

Pentaphragmataceae Family of Campanulales containing about 30 species of perennial herbs lacking alkaloids and a latex system; native to southeast Asia.

Pentaphylacaceae Family of Theales comprising a single species of evergreen shrubs or small trees native to southeast Asia; flowers hermaphrodite, pentamerous, with porous anthers; fruit capsular.

Pentastomata Sole class in the arthropod subphylum Pentastomida *q.v.*

pentastome Pentastomida *q.v.*

Pentastomida Subphylum of vermiform blood-feeding arthropods, endoparasitic in the lungs and nasal passages of reptiles, birds and mammals; body divided into head and annulate abdomen, appendages confined to 4 chitinous hooks on ventral surface of head; mouthparts, circulatory and respiratory systems absent; sexes separate; larval stage with 2 pairs of non-jointed appendages and one or more piercing stylets; contains about 55

tongue worm (Pentastomida)

species in 2 orders, Cephalobaenida and Porocephalida; phylogenetic affinities uncertain, some workers assign the group to the Crustacea on the basis of sperm morphology.

Pentatomidae Shield bugs; family of phytophagous and predatory hemipteran insects comprising about 5000 species, including several economically important agricultural pests; body typically shield-shaped.

Pentazonia Subclass of diplopods (millipedes) comprising 2 superorders, Limacomorpha and Oniscomorpha.

Pentoxylales Extinct group of shrubby or tree-like gymnosperms known from the Jurassic.

peony Paeoniaceae *q.v.*

pepo A type of berry *q.v.* with a hard outer layer derived from the outer fruit wall (epicarp) or from non-carpellary tissue.

pepper Piperaceae *q.v.*

pepper root Brassicaceae *q.v.*

Peracarida Superorder of eumalacostracan crustaceans in which the carapace is usually reduced or absent and the eggs are retained in a ventral brood pouch; diverse in morphology, life history and ecology; contains about 12 000 species in 7 extant orders, Amphipoda, Cumacea, Isopoda, Mictacea, Mysidacea, Spelaeogriphacea and Tanaidacea.

Peramelidae Family of short-haired terrestrial bandicoots from Australia, Tasmania and New Guinea; marsupium opening posteriorly, forelimbs shorter than hindlimbs; habits nocturnal, insectivorous.

bandicoot (Peramelidae)

Peramelina Bandicoots; order of insectivorous terrestrial marsupials found in Australia and New Guinea, comprising 2 families, Peramelidae and Thylacomyidae; marsupium opens posteriorly; chorio-allantoic placenta present but without villi.

perch Percidae *q.v.*

perching bird Passeriformes *q.v.*

Percichthyidae Family containing about 40 species of marine, brackish and freshwater perch-like teleost fishes (Perciformes) widespread in tropical and temperate regions; sometimes included in the family Serranidae; length to about 2 m; important as sport and food-fishes.

Percidae Perches, darters; diverse family of primarily northern hemisphere freshwater teleost fishes (Perciformes); body usually elongate, with 2 dorsal fins; length to 900 mm but often much less; feeding planktotrophic, insectivorous or piscivorous; reproduction oviparous; contains 160 species, some popular with anglers, others fished commercially.

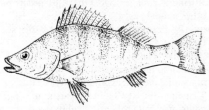

perch (Percidae)

Perciformes Large and diverse order of marine, brackish and freshwater teleost fishes comprising about 7000 species in 150 families; strong fin spines usually present, body scales often ctenoid or absent; adipose fin absent, pelvic fins thoracic or jugular; swim bladder lacking pneumatic duct.

percolation Downward movement of water through porous sediments.

Percophididae Family containing 20 species of deep-water, bottom-dwelling perciform teleost fishes most frequent in the Indian and Pacific Oceans but also present in the Atlantic; body elongate, to 250 mm length, head depressed with large eyes, lower jaw prolonged; 2 dorsal fins present.

Percopsidae Trout-perches; family of small (to 200 mm) North American freshwater teleost fishes (Percopsiformes) containing only 2 species; body slender, compressed, silvery; dorsal and anal fins with 1–2 spines, adipose fin present, caudal fin deeply forked.

Percopsiformes Order of North American freshwater teleost fishes comprising about 8 species and including trout-perches, pirate perch and cave fishes.

peregrine Foreign, non-native; used of organisms transported into an area from outside.

peregrine falcon Falconidae *q.v.*

perennation The survival of plants from year to year with an intervening period of reduced activity.

perennial Used of plants that persist for several years with a period of growth each year; *cf.* annual, biennial.

perfect 1: Pertaining to a hermaphrodite or bisexual flower. 2: Pertaining to fungi or fungal life cycle stages that produce sexual spores.

pergelicolous Living in geloid soils *q.v.* having a crystalloid content below 0.2 parts per thousand; **pergelicole**; *cf.* gelicolous, halicolous, perhalicolous.

perhalicolous Living in haloid soils *q.v.* having a crystalloid content above 2.0 parts per thousand; **perhalicole**; *cf.* gelicolous, halicolous, pergelicolous.

perianth The structure surrounding the inner, reproductive parts of a flower; typically comprising an outer whorl (the calyx) of sepals and an inner whorl (the corolla) of petals but sometimes comprising undifferentiated elements called tepals.

pericontinental sea A shallow sea covering a modern continental shelf, typically less than 200 m in depth; *cf.* epeiric sea.

Peridiniales Large order of predominantly marine thecate dinoflagellates which are typically photosynthetic; some neritic species become so abundant as to discolour the water, producing red tides.

Peridiscaceae Small family of Violales comprising 2 species of trees from tropical South America.

perigean tides The tides of increasing amplitude occurring at the time the moon is nearest the Earth.

perigyny The arrangement of the parts of a flower so that the sepals, petals and stamens are inserted at about the same level as the ovary; *cf.* epigyny, hypogyny.

periodicity The periodic or rhythmic occurrence of an event; also the duration of a single phase of such an oscillation.

periphyton A community of plants, animals and associated detritus adhering to and forming a surface coating on stones, plants and other submerged objects; **periphyte, periphytic.**

Periptychidae Extinct family of primitive ungulates known from the Palaeocene.

Perischoechinoidea Subclass of primitive regular echinoids (Echinoidea) containing a single extant order, Cidaroida *q.v.*, but with many fossil forms known from the Ordovician onwards.

Perissodactyla Odd-toed ungulates; order of large terrestrial herbivorous mammals (Ferungulata) in which the central digit of each foot is dominant and carries most of the weight (mesaxonic); lateral digits present or absent; horns of bone absent; stomach simple; includes 3 families, Equidae (horses), Tapiridae (tapirs) and Rhinocerotidae (rhinoceroses).

Peritricha Large subclass of sessile oligohymenophoran ciliates typified by an inverted bell-shaped body on a long contractile stalk with a prominent ring of cilia around the distal (oral) end; generally found on an aquatic or semiterrestrial host.

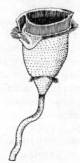

peritrich (Peritricha)

periwinkle 1: Apocynaceae *q.v.* 2: Mesogastropoda *q.v.*

Perkinsea Class of apicomplexan protozoans containing a single species parasitic in American oysters; propagating by flagellate zoospores.

permafrost Permanently frozen subsoil.

permeability 1: The condition of having pores or spaces that permit the passage of fluid molecules. 2: A measure of the freedom of entry of new members into a community or society.

permeant An organism that frequently moves from one community to another.

Permian A geological period of the Palaeozoic (*c.* 290–245 million years B.P.); see geological time scale.

permineralization A process of fossilization by the infiltration and subsequent precipitation of dissolved minerals into the pore spaces of bones, shells and other skeletal parts.

Peronosporales Downy mildews, white rusts, potato blight; order of mainly terrestrial oomycete fungi comprising about 500 species, many of which parasitize flowering plants; characterized by a mycelial thallus with narrow hyphae lacking cross walls.

perpelic Used of plants that live in areas of pure or abundant clay.

perpsammic Used of plants that live in areas of pure or abundant sand.

persimmon Ebenaceae *q.v.*

perturbation A disturbance; any departure of a biological system from a steady state.

Peru Current The northerly extension of the cold Humboldt Current off the Peruvian coast, divided into an inner coastal current and an outer oceanic current by a tongue of warm water from the South Equatorial Countercurrent; see ocean currents.

pessimal Least favourable; used of the levels of environmental factors that are close to the tolerance limits of an organism; *cf.* optimal.

pesticide A chemical agent that kills insects and other animal pests.

peta- (P) Prefix used to denote unit x 10^{15}; see metric prefixes.

Petauridae Family containing about 20 species of small to medium-sized (to 1 m) arboreal marsupial phalangers found in Australia, Tasmania and New Guinea; marsupium opens anteriorly; tail prehensile; habits nocturnal or crepuscular, mainly herbivorous.

petrel 1: Procellariidae *q.v.* 2: Pelecanoididae *q.v.* (diving petrel). 3: Hydrobatidae *q.v.*

petricolous Growing within rock; **petricole.**

petrification A process of fossilization in which tissues are preserved by impregnation with carbonate or silicate minerals; **petrified.**

petrimadicolous Used of an organism living in the surface film of water on rocks; **petrimadicole.**

petrobiontic Living on or amongst rocks or stones; **petrobiont.**

petrochthophilous Thriving on rock-banks; **petrochthophile, petrochthophily.**

petrochthophyte A rock-bank plant.

petrocolous Living in rocky habitats; **petrocole.**

petrodophilous Thriving in boulder fields; **petrodophile, petrodophily.**

petrodophyte A plant inhabiting rocky or stony habitats.

Petromyzoniformes Lampreys; order of jawless vertebrates (Cephalaspidomorphi) with eel-like, naked body; paired fins absent; possessing 7 pairs of external gill openings; adults typically feed on blood of other fishes, but some non-parasitic species do not feed before reproducing; ammocoete larval stage found in fresh water, adults freshwater or anadromus; contains about 30 species in single family, Petromyzonidae.

petrophilous Thriving on rocks or in rocky habitats; **petrophile, petrophily.**

petrophyte A rock plant.

Petrosaviaceae Family of Triuridales comprising only 2 species of small mycotrophic herbs lacking chlorophyll; leaves reduced to alternate scales; single ring of vascular bundles present in stem; native to southeast Asia.

Petrosiida Order of ceractinomorph sponges widely distributed at depths down to 185 m in tropical and warm temperate seas; characterized by a reticulate skeleton of spicule tracts in which the spicules dominate the spongin.

petunia Solanaceae *q.v.*

Pezizales Large order of discomycete fungi distributed worldwide predominantly on soil, dung and plant debris; characterized by cup-shaped, saddle or sponge-shaped apothecia containing asci and interspersed sterile filaments (paraphyses) within a fertile layer (hymenium); the asci open by an operculum or rarely a longitudinal split and eject ascospores forcibly; includes morel.

morel (Pezizales)

pF Back-pull; a measure of the suction force of a soil.

pH The negative logarithm of the hydrogen ion (H^+) concentration, giving a measure of acidity on a scale from 0 (acid) through 7 (neutral) to 14 (alkaline).

Phacidiales Widely distributed order of discomycete fungi typically found as parasites of plants; characterized by the development of apothecia within surrounding sterile tissue (stroma) which ruptures at maturity to expose the fertile hymenium layer; spores violently ejected and disseminated by wind.

Phaeocalpida Order of Phaeodaria in which the shell is small and porcelain-like with one large opening and numerous small pores.

Phaeocystida Order of Phaeodaria in which the skeleton consists of either free spines or spines radiating from a common junction point.

Phaeodaria Class of Actinopoda in which the skeleton, when present, consists of hollow spines and is composed of silica and organic

material; a thick capsular membrane is present in the vegetative stages which has only 3 apertures; typically found in deeper oceanic waters where they feed on small protozoans caught and digested externally.

cormorant (Phalacrocoracidae)

Phaeodendrida Order of Phaeodaria in which the skeleton consists of two valves bearing long branching spines.

Phaeogromida Order of Phaeodaria in which the skeleton is a small diatomaceous or alveolar shell with one large opening.

Phaeogymnocellida Order of naked Phaeodaria lacking a skeleton or surrounded by skeletons from other protozoans.

Phaeophyceae Brown algae; large class containing over 1500 species of predominantly marine seaweeds inhabiting intertidal and sublittoral zones; ranging in form from filamentous to large parenchymatous blades; characterized by their photosynthetic pigments which include chlorophylls *a* and *c*, β-carotene, fucoxanthin and violaxanthin; reproduction mostly sexual involving biflagellate sperm; also treated as a phylum of the Protoctista under the name Phaeophyta.

Phaeophyta The brown algae treated as a phylum of the Protoctista; Phaeophyceae *q.v.*

Phaeosphaerida Order of Phaeodaria in which the skeleton consists of a large latticed shell with wide polygonal meshes.

Phaethontidae Tropic birds; family containing 3 species of oceanic sea birds; feed on fish caught by diving; nest on cliffs.

phage Bacteriophage; a virus of bacteria.

phagocytosis The ingestion of solid particulate matter by a cell.

phagotrophic Feeding by ingesting organic particulate matter; used of cells in the blood or body fluid that ingest foreign particles; **phagotroph, phagotrophy.**

Phalacrocoracidae Cormorants; family containing 33 species of medium-sized water birds (Pelecaniformes), cosmpolitan in coastal marine and freshwater habitats; feed on fishes caught by diving from water surface; breed in colonies, nesting in tree or on ground.

phalanger Phalangeridae *q.v.*

Phalangeridae Phalangers, cuscuses, possums; family containing 9 species of arboreal, diprotodont marsupials found in Australia and New Guinea; marsupium opens anteriorly; tail prehensile; habits nocturnal or crepuscular, mainly herbivorous.

phalarope Phalaropodidae *q.v.*

Phalaropodidae Phalaropes; family containing 3 species of small, aquatic charadriiform shore birds, found in high latitudes of northern hemisphere; feed on crustaceans and insects at water surface; nest in hollows on ground, solitarily or in small colonies.

phalanger (Phalangeridae)

Phallales Stinkhorns; order of gasteromycete fungi in which, at maturity, part of the fruiting body autolyses into a putrid slime containing the spores, which are disseminated by flies.

stinkhorn (Phallales)

Phallostethidae Family of tiny (to 25 mm) fresh- and brackish-water atheriniform teleost fishes found in southeastern Asia; body slender, translucent; pelvic fins reduced or absent in females, modified as asymmetrical copulatory organ in males; reproduction oviparous.

phaneric Pertaining to conspicuous coloration; *cf.* cryptic.

phanerogam A higher plant with reproductive organs conspicuous, in the form of flowers or cones; *cf.* cryptogam.

phanerogamous Reproducing by conspicuous means, used in particular of a plant having conspicuous flowers; **phanerogamy.**

phanerophytes Tall aerial perennial plants, mostly trees or shrubs with their renewal buds at least 250 mm above ground level; subdivided on the basis of bud height above the ground into – nanophanerophyte (up to 2 m), microphanerophyte (2–8 m), mesophanerophyte (8–30 m), megaphanerophyte (greater than 30 m); *cf.* Raunkiaerian life forms.

Phanerozoic That part of geological history in which there is abundant evidence of past life in fossil remains; the aeon comprising the Palaeozoic, Mesozoic and Cenozoic eras (*c.* 570–0 million years B.P.); see geological time scale.

phaoplankton The surface plankton of the upper photic zone; *cf.* knephoplankton, skotoplankton.

Pharetronidia Subclass of calcareous sponges comprising a single order, Inozoida, and characterized by massive reinforcement of the skeleton with additional calcite; usually found in shaded sites such as submarine caves and tunnels.

pharotaxis 1: Navigation by means of landmarks. 2: Movement towards a specific place in response to a learned or conditioned stimulus.

Pharyngobdellida Order of carnivorous hirudinean worms (Annelida) in which a pharynx is present, without jaws; found mostly in freshwater habitats.

Phascolarctidae Koala; monotypic family of arboreal, diprotodont marsupials confined to the forests of eastern Australia; length up to 0.8 m, body covered with soft fur, eyes large, tail absent; marsupium opens posteriorly; habits nocturnal, herbivorous, feeding solely on eucalyptus leaves.

Phasianidae Pheasants, peafowl, jungle fowl, quails; diverse family containing about 185 species of small to large gallinaceous birds

koala (Phascolarctidae)

found worldwide in woodland and grassland habitats; often with colourful plumage in male and cryptic plumage in female; habits gregarious or solitary, monogamous or polygamous, feeding on seeds, fruit and insects; typically nest in hollow in ground; includes the wild ancestor of the chicken.

phasmid Phasmoptera *q.v.*

Phasmoptera Stick insects, leaf insects, walkingsticks, phasmids, spectres; order of phytophagous, nocturnal, arboreal or arbusticolous orthopterodean insects found mainly in wet tropical habitats; contains about 2500 species; camouflage and mimicry highly developed; body commonly elongate and twig-like or flattened and leaf-like, length up to 300 mm; forewings parchment-like or leathery, hindwings forming broad fan, but may be reduced or absent; legs subsimilar; Phasmatodea.

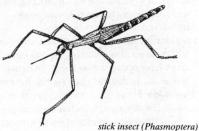

stick insect (Phasmoptera)

pheasant Phasianidae *q.v.*

pheasant shell Archaeogastropoda *q.v.*

phellophilous Thriving on stony or rocky ground; **phellophile, phellophily.**

phellophyte A plant living on gravel or a loose stony substratum.

Phenacodontidae Extinct family of protoungulates, comprising the best known of the condylarths from the Palaeocene to the Eocene; early forms possessed claws, late forms showed more ungulate features including hooves and an elongate third digit.

phenetic Pertaining to overall similarity based on characters selected without regard to evolutionary history.

phenetic method A method of classification based on the criteria of overall morphological, anatomical, physiological or biochemical similarity or difference, with all characters equally weighted and without regard to phylogenetic history; phenetics; *cf.* cladistic method.

phengophilous Thriving in, or having an affinity for, light; **phengophile, phengophily;** *cf.* phengophobous.

phengophobous Intolerant of light; **phengophobe, phengophoby;** *cf.* phengophilous.

phenodeme A local interbreeding population characterized by observed structural and functional properties (phenotype); *cf.* deme.

phenology Study of the temporal aspects of recurrent natural phenomena, and their relation to weather and climate.

phenometry The quantitative measurement of plant growth, mass and leaf area; **phenometric.**

phenotype The sum total of observable structural and functional properties of an organism; *cf.* genotype.

pheromone A chemical messenger secreted by an organism that conveys information to another individual, often eliciting a specific response.

phi grade scale A logarithmic transformation of the Udden grade scale of sediment particle size categories; see sediment particle size.

philodendron Araceae *q.v.*

philopatric Exhibiting a tendency to remain in the native locality; used of species or groups that show little capacity to spread or disperse; **philopatry.**

philothermic Thriving in a warm climate; **philotherm, philothermy.**

Philydraceae Small family of Liliales containing only 5 species of erect perennial herbs from Australia, the Pacific islands and southeastern Asia; flowers with a perianth of 4 petaloid members and a solitary stamen; seeds containing starch, protein and oil.

Phlebobranchia Order of solitary, occasionally colonial, tunicates (Ascidiacea) having internal longitudinal folds in the branchial sac; gonads located in loop of intestine; free-living larva present, but asexual reproduction also occurs by budding and strobilation;

cosmopolitan distribution and found at all depths from littoral to deep water.

Phlebotamidae Sand flies; family of small hairy flies (Diptera) found in moist shady habitats; feed by sucking blood of vertebrates and known to act as vectors of disease.

phlox Polemoniaceae *q.v.*

phobotaxis An avoidance reaction of a motile organism; **phobotactic.**

Phocidae Hair seals; cosmopolitan family of marine mammals (Pinnipedia) containing about 20 species and including grey, hooded, harp, monk, crabeater, Ross, leopard, Weddell, and elephant seals, and the sole freshwater Baikal seal; feed primarily on fishes, squid and crustaceans; ear pinnae absent.

Phocoenidae Porpoises; family containing 6 species of marine mammals (Odontoceta) found in coastal and estuarine waters of the Pacific, Atlantic, and Indian Oceans; head profile lacking distinct beak; feed on squid, crustaceans and fishes.

Phoenicopteridae Flamingos; family containing 6 species of large, long-legged wading birds cosmopolitan in tropical freshwater and marine shallow lakes and lagoons; bill characteristically bent downwards at midlength; feed on plankton and other suspended organic matter filtered out by sweeping movements of bill, with head inverted; breed in colonies, nesting on mounds in mud flats.

sand fly (Phlebotamidae)

Pholididae Gunnels; family containing 14 species of intertidal and shallow marine blennioid teleost fishes (Perciformes), found mostly in rocky habitats; body slender, compressed, to 300 mm in length; dorsal fin long, continuous with caudal and anal; pectorals reduced or absent.

gunnel (Pholididae)

Pholidoskepia Order of solenogastres (Mollusca) characterized by a layer of solid, scale-like calcareous bodies in the mantle and by the absence of true stalked epidermal papillae.

Pholidota Order of placental mammals comprising a single family, Manidae (pangolins).

phoresy A symbiosis in which one organism is merely transported on the body of an individual of a different species.

Phoridae Scuttle flies; family containing over 3000 species of small flies (Diptera) which often have reduced wings and run actively; larvae live in decaying and putrefying matter.

Phoronida Horseshoe worms; phylum or class of tubicolous marine coelomates comprising about 10 species characterized by a tentaculate

flamingo (Phoenicopteridae)

Phoenicopteriformes Order of long-necked, long-legged wading birds comprising a single family, Phoenicopteridae.

Phoeniculidae Small family of birds, the wood hoopoes, usually included within the family Upupidae *q.v.*

pholad Piddock *q.v.*

pholadophyte A plant living in hollows, intolerant of high light intensity.

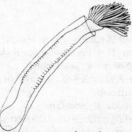
horseshoe worm (Phoronida)

lophophore surrounding the mouth serving respiratory, food collecting and protective functions; gut U-shaped, anus opening outside lophophore; hermaphrodite or with separate sexes, asexual propagation common; planktonic larval stage termed an actinotroch; adults found on hard and soft substrata, or in association with cerianthid anemones; intertidal to 400 m.

phorophyte The host plant of an epiphyte *q.v.*

phosphorescence The emission of light in darkness by the release of absorbed radiation; also the light so produced; *cf.* bioluminescence.

photic zone The surface zone of the sea or a lake having sufficient light penetration for photosynthesis; *cf.* dysphotic zone, euphotic zone.

photoautotrophic Used of organisms, such as green plants, that obtain metabolic energy from light by a photochemical process; **photoautotroph, photoautotrophy**; *cf.* chemoautotrophic.

photocleistogamic Used of a plant in which self-pollination occurs within flowers that remain closed because of insufficient illumination; **photocleistogamy.**

photoepinasty Upward curvature induced by light; **photoepinastic**; *cf.* photohyponasty.

photogenic Light-producing; bioluminescent; **photogenesis.**

photohoramotaxis The directed reaction of a motile organism to colour, or to a light-pattern stimulus; **photohoramotactic.**

photohyponasty Downward curvature induced by light; **photohyponastic**; *cf.* photoepinasty.

photokinesis A change in random movement of an organism in response to a light stimulus; **photokinetic.**

photonasty Growth curvature in response to a diffuse light stimulus; **photonastic.**

photoperiod The light of a light/dark cycle; *cf.* nyctiperiod.

photoperiodic Pertaining to the response of an organism to changes in daylength or to a light/dark cycle.

photophilic Thriving in conditions of full light; **photophile, photophily**; *cf.* photophobic.

photophobic Intolerant of, or avoiding, conditions of full light; **photophobe, photophoby**; *cf.* photophilic.

photophobotaxis A response of an organism to a temporal change in light intensity; positive when stimulated by a decrease in light intensity and negative when stimulated by an increase; **photophobotactic.**

photoplagiotropism Orientation at an oblique angle to incident light; **photoplagiotropic.**

photorespiration A metabolic process involving the uptake of oxygen and release of carbon dioxide, exhibited by many green plants when exposed to light, and which ceases in the dark.

photosynthates The products of carbon dioxide assimilation during photosynthesis.

photosynthesis The biochemical process that utilizes radiant energy from sunlight to synthesize carbohydrates from carbon dioxide and water in the presence of chlorophyll; **photosynthetic**; *cf.* chemosynthesis.

photosynthetically active radiation (PAR) Radiation capable of driving the primary processes of photosynthesis, in the range of wavelengths between 380–710 nm.

phototaxis A directed response of a motile organism towards (positive) or away from (negative) a light stimulus; **phototactic.**

phototropism An orientation response to light; **phototropic.**

Phractolaemidae Monotypic family of small tropical African freshwater teleost fishes (Gonorhynchiformes) found in turbid weedy habitats; body elongate, cylindrical, becoming compressed posteriorly; mouth small but highly protractile bearing only 2 teeth on lower jaw; gas bladder functioning as accessory respiratory organ.

Phragmobasidiomycetes Heterogeneous class of basidiomycotine fungi almost all of which are saprobic; exhibiting a variety of forms of fruiting body from inconspicuous types adhering closely to the substratum to large, pigmented foliose structures; comprises 2 subclasses, Heterobasidiomycetidae and Metabasidiomycetidae.

Phragmophora Order of chaetognaths characterized by a primitive ventral transverse musculature; comprises 2 families, one with 13 pelagic species, the other containing 8 benthic species.

phragmosis The action of closing the entrance to a nest or burrow with the body.

phreatic Pertaining to ground water.

phreaticolous Inhabiting ground water habitats; **phreaticole.**

phreatobiology The study of ground-water organisms.

Phreatoicoidea Small order of mainly subterranean freshwater isopod crustaceans found in Australia, New Zealand, India and South Africa.

pheatophyte A plant that absorbs water from the permanent water table.

phretophilous Thriving in water tanks; **phretophile, phretophily.**

phretophyte A plant inhabiting a water tank.

Phrynophiurida Cosmopolitan order of ophiuroidean echinoderms comprising about 275 species found at all depths from littoral to deep sea; the disk and arms are covered by a thick skin, arms often coiled ventrally or branched (basket stars), spines often ornate; vertebral articulation permits vertical and lateral movement of arms.

Phthiraptera Lice; order of wingless hemipterodean insects comprising 2 groups, Anoplura (sucking lice) and Mallophaga (chewing lice), often classified as distinct orders; lacking eyes and with antennae reduced.

phycobiont The algal partner of a lichen (an algal/fungal symbiosis); *cf.* mycobiont.

phycocoenology The study of algal communities.

phycology The study of algae.

phycophagous Feeding on algae; **phycophage, phycophagy.**

phycophilic Thriving in algae-rich habitats; living on algae; **phycophile, phycophily.**

Phylactolaemata Class of freshwater bryozoans comprising a single order, Plumatellida, in which zooids are cylindrical, monomorphic, with crescent-shaped lophophore bearing 100 or more tentacles; colonies encrusting or gelatinous; typically found attached to tree roots and submerged vegetation.

phyletic Pertaining to a line of direct descent, or a course of evolution.

phylktioplankton Planktonic organisms rendered buoyant by hydrostatic means.

Phyllidae Leaf insects; family containing about 50 species of large, dorsoventrally flattened insects (Phasmoptera) found mainly in southeastern Asia and New Guinea; legs broad with flattened extensions giving entire body a leaf-like appearance.

phyllobiology The study of leaves.

Phyllocarida Subclass of malacostracan crustaceans comprising a single order, Leptostraca *q.v.*

Phyllodocida Order of errant, pelagic and burrowing polychaete worms comprising about 3000 species in 26 families; pharynx cylindrical, muscular, eversible, with or without jaws; mouth surrounded by tentacular cirri; parapodia uniramous or biramous, acicula distinct; includes paddleworms, scaleworms, sea mouse, catworms and ragworms.

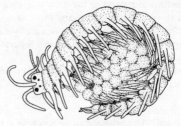

polychaete worm (Phyllodocida)

Phyllomedusidae Family of New World tree frogs (Anura), comprising about 40 species, found in tropical forests from Mexico to Argentina; body usually green dorsally, often brightly coloured on sides; eggs typically laid on vegetation above water, tadpoles aquatic.

phyllophagous 1: Feeding on leaves; **phyllophage, phyllophagy.** 2: Pertaining to plants that obtain nourishment from their leaves.

phyllosoma Zoeal larval stage of palinuran decapod crustaceans.

phyllosphere The microhabitat of a leaf; the immediate surroundings of plant leaves that are influenced by their presence; *cf.* rhizosphere.

Phyllostomatidae Vampire bats, leaf-nosed bats; diverse family containing about 130 species of small to large, New World bats (Microchiroptera); diets include insects, small vertebrates, fruit, nectar and blood.

phyllotaxis The arrangement of leaves on a stem; **phyllotaxy.**

phylogenetic Pertaining to evolutionary relationships within and between groups; **phylogeny.**

phylogenetic relationship Evolutionary relationship; affinity based on recency of common ancestry.

phylogenetic tree Evolutionary tree *q.v.*

phylogeny 1: The evolutionary history of a group or lineage. 2: The origin and evolution of higher taxa; **phylogenetic.**

phylum (phyla) A rank within the zoological hierarchy of classification; the principal category immediately below kingdom.

Phymosomatoida Order of echinoids (Echinacea) found in the Indo-Pacific and Japan, comprising only 2 extant species but having an extensive fossil record.

Physarales Morphologically and developmentally complex order of Myxogastromycetidae characterized by a fruiting body with a dark spore mass and often containing visible lime; spores disseminated by wind.

Physarida The slime mould order Physarales *q.v.*, treated as an order of the protozoan group, Myxogastria.

Physeteridae Sperm whales; family containing 3 species of small to very large (adult length 3–18 m) marine mammals (Odontoceta) widespread in oceanic waters; upper jaw with teeth absent or vestigial; head containing spermaceti organ giving blunt anterodorsal profile; feed mostly on squid, fishes and crustaceans.

sperm whale (Physeteridae)

Physeteroidea Superfamily of whales (Odontoceta) comprising the sperm whales (Physeteridae) and the beaked and bottle-nosed whales (Ziphiidae).

physiognomy The characteristic features or appearance of a plant community or vegetation; **physiognomic**.

physiographic Pertaining to geographical features of the Earth's surface; **physiography**.

physiology Study of the normal processes and metabolic functions of living organisms; **physiological**.

physogastry A condition of excessive enlargement of the abdomen, as found in some insects.

phytal zone That part of a shallow lake bottom supporting rooted vegetation.

phytobenthos 1: A bottom-living plant community. 2: That part of the bottom of a stream or lake covered by vegetation; *cf.* geobenthos.

phytochemistry Chemotaxonomy of plants.

phytochrome A proteinaceous pigment involved in photoperiodic responses and some other photoreactions in plants.

phytocide A chemical used to destroy and control plants; **phytocidal**.

phytocoenology Phytosociology; the study of plant communities.

Phytodiniales Small order of unicellular freshwater dinoflagellates which typically live attached to other aquatic plants by a stalk or basal disk.

phytoedaphon Soil flora; the plant community of the soil.

phytoflagellate Phytomastigophora *q.v.*

phytogenic Arising from, or caused by, plants.

phytogeographical regions The major geographical divisions of the world characterized by floristic composition; the 6 commonly recognized kingdoms are – Antarctic, Australian, Boreal, Neotropical, Palaeotropical and South African; see map on page 294.

phytogeography Study of the biogeography of plants.

Phytolaccaceae Common pokeweed; small family of Caryophyllales comprising about 125 species of often glabrous and succulent herbs best developed in subtropical and tropical areas of the New World.

phytoliths Small particles of opaline silica found in the cell walls of some plants and studied as plant trace fossils.

Phytomastigophora Phytoflagellates; subphylum of Mastigophora commonly treated as algae, comprising all or part of the following algal groups, Euglenophycota, Cryptophyceae, Chrysophyceae, Xanthophyceae, Eustigmatophyceae, Raphidiophyceae, Prymnesiophyceae, and Chlorophyceae.

phytopathology The study of plant disease.

phytophagous Feeding on plants or on plant material; **phytophage**, **phytophagy**.

phytophenology Study of the periodic phenomena of plants, such as flowering and leafing.

phytoplankton Planktonic plant life; typically comprising suspended microscopic algal cells such as diatoms and desmids; **phytoplankter**, **phytoplanktont**; *cf.* zooplankton.

phytopleuston Plants free-floating in aquatic habitats.

Phytosauria Extinct order of crocodile-like reptiles (Archosauria) known from the Triassic.

phytosis A disease caused by a plant parasite.

phytosociology The study of vegetation, including the organization, development, geographical distribution and classification of plant communities.

phytosuccivorous Feeding on sap; sap-sucking; **phytosuccivore**, **phytosuccivory**.

phytotelmata Small pools of water in or on plants, such as the leaf bases of bromeliads or the pitchers of pitcher plants.

phytotelmic Used of organisms that inhabit small pools of water within or upon plants.

phytoteratology The study of malformations and monstrosities in plants; plant teratology.

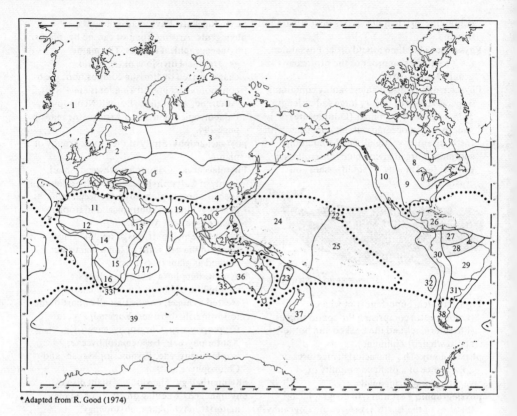

*Adapted from R. Good (1974)

Index

phytogeographical regions

Phytotomidae Plantcutters; family containing 3 species of finch-like passerine birds found in open woodland in South America; bills stout with serrated edges; feed on fruit, leaves and buds; nest in bushes.

phytotoxic Poisonous to plants; **phytotoxicity.**

Picathartidae Small family of passerine birds, the bald crows, usually included in the family Timaliidae *q.v.*

Picidae Woodpeckers, wrynecks; family containing about 200 species of hole-nesting birds (Piciformes) found worldwide in forests, grasslands and deserts; head large, bill typically strong and straight, tongue elongate and extensible; habits solitary, monogamous, non-migratory; feed mostly on insects and fruit; nest in tree holes or ground burrows.

woodpecker (Picidae)

Piciformes Diverse cosmopolitan order of small to medium-sized hole-nesting birds comprising 6 families and including honeyguides, toucans and woodpeckers.

pickerelweed Pontederiaceae *q.v.*

pico- (p) Prefix used to denote unit x 10^{-12}; see metric prefixes.

picoplankton Planktonic organisms between 0.2 and 2.0 μm in diameter.

Picrodontidae Extinct family of prosimian primates known from the Palaeocene.

piddock Pholad; a member of the family Pholadidae in the order Myoida, containing about 100 species of mostly marine bivalves which bore into hard substrata such as chalk, shale or wood.

Pieridae Whites, brimstones; family of medium-sized butterflies (Lepidoptera) with predominantly white or yellow wings; caterpillar with fine hairs, pupa supported by central silk girdle.

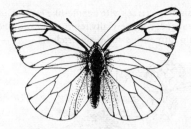

black-veined white (Pieridae)

piezoelectric Pertaining to a crystalline substance the electrical property of which is changed by pressure.

piezotropism A growth movement in response to a compression stimulus; **piezotropic.**

pig Suidae *q.v.*

pigeon Columbidae *q.v.*

pigfish Congiopodidae *q.v.*

pika Ochotonidae *q.v.*

pike Esocidae *q.v.*

pike conger Muraenesocidae *q.v.*

pike-characin Ctenoluciidae *q.v.*

pikehead Luciocephalidae *q.v.*

pilidium Ciliated planktonic larval stage of some nemertean worms.

pilidium

Pilosa Infraorder of edentates (Xenarthra) comprising the sloths (Bradypodoidea), ant-eaters (Myrmecophagoidea) and the extinct ground sloths (Megalonychoidea).

pilot whale Delphinidae *q.v.*

Pimelodidae Family containing about 300 species of nocturnal or crepuscular New World catfishes (Siluriformes) found primarily in freshwaters of South and Central America; body elongate, naked, dorsal fin with stout spine, adipose fin present; 3 pairs of barbels, one pair often elongate; important locally as food-fish and also popular in aquarium trade.

pimpernel Primulaceae *q.v.*

Pinaceae Pines, firs; large family of conifers (Pinatae) from the temperate zone of the northern hemisphere; usually evergreen,

resiniferous trees with needle-like leaves spirally
arranged; commercially important for timber,
pulp, turpentine and resin products.

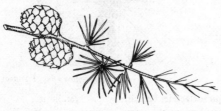

larch (Pinaceae)

pill beetle Byrrhidae *q.v.*

pill bug Oniscoidea *q.v.*

Pilosa Infraorder of edentates (Xenarthra)
comprising the sloths (Bradypodoidea), anteaters (Myrmecophagoidea) and the extinct
ground sloths (Megalonychoidea).

pilot whale Delphinidae *q.v.*

Pimelodidae Family containing about 300
species of nocturnal or crepuscular New World
catfishes (Siluriformes) found primarily in
freshwaters of South and Central America;
body elongate, naked, dorsal fin with stout
spine, adipose fin present; 3 pairs of barbels,
one pair often elongate; important locally as
food-fish and also popular in aquarium trade.

pimpernel Primulaceae *q.v.*

Pinaceae Pines, firs; large family of conifers
(Pinatae) from the temperate zone of the
northern hemisphere; usually evergreen,
resiniferous trees with needle-like leaves
spirally arranged; commercially important for
timber, pulp, turpentine and resin products.

Pinatae Conifers; the largest class of extant
Gymnosperms; usually evergreen shrubs or
trees, branches with short spur shoots and long
shoots; leaves small and scale-like or linear to
needle-like; mature female (pistillate)
strobilus either a cone or drupe-like structure.

pincushion flower Proteaceae *q.v.*

pine Pinaceae *q.v.*

pineapple 1: Bromeliales *q.v.* 2: Coenocarpium
q.v.

pine-cone fish Monocentridae *q.v.*

Pinicae Subdivision of Gymnosperms
(Pinophyta); evergreen or deciduous shrubs
or trees which are usually resiniferous; wood
(xylem) lacks vessels, and stem cortex of minor
importance in mature plant; male (staminate)
structures found in simple strobili, female
(pistillate) structures present in compound
strobili; includes classes Ginkgoatae and
Pinatae.

pink Caryophyllaceae *q.v.*

pinnigrade A form of locomotion by swimming
using flippers as paddles.

Pinnipedia Suborder of carnivores (Ferae)
comprising the sealions, walrus (Otariidae),
and seals (Phocidae); digits fully webbed and
modified with the limbs into paddles.

pinocytosis Active ingestion of fluid by a cell,
by invagination of the cell membrane to form
vesicles; **pinocytotic.**

Pinophyta Gymnosperms; ancient division of
seed-bearing vascular plants extending from
the Devonian to Recent; plants with true
roots, stems and leaves; female (pistillate) and
male (staminate) reproductive structures
borne in clusters or strobili, often forming
cones; pollen sacs present on scales
(microsporophylls) and ovules on scales
(megasporophylls) of cone; pollen usually
fertilizes ovule by entry through the micropyle
pore at its base; no style or stigma present;
extant representatives known in 3
subdivisions, Cycadicae, Gneticae and
Pinicae.

gymnosperm cone (Pinophyta)

pipe snake Aniliidae *q.v.*

pipefish 1: Syngnathidae *q.v.* 2:
Solenostomidae *q.v.*

Piperaceae Peppers; large family of aromatic
herbs (Piperales) and occasionally vines and
epiphytes, widespread in tropical regions in
moist, shady habitats; includes *Piper nigrum*,
from which pepper is made, and the betel nut
tree.

Piperales Order of flowering plants
(Magnoliidae); mostly herbs and shrubs with
reduced flowers lacking a perianth often borne
in dense spikes; ovules usually orthotropic;
fruit a berry or drupe.

Pipidae Family containing 15 species of aquatic
toads (Anura) from tropical west Africa and
South America; body length to 250 mm,
hindlimbs large and webbed, ribs present,

tongue absent; eggs and larvae aquatic; in one genus eggs laid in pockets on back of female, with aquatic larvae or direct development.

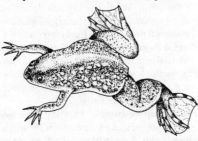

African clawed toad (Pipidae)

pipit Motacillidae *q.v.*

Pipridae Manakins; family containing about 50 species of small Neotropical passerine birds found in wet forest habitats; plumage often colourful; males exhibit lek courtship display; habits solitary to gregarious, arboreal, polygamous, feeding on fruit and insects; nest in trees often over water.

piranha Serrasalmidae *q.v.*

pirate perch Aphredoderidae *q.v.*

Piroplasmia Piroplasms; subclass of apicomplexan protozoans which are usually found as parasites of red blood cells in vertebrates and utilize ticks as vectors; characterized by the lack of a conoid from the apical complex of organelles; includes *Babesia*; also treated as a class, Piroplasmasida, of the protoctistan phylum Apicomplexa.

Pisces Fishes; group of vertebrates that includes all fishes in the broad sense; *cf.* Tetrapoda.

piscicolous Living on or within fishes; **piscicole.**

piscivorous Feeding on fishes; **piscivore, piscivory.**

pistachio Anacardiaceae *q.v.*

pit viper Crotalidae *q.v.*

pitcher plant 1: Nepenthaceae *q.v.* 2: Sarraceniaceae *q.v.*

Pittosporaceae Australian laurel, New Zealand lemonwood; family of Rosales containing about 200 species of often climbing shrubs and trees widely distributed in warm temperate and tropical regions; flowers typically small and bell-like with 5 sepals, petals and stamens, and a superior ovary.

placental mammal Eutheria *q.v.*

Placodermi Class of primitive, heavily armoured, jawed fishes (Gnathostomata) known primarily from the Devonian; body typically depressed, head covered by shield of bony plates; tail heterocercal.

Placodontidae Family of euryapsid reptiles known from the Triassic; most were specialized shellfish-feeders; some were turtle-like in form, others resembled nothosaurs.

Placozoa Phylum of primitive multicellular animals comprising only 2 species found free-living in the littoral zone of warmer seas; characterized by the construction of its 2 body layers, a thin dorsal epithelium of flagellate cells and a thick ventral epithelium of flagellate and glandular cells, with a network of fibre cells running between them; also characterized by the lack of bilateral symmetry of muscular and nervous systems, and of gonads; reproduction usually by binary fission or budding.

placozoan (Placozoa)

Plagiorchiida Order of digenetic trematodes characterized by a single pair of flame cells (excretory organs) in the miracidia larva and by the lack of excretory vessels in the tail of the cercaria larva.

plagiotropism An orientation response at an oblique angle to the vertical; **plagiotropic.**

plaice Pleuronectidae *q.v.*

Planctosphaeroidea Class of hemichordates known only from a planktonic ciliated larva (tornaria) collected from the Bay of Biscay; adult unknown.

plane Platanaceae *q.v.*

planetic Motile; possessing motile or swarming stages; **planetism.**

planidium First instar of bee fly larvae (Diptera: Bombyliidae) parasitic on the eggs and larvae of other insects.

plankter An individual planktonic organism.

plankton Those organisms that are unable to maintain their position or distribution independent of the movement of water or air masses; *cf.* nekton.

planktont An individual planktonic organism.

planktophilous Living or thriving in the plankton; **planktophile, planktophily.**

planktophyte A planktonic plant; a member of the phytoplankton.

planktotrophic Feeding on plankton;

planktotroph, planktotrophy.

Plannipennia Large suborder of advanced Neuroptera *q.v.* comprising the lacewings, dustywings, spongilla flies, owl flies and ant lions.

planogamete A motile gamete; zoogamete; *cf.* aplanogamete.

planomenon All free-living organisms; those organisms not rooted or attached to a substratum; *cf.* ephaptomenon, rhizomenon.

planont A motile gamete, spore or zygote.

Planosol An intrazonal soil with an eluviated surface layer over a strongly compacted eluviated claypan, formed in humid to subhumid climates.

plant 1: Any member of the kingdom Plantae. 2: A multicellular eukaryote organism typically exhibiting holophytic nutrition, lacking locomotion, lacking obvious nervous or sensory organs, and possessing cellulose cell walls.

plant hopper Homoptera *q.v.*

Plantae Plants; kingdom of eukaryotic organisms which typically use light as an energy source via photosynthetic pathways involving the green pigment chlorophyll; mostly non-motile, lacking nervous and excretory systems but possessing cell walls composed largely of cellulose; traditionally including unicellular algae, which may be motile and heterotrophic, and fungi, which lack chlorophyll and absorb their food; comprises 2 subkingdoms, the Thallobionta and the Embryobionta.

Plantaginales Plantain; order of Asteridae containing about 250 species in a single family, Plantaginaceae; mostly herbs with basal leaves; flowers wind-pollinated, or cleistogamous, borne on spikes.

plantain 1: Plantaginales *q.v.* 2: Musaceae *q.v.*

plantcutter Phytotomidae *q.v.*

planticolous Used of organisms that spend most of their active life on or within plants; **planticole.**

plantigrade Walking with the entire sole of the foot in contact with the ground.

planula A free-swimming larval stage of cnidarians with an elongate, radially-symmetrical and ciliated body.

plasmodial slime moulds A group of organisms with phagocytic amoeboid stages which are plasmodial (multinucleate and lacking cell walls) and move by protoplasmic streaming and by filose pseudopodia; produce sedentary fungus-like fruiting bodies containing one to many spores; flagellate stages often present;

found on decaying organic matter in moist microhabitats, and feeding on bacteria, fungi and protistans; classified as a phylum of Protoctista (Myxomycota), as a class of fungal slime moulds (Myxomycetes), or as a class of rhizopod protozoans (Eumycetozoa).

Plasmodiophorea Parasitic slime moulds *q.v.*, treated as a class of rhizopod protozoans.

Plasmodiophoromycetes Parasitic slime moulds *q.v.*, treated as a class of the Myxomycota and classified with the Fungi.

Plasmodiophoromycota Parasitic slime moulds *q.v.*, treated as a phylum of the Protoctista.

plasmogamy Fusion of the cytoplasm of two or more cells without nuclear fusion.

plasmophagous Feeding on body fluid; used of a parasite that freely absorbs the cellular fluid of the host; **plasmophage, plasmophagy.**

plasmotomy The division of a multinucleated cell into multinucleated daughter cells, without accompanying mitosis.

plastogamy The fusion of the cytoplasm of unicellular organisms to form a plasmodium, without fusion of the nuclei.

Platacanthomyidae Spiny dormice; family of arboreal mice with long bushy tails, found in India and southern China; generally included as a subfamily of the Muridae *q.v.*

Platanaceae Plane, sycamore; family of deciduous trees containing 6 species native to the Mediterranean region through to the Himalayas, and to North America; typically with small flowers; fruits are densely hairy achenes or nutlets in globose heads.

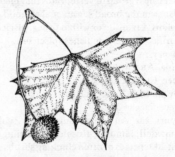

plane tree (Platanaceae)

Platanistidae River dolphins; family containing 4 species of freshwater dolphins widespread in rivers of southern Asia, China and South America; head with long slender beak and distinct forehead bulge; feed mainly on fishes but also take large aquatic invertebrates.

plate tectonics The concept that the Earth's crust is divided into a number of rigid plates that are in motion relative to each other; the plates are formed at the mid-ocean ridges (seafloor spreading) and typically destroyed in deep-sea trenches.

Platyasterida Order of asteroidean echinoderms represented by a single living genus with about 50 species; the group dates from the Lower Ordovician and contains the most primitive of the true asteroids.

Platycephalidae Flatheads; family of bottom-living marine and brackish-water scorpaeniform teleost fishes from the Indo-Pacific and eastern Atlantic; body elongate, cylindrical, to 1.2 m in length; head depressed, spinose and ridged.

Platycopida Order of marine podocopan ostracod crustaceans with a single family, Cytherellidae, containing about 75 species of benthic burrowers and crawlers which are unable to swim; typically abundant on continental shelf and in tropical lagoons; body size between 0.5 and 1.5 mm.

Platyctenida Order of often brightly coloured ctenophores found commonly in coastal waters of tropical and polar regions; containing sessile, creeping and planktonic forms which are often compressed in the oral–aboral axis and may have the anterior part of the gastrovascular system erected to form a creeping sole; comb rows often absent in adults.

Platydesmida Order of slender helminthomorphan diplopods (millipedes) having the sternal sclerites free, the pleural and tergal sclerites fused, and a conspicuous median dorsal suture; contains about 60 species, many of which are colonial, distributed throughout the Holarctic.

Platyhelminthes Flatworms; phylum of dorsoventrally flattened, vermiform animals exhibiting bilateral symmetry and a definite head end but lacking a coelom; other characteristics include the single-celled excretory organs (flame cells), the incomplete digestive system which necessitates the voiding of waste matter through the mouth, and the typically hermaphroditic reproductive system; comprises 3 classes, the mostly free-living Turbellaria, and the wholly parasitic Trematoda and Cestoda.

triclad flatworm (Platyhelminthes)

platypus Ornithorhynchidae *q.v.*

Platyrhinidae Family containing 5 species of small (less than 1 m) rhinobatiform fishes; body disk short and broad, snout rounded, tail slender and demarcated from disk; caudal fin without ventral lobe, axis straight.

Platyrrhini Infraorder of arboreal Neotropical primates (Haplorhini) comprising the marmosets, Callitrichidae, and the New World monkeys, Cebidae.

Platysternidae Big-headed turtle; family containing a single species of carnivorous, freshwater turtle (Testudines) from Asia; jaws serrated, carapace flattened, head too large to be withdrawn into shell.

play Trial-and-error learning especially in young animals.

Plecoglossidae Ayu; monotypic family of anadromous salmoniform teleost fishes found in China and Japan; body stout; teeth feeble, flattened and moveable; adipose fin present; prized locally as food-fish, and cultured in ricefields, ponds and waterways; some are caught using tethered cormorants.

Plecoptera Stoneflies; order of orthopterodean insects found mostly in cryptic habitats under stones and other debris; wings pleated or folded, abdomen 10-segmented bearing paired cerci; larvae usually aquatic, often in running water.

stonefly (Plecoptera)

Plectomycetes Class of ascomycotine fungi in which the asci are scattered through the inner tissues and are typically globose in shape; asci deliquescent or evanescent; typically the fruiting body (ascocarp) lacks an ostiole; comprises 5 orders, Ascophaerales, Elaphomycetales, Eurotiales, Gymnascales and Microascales.

pleiomorphic Exhibiting polymorphism at different stages of the life cycle;

pleiomorphism, pleiomorphous.

pleiotropic Used of a gene that has more than one, apparently independent, phenotypic effect; **pleiotropism, pleiotropy.**

pleioxenous Used of a parasite that is not host-specific, or a parasite that has several hosts during its life cycle; **pleioxeny.**

Pleistocene A geological epoch of the Quaternary period (*c.* 1.6–0.01 million years B.P.); see geological time scale.

Pleistocene refuge An area peripheral to a major ice sheet, or an elevated, ice-free area within an ice sheet (a nunatak) that has been relatively unaltered by glaciation during the Pleistocene and which has a biota once typical of the region as a whole.

Pleocyemata Large suborder of decapod crustaceans in which the gills lack secondary branching; females carry eggs which typically hatch as zoea larvae; comprises 6 infraorders, Astacidea (crayfishes), Anomura (hermit crabs, squat lobsters), Brachyura (crabs), Caridea (shrimps), Stenopodidea (shrimps) and Palinura (lobsters).

pleogamy The maturation and pollination of different flowers on an individual plant at different times; **pleogamic.**

pleometrosis The founding of a colony of social organisms by more than one original female; **pleometrotic;** *cf.* monometrosis.

pleomorphic Polymorphic; assuming various shapes or forms within a species or group, or during a single life cycle.

pleophagous 1: Feeding on a variety of food substances or food species; **pleophage, pleophagy.** 2: Used of a parasite associated with a variety of hosts.

Pleosporales Large order of loculoedaphomycetid fungi which may be parasitic on plants and other fungi, saprobic or lichenized.

plerocercoid Larval stage of tetraphyllidean cestodes, typically found in copepod crustacean hosts.

Plesiadapidae Extinct family of primates with squirrel-like bodies, equal length fore- and hindlimbs, claws on the digits and chisel-like incisors; known from the Palaeocene and Eocene.

Plesiocidaroida Extinct order of small, regular echinoids with very large apical system and ambulacra; known from the Triassic.

plesiomorphic 1: Used of ancestral or primitive characters or character states; *cf.* apomorphic. 2: Having similar shape or structure; **plesiomorph.**

Plesiopidae Roundheads; family of small (to 250 mm) Indo-Pacific perciform teleost fishes found in shallow water over reefs and rocky bottoms; body stout, head rounded with large mouth.

Plesiopora Small order of free-living and tubicolous annelid worms (Oligochaeta).

Plesiosauroidea Plesiosaurs; extinct suborder of aquatic reptiles known from the Jurassic and Cretaceous; body up to 15 m long, limbs paddle-shaped, neck either short (plesiosaurs) or long (pliosaurs).

pliosaur (Pleisiosauroidea)

Plethodontidae Family containing about 210 species of mostly terrestrial salamanders (Caudata); lungs absent; fertilization internal, males producing spermatophores; development variable with evolutionary trend from aquatic larvae to terrestrial eggs and direct development; some forms neotenic; mainly New World distribution, but also represented in Europe.

Pleurocapsales Order of unicellular or multicellular blue-green algae found in marine, freshwater and terrestrial habitats; characterized by an aggregation of cells into chains or filaments and by reproduction involving endospores or exospores.

Pleurodira Side-necked turtles; suborder of testudine reptiles in which the head is withdrawn laterally under the front edge of the carapace; comprises 45 species in 2 families, Pelomedusidae and Chelidae (snake-necked turtles), distributed throughout the southern hemisphere.

Pleurogona Order of ascidiacean tunicates comprising a single suborder, Stolidobranchia, in some classifications treated as a in distinct order.

Pleuronectidae Plaice, halibut, dab, flounder; family of pleuronectoid flatfishes widespread in temperate and tropical waters, in which the eyes are on the right side of the head; body length to 2.5 m; contains about 100 species several of which are important as food-fishes.

Pleuronectiformes Flatfishes; order of bottom-living marine and brackish-water teleost fishes characterized by a strongly compressed body with both eyes on the same side of the head; the pelagic larval stage is symmetrical, undergoing metamorphosis in adopting benthic mode of life; comprises 7 families in 3 suborders; many very important commercially, including flounders, sole, plaice, halibut, turbot, dab.

Pleuronectoidei Flounders; suborder of marine and brackish-water flatfishes comprising 4 families (Citharidae, Bothidae, Pleuronectidae and Scophthalmidae) with about 300 species; median fins lacking fin spines; pectoral fins present; vomerine and palatine teeth absent; typically use body undulations for locomotion.

Pleurostomatida Order of gymnostome ciliates comprising a single family of voracious carnivores widely distributed in aquatic habitats; characterized by a slit-like cytostome positioned on the edge of a laterally compressed body.

Pleurotremata Older name for the sharks; Selachii *q.v.*

pleustohelophyte A plant which floats at the surface of a water body but which also has emergent structures.

pleuston Aquatic organisms that remain permanently at the water surface by their own buoyancy, normally positioned partly in the water and partly in the air; **pleustonic**, **pleustont**; *cf.* neuston.

pleustophyte A free-floating macroscopic plant.

Pliocene A geological epoch within the Tertiary period (*c.* 5.0–1.6 million years B.P.); see geological time scale.

pliosaur Plesiosauroidea *q.v.*

Ploceidae Weaver finches, sparrows; family containing about 115 species of small Old World passerine birds found in forest, grassland and open arid habitats of Africa and southeast Asia; bill conical and robust; habits gregarious, arboreal to terrestrial, monogamous to polygamous or polyandrous, feeding mainly on seeds and insects; nest colonial to solitary, woven from grasses, usually placed in tree; members of one African subfamily are nest parasites.

ploidy The number of sets of chromosomes present; *cf.* monoploid, haploid, diploid, triploid, tetraploid, polyploid.

Ploima Large order of monogonatan rotifers containing both marine and freshwater forms, the latter often producing resistant resting eggs which ensure the survival of the rotifers during periods of desiccation; characterized by a circumapical ciliated band on the corona and by the presence of 1 or 2 toes and pedal glands on the foot.

plotophyte A floating plant, usually having special flotation structures.

Plotosidae Catfish eels; family containing 30 species of marine, brackish and freshwater catfishes (Siluriformes) found in coastal waters of the Indo-Pacific; body elongate and tapering, naked; first dorsal fin short with serrate spine, second long and contiguous with caudal and anal; 4 pairs of barbels present.

plover Charadriidae *q.v.*

plum Rosaceae *q.v.*

Plumatellida Sole order of bryozoan class, Phylactolaemata *q.v.*

Plumbaginales Devil's herb, statice; small but cosmopolitan order of Caryophyllidae comprising the family Plumbaginaceae containing about 400 species of herbs or low shrubs, many xerophytic or maritime; characterized by flowers with 5 sepals, 5 petals, 5 stamens and a superior ovary with 5 styles.

acantholimon (Plumbaginales)

pluriparous 1: Producing several offspring within a single brood; multiparous; *cf.* biparous, uniparous. 2: Having produced more than one previous brood.

plurivorous Feeding on a variety of different food sources or food species; **plurivore**, **plurivory**.

pluteus Free-swimming, planktonic larval stage of sea urchins (echinopluteus) and brittle stars (ophiopluteus) in which the long larval lobes bearing bands of cilia are supported by skeletal rods.

pluvial 1: Pertaining to, or resulting from, the action of rain or precipitation. 2: Used of a geological period, or the climate, characterized by abundant rainfall.

pluvial periods Periods during the Quaternary *q.v.* characterized by excessive rainfall and the formation of great lakes in tropical and subtropical regions.

pluviofluvial Pertaining to the combined action or effects of rainfall and streams.

pluviophilous Thriving in conditions of abundant rainfall; **pluviophile, pluviophily**; *cf.* pluviophobous.

pluviophobous Intolerant of conditions of abundant rainfall; **pluviophobe, pluviophoby**; *cf.* pluviophilous.

pneumatophore 1: Gas-filled apical zooid in colonial hydrozoan cnidarians (Siphonophora) that acts as a float. 2: Specialized aerial roots through which gaseous exchange can take place; found in mangroves and other plants in waterlogged or compacted soils.

pneumotaxis The directed response of a motile organism towards (positive) or away from (negative) a stimulus of dissolved carbon dioxide or other gas; **pneumotactic.**

pneumotropism An orientation response to a stimulus of dissolved carbon dioxide or other gas; **pneumotropic.**

Poaceae Grasses; cosmopolitan family of Cyperales containing about 8000 species of mostly herbs, occasionally woody plants (bamboo), which tend to accumulate silica; stems mostly round and hollow, leaves usually narrow and parallel-veined; flowers small and arranged in spikelets which are in turn arranged in spikes; pollen grains with a single pore; includes many domesticated species providing staple foods for man, such as wheat, rice, millet, maize, oats, sugarcane, sorghum, rye, barley and bluegrass; formerly known as Graminae.

oats (Poaceae)

poacher Agonidae *q.v.*

pocket gopher Geomyidae *q.v.*

pocket mouse Heteromyidae *q.v.*

Podargidae Frogmouths; family containing 13 species of solitary nocturnal birds (Caprimulgiformes) found in forest and bushland from India through to Australia; head large, bill broad and hooked with wide gape; feed on insects and small mammals; nest in trees.

frogmouth (Podargidae)

Podaxales Order of gasteromycete fungi containing a single species found in warm sandy habitats, frequently on ants' nests, which are penetrated by mycelium; fruiting body produced on surface.

Podicipedidae Grebes; family containing 20 species of specialized aquatic birds; legs located posteriorly for efficient swimming and diving, but locomotion on land clumsy; sexes similar; feed on fish, aquatic invertebrates and plant material; breed on freshwater lakes and ponds, overwintering along coasts.

Podicipediformes Order of neognathous birds containing a single cosmopolitan family, Podicipedidae (grebes).

Podocarpaceae Family containing over 100 species of evergreen, resiniferous conifers (Pinatae) native mainly to tropical and subtropical mountains of southern hemisphere; leaves scale- to needle-like; mature pistillate cones small and drupe-like.

Podocopa Diverse subclass of ostracod crustaceans found in marine, freshwater and, rarely, terrestrial habitats; mostly benthic creepers and burrowers; contains 2 orders, Platycopida and Podocopida.

Podocopida Order of podocopan ostracod crustaceans found in marine, brackish and freshwater habitats; mostly benthic creepers and crawlers, some living on algae or commensal on crustaceans; includes about 30 families and more than 5000 species.

Podostemales Small order of Rosidae containing over 200 species of aquatic herbs producing tiny aerial flowers; found in fast rivers with stony beds in tropical regions;

species of family Podostemaceae have a thallus-like form anchored by roots.

Podzol soil A zonal soil with acidic layer of litter and duff over a highly acidic leached A-horizon and dark brown illuvial acidic lower horizon with translocated deposits of humus and iron oxide, formed in cool temperate to temperate, humid climates, under coniferous or mixed coniferous/deciduous forest or heath.

podzolization A soil-forming process in regions of high precipitation, good internal soil drainage, and low calcium carbonate levels; the soil becomes more acidic and iron and aluminium sesquioxides, as well as humus colloids, are translocated down the soil profile by percolating water.

Poeciliidae Live-bearers, guppies; family containing 140 species of small viviparous fresh- and brackish-water cyprinodontiform teleost fishes widespread in South and Central America, and also found in parts of North America; anal fin of male modified as gonopodium for sperm transfer; sexual dimorphism often well developed.

Poecilosclerida Large order of ceractinomorph sponges found in all seas and from the intertidal to depths in excess of 7000 m; characterized by a skeleton of megascleres and spongin, varying in body form from encrusting to branching or massive; extensive fossil record from Cambrian to recent.

Poecilostomatoida Order of marine copepod crustaceans containing about 1350 species of free-living predatory or surface-feeding forms, commensals and parasites, typically with powerful grasping maxillipeds and short antennae in males.

Poeobiida Monotypic order of aberrant pelagic polychaete worms; body compressed, transparent, lacking segmentation; prostomium and peristomium indistinct, bearing branchiae and tentacular palps.

Pogonophora Beard worms; phylum of sedentary marine worms inhabiting cylindrical tubes typically buried vertically in fine ocean sediments; depth range 20–4000 m; body extremely slender, length 50–800 mm, divided into 4 regions – head bearing tentacles, mesosoma with or without an oblique ridge (bridle), long metasoma containing gonads, and a terminal opisthosoma armed with setae; mouth, gut and anus absent, nutrients absorbed through tentacles; sexes separate; contains about 100 species in 2 classes, Frenulata and Afrenulata.

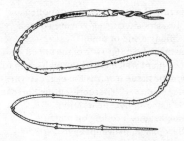

beard worm (Pogonophora)

poic Pertaining to meadows.

poikilohaline Used of organisms having body fluids that conform to external changes in salinity.

poikilosmotic Used of organisms that have an internal osmotic pressure that conforms to the osmotic pressure of the external medium; *cf.* homoiosmotic.

poikilothermic Cold-blooded; used of an organism having no mechanism for internal temperature regulation so that body temperature fluctuates with that of the immediate environment; **poikilotherm**, **poikilothermy**; *cf.* homoiothermic.

poikilothermy The condition of being poikilothermic *q.v.*

poinsettia Euphorbiaceae *q.v.*

poison ivy Anacardiaceae *q.v.*

poison oak Anacardiaceae *q.v.*

polar Pertaining to the areas within the Arctic and Antarctic Circles; characteristic of regions around the geographical poles; *cf.* temperate, tropical.

polecat Mustelidae *q.v.*

Polemoniaceae Phlox; widely distributed family of Solanales containing nearly 300 species of mostly herbs with glandular hairs producing an often mephitic odour; flowers in crowded cymes, each flower with 5 sepals, petals and stamens, and a superior ovary.

poleophilous Thriving in urban habitats; **poleophile, poleophily**; *cf.* poleophobous.

poleophobous Intolerant of urban habitats; **poleophobe, poleophoby**; *cf.* poleophilous.

Polioptilidae Gnatcatchers; family containing 13 species of small, active passerine birds found in the New World; bills long and slender for feeding on insects; live in scrub, nest in trees.

pollarding Severe pruning in which all the younger branches of a tree are cut back almost to the trunk; *cf.* coppicing.

pollen The microspores of flowering plants containing the male gametophyte.

pollen rain The total amount of pollen and spores deposited in a given area over a specified period of time.

pollen tube An outgrowth of the pollen grain of a seed plant which grows through compatible stigmatic tissue towards the egg, carrying the male gametes with it; representing the reduced male gametophyte generation.

pollenophagous Feeding on pollen; **pollenophage, pollenophagy.**

pollination Transference of pollen from the anther to the receptive area of a flower; used loosely to mean fertilization of a seed plant.

pollution The contamination of a natural ecosystem, especially with reference to the activity of man; **pollutant.**

polyacmic Exhibiting many abundance peaks per year; *cf.* diacmic, monacmic.

polyandrous Used of a female that mates with several males; **polyandry;** *cf.* monandrous.

polybrominated biphenyls PBBs *q.v.*

polycarpic Producing fruit or spores more than once during a life cycle; *cf.* monocarpic.

Polychaeta Bristleworms; class of marine, brackish and freshwater annelid worms comprising about 8000 species in 85 families; body length from 1 mm to 3 m, segments typically with paired lobes (parapodia) bearing variety of chaetae and bristles; exhibit diverse errant, sedentary, tubiculous, pelagic and parasitic habits with associated morphological adaptations.

scaleworm (Polychaeta)

polychlorinated biphenyls PCBs *q.v.*

Polycladida Large order of mostly predatory turbellarians found in coastal habitats and occasionally in plankton, rarely in fresh water; characterized by a large, flattened body, a main median intestine with numerous lateral branches, and many small testes and ovaries.

Polycystina Class of Actinopoda characterized by a skeleton made of silica and by the presence of a capsular membrane in the vegetative stages, which is regularly perforated; shell architecture depends on the arrangement of axopodia; found in all oceanic waters; includes 2 orders, Nassellarida and Spumellarida.

Polydesmida Diverse order of blind helminthomorphan diplopods (millipedes); body compact, able to roll into a loose coil or spherical ball; habits commonly nocturnal and fossorial, but order also includes arboreal, troglophilic, and active ground-living carnivorous species; the largest order of millipedes.

polydomic Used of colonies of social insects that occupy more than one nest; *cf.* monodomic.

polyethism The division of labour within a society on the basis of morphological castes (caste polyethism) or age (age polyethism).

Polygalaceae Milkwort; cosmopolitan family containing about 750 species of herbs and woody plants often with extrafloral nectaries; flowers bisexual but irregular with 5 sepals (the 2 inner sepals being petaloid), usually 8 stamens and a superior ovary.

Polygalales Order of Rosidae consisting of 7 families of woody and herbaceous plants, the largest of which are the Malpighiaceae, Polygalaceae and Vochysiaceae.

polygamous 1: Pertaining to the condition in which an individual has more than one sexual partner; **polygamy;** *cf.* monogamous. 2: Used of a plant having perfect and imperfect flowers together on the same individual.

Polygonales Buckwheat, dock, rhubarb, knotweeds, sorrels, sulphur plant; small order of Caryophyllidae consisting of a single family, the Polygonaceae, of mostly herbaceous or woody plants chiefly from northern temperate regions; characterized by small, perfect flowers usually with 3–6 perianth segments, up to 9 stamens and a superior ovary.

persicara (Polygonales)

polygoneutic Producing several broods per year or season; **polygoneutism;** *cf.* monogoneutic, digoneutic, trigoneutic.

Polygordiida Order of deposit-feeding polychaete worms found on intertidal and subtidal coarse sediments, comprising a single family and about 15 species; body segmentation indistinct; prostomium and peristomium with pair of short tentacles; pharynx eversible; parapodia and chaetae absent.

polygyny 1: The mating of a single male with several females. 2: The presence of many queens within a single colony of social insects; *cf.* monogyny, oligogyny.

polyhaline Pertaining to brackish water having a salinity between 10 and 17 parts per thousand; or to sea water having a salinity greater than 34 parts per thousand; *cf.* mesohaline, oligohaline.

polyhalophilic Thriving in a wide range of salinities; **polyhalophile**, **polyhalophily**.

Polyhymenophora Class of ciliates with sparse body ciliature but a highly developed oral ciliature comprising numerous membranelles; found most commonly as free-swimming forms in aquatic and soil habitats, although some are endocommensals; contains a single subclass, Spirotricha.

polymictic Used of a lake having no persistent thermal stratification, which is continually circulating with only brief periods of stability: *cf.* mictic.

Polymixiidae Beardfishes; family containing 3 species of mesopelagic marine teleost fishes widespread in tropical and temperate seas; body deep, compressed, to 250 mm length; eyes well developed; pair of long chin barbels present.

Polymixiiformes Monofamilial order of pelagic marine teleost fishes sometimes included in the Beryciformes.

Polymorphida Order of palaeacanthocephalan thorny-headed worms that are parasites of amphibians, reptiles, birds and mammals.

polymorphism 1: The co-occurrence of several different forms. 2: The co-existence of two or more discontinuous, genetically determined, segregating forms in a population, where the frequency of the rarest type is not maintained by mutation alone; *cf.* dimorphic, monomorphic.

Polynemidae Threadfins; family containing 35 species of small to large (to 2 m), tropical and subtropical, coastal marine and brackish-water teleost fishes (Perciformes); snout pointed, mouth subterminal, 2 dorsal fins present; pectorals subdivided, with several elongate rays; feed on pelagic crustaceans and fishes.

Polynemoidei Suborder of marine and brackish-water perciform teleost fishes comprising a single family, Polynemidae (threadfins).

Polynesia An assemblage of a large number of small oceanic islands scattered over the eastern Pacific Ocean, extending northwards to include Hawaii and eastwards to Easter Island, sometimes also including New Zealand; used loosely as a geographical area and ethnological unit to include both eastern Melanesia *q.v.* and Micronesia *q.v.*

Polynesian subkingdom A subdivision of the Palaeotropical kingdom *q.v.*

polynya An expanse of open water in the middle of sea ice.

Polyodontidae Paddlefishes; family of primitive freshwater fishes, allied to sturgeons; body elongate, mostly naked, length to 2 m, snout paddle-shaped; contains 2 species, one found in North America, the other in China.

polyoestrous Having a succession of breeding periods in one sexual season; *cf.* anoestrus, dioestrus, monoestrous.

polyoxybiotic Used of organisms requiring abundant free oxygen.

polyp The individual, soft-bodied, sedentary form of cnidarians; typically a cylindrical trunk with an apical mouth surrounded by tentacles; attached basally to substratum in solitary forms or to branching tubular system in colonial forms.

Polyphaga The largest suborder of beetles (Coleoptera) comprising about 140 families, including Staphylinidae (rove beetles), Lucanidae (stag beetles), Scarabaeidae (dung beetles), Buprestidae (jewel beetles), Elateridae (click beetles), Lampyridae (fireflies, glow worms), Cantharidae (soldier beetles), Coccinellidae (ladybird beetles), Tenebrionidae (darkling ground beetles), Meloidae (blister beetles), Cerambycidae (long-horned beetles), Bruchidae (bean and pea weevils), Chrysomelidae (leaf beetles) and Curculionidae (weevils).

polyphagous Feeding on a wide variety of different foods or food species; **polyphage**, **polyphagy**; *cf.* monophagous, oligophagous.

polyphyletic Derived from two or more distinct ancestral lineages; used of a group comprising taxa derived from two or more different ancestors; **polyphyly**; *cf.* monophyletic.

Polyplacophora Chitons, coat-of-mail shells; class of marine, free-living molluscs mostly found in shallow waters and characterized by a dorsal shell consisting of 8 overlapping, calcareous plates; typically with broad, ventral foot used for movement over or attachment to the substratum, surrounded by a groove containing 6 to many pairs of gills (ctenidia); jaws lacking, radula usually consisting of regular transverse rows of teeth; all living chitons belong to the subclass, Neoloricata.

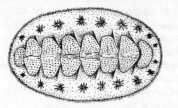

chiton (Polyplacophora)

polyplanetic Having several motile stages and intervening resting phases during the life cycle; **polyplanetism**; *cf.* diplanetic, monoplanetic.

polyploid Having more than two sets of homologous chromosomes; **polyploidy**; *cf.* ploidy.

Polypteridae Reedfishes, bichirs; family containing 11 species of African freshwater fishes found in weedy lakes and rivers; body elongate, subcylindrical, to almost 1 m in length, bearing robust ganoid scales; gas bladder serves as accessory respiratory organ; dorsal fin comprising many small finlets, pelvics absent; larvae with external gills; adults predatory on fishes and amphibians.

polysaccharide A polymer comprising monosaccharides (simple sugars) or their derivatives; function mainly as structural elements (such as cellulose in plant cell walls) or as energy storage molecules (such as starch).

polyspermy The penetration of two or more sperms into a single ovum at the time of fertilization; **polyspermic**; *cf.* dispermy, monospermy.

polystenohaline Used of organisms that only inhabit oceanic waters of relatively constant high salinity; *cf.* euryhaline, holeuryhaline, oligohaline, stenohaline.

polythermic Tolerating relatively high temperatures; *cf.* oligothermic.

polytokous Having many offspring per brood; fruiting many times during the life cycle; *cf.* ditokous, monotokous, oligotokous.

polytopic Occurring or arising independently in two or more separate localities or geographical areas; **polytopism**, **polytopy**; *cf.* monotopic.

Polytrichidae Subclass of mosses (Bryopsida) in which plants are differentiated into above- and below-ground portions, with erect,

sparsely branching stems, a well developed conducting system and complex leaf structure; grow as individual plants up to 800 mm in height or as dense mats.

polytrophic Feeding on a variety of different food substances or food species; **polytrophy.**

Polyxenida Order of minute (to 4 mm) diplopods (millipedes) having thin flexible body, lacking mineralization; typically active, colonial, prefering arid habitats; contains about 90 species having primarily tropical distribution.

bristly millipede (Polyxenida)

Polyzoa Bryozoa *q.v.*

Polyzoniida Cosmopolitan order of helminthomorphan diplopods (millipedes) having smooth polished body typically without median suture; contains about 25 species, inhabiting moist leaf litter and humus.

Pomacentridae Damselfishes; family containing about 200 species of primarily tropical marine, reef-dwelling perciform teleost fishes, also occasionally found in shallow brackish and fresh waters; small (to 350 mm) and colourful, body deep and compressed; very popular amongst aquarists.

Pomadasyidae Haemulidae *q.v.*

Pomatomidae Bluefishes; family containing 3 species of large (to 1 m) predatory, pelagic, tropical and warm temperate, perciform marine teleost fishes; body elongate and robust, jaws and teeth strong; 2 dorsal fins present; popular as sport and food-fishes.

pome A type of fleshy false fruit (pseudocarp) such as the apple, in which the true fruit (core) is surrounded by succulent tissues derived from the enlarged receptacle of the flower.

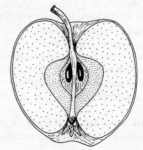

apple (pome)

pomegranate Punicaceae *q.v.*

Pomeranian glaciation A subdivision of the Weichselian glaciation *q.v.*

pomfret Bramidae *q.v.*

pomology The study and practice of fruit cultivation.

pompano Carangidae *q.v.*

Pompilidae Spider wasps; family of slender solitary wasps (Hymenoptera) often with black or metallic blue bodies and orange or yellow wings; females paralyse spiders then lay one egg on each prey which is usually left in a burrow; contains approximately 2500 species.

pond A small natural or artificial body of standing fresh water, intermediate in size between a pool and a lake, usually with negligible current and having more or less continuous vegetation from the marginal land areas into the water.

pond skater Gerridae *q.v.*

pond snail Basommatophora *q.v.*

pondweed Potamogetonaceae *q.v.*

Pongidae Great apes; family of large, primarily herbivorous, arboreal or terrestrial primates (Catarrhini) found in the Oriental and Ethiopian regions; body robust, forelimbs longer than hindlimbs; tail and ischial callosities absent; contains the orang-utan, gorilla and chimpanzee.

chimpanzee (Pongidae)

Pontederiaceae Water hyacinth, pickerelweed, mud plantain; family of Liliales comprising about 30 species of aquatic or semiaquatic herbs with floating or emergent leaf blades; flowers with 6 stamens; widespread in warm regions and extending into the north temperate zone.

pontic Pertaining to the deep sea.

pontophilous Thriving in the deep sea; **pontophile, pontophily.**

ponyfish Leiognathidae *q.v.*

poocolous Living in meadows; **poocole.**

poophilous Thriving in meadows; **poophile, poophily.**

poophyte A meadow plant.

poplar Salicales *q.v.*

poppy Papaveraceae *q.v.*

population All individuals of one species occupying a defined area and usually isolated to some degree from other similar groups.

population biology The study of the spatial and temporal distribution of organisms.

population dynamics The study of changes within populations and the factors that cause or influence those changes.

population genetics The study of gene frequencies and selection pressures in populations.

population size The number of items or individuals in a finite population.

population structure The composition of a population according to age and sex of the individuals.

porbeagle Lamnidae *q.v.*

porcupine 1: Hystricidae *q.v.*, Old World porcupines. 2: Erethizontidae *q.v.*, New World porcupines.

porcupine fish Diodontidae *q.v.*

pore volume The volume of air-filled space in a soil.

porgy Sparidae *q.v.*

poricidal dehiscence Spontaneous release of the contents of a ripe fruit through pores.

Porifera Sponges; phylum of aquatic animals characterized by an internal canal system through which water is propelled by flagella borne singly on choanocyte cells typically located in special chambers; water enters inhalent spaces through ostia in the outer covering, pores in the pinacocyte-lined walls of these inhalent spaces allow the water to percolate into the mesohyl (area containing the skeletal elements and other cell types of the sponge); water leaves sponge via the choanocyte chambers, the exhalent canal system and a set of larger openings, or oscules; skeleton made up of spicules of varying shape and chemical composition.

sponge (Porifera)

Porocephalida Order of pentastomes
containing 38 species in 2 superfamilies, one
parasitic in mammals, the other in reptiles
(crocodiles, monitor lizards, snakes, turtles);
in mammals, pentastomes found as adults only
in nasal passages of Canidae, with juveniles
infesting variety of intermediate hosts
including horses, camels, pigs, hares,
ruminants and occasionally man; body length
up to 90 mm; head flattened; hooks arranged
in transverse row.

porosity The ratio of the volume of pore space
to the total volume of sediment.

porosphere The habitat or environment
provided by a porous medium such as soil, that
is more or less solid but with internal spaces
(interstices) and surfaces.

Porphyridiales Order of marine and freshwater
red algae (Rhodophyceae) with typically
single-celled thalli, or rows of cells held
together by mucilage.

porpoise Phocoenidae *q.v.*

Portuguese man o'war Siphonophora *q.v.*

Portulaceae Moss rose, bitterroot; family of
Caryophyllales containing about 500 species
of often succulent herbs with capsular fruits;
cosmopolitan but best developed in western
North America and the Andes; flowers
bisexual, typically borne on cymes and with 2
sepals, 4 to many petals, 1 or 2 whorls of 5 or
several stamens and, commonly, a superior
ovary.

Posidoniaceae Family of Najadales consisting
of 3 species of glabrous, rhizomatous herbs
growing submerged in sea water or on rocks
that may be exposed at low tide.

possum Phalangeridae *q.v.*

post coitum Occurring after mating.

pot worm Enchytraeidae *q.v.*

potamic Pertaining to rivers, or transport by
river currents.

potamicolous Living in rivers; **potamicole.**

potamodromous Used of an organism that
migrates only within fresh water; *cf.*
anadromous, catadromous, oceanodromous.

Potamogetonaceae Pondweeds; cosmopolitan

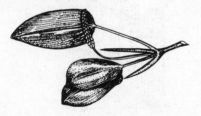

pondweed (Potamogetonaceae)

family of Najadales comprising about 100
species of glabrous herbs with creeping
rhizomes, erect leafy branches and small,
wind-pollinated flowers, found in freshwater
habitats.

potamology The study of rivers.

potamophilous Thriving in rivers; **potamophile,
potamophily.**

potamophyte A river plant.

potamoplankton Planktonic organisms of slow-
moving streams and rivers.

Potamotrygonidae River rays; family
containing 20 species of medium-sized
myliobatiform fishes confined to freshwater
habitats of tropical South America; body disk
oval to circular, tail slender and filamentous
with serrated dorsal spine; dorsal and caudal
fins absent.

potamous Pertaining to the lower reaches of
rivers and streams; *cf.* rhithrous.

potato Solanaceae *q.v.*

potato blight Peronosporales *q.v.*

potter wasp Eumenidae *q.v.*

Pottiales Small cosmopolitan order of mosses
(Bryidae) containing generally small plants
that grow as tufts on soil, rock and trees; stems
usually having terminal sporophyte, the apical
cell of the stem with 3 cutting faces; erect spore
capsule typically borne on long stalk.

pottoo Nyctibiidae *q.v.*

pot worm Enchytraeidae *q.v.*

powdery mildew Erysiphales *q.v.*

Praecardioida Extinct order of thin-shelled,
bivalve molluscs known from the Ordovician
to the Carboniferous.

prairie dog Sciuridae *q.v.*

Prairie soil A zonal soil with a dark greyish
brown to black moderately acidic A-horizon
grading through brown into a light-coloured
parent material, formed in temperate to cool
temperate humid climates under tall grass
vegetation; Brunizem.

praniza Parasitic larval stage of gnathiidean
isopods (Crustacea).

Prasinomonadida Prasinophyceae *q.v.* treated
as an order of the protozoan class,
Phytomastigophora.

Prasinophycales Prasinophyceae *q.v.* treated
as a class of the protoctistan phylum,
Chlorophyta.

Prasinophyceae Class of mostly flagellate
chlorophycote green algae found in marine,
brackish and freshwater habitats;
characterized by a covering of polysaccharide
scales over the cells and flagella; comprises 5
orders, Pedinomonadales, Monomastigales,

Pyramimonadales, Pterospermatales and Chlorodendrales; also treated as a class of the protoctistan phylum Chlorophyta under the name Prasinophycales, and as an order of the protozoan class Phytomastigophora, under the name Prasinomonadida.

Prasiolales Small order of terrestrial, freshwater and marine green algae in which the body form varies from filamentous to a parenchymatous cylinder with attachment rhizoids; cells uninucleate with an axial, stellate chloroplast.

pratal Pertaining to grassland and meadowland.

pratincole Glareolidae *q.v.*

pratinicolous Living in grassland and meadowland; **pratinicole.**

prawn Dendrobranchiata *q.v.* and Pleocyemata *q.v.*; shrimp.

prayer plant Marantaceae *q.v.*

praying mantis Mantodea *q.v.*

Preboreal period A period of the Blytt–Sernander classification *q.v.* (*c.* 9000–10 000 years B.P.), characterized by birch and pine vegetation and by relatively colder and wetter conditions than in the Boreal period.

Precambrian A geological era ending about 570 million years B.P. comprising the Proterozoic, Cryptozoic, Archaeozoic and Azoic eons; see geological time scale.

precipitation Rainfall; also including snow, hail and sleet.

precocial Used of offspring that exhibit a high level of independent activity from birth; praecocial; *cf.* altricial.

precocious Occurring particularly early in development.

predation The consumption of one animal (the prey) by another (the predator); also used to include the consumption of plants by animals, and the partial consumption of a large prey organism by a smaller predator (micropredation).

predator An organism that feeds by preying on other organisms, killing them for food; **predation.**

prehensile Adapted for gripping.

Prenebraskan glaciation An early glaciation of the Quaternary Ice Age *q.v.* in North America.

preservation The maintenance of individual organisms, populations or species by planned management and breeding programmes; *cf.* conservation.

presocial Pertaining to a social group, or the condition in which members display some degree of social behaviour but fall short of full eusocial *q.v.* behaviour.

prevalence of infection The number of individuals of a host species infected by a particular parasite species divided by the number of hosts examined, usually expressed as a percentage.

prevernal Pertaining to early spring.

prey An animal or animals killed and consumed by a predator.

prey set The total range of prey items utilized by a predator.

Priacanthidae Bigeye; family containing 15 species of nocturnal predatory bottom-living tropical and subtropical marine perciform teleost fishes; body deep, compressed, usually bright red in colour; eyes large, lower jaw prolonged; pelvic fin joined to body by flap of skin.

Priapulida Small phylum of generally rare marine worms known from less than 10 species found burrowing within sediments from intertidal to abyssal depths; characterized by an unsegmented cylindrical body typically divided into an anterior portion, the introvert, which bears the mouth and is invaginable into the abdomen, an abdomen bearing the anus and urogenital pores posteriorly, and a postanal tail; sexes separate, fertilization generally external; larval stages covered by a cuticular lorica.

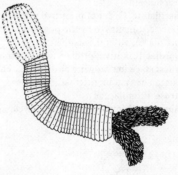

priapulid (Priapulida)

prickleback Stichaeidae *q.v.*

pricklefish Stephanoberycidae *q.v.*

primary consumer A heterotrophic organism that feeds directly on a primary producer.

primary producer An organism that synthesizes complex organic substances from simple inorganic substrates.

primary production The assimilation of organic matter by autotrophs; *cf.* gross primary production, net primary production.

primary sexual characters Those differences between the sexes that relate to the reproductive organs and gametes; *cf.* secondary sexual characters.

Primates Order of arboreal and terrestrial mammals that includes lemurs, tarsiers, monkeys, gibbons, apes, and man; eyes well developed, with binocular vision; digits usually prehensile, pollex and hallux opposable; typically produce single altricial young; many exhibit strong social organization; cosmopolitan in distribution but most abundant in the tropics; comprises 2 suborders, Strepsirhini (prosimians) and Haplorhini (monkeys, apes).

gorilla (Primates)

primeval Pertaining to earliest times, or the beginning of the Universe.

primitive 1: Early; simple; poorly developed; unspecialized. 2: Used of characters or character states present in ancestral forms; plesiomorphic.

Primociliatida Order of primitive gymnostome ciliates distinguished by the possession of 2 or more nuclei, all of the same kind.

primordial Primitive; primary; used of the earliest stage in ontogeny or development of an organ or system; original.

primrose Primulaceae *q.v.*

Primulaceae Primrose, pimpernel, cyclamen, cowslip, loosestrife, primula; large family of mostly herbs containing about 1000 species found in cold and temperate regions of northern hemisphere; flowers typically with a 5-part calyx, a 5-lobed corolla, 5 stamens, 1 style and a superior ovary.

Primulales Order of Dilleniidae containing about 2100 species of herbs or woody plants in 3 families, Myrsinaceae, Primulaceae and Theophrastaceae.

Prionodonta Superorder of bivalve molluscs comprising a single order, Arcoida *q.v.*

Pristidae Sawfishes; family containing 9 species of large (to 6 m) tropical and subtropical shallow marine, brackish and freshwater elasmobranch fishes (Pristiformes) in which the snout is prolonged into a long blade bearing a series of marginal teeth; body only moderately depressed, pectoral fins weakly expanded, dorsals and caudal large; feed on fishes and bottom invertebrates using snout to forage in sediment or slash small schooling fishes; reproduction viviparous, without placenta.

Pristiformes Order of large elasmobranch fishes (Batoidea) comprising a single family, Pristidae (sawfishes).

Pristiophoridae Saw sharks; family containing 5 species of medium-sized (to 2 m) pristiophoriform elasmobranch fishes confined to the Indo-Pacific; body elongate and subcylindrical, head depressed with long blade-like snout bearing irregular saw-teeth and long ventral barbels; large dorsal fins without spines, anal fin absent.

Pristiophoriformes Order of squalomorph sharks containing a single family, Pristiophoridae.

privet Oleaceae *q.v.*

proaposematic coloration Warning coloration as exhibited by many venomous or unpalatable animals; *cf.* aposematic coloration.

Proboscidea Order of terrestrial mammals (Ferungulata) comprising the elephants and their extinct relatives; typically large body size

primrose (Primulaceae)

mammoth (Proboscidea)

and showing development of a trunk from the nose and upper lip; teeth reduced, upper jaw often with tusks; includes elephants and mammoths (Elephantidae), mastodons (Mammutidae) and the Deinotherioidea, Moeritherioidea and Gomphotheriidae.

procaryotic Prokaryotic *q.v.*

Procaviidae Hyraxes; family of small terrestrial or arboreal mammals (Hyracoidea) comprising about 10 species; forefeet with 4 digits, soles of hindfeet bearing large elastic adhesive pads to assist in climbing; habits diurnal to nocturnal, colonial, herbivorous; distributed through Ethiopian region to Mediterranean.

hyrax (Procaviidae)

Procellariidae Petrels, fulmars, shearwaters; cosmopolitan family containing about 65 species of medium-sized to large, oceanic sea birds (Procellariiformes) specialized for dynamic soaring close to sea surface; feed on fishes, squid and crustaceans; breed in colonies on islands, having a single egg per clutch.

Procellariiformes Order of neognathous sea birds specialized for soaring flight close to sea surface; comprises 4 families, and includes albatrosses, fulmars and petrels.

Prochlorophycota Division of photosynthetic, alga-like, unicellular, moneran microorganisms found in association with ascidians in tropical seas; characterized by the presence of chlorophyll *a* and *b* but without phycobiliproteins, and by the formation of photosynthetic lamellae from pairs or stacks of thylakoids; Chloroxybacteria.

procryptic Pertaining to coloration and behaviour that affords protection against enemies; **procrypsis**; *cf.* anticryptic.

Procyonidae Raccoons, coatis, red panda;

kinkajou (Procyonidae)

family of arboreal mammals (Carnivora) comprising about 15 species found in the New World and in the Palaearctic (Himalayan red panda); diets include shoots, roots, fruit, small animals and carrion.

producer An organism that synthesizes complex organic substances from simple inorganic substrates; primary producer.

production 1: Gross production, the actual rate of incorporation of energy or organic matter by an individual or population. 2: Net production, that part of assimilated energy converted into biomass through growth and reproduction by an individual or population per unit time.

productivity 1: The potential rate of incorporation or generation of energy or organic matter by an individual or population; rate of carbon fixation; *cf.* production. 2: Often used loosely for the organic fertility or capacity of a given area or habitat.

proepisematic Pertaining to a character or trait that aids social recognition; *cf.* episematic.

profile The vertical sequence of distinct horizontal layers of a soil, sediment or vegetation.

profundal Pertaining to a deep zone of a lake below the level of effective light penetration.

progamous Occurring or existing before fertilization.

progenesis The precocious sexual maturation of an organism that is still at a morphologically juvenile stage.

progeny The offspring of a single mating or of an asexually reproducing individual.

Prokaryotae Superkingdom comprising viruses and the typical moneran forms; predominantly single-celled microorganisms or infectious agents of cells, lacking an organized nucleus and surrounding nuclear envelope in their cells; *cf.* Eukaryotae.

prokaryotic Used of organisms lacking a discrete nucleus separated from the cytoplasm by a membrane, as in bacteria and blue-green algae; **prokaryote, procaryotic**; *cf.* eukaryotic.

Prolecithopora Order of marine and freshwater turbellarians found on mud, algae or interstitially in sediment; characterized by a plicate or variable pharynx, a simple sac-like gut and a common genital atrium receiving both male and female systems.

prominent Notodontidae *q.v.*

promiscuous Pertaining to the mating system in which neither sex is restricted to a single mate.

pronghorn Antilocapridae *q.v.*

propagation 1: Vegetative increase. 2: Sexual

or asexual multiplication; **propagate.**

propagule 1: Any part of an organism, produced sexually or asexually, that is capable of giving rise to a new individual. 2: The minimum number of individuals of a species required for colonization of a new or isolated habitat.

prophylactic Pertaining to the prevention of a disease; **prophylaxis.**

Proplicastomata Small order of free-living turbellarians known only from a single species found off Greenland at a depth of 180 m; characterized by an elongate, plicate pharynx, sac-like gut and by the absence of a statocyst.

Prorocentrales Small order of typically marine unicellular dinoflagellates.

Prosauropoda Extinct order of large, herbivorous dinosaurs (Saurischia) which were partly bipedal; known mainly from the Triassic.

Proscyllidae Family containing 7 species of small (to 1 m), tropical, bottom-dwelling, carcharhiniform cat sharks, found mostly in deep water, to depths of 600 m.

Proseriata Order of predominantly marine free-living turbellarians widely distributed in intertidal habitats, as well as occurring in deeper coastal waters and interstitially; characterized by a plicate pharynx, an unbranched gut, and a distinctive type of statocyst.

prosimian Strepsirhini *q.v.*

Prosimii Subgroup of primates usually known as the Strepsirhini in modern classifications; formerly used for a subgroup comprising the more primitive lemurs, lorises and tarsiers.

Prosobranchia Snails; subclass of gastropod molluscs found in marine, freshwater and terrestrial habitats, and occasionally as parasites; characterized by a typically twisted body enclosed in a spirally coiled shell, by the position of the mantle cavity opening above or near the head and by the muscular creeping foot which usually bears a horny or calcareous plate (operculum) to close the aperture of the shell upon retraction; comprises 3 orders, Archaeogastropoda, Mesogastropoda and Neogastropoda.

Prosopora Small order of aquatic annelid worms (Oligochaeta); typically with paired testes.

Prosotheca Pot worms; small order of oligochaete worms comprising a single family, the Enchytraeidae; usually included within the order Haplotaxida.

Prostigmata Subgroup of acariform mites

(Acari) comprising about 14 000 species in 130 families; exceedingly diverse in terrestrial, freshwater and marine habitats, including fungal and algal feeders, saprophages, phytophages and predators, as well as parasites of vertebrates and invertebrates.

Prostomatida Order of gymnostome ciliates possessing a polyploid macronucleus, and an apical or subapical cytostome surrounded by unspecialized ciliature.

protandrous Pertaining to a hermaphrodite organism that assumes a functional male condition during development before reversal to a functional female state; **protandry**; *cf.* protogynous.

protaspis Planktonic, first larval stage of trilobites, in which the body was covered by a single dorsal carapace.

Protea Proteaceae *q.v.*

Proteaceae Protea, firewheel tree, pincushion flower, silky oak; large family of over 1000 species of evergreen shrubs or trees often possessing characteristic 3-celled hairs; widespread in tropical and subtropical regions, especially South Africa and Australia; flowers usually bisexual with 4 perianth segments, 4 long stamens bearing spoon-shaped anthers, and a superior ovary; flowers grouped into large showy inflorescences.

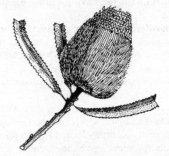

banksia (Proteaceae)

Proteales Order of Rosidae containing only 2 families of mostly woody plants, Proteaceae and Elaeagnaceae.

Proteidae Mud puppies, olm; small family of aquatic salamanders (Caudata); body length to 400 mm; metamorphosis incomplete; fore- and hindlimbs, lungs, and external gills present; fertilization internal, males producing spermatophores; eggs typically laid under stones in streams; contains 5 North American species and a single subterranean species from Europe.

protein Large complex organic molecule composed of one or more polypeptide chains each of which consists of a sequence of amino acids.

protelean Used of an organism that is parasitic when immature and free-living as an adult.

Proteocephalidea Order of tapeworms parasitic in the gut of fishes, amphibians and reptiles, and utilizing copepod crustaceans as intermediate hosts; characterized by a scolex with 4 suckers commonly covered with minute spines, the presence of a neck and by the segmented body.

Proteromonadida Small order of zooflagellates with 2 or 4 flagella, one of which is recurrent and arises from the anterior pole of the body; found as parasites in the guts of amphibians, reptiles and mammals.

Proterozoic A geological period (*c.* 2400–570 million years B.P.); the late Precambrian era; Algonkian, Agnotozoic; see geological time scale.

Proteutheria Suborder of insectivores comprising the modern tree shrews (Tupaiidae) and a variety of extinct forms known from fossils in the Cretaceous.

prothallus The free-living gametophyte generation of ferns and certain other lower vascular plants.

prothetely Accelerated sexual maturation in insects resulting in the production of an imago exhibiting larval or pupal characters; **prothetelic.**

Protista Kingdom of eukaryotic microorganisms comprising protozoans, some algae and fungi with flagellated spores which are classified in up to 30 separate phyla; typically possessing a true flagellum (9 + 2 undulipodium); mitochondria and plastids often present.

protistan Pertaining to the eukaryotic organisms of the kingdom Protista, including some algae, fungi and protozoans; formerly used only of unicellular eukaryotic organisms; *cf.* moneran.

protistology The study of protistans.

Protoalcyonaria Order of deep-water solitary octocorals that reproduce sexually and possess only one type (morph) of polyp.

Protobranchia Subclass of marine bivalve molluscs containing about 500 species typically found living in the sediment and feeding on detritus; characterized by 2 shell-closing muscles and by protobranch gills consisting of simple filaments with plate-like leaflets on both sides; comprises 2 orders, Nuculoida and Solemyoida.

Protococcidiida Small, primitive order of coccidians in which merogony is absent; found in marine annelids.

Protoctista Kingdom of eukaryotic organisms which are defined by the lack of a blastula, an embryo or spores in the life cycle and by the lack of true flagella (9+2 undulipodia).

Protodrilida Order of interstitial polychaete worms found on coarse, shallow-marine sediments, comprising about 45 species in 2 families; body depressed and ciliated; prostomium and peristomium with pair of tentacles, pharynx non-eversible; parapodia reduced.

Protogastrales Order of gasteromycete fungi containing 3 species each in a separate family; characterized by a small fruiting body with a single hymenial cavity; typically found growing on plant roots and rabbit dung.

protogynous Pertaining to a hermaphroditic organism that assumes a functional female condition first during development before changing to a functional male state; **protogyny**; *cf.* protandrous.

Protomedusae Diverse class of extinct primitive cnidarians having bodies swollen into radial pouches separated by radial grooves; known from the Precambian to Ordovician.

Protomycetales Order of hemiascomycete fungi comprising a single worldwide family of parasites causing galls or other lesions in tissues of flowering plants.

protonema The typically filamentous juvenile form of a moss or liverwort which develops on germination of a spore.

protonymph Immature stage of mites (Acari) possessing 4 pairs of legs, and relatively simple chaetotaxy and genitalia; formed from the larval stage that has only 3 pairs of legs; proteronymph.

Protopteridae African lungfish; family containing 4 species of freshwater fishes found in weedy sluggish backwaters of central Africa; body elongate, cylindrical with paired swim bladders functioning as a lung; pectoral and pelvic fins long and filamentous; larvae with 4 pairs of feathery external gills; adults aestivate in dry season.

African lungfish (Protopteridae)

Protostelia Subclass of slime moulds (Eumycetozoa) which range from uninucleate amoeboid forms to multinucleate reticulate plasmodia, found on dead plants, dung and decaying organic matter; characterized by minute, stalked fruiting bodies bearing usually 1–4 spores.

Protosteliomycetes The Protostelia *q.v.* treated as a group of plasmodial slime moulds (Myxomycota).

protostomes Group of invertebrate animals linked by embryological characters such as spiral cleavage and formation of a coelom by splitting of the mesoderm (schizocoelic); includes annelids, molluscs and arthropods.

Prototheria Subclass of egg-laying mammals comprising a single Recent order, Monotremata, with 2 living families, Ornithorhynchidae (platypus) and Tachyglossidae (echidna); primitive features include retention of cloaca, unfused uteri, shell glands, and abdominal testes; female incubates eggs, the young feeding on milk from mammary glands that lack nipples.

Protoungulata Superorder of placental mammals (Ferungulata) including the earliest known ungulates from the late Cretaceous, which began to specialize for a herbivorous diet; comprises the orders Condylarthra, Litopterna, Astrapotheria, Notoungulata and Tubulidentata.

prototrophic 1: Autotrophic *q.v.*; protrophic. 2: Obtaining nourishment from a single source only; **prototrophism.** 3: Used of microorganisms capable of growth on simple sugars and salts without specific additional nutrient requirements.

Protozoa Diverse group of unicellular, eukaryotic organisms typically found as free-living heterotrophs in all types of habitat and as symbionts and parasites of other organisms; typically possess a single nucleus, sometimes with 2 or many nuclei; reproduction usually by simple binary fission although more elaborate cycles involving sexual phenomena occur in many protozoan groups; some forms are sessile, attached to the substratum by a stalk, some construct lorica or test to live in, others are colonial; they may be classified within the kingdom Protista, or as a subkingdom of the kingdom Animalia; includes at least 7 phyla, Apicomplexa, Ascetospora, Ciliophora, Labyrinthulata, Microspora, Myxozoa and Sarcomastigophora.

protozoea The early zoeal phase of larval development found in crustaceans.

protozoology The study of Protozoa.

protozoophilous Pollinated by Protozoa or other microscopic animals, as in some aquatic plants; **protozoophily.**

provisioning Providing food for young.

Protura Class of primitively wingless, white insects which lack eyes and antennae and have reduced mouthparts; body elongate with 3 pairs of jointed legs on the thorax and 3 pairs of rudimentary limbs (styli) on the abdomen; contains about 120 species found worldwide under bark, stones or in rotting vegetation.

proturan (Protura)

prowfish Pataecidae *q.v.*

Prunellidae Hedge sparrows; family containing 12 species of small Eurasian passerine birds found in woodland, meadow, and montane habitats; bill slender and pointed, wings short and rounded; habits solitary, terrestrial, often migratory, feeding on ground insects, seeds and fruit; nest on ground, in crevices or in bushes.

Prymnesiales Order of mostly marine, planktonic flagellate algae (Prymnesiophyceae) which are covered in unmineralized organic scales.

Prymnesiophyceae Haptophytes; class of mostly marine flagellate chromophycote algae typically unicellular and possessing 2 equal flagella; characterized by the presence of chlorophylls *a* and *c*, and xanthophyll pigments which produce a golden brown coloration; usually covered with organic scales, which may be calcified to form coccoliths around the cell; comprises 4 orders, Coccosphaerales, Isochrysidales, Pavlovales and Prymnesiales; also treated as a phylum of Protoctista under the name Haptophyta, and as an order of the protozoan class Phytomastigophora under the name Haptomonadida.

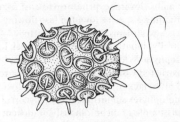

haptophyte (Prymnesiophyceae)

psamathophilous Thriving in the strandline of a sandy seashore; **psamathophile, psamathophily.**

psamathophyte A strandline plant of a sandy seashore.

Psamminida Widely distributed order of xenophyophores distinguished by the general rigidity of the body and by the absence of organic threads in the test.

psammobiontic Living in sand; living interstitially between or attached to sand particles.

Psammodrilida Order of small (to 9 mm) interstitial neotenic polychaete worms that inhabit mucus tubes covered with sand grains in shallow water; typically feeding on benthic diatoms; body segmentation indistinct; prostomium and peristomium fused, naked; pharynx non-eversible; parapodia reduced.

psammofauna Those animals associated with a sandy substratum, or living within a sandy area.

psammolittoral Pertaining to a sandy shore.

psammon Those organisms living in, or moving through, sand; the interstitial flora and fauna; *cf.* benthos.

psammophilous Thriving in sandy habitats; **psammophile, psammophily.**

psammophyte A plant growing or moving in unconsolidated sand.

psammosere An ecological succession commencing on an area of unconsolidated sand.

Psettodidae Family of primitive marine flatfishes (Pleuronectiformes) containing 3 species found in west Africa and the Indo-

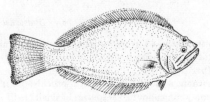

flatfish (Psettodidae)

Pacific; body length to 700 mm; median fins spinous, pectorals present; vomerine and palatine teeth present; typically use caudal fin for locomotion.

pseudaposematic coloration Warning coloration which takes the form of a bluff, as in Batesian mimicry; *cf.* aposematic coloration.

pseudepisematic Pertaining to a character or trait aiding recognition but involving deception; *cf.* **episematic**.

Pseudidae Family containing 4 species of fully aquatic frogs (Anura) found in tropical lowlands of South America; hindlimbs strong, feet webbed; tadpoles extremely large, up to 250 mm in length.

Pseudocarchariidae Crocodile shark; monotypic family of small (about 1 m) and little-known lamniform elasmobranch fishes; head and snout conical, eyes large; distribution circumtropical, in oceanic waters from surface to 300 m.

pseudocarp False fruit; a fruit incorporating tissues other than those derived from the gynoecium *q.v.*

Pseudochromidae Dottybacks; family of small colourful, slender-bodied Indo-Pacific perciform teleost fishes found in shallow waters over reefs and rocky bottoms; dorsal fin elongate bearing only 3 spines.

pseudocopulation A mode of pollination in which a male insect attempts to copulate with a flower that resembles a female insect, and in the process transfers pollen from one flower to another.

pseudogamy The activation of an egg by a male gamete that degenerates without its nucleus fusing with that of the egg.

pseudomycorrhizal association A symbiosis between a fungus and a higher plant in which the fungus is parasitic.

Pseudophyllidea Widely distributed order of tapeworms parasitic in fishes, amphibians, birds and mammals, and often utilizing crustaceans as intermediate hosts; characterized by a scolex with dorsal and ventral grooves (bothria) which may be fused along their lengths.

Pseudoplesiopidae Small family of marine perciform teleost fishes with an elongate body, large eyes and rounded fins; commonly found on coral reefs between Australia and Indonesia.

Pseudorajidae Family containing 4 species of small (to 0.5 m) rajiform fishes found in the tropical western Atlantic at depths of 250–

1000 m; dorsal fin absent, caudal fin bearing both dorsal and ventral lobes.

Pseudoscorpiones False scorpions; order of small (to 7 mm) predatory terrestrial arthropods resembling tiny scorpions but without slender tail and sting; chelicerae large and chelate, often bearing a spinneret used to construct silk cocoons; pedipalps often with venom gland; contains about 2000 species, most abundant in tropical regions; Chelonethida.

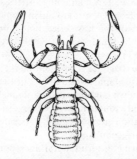

false scorpion (Pseudoscorpiones)

Pseudoscorpionida Pseudoscorpiones *q.v.*
pseudosematic Used of a character or trait that serves as a false danger signal or warning.
Pseudotriakidae False cat sharks; family containing 2 species of large (to 3 m) carcharhiniform elasmobranch fishes, found in deep waters (300–1500 m) of boreal to tropical seas.
pseudozoea Early free-swimming larval stage of some stomatopods possessing biramous antennules, 2 pairs of uniramous thoracic limbs and 4 or 5 pairs of functional pleopods.
psicolous Living in prairie or savannah habitats; **psicole.**
psilic Pertaining to savannah or prairie communities or habitats.
psilicolous Living in savannah or prairie habitats; **psilicole.**
Psilidae Rust flies; family containing about 200 species of small slender-bodied flies (Diptera) usually found in habitats shaded by dense vegetation; larvae feed on plant stems and roots, and some, such as the carrot rust fly, are important horticultural pests.
psilopaedic Used of birds that are naked when hatched; *cf.* ptilopaedic.
psilophilous Thriving in prairie or savannah habitats; **psilophile, psilophily.**
psilophyte A prairie or savannah plant.
Psilophytopsida Psilotophyta *q.v.*
Psilorhynchidae Family of tiny (to 50 mm)

Indian freshwater cypriniform teleost fishes containing 3 species found in fast-flowing rivers and streams; body deep, flattened ventrally, mouth subterminal with thick lips.

Psilotophyta Whiskferns; division of simply constructed vascular plants lacking true roots and often lacking leaves; usually epiphytic; gametophytes subterranean and mycorrhizal; stems simple or dichotomously branched; sporangia large, each having 2 or 3 chambers; homosporous; consists of 2 genera in a single order, Psilotales; Psilophytopsida is an alternative name.

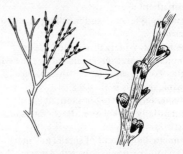

whiskfern (Psilotophyta)

Psittacidae Parrots, cockatoos, macaws; large family of small to medium-sized birds with strong hooked bills and fleshy tongues; plumage often brightly coloured; habits gregarious, monogamous, sometimes nocturnal; feed mainly on fruit, also on pollen and nectar; contains about 340 species, widely distributed in tropical regions of Australia and South America.

parrot (Psittacidae)

Psittaciformes Diverse order of mainly arboreal birds comprising a single family, Psittacidae (parrots).
Psocoptera Book lice, bark lice, psocids; order containing about 2600 species of small (1–10 mm) hemipterodean insects found in diverse

habitats such as leaf litter, beneath bark, stones and other debris, and in stored products where they may become pests; compound eyes larger in males than females; wing venation reduced; polymorphism common, involving reduction or loss of wings in either or both sexes; parthenogenesis obligatory or facultative; psocids feed mainly on microflora.

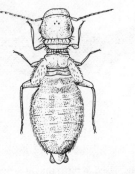

psocid (Psocoptera)

Psophiidae Trumpeters; family containing 3 species of medium-sized wading birds (Gruiformes) found in wet forests of South America; neck and legs long, wings small; feed on insects and plant material; nest in tree holes.

Psychodidae Sand flies; family containing about 500 species of dipteran insects; adult body small with a dense covering of hairs; mandibles typically absent; one bloodsucking group is responsible for transmitting leishmaniasis in man.

psychophilous Pollinated by diurnal Lepidoptera; **psychophile, psychophily.**

psychric Pertaining to low temperatures or cold habitats.

psychrocleistogamic Used of plants in which self-pollination takes place within flowers that remain closed because of abnormally cold conditions.

psychrokliny Growth behaviour influenced by low temperatures.

Psychrolutidae Family containing 7 species of small (to 60 mm) marine sculpins (Scorpaeniformes) found in the Pacific Ocean; body naked, compressed posteriorly, rounded anteriorly with large head and mouth; skin loose and spinulose.

psychrophilic 1: Thriving at low temperatures; **psychrophile, psychrophily.** 2: Used of microorganisms having an optimum for growth below 20°C.

psychrophyte A plant living in relatively cold polar or alpine habitats or on a cold substratum.

Psyllidae Jumping plant lice, psyllids; family containing about 1300 species of small, jumping insects (Homoptera) resembling cicadas; usually host-specific on various plants, feeding by sucking sap; nymphs live in galls or free on the plant, often producing honeydew or protective wax.

ptenophyllophilous Thriving in deciduous forest; **ptenophyllophile, ptenophyllophily.**

ptenophyllophyte A deciduous woodland plant.

ptenothalophilous Thriving in deciduous thickets; **ptenothalophile, ptenothalophily.**

ptenothalophyte A deciduous thicket plant.

Pteraspidimorphi Class of eel-like, jawless vertebrates (Agnatha) which lack paired fins, possess 2 semicircular canals and lack a true bony skeleton; includes the fossil pteraspids and the hagfish (Myxiniformes).

pteridology The study of ferns.

Pteridophyta Ferns; commonly classified under the name Filicophyta *q.v.*; pteridophytes.

Pteridophytic Palaeophytic *q.v.*; the age of ferns.

Pteridospermales Seed ferns; extinct order of primitive seed plants (Pinophyta) that flourished during the Carboniferous.

Pterioida Pearl oysters, hammer oysters; order comprising about 100 species of mainly marine pteriomorphian bivalve molluscs; characterized by detailed shell and hinge structure, and by the posterior shell closing muscle being larger than the anterior.

Pteriomorphia Large subclass of mainly marine, sedentary bivalve molluscs often attached to the substratum by byssus apparatus; comprises the Eupteriomorphia, Isofilobranchia and Prionodonta.

Pterobranchia Class of littoral to abyssal, tubicolous, tentaculate hemichordates; groups of individuals (zooids) termed a coenecium; body sac-like, gut U-shaped, mesosome with anteroventral mouth and up to 9 pairs of dorsal tentaculate arms; nerve chord absent, coelom confined to tentacles and metasome; development takes place without free larval stage, and asexual reproduction by budding is common.

Pteroclididae Sandgrouse; family containing about 15 species of small grouse-like terrestrial columbiform birds found in grassland and deserts of Africa and Eurasia; habits gregarious, monogamous, solitary when breeding, feeding mainly on fruit and insects.

pteropod ooze A calcareous pelagic sediment containing at least 30% calcium carbonate in the form of skeletal remains, predominantly of pteropod molluscs; *cf.* ooze.

Pteropodidae Family of large tree-dwelling fruit bats found in the Old World tropics; contains about 150 species in 2 subfamilies, one frugivorous (flying foxes), the other mainly nectivorous (long-tongued fruit bats).

pteropods Marine gastropod molluscs belonging to the orders Gymnosomata *q.v.* and Thecostomata *q.v.*

Pterosauria Pterosaurs; extinct order of flying reptiles most widespread during the Jurassic; probably lived around sea coasts feeding on fishes.

pterosaur (Pterosauria)

Pterospermatales Small order of marine flagellates (Prasinophyceae) characterized by the production of spheroidal cyst-like cells (phycomata) during the life cycle.

Pterygota Subclass of winged or secondarily wingless insects comprising all but a tiny minority of insect species; adult, terminal, or reproductive stage typically bearing 1 or 2 pairs of wings on meso- and metathorax; in many groups the wings are modified, in either or both sexes, as for example the shield-like elytra of beetles (Coleoptera) or the balancing halteres of flies (Diptera), or may be absent; fertilization usually internal; comprises 2 infraclasses, Paleoptera and Neoptera.

Pthiridae Monogeneric family of sucking lice (Anoplura) comprising 2 species, one found on gorillas, the other on man (the pubic or crab louse).

Ptilichthyidae Quillfish; monotypic family of northeastern Pacific, blennioid teleost fishes (Perciformes); body extremely slender, naked, to 350 mm in length; dorsal fin bearing numerous (up to 90) isolated hooked spines; pelvic fins and caudal fin absent.

Ptilonorhynchidae Bowerbirds; family containing 18 species of medium-sized passerine birds found in forests of Australia and New Guinea; males court females in forest clearings or in a specially constructed bower of twigs decorated with coloured objects; habits solitary to gregarious, non-migratory, monogamous to polygamous, feeding on ground invertebrates and fruit; the shallow nest is usually placed in a tree.

bower bird (Ptilonorhynchidae)

ptilopaedic Used of birds that are covered with down when hatched; *cf.* psilopaedic.

Ptinidae Spider beetles; family containing about 700 species of small spider-like beetles (Coleoptera); larvae white and fleshy with tiny legs; common pests in homes and warehouses, occurring naturally in the nests of birds and mammals, and in ant nests.

Ptychodactiaria Small order of deep-water solitary zoantharians resembling sea anemones; possess a definite base but no basilar muscles.

puberty The onset of sexual maturity.

puerulus Late (megalopal) larval stage of palinuroid decapod crustaceans.

puff ball 1: Lycoperdales *q.v.* 2: Sclerodermatales *q.v.*

puffbird Bucconidae *q.v.*

puffer Tetraodontidae *q.v.*

puffin Alcidae *q.v.*

Pulmonata Snails; subclass of predominantly terrestrial and freshwater gastropod molluscs generally possessing a spirally coiled shell typically without an operculum; characterized by the modification of the mantle cavity into a vascularized, pulmonary cavity for aerial

garden snail (Pulmonata)

respiration; divided into 4 orders,
Archaeopulmonata, Basommatophora,
Stylommatophora and Systellommatophora.

puma Felidae *q.v.*

pumpkin Cucurbitaceae *q.v.*

Punicaceae Pomegranate; family of Myrtales
containing only 2 species of often thorny
shrubs or trees; flowers brilliant scarlet in
colour, with 5–8 sepals and petals, and many
stamens; forming a fruit with a tough, leathery
rind enclosing many seeds, each with a fleshy
aril.

punky Ceratopongonidae *q.v.*

pupa The life cycle stage during which the larval
form is reorganized to produce the final, adult
form; commonly an inactive stage enclosed
within a hard shell (chrysalis) or silken
covering (cocoon).

pupate To change from larva to pupa.

pupigerous Containing a pupa.

pupiparous Producing offspring already at the
pupal stage, or larvae ready to pupate.

pupivorous Feeding on pupae; **pupivore,
pupivory.**

purple glory tree Melastomataceae *q.v.*

purple loosestrife Lythraceae *q.v.*

putrefaction The decomposition of organic
matter with incomplete oxidation, producing
methane and other gaseous products subject
to further oxidation; *cf.* decay, mouldering.

puya Bromeliales *q.v.*

pycnocline A zone of marked density gradient.

Pycnogonida Sea spiders; class of marine
chelicerate arthropods containing about 1000
species found in benthic habitats from the
intertidal to abyssal depths, commonly in
association with hydroids, sea anemones and
algae; body typically small with long legs; feed
as succivorous predators; Pantopoda.

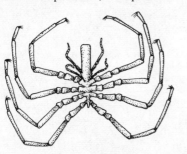

sea spider (Pycnogonida)

Pycnonotidae Bulbuls; family containing about
120 species of small Old World passerine birds
found in woodland and scrubland habitats
from Africa to southeast Asia and Japan;

habits gregarious, sedentary, arboreal, non-
migratory, feeding on fruit and insects; nest in
tree or bush; some species have been
introduced into the New World.

Pygasteroida Extinct order of irregular
echinoids known from the Jurassic and
Cretaceous.

pygmy phalanger Burramyidae *q.v.*

Pygopodidae Snake lizards; family containing
about 30 species of limbless lizards (Sauria)
found in Australia and New Guinea; body
snake-like with large scales; length to 250 mm;
forelimbs absent, hindlimbs reduced to minute
flaps; habits mostly terrestrial, nocturnal and
insectivorous, some species fossorial.

Pyralidae Family of small to large, often
slender, moths (Lepidoptera) the larvae of
which are phytophages, scavengers, or rarely
parasites on other insects; contains over 20 000
species including many important agricultural
pests that bore into grain, or damage leaves of
trees and other crops.

Pyramimonadales Order of aquatic flagellates
(Prasinophyceae) that are covered with one to
several layers of organic scales and possess 2–4
flagella.

Pyrenomycetes Class of ascomycotine fungi,
characterized by unitunicate, sometimes
evanescent, asci and by an ascocarp which
develops from an initial coiled hypha
(ascogonium); the ascogonium becomes
enveloped by other hyphae which form the
wall of the ascocarp.

Pyrocystales Small order of pelagic marine,
photosynthetic unicellular dinoflagellates
many of which are intensely bioluminescent.

Pyrolaceae Wintergreen; small family of
Ericales containing about 45 species of
strongly mycotrophic perennial herbs or half
shrubs; commonly growing in acid soils in cool
regions of the northern hemisphere.

wintergreen (Pyrolaceae)

pyrophilous Thriving on ground that has
recently been scorched by fire; **pyrophile,
pyrophily.**

pyrophobic Intolerant of the soil conditions produced by fire; used of a plant that is unable to re-establish itself following a fire; **pyrophobe**, **pyrophoby**.

pyrophyte A plant resistant to permanent damage by fire.

Pyrosomida Order of compound, barrel-shaped, pelagic tunicates (Thaliacea) in which the colony may be several metres in length and strongly bioluminescent; zooids have inhalent siphons on external wall and exhalent siphons opening into chamber of barrel, the anterior end of which is closed; colony formed by budding from primary sexually produced zooid, larval stage absent.

Pyrotheria Extinct order of elephant-like South American ungulates (Ferungulata) known from the Oligocene.

pyroxylophilous Thriving on burnt wood; **pyroxylophile**, **pyroxylophily**.

pythmic Pertaining to lake bottoms.

python Boidae *q.v.*

q

quadrat 1: A delimited area for sampling flora or fauna; usually taken randomly within the study area and typically consisting of a 1 metre square frame. 2: The sampling frame itself.

quadrature The time at which the sun and moon are approximately at right angles with respect to the Earth; *cf.* syzygy.

quadrivoltine Having four broods per year or per season; *cf.* voltine.

quagga Equidae *q.v.*

quail Phasianidae *q.v.*

qualitative Descriptive; non-numerical.

quantitative Numerical; based on counts, measurements, ratios or other values.

Quaternary A geological period of the Cenozoic (*c.* 1.6 million years B.P. to present), comprising the Pleistocene and the Holocene; see geological time scale.

Quaternary Ice Age A period in the Quaternary when glaciation was much more widespread than at present; Pleistocene glaciation.

queen substance A pheromone or combination of pheromones produced by a queen to control the reproductive behaviour of workers and other queens; trans-9-keto-2-decenoic acid is the most potent of the set of such pheromones used by a queen.

queenright Used of a colony of honeybees that contains a functional queen.

Questida Monofamilial order of polychaete worms.

quiescence 1: A temporary resting phase characterized by reduced activity or cessation of development; **quiescent**. 2: The condition of seeds that have not germinated because one or more environmental variables is unfavourable; *cf.* dormancy.

Quiinaceae Small family of Theales, native to tropical America, containing about 40 species of mostly trees or shrubs with mucilage channels.

quillfish Ptilichthyidae *q.v.*

quillwort Isoetales *q.v.*

Estimated duration of Glacial Period (1000 years)	The Alpine area	British Isles	North Germany and Poland	North America
70	Würm 3 Würm 2 Würm 1	Devensian	Weichselian: Pomeranian Frankfurt Brandenburgh	Wisconsin Iowan
Interglacial Period	Riss-Würm	Ipswichian	Eemian	Sangamonian
100	Riss	Wolstonian	Warthe Saale	Illinoian
'Great' Interglacial Period	Mindel-Riss	Hoxnian	Holsteinian	Yarmouthian
90	Mindel	Anglian	Elsterian	Kansan
Interglacial Period	Günz-Mindel	Cromerian		Aftonian
100	Günz	Beestonian	Elbe	Nebraskan
Interglacial Period		Pastonian		
260	Donau	Baventian		Pre-Nebraskan

chronology of the Quaternary Ice Age

r

r selection Selection favouring a rapid rate of population increase, typical of species that colonize short-lived environments or of species that undergo large fluctuations in population size; *cf. K* selection.

rabbit Leporidae *q.v.*

rabbitfish 1: Siganidae *q.v.* 2: Chimaeriforms *q.v.*

raccoon Procyonidae *q.v.*

race A subgroup of a species characterized by conspicuous physiological (physiological race), biological (biological race), geographical (geological race) or ecological (ecological race) properties.

racehorse Congiopodidae *q.v.*

raceme An inflorescence in which the main axis continues to grow, producing flowers laterally the youngest flowers are apical or central; racemose inflorescence.

raceme

rachion The zone on the shore of a lake subject to pronounced wave action.

Radiata Cnidaria *q.v.*

radicicolous Growing on or in roots; **radicicole.**

radicivorous Feeding on roots; **radicivore, radicivory.**

Radiolaria Loose assemblage of marine pelagic

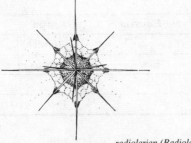

radiolarian (Radiolaria)

protozoans (Actinopoda) comprising the classes Polycystina and Phaeodaria; may also include the Acantharia.

radiolarian ooze A pelagic siliceous sediment comprising at least 30% siliceous material, predominantly in the form of radiolarian tests; *cf.* ooze.

radiotropism A growth effect produced by radiation; **radiotropic.**

radish Brassicaceae *q.v.*

Rafflesiaceae Family of plants lacking chlorophyll which are endoparasitic in the roots or shoots of other plants and vegetatively resemble a fungal mycelium; flowers typically produced terminally on short fleshy shoots; includes *Rafflesia arnoldi* from Sumatra which has the largest flower of any flowering plant.

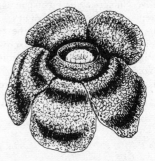

Rafflesiales Small order of Rosidae containing about 60 species of fleshy parasitic herbs lacking chlorophyll, often with only the flowers or inflorescence emerging from the host.

rafting Dispersal of terrestrial organisms across water on floating objects.

ragfish Icosteidae *q.v.*

ragweed Asterales *q.v.*

ragworm Phyllodocida *q.v.*

rail Rallidae *q.v.*

rail babblers Orthonychidae *q.v.*

rain forest A tropical forest having an annual rainfall of at least 254 cm (100 inches).

rain shadow An area on the leeward side of a mountain or range of mountains having a significantly lower rainfall than an area on the windward side.

rain shadow desert An area of desert or arid land in the rain shadow on the leeward side of a mountain range.

rainbowfish Melanotaeniidae *q.v.*

rainfall The amount of water precipitated from the atmosphere over a given period of time.

Rajidae Family containing about 175 species of shallow littoral to abyssal rajiform fishes; body disk subcircular to rhomboidal, tail slender;

dorsal fins close to tip of tail, caudal fin small to minute.

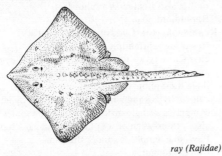

ray (Rajidae)

Rajiformes Skates, rays; order containing about 200 species of bottom-dwelling elasmobranch fishes (Batoidea); body strongly depressed, length to 2 m, pectoral fins forming subcircular to rhomboidal disk, typically broader than long; tail slender to whip-like, dorsal and caudal fins small or absent; habits littoral to abyssal, feeding on benthic invertebrates and fishes; reproduction oviparous, egg capsules having adhesive filaments for attachment.

Rallidae Rails, coots, gallinules; diverse family containing about 140 species of small to medium-sized gruiform birds found in wet and woodland habitats worldwide; most are terrestrial or aquatic with weak flight, several island species are flightless; habits solitary, secretive, often crepuscular, monogamous, feeding on variety of plant and animal material; nest on ground, on water, or in bushes.

coot (Rallidae)

ramapithecines Extinct group of Miocene apes that lived about 14–8 million years ago, known from fossil fragments found in Africa, the Near East and Asia; hominid affinities uncertain, having jaw and teeth features that link them to chimpanzees and gorillas as well as to man.

ramet A member or modular unit of a clone, that may follow an independent existence if separated from the parent organism; *cf.* genet, ortet.

ramicolous Living on twigs and branches; **ramicole.**

ramification A branch or dichotomy; branching.

Ramphastidae Toucans; family of medium-sized to large Neotropical forest birds (Piciformes); bill variable, small to very large; habits gregarious, non-migratory, monogamous, feeding on fruit and small animals; nest in tree hole; contains about 30 species found from Mexico to Argentina.

toucan (Ramphastidae)

ram's horn Sepioidea *q.v.*

ramshorn snail Basommatophora *q.v.*

range 1: The limits of the geographical distribution of a species or group. 2: Home range *q.v.*

Rangoon creeper Combretaceae *q.v.*

Ranidae Large family of riparian frogs (Anura); most species with aquatic eggs and larvae found in still waters, some in streams, others with terrestrial eggs and direct development; tadpoles with denticles, beak and single sinistral spiracle, barbels absent; contains about 590 species, including the edible frog; distribution largely cosmopolitan except for much of Australia and parts of South America.

ranivorous Feeding on frogs; **ranivore, ranivory.**

Ranunculaceae Buttercup, larkspur, hellebore, delphinium, anemone, clematis; diverse family of about 2000 species including many ornamental garden plants; mostly herbaceous perennials with typically bisexual flowers in which the carpels are usually distinct, forming simple pistils with several to many marginal

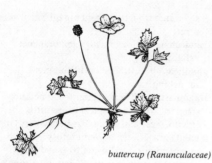

buttercup (Ranunculaceae)

ovules; widely distributed in temperate and boreal regions.

Ranunculales Large order of dicotyledons (Magnoliidae) containing about 3200 species of herbs, shrubs or vines with variable flower morphology.

rapacious Feeding on prey.

Rapateaceae Small family of Commelinales containing nearly 100 species of often coarse perennial herbs with basal, parallel-veined leaves; flowers with 3 petals, 3 sepals, 6 stamens and 3 united carpels; native to tropical South America.

rape Brassicaceae *q.v.*

Raphidae Dodos; family containing 3 species of large flightless pigeons (Columbiformes) from the Mascarene Islands in the Indian Ocean; extinct since the end of the 17th century as result of the activitites of man.

dodo (Raphidae)

Raphidiidae Snakeflies; family containing about 100 species of mostly predatory neuropteran insects that feed on soft-bodied plant lice or on dead invertebrate remains; characterized by an elongate and highly mobile thorax.

Raphidioideae Order of pennate diatoms (Pennales) possessing a primitive type of raphe; also treated as a suborder, Raphidioidineae.

Raphidophyceae Class of predominantly freshwater chromophycote algae; characterized by biflagellate cells, with a hairy anterior flagellum and naked posterior flagellum; photosynthetic pigments include chlorophylls *a* and *c* and various xanthophylls; comprises a single order, Raphidomonadales; also treated as an order of the protozoan class Phytomastigophora, under the name Chloromonadida.

raptorial Adapted for seizing prey.

rasorial Adapted for scratching or scraping the ground.

raspberry Rosaceae *q.v.*

rat 1: Cricetidae *q.v.* (New World rats). 2: Muridae *q.v.* (Old World rats).

rat chinchilla Abrocomidae *q.v.*

rat opossum Caenolestidae *q.v.*

ratfish Chimaeriformes *q.v.*

ratites A loose assemblage of flightless, running birds which lack a keel on the sternum.

rattail Macrouridae *q.v.*

rattlesnake Viperidae *q.v.*

Raunkiaerian leaf size classes A system of classifying leaf sizes according to surface area; the 6 categories are leptophyll (up to 25 mm^2), nanophyll (25–225 mm^2), microphyll (225–2025 mm^2), mesophyll (2025–18 225 mm^2), macrophyll (18 225–164 025 mm^2), megaphyll (over 164 025 mm^2); the category notophyll has been added for a small mesophyll (2025–4500 mm^2).

Raunkiaerian life forms A system of classification of the life forms of plants based on the type and position of the renewal buds with respect to ground level; categories include chamaephyte, cryptophyte, geophyte, helophyte, hemicryptophyte, hydrophyte, phanerophyte, therophyte.

raw data Data which have not been organized numerically or treated statistically.

ray Rajiformes *q.v.*

razor shell Veneroida *q.v.*

reaction Any change in the activity of an organism in response to a stimulus.

reaction time The time interval between a stimulus and the response to that stimulus.

realized niche That part of the fundamental niche *q.v.* actually occupied by a species.

realm The largest of the biogeographical units, encompassing major climatic or physiographic zones.

recapitulation The repetition of a sequence of ancestral adult stages in the embryonic or juvenile stages of descendants.

Recent Holocene *q.v.*

recent Still existing; extant; *cf.* fossil.

recessive An allele that is not expressed in the phenotype except when homozygous; *cf.* dominant.

recessiveness The failure of one allele (recessive) to be expressed over another (dominant) in the phenotype of the heterozygote; *cf.* dominance.

recombination Any process that gives rise to a new combination of hereditary determinants, such as the reassortment of parental genes during meiosis through crossing over.

recretion The elimination of mineral salts from plants in the same form in which they were taken up.

recruitment The influx of new members into a population by reproduction or immigration.

rectilinear Used of growth or movement that proceeds in a straight line.

Recurvirostridae Avocets, stilts; family containing 11 species of medium-sized, long-legged, charadriiform shore birds, cosmopolitan in tropical and temperate coastal and marsh habitats; bill characteristically long and slender, upcurved or straight; habits gregarious, often migratory, monogamous, feeding on invertebrates and small vertebrates.

red algae Rhodophycota *q.v.*

red bud Caesalpiniaceae *q.v.*

red clay A pelagic sediment comprising an accumulation of volcanic and aeolian dust with less than 30% biogenic material.

Red Desert soil A zonal soil with a light reddish brown A-horizon overlying a heavy reddish alkaline B-horizon above calcareous material formed under warm temperate to dry tropical climates.

red:far red ratio The proportion of red light to far red light in the radiation spectrum available to a plant.

red light Solar radiation in the spectral range 620–680 nm.

red mangrove Rhizophorales q.v.

red mud A well oxidized terrigenous marine sediment having a red coloration produced by ferric oxide.

red mullet Mullidae *q.v.*

red panda Procyonidae *q.v.*

red tide A reddish brown coloration of sea water caused by a bloom of dinoflagellates, usually belonging to the Peridiniales *q.v.*

Reddish Brown lateritic soil A zonal soil, highly weathered, with a thin litter and duff layer, reddish brown slightly acidic A-horizon over a red or reddish brown clayey, strongly acidic, B-horizon, formed in forested warm temperate and tropical regions.

Reddish Brown soil A zonal soil with a reddish brown surface layer over a darker red horizon and a light calcareous lower horizon, formed in warm, temperate to tropical, semi-arid climates beneath short grass and shrub vegetation.

Reddish Chestnut soil A zonal soil with a thick red to dark brown A-horizon over a heavy reddish brown B-horizon and a calcareous lower horizon, formed in warm temperate, semi-arid climates beneath mixed grasses and shrub vegetation.

Reddish Prairie soil A zonal soil with a dark reddish brown acidic A-horizon grading through a reddish brown soil to the parent material, formed in warm to temperate, humid to subhumid, climates under tall grasses.

redfish Scorpaenidae *q.v.*

red-hot poker Aloeaceae *q.v.*

redia Sac-like larval stage of a digenean trematode parasite found within the molluscan intermediate hose, that asexually produces infective cercariae *q.v.*.

redox potential (Eh) Oxidation–reduction potential; a measure of the tendency of a given system to act as an oxidizing (electron acceptor) or reducing (electron donor) agent.

reducer Any heterotrophic organism responsible for degrading or mineralizing organic matter.

reduction The process of producing or evolving simpler structures from more complex ones.

reduction division The first division of meiosis in which the chromosome complement is halved.

Reduviidae Assassin bug, kissing bug; family

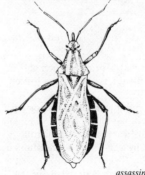

assassin bud (Reduviidae)

containing over 4000 species of bugs (Heteroptera); most are predators of other arthropods which are seized with the front legs; some are blood suckers of mammals and birds and may act as vectors of diseases such as Chagas'.

Red-Yellow podzolic soil An acidic zonal soil with little or no litter, a light coloured leached A-horizon over a red to yellow clayey B-horizon, formed in humid warm temperate or tropical climates under coniferous or deciduous forest.

reedfish Polypteridae *q.v.*

reedmace Typhaceae *q.v.*

refection The habit of an animal that eats its own faeces.

reflex A rapid innate stereotyped response to a stimulus involving the central nervous system.

reforestation The process of re-establishing a forest on previously cleared land; *cf.* afforestation.

refuge An area in which prey may escape from or avoid a predator.

refugium An isolated habitat that retains the environmental conditions that were once widespread.

reg A stony desert.

Regalecidae Oarfishes; family containing 3 species of little-known mesopelagic lampridiform teleost fishes; body ribbon-like and naked, to 7 m in length; pelvic fins long and slender, dorsal fin elongate running entire length of body; teeth and anal fin absent; thought to feed mainly on euphausiid crustaceans.

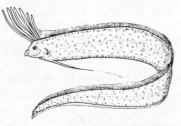

oarfish (Regalecidae)

regeneration Renewal or restoration of structures or tissues, typically after loss or damage.

region 1: Any major world area supporting a characteristic biota; biogeographical region. 2: A major world area and climatic zone characterized by a 50–75% endemic flora or fauna.

regolith All loose earth material above the underlying solid rock; more or less equivalent to the term soil *q.v.*

Regosol An azonal soil without pronounced horizons formed on unconsolidated soft mineral deposits such as loess or sand.

regression A retreat of the sea from a land area; *cf.* transgression.

Regulidae Tit warblers; family of tiny passerine birds with pointed bills, found in scrub and heathland; commonly included in the family Sylviidae.

reindeer Cervidae *q.v.*

rejecta That part of the total food or energy intake of an individual that is not utilized for production and respiration, and is lost as egesta *q.v.* and excreta *q.v.*

relative age The geological age of a fossil or rock expressed in relation to other fossils or events rather than as an actual number of years; *cf.* absolute age.

relative dating A method of geological dating in which only the relative age of a rock is used, rather than the absolute age *q.v.*; *cf.* absolute dating.

relative humidity The ratio of the actual water vapour present to that of total saturation at the existing temperature, expressed as a percentage; *cf.* absolute humidity, saturation deficit.

releaser A stimulus that serves to initiate instinctive activity.

relict species Persistent remnants of formerly widespread fauna or flora existing in certain isolated areas or habitats.

rem Abbreviation of röntgen equivalent man; defined as the equivalent dose of ionizing radiation having the same biological effect as a single röntgen unit of X-rays.

Remipedia Class of small cavernicolous crustaceans characterized by an elongate body bearing about 30 pairs of biramous trunk limbs; mostly predatory feeding on other crustaceans; about 6 species are known at present, all from inland marine caves.

remora Echeneidae *q.v.*

Rendzina soil An intrazonal soil with a dark friable upper A-horizon, a calcareous lower A-horizon over a light grey to yellow calcareous B-horizon, derived from calcium-rich parent material in humid and semi-arid climates under grasses or mixed grass and forest vegetation.

Rensch's laws 1: That in cold climates races of mammals have larger litters and birds larger clutches of eggs than races of the same species in warmer climates. 2: That birds have shorter wings and mammals shorter fur in warmer climates than in colder ones; wing rule. 3: That races of land snails in colder climates have brown shells and those in warmer climates have white shells. 4: That the thickness of snail shell is positively associated with strong sunlight and arid conditions.

replete Fully nourished; well fed.

reproduction The act or process of producing offspring.

reproductive effort That proportion of metabolic resources devoted to reproduction.

reproductive success The number of offspring of an individual surviving at a given time.

Reptantia The lobster-like and crab-like decapod crustaceans; formerly used as a suborder of the Decapoda but now incorporated within the suborder Pleocyemata *q.v.*

Reptilia Class of primitively quadripedal, pentadactyl, tailed vertebrates comprising 3 extant subclasses, Anapsida (turtles, tortoises), Lepidosauria (lizards, snakes) and Archosauria (crocodiles); epidermal scales present; fertilization internal, eggs amniotic, development direct; most species oviparous, some ovoviviparous; abundant fossil record, including dinosaurs, pterosaurs, hadrosaurs and ichthyosaurs.

triceratops (Reptilia)

repugnatorial Adapted for defence or offence.

requiem shark Carcharhinidae *q.v.*

Resedaceae Mignonette; small family of Capparales containing about 70 species of herbs restricted to the northern hemisphere; flowers usually arranged in racemes, with 2–8 sepals, 0–8 petals, 3 to many stamens and a superior ovary.

reservoir host A host that carries a pathogen without detriment to itself and which serves as a source of infection for other host organisms.

residence time That time a given substance remains within a system.

resiniferous Producing resin (a mixture of compounds including polymerized acids, esters and terpenoids), which is commonly exuded by plants when tissues are damaged.

resource Any component of the environment that can be utilized by an organism.

respiration 1: The exchange of gases between an organism and its surrounding medium; breathing. 2: Intracellular oxidation of organic molecules to release energy and typically to produce carbon dioxide and water.

respiratory quotient The ratio of the volume of oxygen consumed to the volume of carbon dioxide released.

response Any change in an organism or a behaviour pattern as the result of a stimulus.

restibilic Perennial.

Restionaceae Southern hemisphere grasses; family of about 400 species of perennial herbs often accumulating silica and with reduced leaf blades; widely distributed on nutrient-poor soils in the southern hemisphere.

Restionales Order of Commelinidae comprising 4 families of herbs with more or less grass-like or rush-like leaf blades and small inconspicuous flowers lacking nectar and wind-pollinated or self-pollinated; producing seeds with abundant mealy or starchy endosperm.

Retortamonadida Order of small parasitic zooflagellates with 2 or 4 flagella, one of which is deflected posteriorly across the cytostome; all species feed on bacteria in the guts of their hosts and are transmitted from host to host in an encysted form.

retrogressive Pertaining to a process of dedifferentiation or reversal to a simpler state or form; **retrogression.**

Retropinnidae Family containing 6 species of freshwater or anadromous salmoniform teleost fishes found in New Zealand and southeastern Australia; body slender and compressed, resembling the smelts; adipose fin small, dorsal fin positioned above anal.

Retziaceae Family of Gentianales containing a single species of erect, evergreen shrub confined to the Cape Province in South Africa.

reversion Return towards an ancestral condition, or from a cultivated to a wild state.

revivescence Emergence from hibernation or some other quiescent state.

Rhabditia Large subclass of secernentean nematodes containing free-living, microbivorous terrestrial forms and many parasites of both vertebrates and invertebrates; characterized by the complex valve consisting of 3 muscular, cuticle-lined lobes, in the hind part of the oesophagus; comprises 3 orders, Ascaridida, Rhabditida and Strongylida.

Rhabditida Large order of rhabditian nematodes comprising free-living forms which typically feed on bacteria and other microorganisms, and parasitic forms found in insects, snails, other invertebrates and vertebrates.

Rhabdocoela Group of mostly free-living turbellarians characterized by a straight gut and comprising the orders Catenulida, Macrostomida, Haplopharyngida and Neorhabdocoela; not used in modern classifications.

Rhabdodendraceae Small family of Rosales containing 3 species of shrubs native to tropical South America.

Rhadomonadales Order of colourless, non-phagotrophic euglenoid flagellates with a single emergent flagellum.

Rhabdopleurida Order of colonial pterobranchs in which the individual zooids are attached to the stolon by a contractile stalk; single pair of arms on the mesosoma; gonads unpaired.

Rhacophoridae Family of Old World tree-frogs (Anura) having adhesion disks on the digits; eggs and larvae typically aquatic; some deposit eggs on vegetation above still water, others have abbreviated development; contains about 95 species found from Africa and Madagascar through to Indo-Australian region.

Rhamnaceae Buckthorn, deer brush, anchor plant; large family containing about 900 species of often thorny woody plants of widespread distribution especially in warm regions; characterized by a cymose inflorescence composed of small flowers usually with 4–5 sepals, small petals and stamens.

Rhamnales Order of Rosidae consisting of 3 families of mucilage-producing and usually woody plants, Leeaceae, Rhamnaceae and Vitaceae.

Rhamphichthyidae Small family of freshwater knifefishes (Siluriformes) from South America; body long, tapering posteriorly; anal fin well developed.

Rhaphidophoridae Cave crickets; small family of wingless crickets (Ensifera) with very long, slender legs and antennae; found in moist environments, especially caves.

rhea Rheidae *q.v.*

Rheidae Rheas; family containing 2 species of large (to 1.5 m tall) South American ratites (Palaeognathae) found in open grassland and brush; legs long and powerful, bearing 3 digits; neck elongate, head feathered; habits cursorial, gregarious, polygamous, feeding on seeds, grasses and insects.

Rheiformes Order of ratite birds comprising a single family, Rheidae.

rheocrene A flowing spring.

rheology The study of fluid movements; that aspect of limnology devoted to running waters.

rheophilous Thriving in or having an affinity for running water; **rheophile, rheophily;** *cf.* rheophobous.

rheophobous Intolerant of running water; **rheophobe, rheophoby;** *cf.* rheophilous.

rheophyte A plant inhabiting running water.

rheoplankton Planktonic organisms associated with running water.

rheopositive Sensitive to currents; used of organisms showing behavioural and other responses to air or water currents.

rheoreceptive Sensitive to water or air currents.

rheotaxis A directed response of a motile organism to a water or air current, either into the current (positive rheotaxis) or with the current (negative rheotaxis); **rheotactic.**

rheotrophic Pertaining to organisms that obtain nutrients largely from percolating or running water.

rheotropism An orientation response to a water or air current; **rheotropic.**

rheoxenous Used of organisms that occur only occasionally in running water; **rheoxene.**

rhexigenous Produced as a result of breakage or rupture.

Rhinatrematidae Family containing 8 species of primitive terrestrial caecilians (Gymnophiona) found in tropical South America; body small (to about 300 mm) and covered with dermal scales, tail present; larvae aquatic bearing branched gills.

Rhincodontidae Whale shark; monotypic family of very large (to 20 m) inoffensive orectolobiform elasmobranch fishes (carpet sharks), the largest of all fish species; body cylindrical with flattened head; gill openings large bearing internal filter screens; mouth terminal, teeth small and weakly

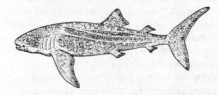

whale shark (Rhincodontidae)

differentiated; reproduction oviparous; circumtropical in pelagic waters, filter feeding on planktonic crustaceans, squid and small fishes.

Rhinobatidae Family containing about 35 species of rhinobatiform fishes in which the body disk is elongate, and the tail moderately slender but not demarcated from disk; length to 1.5 m; caudal fin without distinct ventral lobe, axis upturned.

Rhinobatiformes Guitarfishes; order of bottom-living, shallow water elasmobranch fishes (Batoidea); anterior body and pectoral fins discoidal, snout prolonged, tail slender, pelvic fins small; habits sluggish, feeding on fishes, crustaceans, molluscs and other benthic invertebrates; reproduction viviparous; comprises about 50 species in 3 families.

Rhinocerotidae Rhinoceroses; family containing 5 species of massive terrestrial mammals (Perissodactyla) found in a variety of open savannah and forest habitats of the Oriental and Ethiopian regions; head elongate with 1 or 2 median horns composed of dense horny fibres, but lacking bone; skin very thick with sparse stiff hairs; habits typically nocturnal, solitary, herbivorous, grazing or browsing on vegetation.

rhinoceros (Rhinocerotidae)

Rhinochimaeridae Family containing 8 species of small to medium-sized (to 1.5 m) deep-water (700–2000 m) holocephalan fishes; body compressed and naked, snout pointed, tail tapering; dorsal fins contiguous, caudal narrow with straight axis.

Rhinocryptidae Tapaculos; family containing about 30 species of small to medium-sized terrestrial passerine birds that inhabit forests and grassland of the Neotropical region; body compact, bill short and pointed, wings short and rounded, flight poor; habits mostly solitary, cursorial, feeding on insects and seeds; nest in burrow, or in domed nest on ground or in tree or rock crevice.

Rhinodermatidae Family containing 2 species of small (to 30 mm) terrestrial frogs (Anura)

from temperate rainforests of Chile and Argentina; eggs terrestrial, tadpoles aquatic in one species, brooded in vocal sac of male in the other; sometimes included in the family Leptodactylidae.

Rhinolophidae Horseshoe bats; family containing about 110 species of insectivorous microchiropteran Old World leaf-nosed bats, widespread in Ethiopian, Oriental and Australian regions; habits colonial or solitary, some species hibernate, others are migratory.

Rhinophoridae Woodlouse flies; small family of flies (Diptera) the larvae of which are endoparasitic on terrestrial isopods (woodlice).

Rhinophrynidae Monotypic family of fossorial toads (Anura) found in Central America; body length to 75 mm, head very small, limbs short and specialized for digging, ribs absent; larvae aquatic.

Rhinopomatidae Mouse-tailed bats; family containing 3 species of small insectivorous microchiropteran bats found in arid habitats from North Africa to India.

Rhinopteridae Cow-nosed rays; family containing about 10 species of tropical and subtropical, estuarine and coastal marine, myliobatiform elasmobranch fishes; body disk broader than long and demarkated from head, pectoral fins prolonged anteriorly as pair of lobes in front of snout; tail filamentous distally, caudal fin absent; mouth with series of grinding plates; feed on benthic molluscs and crustaceans dug up using pectoral fins.

Rhipidistia Extinct group of tassel-finned fishes (Crossopterygia) with 2 dorsal fins, and stalked or lobate pectoral and pelvic fins; nares internal; known from the Devonian to the Permian.

rhithrous Pertaining to the upper reaches of a river; *cf.* potamous.

rhizanthous Used of plants producing flowers that appear to arise directly from the roots.

rhizobenthos Those organisms rooted in the substratum; **rhizobenthic**; *cf.* benthos.

rhizocarpous 1: Used of plants having perennial roots and annual stems and foliage. 2: Pertaining to perennial herbs that produce flowers and fruit below ground level as well as above ground.

Rhizocephala Order of exclusively parasitic barnacles (Cirripedia) found only on crustacean hosts; body highly modified, consisting of a branching rootlet system penetrating the host and an external sac containing the reproductive organs; larval development typically includes nauplius and cyprid stages; contains about 250 species.

Rhizochloridales Order of naked, amoeboid xanthophyceaen algae found in both freshwater and marine habitats; includes solitary and colonial forms, characterized by the lack of flagella.

rhizogenic Producing roots; stimulating root formation or growth; **rhizogenesis.**

Rhizomastigida Order of mostly free-living zooflagellates which possess both flagella and pseudopodia; typically found in stagnant fresh waters.

rhizomenon All organisms rooted into the substratum; *cf.* ephaptomenon, planomenon.

Rhizomyidae Mole rats, bamboo rats; family containing 18 species of fossorial myomorph rodents from Ethiopian and eastern Oriental regions; body fusiform, limbs short, eyes reduced, tail present; typically burrow using incisor teeth or feet.

rhizophagous 1: Feeding on roots; **rhizophage, rhizophagy.** 2: Used of a plant that obtains nourishment through its own roots.

rhizophilous Growing on roots; **rhizophile, rhizophily.**

Rhizophorales Red mangrove; order of Rosidae comprising the family Rhizophoraceae, which contains about 100 species of tanniferous trees and shrubs including several mangrove species; widely distributed in tropical regions; flowers hermaphrodite, with 3–16 sepals and petals, and 8 to many stamens.

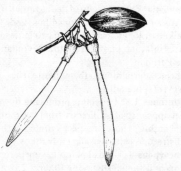

mangrove droppers (Rhizophorales)

rhizopleustohelophyte A plant rooted in the bank or bottom sediment of shallow water with a floating stem and emergent leaf-bearing shoots.

Rhizopoda Cosmopolitan superclass of Sarcodina distinguished on the basis of their type of pseudopodia and by their lack of

axopodia and internal skeletons; includes amoebae and foraminiferans; also treated as a distinct phylum of Protoctista.

rhizosphere The soil immediately surrounding plant roots that is influenced structurally or biologically by the presence of such roots; root zone; *cf.* phyllosphere.

Rhizostomeae Order of medusae (Scyphozoa) found primarily in shallow waters of the tropical Indo-Pacific Ocean; characterized by a domed or discoidal bell and by the fusion of the frilled edges of the oral arms over the mouth so that many suctorial mouths are formed.

rhizotaxis The spatial arrangement of roots; **rhizotaxy.**

Rhodochaetales Order of marine red algae containing a single species with a branched filamentous body form.

rhododendron Ericaceae *q.v.*

Rhodophyceae The only class of the division Rhodophycota *q.v.*; comprises 2 subclasses, Bangiophycideae and Florideophycideae.

Rhodophycota Red algae; diverse assemblage of eukaryotic algae containing forms ranging from unicells to those with a true parenchymatous thallus, characterized by reddish plastids containing chlorophyll *a*, phycobiliproteins and carotenoids; reproduction oogamous, the egg cell being produced inside an oogonium which is receptive to the non-motile male gamete; classified either as a division of eukaryotic algae or as a phylum of Protoctista, under the name Rhodophyta; comprises the Bangiophycideae and the Florideophycideae.

red alga (Rhodophycota)

Rhodophyta Red algae; the Rhodophycota *q.v.* treated as a phylum of Protoctista.

Rhodymeniales Order of typically crustose or erect and frondose red algae found in marine habitats.

Rhoipteleaceae Family of Juglandales comprising a single species of aromatic tree confined to Vietnam and southern China.

Rhombifera Extinct class of echinoderms (Pelmatozoa) with biserial brachioles; known from the Lower Ordovician to the Upper Devonian.

Rhombozoa Class of mesozoans found as endoparasites in the renal organs of cephalopod molluscs; characterized by a body organized into a superficial layer of somatic cells enclosing a medullary, axial cell; form 2 different kinds of embryos, nematogens and rhombogens, according to whether the host is juvenile or adult.

rhoophilous Thriving in creeks; **rhoophile, rhoophily.**

rhoophyte A creek plant.

Rhopalosomatidae Cricket wasps; family of slender solitary wasps (Hymenoptera) which have parasitoidal larvae living and feeding in sacs attached to the abdomens of crickets; contains about 20 species.

R-horizon The bedrock underlying a soil profile; D-horizon; see soil horizons.

rhubarb Polygonales *q.v.*

rhyacophilous Thriving in, or having an affinity for, torrents; **rhyacophile, rhyacophily.**

rhyacophyte A plant living in a torrent.

Rhynchobatidae Family containing 5 species of rhinobatiform fishes; length up to 2 m, body disk elongate, tail stout and not demarcated from disk; caudal fin with distinct ventral lobe.

Rhynchobdellae Order of freshwater and marine leeches parasitic or predacious on a wide range of vertebrate and invertebrate hosts; pharynx forming protractile piercing proboscis, jaws absent; contains about 250 species in 3 families; one family, Piscicolidae (fish leeches), contains mostly marine forms.

Rhynchocephalia Order of primitive lepidosaurian reptiles represented by a single extant family, the monotypic Sphenodontidae (tuatara).

tuatara (Rhynchocephalia)

Rhynchocoela Alternative name for the Nemertea *q.v.*

Rhynchodida Order of hypostomatan ciliates found on the gills of freshwater and marine invertebrates; small rostrate forms with a single anterior sucking tube or tentacle which functions in ingestion; body nearly naked; reproduction generally by budding.

Rhynchonellida Order of articulate brachiopods containing about 30 species; shell typically impunctate; spicules absent from soft parts: found in shallow water to 3000 m, attached to hard substrata.

rhythm Any cyclical variation in the intensity of a behavioural or physiological activity; **rhythmicity.**

ribbon worm Nemertea *q.v.*

ribbonfish Trachipteridae *q.v.*

ribonucleic acid RNA *q.v.*

ribosomal RNA The structural RNA of the ribosomes; rRNA.

rice Poaceae *q.v.*

ricefish Oryziatidae *q.v.*

Richter scale A logarithmic scale for measuring the magnitude of seismic disturbances; ranging from 1.5 for the smallest detectable tremor to 8.5 for a devastating earthquake.

Ricinulei Monofamilial order of small (to 15 mm) ground-living predatory arthropods (Arachnida) containing about 30 species found in tropical West Africa and America; cephalothorax and abdomen joined by a broad waist; chelicerae forming scorpion-like pincers; respiration by means of tracheae.

Rickettsias Group of mostly small, rod-like bacteria which possess typical bacterial cell walls and multiply by binary fission inside host cells.

ricochetal Type of jumping or bouncing locomotion involving the storage of energy in elastic tendons in one phase of the movement and its subsequent release during the power stroke.

right whale Balaenidae *q.v.*

rill erosion The formation of many small channels due to erosion by surface water, especially on cultivated land.

ring species The graded series of populations along an extensive cline *q.v.* that curves round on itself so that the populations at the extremes overlap but are unable to interbreed successfully.

ringed worm Annelida *q.v.*

ringworm Gymnascales *q.v.*

riparian Pertaining to, living or situated on, the banks of rivers and streams.

ripicolous Living on the banks of rivers and streams; **ripicole.**

Riss glaciation A glaciation of the Quaternary Ice Age *q.v.* in the Alpine area with an estimated duration of 100 000 years.

Riss–Würm interglacial The interglacial period preceding the Würm glaciation in the Alpine area; see Quaternary Ice Age.

rival Pertaining to a stream or rivulet.

river dolphin Platanistidae *q.v.*

river ray Potamotrygonidae *q.v.*

riverine Pertaining to a river; formed by the action of a river.

RNA Ribonucleic acid; a polynucleotide that governs protein synthesis in a cell, existing in a variety of forms serving particular functions, such as messenger RNA, ribosomal RNA, transfer RNA.

roadrunner Cuculidae *q.v.*

robber fly Asilidae *q.v.*

rock crawler Grylloblattaria *q.v.*

rock jumper Thysanura *q.v.*

rock oyster Hippuritoida *q.v.*

rock rat Echimyidae *q.v.*

rockfish Scorpaenidae *q.v.*

Rodentia Diverse order of mostly small, herbivorous or omnivorous, terrestrial, arboreal or semiaquatic mammals (Glires); comprises about 1650 species in 33 families arranged in 3 suborders, Sciuromorpha (squirrels, beavers), Myomorpha (rats, mice, jerboas, hamsters, gerbils) and Hystricomorpha (porcupines, guinea-pigs); the largest order of Mammalia; incisor teeth chisel-like, continuously growing and separated from cheek teeth by broad gap (diastema).

gerbil (Rodentia)

rodenticide A chemical used to kill rodents.

rogue To remove undesirable weeds, diseased plants or genetic variants (rogues) from a population.

roller Coraciidae *q.v.*

Rondeletiidae Whalefishes; cosmopolitan family containing 2 species of small (to 100 mm) bathypelagic beryciform teleost fishes; body devoid of scales but bearing vertical rows of small papillae; head large, jaws with many minute teeth.

ronquil Bathymasteridae *q.v.*

rookery A breeding or nesting place for some mammals and birds.

root parasitism A condition found in a semi-parasitic plant, the roots of which penetrate the roots of the host plant and remove elaborated food material.

rorqual Balaenopteridae *q.v.*

Rosaceae Rose, hawthorn, bridal wreath, meadow sweet, strawberry, mountain ash, cherry, peach, apricot, plum, apple, pear, blackberry, raspberry; large family of herbs, shrubs or trees containing about 3000 species nearly cosmopolitan in distribution and including many species cultivated as edible fruits or as ornamentals; bisexual flowers arranged in racemes or cymes and with 5 sepals and petals, many stamens and, usually, a superior ovary.

boysenberry (Rosaceae)

Rosales Large order of Rosidae containing 24 families of chemically and anatomically diverse, usually terrestrial plants which lack milky juice and internal phloem.

rose Rosaceae *q.v.*

rosemary Lamiaceae *q.v.*

Rosidae Large subclass of dicotyledons producing various sorts of repellents, frequently alkaloids or triterpenoid compounds; flowers variable, usually with separate petals; contains nearly 60 000 species arranged in 114 families and 18 orders; majority of species belonging to Euphorbiales, Fabales, Myrtales, Rosales and Sapindales.

Rostroconchia Extinct class of bilaterally symmetrical, bivalved, marine molluscs known from the Lower Cambrian to the Permian; shell with distinct anterior gape from which the foot probably emerged; shell without functional dorsal hinge.

rot Decay; decomposition.

Rotatoria Rotifera *q.v.*

Rotifera Rotifers, wheel animalcules; phylum of microscopic pseudocoelomate animals found in aquatic habitats and occasionally in soil and damp moss; characterized by a typically unsegmented body covered by a cuticle which may be thickened into a lorica consisting of plates, and by the absence of a muscular body wall; pharynx (mastax) with internal jaws (trophi); excretory organs comprising pair of flame bulb protonephridia; many organs are syncytial but the nuclear number in adults is constant; sexes separate, although males may be rare and parthenogenesis is common; comprises 3 classes, Bdelloidea, Monogonata and Seisonidea; formerly known as Rotatoria.

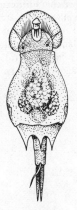

rotifer (Rotifera)

roundhead Plesiopidae *q.v.*

roundworm Nematoda *q.v.*

rove beetle Staphylinidae *q.v.*

Roveacrinida Extinct order of small, free-swimming, pelagic sea lilies (Crinoidea); known from the Triassic to the Cretaceous.

royal poinciana Caesalpiniaceae *q.v.*

rRNA Ribosomal RNA *q.v.*

r-selected species A species characteristic of variable or unpredictable environments; typically with rapid development, high innate capacity for increase, early reproduction, small body size and semelparity; opportunistic species; *r*-strategist; *cf.* K-selected species.

rubber Euphorbiaceae *q.v.*

Rubiaceae Madder, coffee, gardenia; large family containing about 6500 species of mostly tropical and subtropical, woody or herbaceous plants producing a wide range of repellent chemicals including iridoids and alkaloids; flowers usually bisexual and regular, commonly with 4 or 5 sepals, petals and stamens, and an inferior ovary.

Rubiales Order of Asteridae containing 2 families of woody, or sometimes herbaceous, plants commonly producing iridoid compounds; leaves commonly opposite and flowers sympetalous but sometimes reduced; includes Rubiaceae and Theligonaceae.

rubicolous Living on brambles; **rubicole.**

rubrification The appearance of a red coloration in a soil due to oxidative weathering and the formation of haematite.

ruderal 1: Pertaining to or living amongst rubbish or debris, or inhabiting disturbed sites. 2: A plant inhabiting such a habitat.

Rudimicrosporea Subclass of minute microsporidian protozoans which are all hyperparasitic inside gregarines parasitizing mostly marine annelids; characterized by a rudimentary extrusion apparatus on the spores.

rugophilic Thriving on rough surfaces or in surface depressions; **rugophile, rugophily.**

Rugosa Extinct order of solitary or compound corals (Zoantharia); large, diverse group known from the Cambrian to the Upper Permian; characterized by bilateral symmetry, with metasepta inserted between only 4 of the 6 primary septa.

rugotropism An orientation with respect to a surface depression or roughness; **rugotropic.**

Ruminantia Suborder of cloven-hooved herbivorous ungulates (Artiodactyla) that possess a 4-chambered stomach and chew the cud (ruminate); feet with 2 or 4 digits and well developed hooves; horns or antlers typically present; comprises 5 families, Tragulidae (chevrotains), Antilocapridae (pronghorn), Giraffidae (giraffe), Cervidae (deer) and Bovidae (cattle, goats, sheep, antelopes).

Runcinoidea Order of small, marine opisthobranch molluscs, most of which are herbivores found in shallow water; characterized by an elongate, vermiform body which often lacks a shell and cephalic tentacles.

runner An elongate creeping stem which arises from an axillary bud and gives rise to plantlets at the nodes, found commonly in rosette plants.

runoff That part of precipitation that is not held in the soil but drains freely away.

rupestral Pertaining to or living on walls or rocks.

rupicolous Living on walls or rocks; **rupicole.**

Ruppiaceae Ditch grass; cosmopolitan family of Najadales comprising a single variable species of glabrous herb submerged in alkaline or brackish water; flowers minute, borne in short, terminal racemes; pollen grain elongate, pollination occurring at water surface.

rush Juncaceae *q.v.*

Russian olive Elaeagnaceae *q.v.*

rust fly Psilidae *q.v.*

rust fungus Uredinales *q.v.*

rut A period of sexual excitement in a mammal.

Rutaceae Citrus family; orange, lemon, lime, grapefruit, common rue; family of Sapindales containing about 1500 species of mostly woody, aromatic plants generally producing ethereal oils; widely distributed especially in warm regions; flowers typically with 4 or 5 sepals and petals, 8 or 10 stamens and a superior ovary.

Rutoceratida Extinct order of cephalopods (Nautiloidea) in which the siphuncle is small and located near the centre of the chambers; known from the Lower Devonian to Middle Jurassic.

rye Poaceae *q.v.*

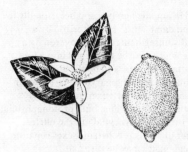

lemon (Rutaceae)

Rynchopidae Skimmers; family containing 3 species of tropical charadriiform sea birds with long blade-like bill used to catch fish during skimming flight over water; habits gregarious, often crepuscular or nocturnal.

rypophagous Feeding on putrid matter or refuse; **rypophage**, **rypophagy**.

S

Saale glaciation A glaciation of the Quaternary Ice Age *q.v.* in northern Germany and Poland with an estimated duration (together with the Warthe glaciation) of 100 000 years.

Sabellida Fanworms; order of sedentary and tubicolous polychaete worms comprising about 800 species in 4 families; prostomium and peristomium fused, supporting a crown of tentacles or branchiae around the mouth used for suspension feeding or respiration; tube construction carried out by anterior thoracic collar.

sedentary polychaete (Sabellida)

Sabiaceae Small family of tanniferous trees, shrubs or woody vines found mainly in southeast Asia and tropical America.

sable Mustelidae *q.v.*

sablefish Anoplopomatidae *q.v.*

sabretooth Evermannellidae *q.v.*

sabulicolous Living in sand; **sabulicole.**

sabuline Sandy; used especially of species growing on coarse sand.

sac fungus Ascomycotina *q.v.*

Saccifoliaceae Family of Gentianales containing a single species of subshrub native to southern Venezuela.

Saccopharyngidae Family of bizarre bathypelagic anguilliform teleost fishes (gulper-eels) possessing a huge mouth and many sharp curved teeth; body smooth, length to about 1.8 m, eyes set close to the snout; tail long and tapering.

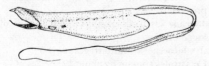

gulper-eel (Saccopharyngidae)

Saccopharyngoidei Deep-sea gulper-eels; suborder of bizarre carnivorous bathypelagic anguilliform teleost fishes exhibiting loss of gas bladder, scales, ribs, and reduction of eyes, gill openings and caudal fin; tail long and tapering, mouth very large; comprises about 10 species in 3 families.

Sacoglossa Order of generally small, often cryptically coloured, opisthobranch molluscs which typically feed on algae; characterized by the retention of older radula teeth in a ventral sac off the muscular, sucking pharynx.

sacoglossan (Sacoglossa)

sac-winged bat Emballonuridae *q.v.*

saddle oyster Ostreoida *q.v.*

sage Lamiaceae *q.v.*

sagebrush Asterales *q.v.*

Sagittariidae Secretary bird; monotypic family of large long-legged raptors (Falconiformes) found on open plains of southern and eastern Africa; wings broad; habits solitary, monogamous, mostly terrestrial, running to catch prey; feeds on vertebrates, especially snakes, and insects; nests in bush or tree.

secretary bird (Sagittariidae)

Sagittoidea The single class of the phylum Chaetognatha *q.v.*

sailfish Istiophoridae *q.v.*

salamander Salamandridae *q.v.*

Salamandridae Newts, salamanders; diverse family containing about 40 species of aquatic or amphibious Holarctic salamanders

(Caudata); body length to 200 mm; metamorphosis complete; forelimbs, hindlimbs and lungs present; fertilization internal, males producing spermatophores; eggs typically aquatic, the larvae bearing large gills although some forms have terrestrial eggs and direct development.

salamander (Salamandridae)

Salangidae Icefishes; family containing 12 species of mostly small anadromous salmoniform teleost fishes from the Oriental region; body slender, transparent, subcylindrical, scales and pelvic axillary process absent; adipose fin present.

Salenioida Order of small (to 20 mm) echinoids (Echinacea) comprising about 13 extant species found mostly at continental slope depths; dentition stirodont, oral membrane with gill notches, anus displaced to one side by large apical plate; diverse fossil record in the Mesozoic.

Salicales Willow, poplar, cottonwood; order of Dilleniidae comprising the family Salicaceae which contains about 300 species of dioecious woody plants, best developed in northern temperate regions; flowers borne in unisexual catkins and primarily wind-pollinated.

Salientia Older name for the Anura *q.v.*

saline Salty; pertaining to soil or water rich in soluble salts.

salinity A measure of the total concentration of dissolved salts in sea water usually measured in parts per thousand (ppt); *cf.* chlorinity.

salinization The process by which soluble salts accumulate in soil.

salmon Salmonidae *q.v.*

salmon shark Lamnidae *q.v.*

Salmonidae Salmon, trout, chars, graylings, whitefishes; family containing 70 species of northern hemisphere, anadromous marine or freshwater, salmoniform teleost fishes; body compressed, to 1.5 m in length; adipose fin

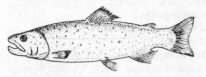

salmon (Salmonidae)

present, dorsal fin arising above or anterior to pelvics; pelvic axillary process present; commonly migrate upstream to spawn; salmonids are very important food-fishes widely exploited in pisciculture.

Salmoniformes Large order of soft-rayed marine and freshwater teleost fishes possessing many primitive or generalized features; swim bladder lacking connection to inner ear; adipose fin often present; comprises about 270 species in 4 suborders, Esocoidei (pikes, mudminnows), Salmonoidei (salmon, trout, chars), Argentinoidei (herring smelts, deep-sea smelts, barreleyes, slickheads, tubeshoulders) and Giganturoidei (giganturids).

Salmonoidei Suborder of marine, freshwater and anadromous teleost fishes comprising about 145 species in 7 families; including salmon, trout, chars, graylings, whitefishes, smelts, galaxiids and icefishes; adipose fin usually present; dorsal fin originating close to midpoint of body.

salp Thaliacea *q.v.*

Salpida Order of pelagic tunicates (Thaliacea) that exhibit alternation of solitary, sexually reproducing (gonozooids) and compound, asexually produced forms, (ovozooids); zooids barrel-shaped with anterior inhalent and posterior exhalent openings producing locomotory water current; gonozooid hermaphrodite, larval stage absent, development viviparous.

salsuginous Pertaining to, or living in, habitats inundated by salt or brackish water; **salsugine.**

salt lake An inland water body having a high salinity due to an excess of evaporation over precipitation.

salt marsh A flat poorly drained coastal swamp inundated by most high tides.

saltation 1: To move by leaping or bounding; saltatory. 2: A drastic and sudden mutational change; an abrupt evolutionary change.

saltatorial Pertaining to, or adapted for, leaping or bounding locomotion; saltatory.

Salticidae Jumping spiders; family containing about 4000 species of active predatory spiders (Araneae) with large eyes; prey jumped on from a distance, the jumping mechanism involving extension of legs by internal body pressure.

saltigrade Used of a leaping or hopping form of locomotion.

Salvadoraceae Mustard tree; family of Celastrales containing about 12 species of woody plants sometimes producing mustard

oils; native from Africa through to southeast
Asia.

Salviniales Water ferns; order of small ferns
(Filicopsida) producing rhizomes which creep
on the surface of the water, and hanging roots;
typically heterosporous forming
microsporangia and megasporangia in
separate walled sporocarps.

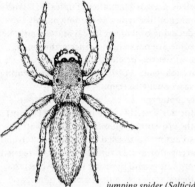

jumping spider (Salticidae)

samara A type of achene or nut in which the
wall of the fruit (pericarp) is produced into a
membranous wing to aid dispersal of the seed
by wind.

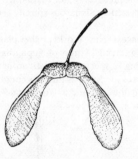

sycamore fruit (samara)

samphire Apiaceae *q.v.*

sample Any subset of a population; a
representative part of a larger unit used to
study the properties of the whole.

sand Cohesionless sediment particles of size
range 2.0–0.0625 mm; subdivided into very
coarse sand (2.0–1.0 mm), coarse sand (1.0–
0.5 mm), medium sand (0.5–0.25 mm), fine
sand (0.25–0.125 mm) and very fine sand
(0.125–0.0625 mm); see sediment particle size.

sand cricket Stenopelmatidae *q.v.*

sand dollar Clypeasteroida *q.v.*

sand eel 1: Hypoptychidae *q.v.* 2:
Gonorhynchidae *q.v.*

sand flea Tungidae *q.v.*

sand fly 1: Psychodidae *q.v.* 2: Phlebotamidae
q.v.

sand lance Ammodytidae *q.v.*

sand stargazer Dactyloscopidae *q.v.*

sand tiger shark Odontaspididae *q.v.*

sandalwood Santalaceae *q.v.*

sandbar A submerged ridge of alluvial sand in
shallow water.

sandfish 1: Trichodontidae *q.v.* 2:
Leptoscopidae *q.v.*

sandgrouse Pteroclididae *q.v.*

sandhopper Amphipoda *q.v.*

sandperch Mugiloididae *q.v.*

sandpiper Scolopacidae *q.v.*

sand-smelt Atherinidae *q.v.*

Sangamonian interglacial The interglacial
period preceding the Iowan glaciation in
North America; see Quaternary Ice Age.

sanguicolous Living in blood; **sanguicole.**

sanguivorous Feeding on blood; **sanguivore,
sanguivory.**

Santalaceae Sandalwood, false toadflax;
cosmopolitan family containing about 400
species of hemiparasitic herbs and shrubs that
attach to roots of other plants; flowers small,
typically with 3–6 fused tepals and adnate
stamens.

Santalales Order of Rosidae containing a
diverse array of trees, or hemiparasitic (with
chlorophyll) or fully parasitic (lacking
chlorophyll) plants attached to trees, shrubs
or herbs; leaves simple, often evergreen,
sometimes reduced to scales; contains about
2000 species in 10 families, including the
mistletoes (Viscaceae).

Sapindaceae Goldenrain tree, soapberry, litchi;
large family containing about 1500 species of
mostly woody plants often with anomalous
stem structure, widespread mainly in warm
regions; flowers usually with 4 or 5 sepals and
petals, 4, 5, 10 or many stamens and a superior
ovary.

Sapindales Order of Rosidae containing about
5400 species of chemically diverse, mostly
woody, plants arranged in 15 families, the
largest of which are the Sapindaceae and
Rutaceae.

sapodilla Sapotaceae *q.v.*

Sapotaceae Star apple, sapodilla; large family
of Ebenales containing about 800 species of
trees or shrubs with latex sacs; widespread
chiefly in tropical regions; provides some
edible fruits and produces gums, including
chicle; ovary superior, fruit typically a berry.

saprobe 1: A saprophytic organism. 2: An

organism thriving in water rich in organic matter.

saprobic 1: Pertaining to water rich in decaying organic matter. 2: Saprophytic *q.v.*

saprobiontic Saprophagous *q.v.*; **saprobiont.**

saprogenic Causing, or caused by, the decay of organic matter.

Saprolegniales Water moulds; order of oomycete fungi; mostly saprophytic in wet soil or aquatic habitats, occasionally parasitic.

sapromyiophilous Pollinated by dung flies; **sapromyiophile, sapromyiophily.**

sapropelic Used of organisms inhabiting mud rich in decaying organic matter; **sapropel.**

saprophagous Feeding on dead or decaying organic matter; **saprophage, saprophagy;** *cf.* biophagous.

saprophilous 1: Thriving in humus-rich substrata. 2: Feeding on decaying organic matter; **saprophile, saprophily.**

saprophytic Used of a plant obtaining nutrients from dead or decaying organic matter; **saprobic, saprophyte.**

saprophytophagous Feeding on decomposing plant material; **saprophytophage, saprophytophagy.**

saproplankton Plankton organisms inhabiting water rich in decaying organic matter.

saprotrophic Obtaining nourishment from dead or decaying organic matter; **saprotroph, saprotrophy.**

saproxylobios Those organisms living in or on rotting wood.

saprozoic Used of an animal feeding upon decaying plant or animal matter in the form of dissolved organic compounds; **saprozoite.**

Sarcodina Subphylum of Sarcomastigophora; protozoans possessing some kind of pseudopodia, most of which are free-living in soil and aquatic habitats; body typically naked but sometimes with external tests or internal skeletons; cysts are common and sexual reproduction when present usually involves flagellated gametes; comprises 2 superclasses, Actinopoda and Rhizopoda.

Sarcolaenaceae Small family of usually evergreen woody plants, often with stellate hairs; confined to Madagascar.

Sarcomastigophora Diverse phylum of protozoans comprising over 18 000 species which are widespread in aquatic habitats and soil, as well as in symbiotic or parasitic association with other animals and plants; possessing either flagella, pseudopodia or both and a single type of nucleus; asexual reproduction typically involves binary fission, and sexual reproduction, when present, involves fusion of gametes; typically heterotrophic, although some classifications include certain groups of autotrophic flagellated algae in this phylum; comprises 3 subphyla, Mastigophora, Opalinata and Sarcodina.

Sarcophagidae Flesh flies; cosmopolitan family containing about 1400 species of medium-sized grey or black flies (Diptera); larvae maggot-like typically feeding mainly on decaying organic or faecal matter, although some are parasitic on other arthropods, earthworms or molluscs.

sarcophagous Feeding on flesh; flesh-eating; **sarcophage, sarcophagy.**

Sarcopterygii Lobe-finned fishes; group formerly used for the lungfishes (Dipnoi) and the crossopterygian fishes (coelacanths); sometimes also used to include the tetrapods.

Sarcoptiformes Diverse suborder of free-living, predatory and parasitic mites (Acari) that includes cheese mites, scabies mites, and several families of oribatids (beetle or moss mites).

sardine Clupeidae *q.v.*

Sargentodoxaceae Family of Ranunculales containing a single twining woody vine species native to China, Laos and Vietnam; characterized by numerous spirally arranged pistils.

Sarraceniaceae New World pitcher plants, cobra plant; small family of Nepenthales containing 15 species of insectivorous herbs with leaves modified as pitchers containing a digestive liquid; found in North and South America.

New World pitcher plant (Sarraceniaceae)

Sarsostraca Subclass of branchiopodan crustaceans comprising a single extant group, the fairy shrimps (Anostraca *q.v.*).

sathrophilous Thriving on humus or decaying organic matter; **sathrophile, sathrophily.**

sathrophyte A plant living on humus; **sathrophytic.**

saturation deficit A measure of humidity, derived by subtracting the actual water vapour pressure from the maximum possible vapour pressure at a given temperature; *cf.* relative humidity.

Saturniidae Atlas moths, emperor moths; family containing about 1000 species of large moths (Lepidoptera) having conspicuous eye spots and banded markings on the wings; larvae of some species are economically important pests, causing damage to the foliage of various trees; pupal cocoons of some species are used for the production of silk.

Sauria Lizards; large and diverse suborder of reptiles (Squamata) containing 3300 species in 17 families; body length to about 3 m; habits mostly terrestrial, diurnal or nocturnal, a few species fossorial or semiaquatic; limbs may be reduced in burrowing forms; feeding as herbivores, insectivores or as predators of small vertebrates; reproduction oviparous or ovoviviparous.

Saurischia Order of dinosaurs characterized by the lizard-type hip; includes large bipedal carnivores (Theropoda) and large quadripedal herbivores (Sauropoda).

saurochorous Dispersed or distributed by lizards or snakes; saurophilous; **saurochore, saurochory.**

saurophilous Saurochorous *q.v.*

Sauropoda Group of saurischian dinosaurs containing many large, quadripedal herbivores; includes *Diplodocus*, the largest known terrestrial animal; known from the Jurassic and Cretaceous.

diplodocus (Sauropoda)

Sauropterygia Extinct order of reptiles (Archosauria) comprising the plesiosaurs (Plesiosauroidea) and nothosaurs (Nothosauria).

Saururaceae Small family of aromatic rhizomatous herbs (Piperales) occurring in eastern Asia and North America.

saury Scomberesocidae *q.v.*

savannah The tropical and subtropical grassland biome, transitional in character between grassland or desert and rain forest, typically having drought-resistant vegetation dominated by grasses with scattered tall trees; Tropical grassland biome; savanna.

saw shark Pristiophoridae *q.v.*

sawfish Pristidae *q.v.*

sawfly Symphyta *q.v.*

saxatile Living or growing among rocks.

saxicavous Rock-boring.

saxicolous Living or growing on or among rocks or stones; **saxicole, saxicoline.**

Saxifragaceae Saxifrage, astilbes, foam flower, bronze leaf; large family of Rosales containing herbs, often with red root tips and basal rosettes of leaves; best developed in the cold and temperate northern hemisphere; flowers usually bisexual with 5 sepals and petals, and 10 stamens.

saxifrage (Saxifragaceae)

saxifragous Living in rock crevices; **saxifrage.**

scabious Dipsacaceae *q.v.*

scad Carangidae *q.v.*

scale insect Coccoidea *q.v.* (Homoptera).

scaled blenny Clinidae *q.v.*

scaleworm Phyllodocida *q.v.*

scallop Ostreoida *q.v.*

scaly anteater Manidae *q.v.*

scaly-tailed squirrel Anomaluridae *q.v.*

scansorial Climbing or adapted for climbing; scandent.

Scaphopoda Tusk shells, tooth shells; class of exclusively marine, benthic molluscs found partly embedded in the sediment from the sublittoral to abyssal depths; characterized by a bilaterally symmetrical body with a curved, tubular calcareous shell open at both ends; head rudimentary with a proboscis used for

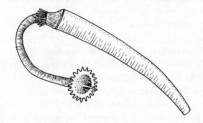

tusk shell (Scaphopoda)

capturing foraminiferans; foot protrusible and often modified for burrowing; possessing a radula, used for crushing shells of prey; comprises 2 orders, Dentaliida and Gadilida.

Scarabaeidae Chafers, dung beetles; family containing about 25 000 species of beetles (Coleoptera) the adults and larvae of which feed mainly on dung and carrion; larvae also found sometimes on roots and decaying vegetation; larvae typically C-shaped.

Scaridae Parrotfishes; family containing 70 species of colourful tropical and warm temperate marine labroid teleost fishes (Perciformes) in which the jaws form a non-protractile parrot-like beak used to scrape algae and coral from hard surfaces; body length to 1.2 m.

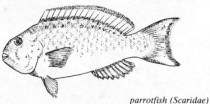

parrotfish (Scaridae)

scarification Any chemical or physical treatment of a hard seed which renders it permeable.

scat 1: Scatophagidae *q.v.* 2: An animal faecal dropping.

scatology The study of animal faeces; **scatological.**

Scatophagidae 1: Dung flies; family of small to large hirsute flies (Diptera) that feed mainly on dung; larvae phytophagous or scatophagous, often found in meadows. 2: Scats; small family of Indo-Pacific coastal marine and brackish-water teleost fishes (Perciformes) that feed mainly on bottom detritus and are often abundant around sewage outfalls; body deep, compressed, with high dorsal profile; 2 dorsal fins present; popular in the aquarium trade.

scatophagous Feeding on faecal matter or dung; **scatophage, scatophagy.**

scavenger Any organism that feeds on carrion or organic refuse.

Scenopinidae Window flies; cosmopolitan family containing about 250 species of small flies (Diptera) having a dark oblong body lacking bristles, commonly found on vegetation or on windows of houses.

Scheuchzeriaceae Family of Najadales comprising a single species of cyanogenic

rhizomatous herb found in bogs or other very wet habitats; typically possessing a bract beneath each flower; native to cooler parts of the northern hemisphere.

Schilbeidae Family containing 40 species of small (to 350 mm) Old World freshwater catfishes (Siluriformes) ranging from Africa to southeast Asia; body slender and compressed, dorsal fin short with single spine, or absent; anal fin very long; 2–4 pairs of barbels present.

Schindleriidae Family containing 2 species of small (to 25 mm) neotenic marine perciform teleost fishes found in the north Pacific Ocean; adult body transparent retaining many larval characters.

Schisandraceae Small family of Illiciales containing about 50 species of glabrous, aromatic, evergreen or deciduous woody vines.

Schistostegales Order containing a single species of mosses (Bryidae) found in low-light habitats; protonema persistent, with light-refracting properties.

schizocarp A dry fruit developed from 2 or more one-seeded carpels that divide into separate units at maturity.

schizogenesis Reproduction by multiple fission; **schizogenetic.**

schizogony A type of multiple fission in which the parent organism (schizont) divides to form many new individuals (merozoites).

Schizomida Order of small (2–15 mm) predatory terrestrial arthropods (Arachnida); mesosoma bearing single pair of book lungs, metasoma with short terminal flagellum; chelicerae possessing fang, pedipalps spinose and raptorial; contains about 80 species, typically found in leaf litter of the tropics and subtropics; Schizopeltida.

Schizomycetes A group name formerly used for the bacteria when classified as fungi.

Schizopeltida Schizomida *q.v.*

schizophyte A plant that reproduces by fission.

Schizopyrenida Order of naked lobose amoebae (Gymnamoeba) typically with a flagellate phase in the life cycle; widely distributed in aquatic habitats and soil.

school To aggregate and swim together in an organized manner; also the swimming group itself.

Sciaenidae Drums, croakers; family containing about 200 species of coastal marine, brackish and freshwater, teleost fishes (Perciformes) widespread in tropical and temperate regions, that derive their common names from the ability to produce sounds by resonating the gas bladder; body elongate, weakly compressed, to 2 m in length; chin barbels may be present; 2 dorsal fins, the second elongate.

Scincidae Skinks; very large cosmopolitan family of terrestrial lizards comprising more than 1000 species; body length to 650 mm, limbs usually small, reduced or absent in some fossorial forms; habits diurnal, insectivorous, some species herbivorous; reproduction oviparous or viviparous.

Sciomyzidae Family containing about 500 species of small to large brown flies (Diptera) in which the larvae prey on aquatic or terrestrial molluscs.

scion That part or structure transplanted from one plant to another during a graft.

sciophilous Thriving in shaded situations, or in habitats of low light intensity; **sciophile, sciophily.**

sciophyllous Having leaves that can tolerate shaded situations; skiophyllous; *cf.* heliophyllous.

sciophyte A plant growing in shady situations, or in habitats of low light intensity; *cf.* heliophyte.

scissiparous Used of multicellular organisms that reproduce by fission.

Sciuridae Squirrels, marmots, prairie dogs; large cosmopolitan family containing about 250 species of diurnal, arboreal or terrestrial mammals (Rodentia); mostly herbivorous, feeding primarily on nuts, seeds and leaves, but some also taking invertebrates.

squirrel (Sciuridae)

Sciuromorpha Suborder of rodents comprising 7 families, Aplodontidae (mountain beaver), Sciuridae (squirrels), Geomyidae (pocket gophers), Heteromyidae (kangaroo rats), Castoridae (beavers), Anomaluridae (scaly-tailed squirrels) and Pedetidae (springhaas).

Scleractinia Stony corals, true corals; large, diverse order of colonial zooantharians found mainly in shallow tropical seas between 30° N and 30° S; play central role in the formation of coral reefs by producing massive calcareous (aragonitic) exoskeletons into which the

polyps can usually retract; colony growth takes place by asexual budding; formerly known as Madreporaria; over 2500 extant species species are known in addition to a large number of fossils.

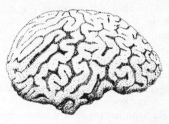

brain coral (Scleractinia)

Sclerodermatales Puff balls; order of gasteromycete fungi producing subglobular fruiting bodies; typically found growing in mycorrhizal association.

Sclerospongiae Class of sponges which secrete a basal skeleton of crystalline calcium carbonate; the living tissues are organized like those of demosponges and contain siliceous and spongin skeletal elements.

Scolecomorphidae Small family of tropical eastern African terrestrial caecilians (Gymnophiona) comprising a single genus and 6 species; body length to 450 mm, scales and tail absent; reproduction viviparous.

scolite A fossil worm-burrow.

Scolopacidae Snipe, sandpipers, curlews, turnstones; family containing about 85 species of small to large charadriiform shore birds cosmopolitan in freshwater and marine marsh and shore habitats, also occasionally in dry grassland and woodland; bill usually long and slender, straight or curved; habits gregarious, often migratory, monogamous, feeding mainly on invertebrates.

snipe (Scolopacidae)

Scolopendrida Order containing about 500 species of small to large (to 300 mm) epimorphan chilopods found in soil, leaf litter or under bark and stones; ocelli present or absent; trunk composed of 25 or 27 segments bearing 21 or 23 pairs of legs; genital apparatus concealed; body often with conspicuous coloration.

Scolopendromorpha Scolopendrida *q.v.*

Scolytidae Bark beetles; family of small, cylindrical beetles (Coleoptera) which excavate an egg chamber beneath tree bark; larvae fleshy and legless tunnelling away from chamber, through bark or wood.

Scomberesocidae Sauries, skippers; family containing 4 species of small (to 35 mm) surface-living marine beloniform teleost fishes; body elongate, slender, compressed; jaws prolonged into narrow beak bearing many very small teeth, upper jaw shorter than lower; typically active zooplankton feeders that commonly skip and jump at the surface in large schools.

Scombridae Tunas, mackerels; widespread tropical to temperate family of active, predatory, oceanic perciform teleost fishes; body spindle-shaped, to about 4 m in length; jaw teeth well developed, dorsal and anal fins with finlets; feed mainly on squid and other fishes, although smaller species are selective plankton feeders; contains 45 species many of which are important commercially and prized as sport fish.

mackerel (Scombridae)

Scombroidei Suborder of medium-sized to large, active pelagic predatory perciform teleost fishes comprising about 100 species in 6 families and including mackerels, tunas, billfishes, swordfish and cutlassfishes.

Scopelarchidae Pearleyes; family containing 17 species of pelagic predatory deep-sea myctophiform teleost fishes having telescopic eyes containing iridescent connective tissue; body compressed, mouth large; photophores present or absent; some forms hermaphroditic.

Scopelosauridae Family containing 5 species of deep-sea myctophiform teleost fishes widespread in tropical and temperate oceans; body slender, subcylindrical becoming compressed posteriorly, head depressed; mouth and eyes large, teeth small and sharp;

adipose fin present; photophores and swim bladder absent.

Scophthalmidae Turbot, brill, topknot; family of North Atlantic and Mediterranean flatfishes (Pleuronectiformes) in which the eyes are located on the left side; both pelvic fins well developed.

scopulous Pertaining to crags and steep overhanging cliffs.

Scorpaenidae Scorpionfishes, rockfishes, redfishes, turkeyfishes; family of bottom-living tropical and temperate marine scorpaeniform teleost fishes widespread in nearshore and shelf waters; body robust, to 1 m in length, head large and spinose; dorsal and anal fins spinose, fin spines venomous; reproduction oviparous, one genus viviparous; contains about 330 species, some important as food-fish; the turkeyfishes (subfamily Pteroinae) are popular with aquarists.

Scorpaeniformes Diverse order of marine, brackish and freshwater teleost fishes characterized by a large stout head, and body covered with spines or bony plates; comprises about 1000 species in 21 families, including scorpionfishes, rockfishes, stonefishes, sea robins, velvetfishes, sablefishes, greenlings, combfishes, flatheads, pigfishes, sculpins, poachers, lumpfishes and snailfishes.

Scorpiones Scorpions; order of terrestrial arthropods (Arachnida) most of which are ground-dwelling nocturnal predators feeding on other arthropods; a few are arboreal or cavernicolous; length 10–180 mm, body divided into a carapace-bearing cephalothorax and slender segmented abdomen; pedipalps large and chelate; mesosoma bearing pectines and 4 pairs of book lungs; metasoma armed with conspicuous terminal sting; contains about 1200 species.

scorpion (Scorpiones)

scorpion fly Mecoptera *q.v.*

scorpionfish Scorpaenidae *q.v.*

scotophilic Skotophilic *q.v.*; **scotophile**, **scotophily**.

scotophyte A plant living in darkness.

scototaxis Skototaxis *q.v.*; **scototactic**.

scototropism Skototropism *q.v.*; **scototropic**.

scouring rush Equisetophyta *q.v.*

screamer Anhimidae *q.v.*

screech beetle Hygrobiidae *q.v.*

screw pine Pandanales *q.v.*

Scrophulariaceae Foxglove, snapdragon, mullein, calceolaria, musk, toadflax, speedwell, lousewort, Indian paintbrush, antirrhinum; cosmopolitan family containing about 4000 species of mostly herbs, some of which are hemiparasitic or fully parasitic; fruit usually a 2-celled capsule with numerous seeds, rarely a berry.

foxglove (Scrophulariaceae)

Scrophulariales Order of Asteridae containing 12 families characterized by a sympetalous irregular corolla bearing fewer functional stamens than corolla lobes.

SCUBA An acronym of self-contained underwater breathing apparatus; aqualung.

scud Amphipoda *q.v.*

sculpin Common name of various fishes of the families Cottidae, Icelidae, Cottocomephoridae, Cottunculidae and Psychrolutidae.

Scuticociliatida Order of hymenostome ciliates abundant in marine and brackish habitats; either free-living or found in association with typically invertebrate hosts.

Scutigerida Small order of mostly tropical anamorphan chilopods; eyes large and multifaceted, antennae very long with up to 400 minute articles; trunk composed of 15 pedigerous segments covered by 8 large tergal plates; legs elongate, for rapid cursorial locomotion.

Scutigeromorpha Scutigerida *q.v.*

scuttle fly Phoridae *q.v.*

Scyliorhinidae Cat sharks; widespread family containing 85 species of small to medium-sized (to 1.5 m), mostly benthic, carcharhiniform elasmobranch fishes; found from intertidal to 2000 m, feeding on fishes, crustaceans, molluscs and echinoderms; reproduction oviparous to fully viviparous.

scyphistoma Sessile, polypoid larval stage of some jellyfishes (Scyphozoa) that may undergo asexual budding to produce further scyphistomae or transverse fission (strobilation) to form free-living medusae.

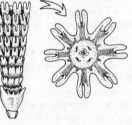

scyphistoma

Scyphostegiaceae Family of Violales comprising a single species of small glabrous tree found in Borneo; flowers borne in spikes, with 6 tepals, stamens forming a column and tiny carpels.

Scyphozoa Jellyfish; class of exclusively marine cnidarians in which the medusoid stage is dominant; the attached polyp small, with single whorl of tentacles around the mouth; medusae produced by budding and subsequent transverse fission (strobilation); medusae typically flat or bell-shaped jellyfish, usually free-swimming but occasionally attached, with marginal tentacles and a median mouth; comprises 4 orders, Coronatae, Rhizostomeae, Semaeostomeae and Stauromedusae.

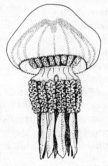

jellyfish (Scyphozoa)

Scytopetalaceae Small family of Theales containing about 20 species of glabrous woody plants, all native to Africa.

Scytosiphonales Small order of brown algae exhibiting alternation of a large parenchymatous sporophyte and a small pseudoparenchymatous, encrusting, gametophyte generation.

sea 1: A geographical division of an ocean. 2: An inland body of salt water.

sea anemone Actiniaria *q.v.*

sea bass Serranidae *q.v.*

sea bream Sparidae *q.v.*

sea breeze A light wind blowing off the sea on warm days when the land surface is warmer than the sea surface; *cf.* land breeze.

sea buckthorn Elaeagnaceae *q.v.*

sea cow Sirenia *q.v.*

sea cucumber Holothuroidea *q.v.*

sea daisy Concentricycloidea *q.v.*

sea devil Ceratiidae *q.v.*

sea fan Gorgonacea *q.v.*

sea gooseberry Ctenophora *q.v.*

sea grass Zosteraceae *q.v.*

sea hare Anaspidea *q.v.*

sea holly Apiaceae *q.v.*

sea horse Syngnathidae *q.v.*

sea lettuce Ulvales *q.v.*

sea lily Crinoidea *q.v.*

sealion Otariidae *q.v.*

sea moth Pegasidae *q.v.*

sea mouse Phyllodocida *q.v.*

sea pansy Pennatulacea *q.v.*

sea pen Pennatulacea *q.v.*

sea perch Embiotocidae *q.v.*

sea robin Triglidae *q.v.*

sea slug Nudibranchia *q.v.*

sea snake Hydrophiidae *q.v.*

sea snow The settlement of particulate organic detritus and living organisms down the water column.

sea spider Pycnogonida *q.v.*

sea squirt Ascidiacea *q.v.*

sea star Asteroidea *q.v.*

sea toad Chaunacidae *q.v.*

sea turtle Cheloniidae *q.v.*

sea urchin Echinoidea *q.v.*

sea wasp Cubozoa *q.v.*

sea whip Gorgonacea *q.v.*

seafloor spreading The formation and lateral displacement of oceanic crust by upwelling of material from the Earth's mantle along mid-ocean ridges.

seal 1: Otariidae *q.v.* (fur seal). 2: Phocidae *q.v.* (hair seal).

seamount An elevated area of limited extent rising 1000 m or more from the surrounding ocean floor; guyot.

search image A model for visual comparison, formed by individual predators which select certain prey types and ignore others.

Searsiidae Tubeshoulders; family containing 30 species of mostly small deep-sea salmoniform teleost fishes closely resembling the slickheads (Alepocephalidae); possessing sac of luminous fluid behind the gill cover, opening to outside via a tubular papilla.

season Any of the major climatic periods of the year; *cf.* aestival, hibernal, hiemal, serotinal, vernal.

seasonal Showing periodicity related to the seasons; used of vegetation exhibiting pronounced seasonal periodicity marked by conspicuous physiognomic changes.

Secernentea Class of nematodes almost all of which are terrestrial, although some occur in fresh water; most are parasitic, with over 3000 species found in animal hosts and over 2000 species found in plant hosts; characterized, in general, by a pore-like external opening of the cephalic chemosensory organs (amphids), 3 oesophageal glands, and sensory glands (phasmids) in the tail area; comprises 3 subclasses, Diplogasteria, Rhabditia and Spiruria.

secondary consumer Any heterotrophic organism that feeds directly on a primary consumer *q.v.*

secondary production The assimilation of organic matter by a primary consumer; *cf.* gross secondary production, net secondary production.

secondary sexual characters Those differences between the sexes that relate to structures other than the reproductive organs and gametes; *cf.* primary sexual characters.

secondary succession An ecological succession resulting from the agricultural or other activities of man, or that follows destruction of all or part of an earlier community.

secretary bird Sagittariidae *q.v.*

secretion 1: Any substance or product elaborated and released by a cell or gland to perform a specific function. 2: The process of elaboration and release; secretory; *cf.* excretion.

secular 1: Pertaining to periods of hundreds of years; *cf.* decennial, millennial. 2: Used of a process persisting for an indefinitely long period.

Sedentaria Taxon formerly used to group together all sedentary and tubicolous

PARTICLE GRADE SCALES				
UDDEN (mm)	PHI VALUE	WENTWORTH		
		BOULDER		
256.0	−8.0			
		COBBLE	G	
64.0	−6.0		R	
		PEBBLE	A	
			V	
4.0	−2.0		E	
		GRANULE	L	
2.0	−1.0			
		VERY COARSE		
1.0	0.0			
		COARSE		
0.5	1.0		S	
		MEDIUM	A	
			N	
0.25	2.0		D	
		FINE		
0.125	3.0			
		VERY FINE		
0.0625	4.0			
		SILT	M	
0.0039	8.0		U	
		CLAY	D	
0.00024	12.0			
		COLLOID		

sediment particles sizes

polychaete worms; *cf.* Errantia.

sedentary Attached to the substratum; not free-living.

sedge Cyperaceae *q.v.*

sediment Particulate matter that has been transported by wind, water or ice and subsequently deposited, or that has been precipitated from water; sedimentation.

sediment particle sizes See table for particle size classification and terminology.

sedimentary rock Rock formed by deposition of particulate matter transported by wind, water or ice, or by precipitation from solution in water under normal surface temperatures and pressures, or by the aggregation of inorganic material from skeletal remains; *cf.* igneous rock, metamorphic rock.

seed The dispersal and reproductive structure that develops from the fertilized ovule in higher plants, comprising an embryo with food supply either in the endosperm or in the cotyledons.

seed bank The store of dormant seeds buried in soil.

seed fern Pteridospermales *q.v.*

seedling stratum The lowest layer of a stratified forest community comprising germinating plants.

seedsnipe Thinocoridae *q.v.*

segmented spider Any of the trapdoor spiders in the suborder Mesothelae *q.v.*

segregation, law of Mendel's first law; that in sexual organisms the two members of an allele pair or a pair of homologous chromosomes separate during gamete formation and that each gamete receives only one member of the pair.

seiche A standing wave oscillation of an enclosed or partially enclosed water body that continues after the cessation of the original generating force, such as wind or heavy rain.

seismonasty A growth movement of a plant in response to a non-directional shock or mechanical vibration stimulus; **seismonastic, seismonic.**

seismotaxis A directed response of a motile organism to mechanical vibration or a shock stimulus; **seismotactic.**

seismotropism An orientation response to mechanical vibration or a shock stimulus; **seismotropic.**

Seisonidea Class of exclusively marine rotifers found as epizoites on the gills of the crustacean *Nebalia*; characterized by a large body ending in a proximal adhesive disk and by the possession of fully developed males which produce spermatophores; comprises a single order, Seisonida.

Selachii Group of cartilaginous fishes containing all the sharks; Pleurotremata.

Selaginellales Spike mosses; small terrestrial plants usually with creeping stems bearing spirally arranged scale-like or needle-like leaves; roots dichotomously branching; sporangia borne in cones.

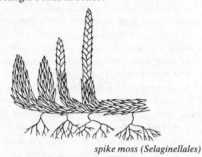

spike moss (Selaginellales)

selection 1: Non-random differential reproduction of different genotypes in a population. 2: Natural selection *q.v.*

selective breeding The selection by man of particular genotypes in a population because they exhibit desired characters; artificial selection.

selenophyte A plant adapted to, or tolerating, high selenium levels in the soil, frequently used as an indicator of this particular soil type.

selenotropism An orientation response to moonlight; **selenotropic.**

self-compatible Used of a plant that can be self-fertilized.

self-fertilization The union of male and female gametes produced by the same individual; *cf.* cross fertilization.

selfing Self-fertilizing or self-pollinating.

self-pollination Transfer of pollen from anther to stigma of the same flower or to another flower on the same plant; *cf.* cross pollination.

self-sterility The condition of hermaphroditic organisms that cannot produce viable offspring by self-fertilization.

Seligeriales Small order of mosses (Bryidae); small to medium-sized plants typically forming tufts or cushions on rocks.

SEM Scanning electron microscope.

Semaeostomeae Order of Scyphozoa found commonly in coastal waters of all seas and including the typical disk-shaped jellyfish of temperate regions; characterized by a simple mouth opening with 4 broad, lobed oral arms.

sematic Pertaining to or serving as a signal or danger warning.

semelparous Pertaining to organisms that have only one brood during the lifetime; **semelparity** (big-bang reproduction); *cf.* iteroparous.

semestrial Pertaining to periods of 6 months; half-yearly.

semidiurnal Having a period of about half a lunar day; equal to 12.42 h.

Semionotiformes Order of primitive bony fishes with a single extant family, Lepisosteidae (gars).

semiotics The study of communication.

semiparasitic Pertaining to an organism (partial parasite) that derives only part of its nourishment from its host, or that lives for only part of its life cycle as a parasite; **semiparasite.**

sempervivum Crassulaceae *q.v.*

senescence The biological process of ageing; **senescent.**

senility The degenerative condition of old age; post-reproductive; **senile.**

senna Caesalpiniaceae *q.v.*

sensibility The capacity to perceive a stimulus.

sensitivity The capacity of an organism to respond to stimuli; irritability; **sensitive.**

sensory Pertaining to a sense organ or sensation.

sepal Each element in the outermost whorl of a typical flower, probably representing a leaf modified as part of the protective outer layer (the calyx) of a flower.

sepicolous Living in hedgerows; **sepicole.**

Sepioidea Cuttlefish, Ram's horn; order of benthic or near bottom cephalopod molluscs characterized by 8 arms and 2 tentacles which are used to capture prey; shell typically internal, forming the cuttlebone, but may be spirally coiled or absent.

cuttlefish (Sepioidea)

septentrional Northern, northerly.

septic Heavily polluted; used of a habitat or zone of fresh water rich in decomposing organic matter, high in carbon dioxide and very low in dissolved oxygen.

Septobasidiales Order of phragmobasidiomycete fungi containing about 200 species which grow on stems and leaves of woody plants; producing fruiting bodies usually adhering close to surface of host; many species are involved in complex associations with scale insects.

sequoia Taxodiaceae *q.v.*

seral stages The developmental stages of an ecological succession not including the climax community.

sere A succession of plant communities in a given habitat leading to a particular climax association; a stage in a community succession; **seral.**

seriemas Cariamidae *q.v.*

serir A stony desert.

serodeme A local interbreeding population characterized by immunological properties *cf.* deme.

serology The study of antigens and antibodies; **serological.**

serotaxonomy Taxonomy based on serological characters.

serotinal Pertaining to late summer.

serotinous Coming late; used particularly with reference to late flowering plants, and for behaviour that occurs late in the day, or late in the season.

Serpentes Snakes; large cosmopolitan suborder of limbless reptiles comprising about 2250 species in 14 families; body length up to 10 m, forelimbs and girdle absent, hindlimbs and girdle vestigial or absent; tongue forked and extensible; left lung small or absent; habits terrestrial, arboreal or aquatic, carnivorous, many species venomous; reproduction oviparous or ovoviviparous; Ophidia.

serpentinophilous Thriving in habitats rich in the mineral serpentine; **serpentinophile, serpentinophily.**

serpulite A fossil polychaete tube.

Serranidae Sea basses, grouper; family containing about 370 species of primarily predatory demersal marine teleost fishes (Perciformes) widespread in inshore tropical and temperate waters, but also found in brackish and occasionally freshwater habitats; body robust to elongate, length to 3 m; 1 or 2 dorsal fins presnt; mouth large and protractile; many species hermaphroditic.

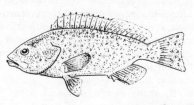

grouper (Serranidae)

Serrasalmidae Piranhas, silver dollars; family containing 65 species of small to medium-sized South American freshwater characiform teleost fishes; dorsal fin situated at or behind mid-body, adipose fin present; dentition variable; piranhas are predators feeding on fishes and mammals, others herbivorous on fruits and seeds.

Serrivomeridae Family containing 10 species of pelagic deep-sea anguilliform teleost fishes resembling duckbill eels; body smooth, tapering posteriorly, jaws prolonged.

sesame Pedaliaceae *q.v.*

sessile Non-motile; permanently attached at the base.

seston The total particulate matter suspended in water.

sewage Liquid or solid waste matter channelled through sewers.

sex The sum of all structural, functional and behavioural characteristics distinguishing males, females and hermaphrodites; **sexual.**

sex chromosome A chromosome or group of chromosomes of eukaryotic organisms, represented differently in the sexes, responsible for the genetic determination of sex.

sex linkage The location of a gene or character on a sex chromosome.

sex mosaic A phenotype that manifests characters of both the male and the female.

sex ratio The relative number of males and females in a population, expressed as the number of males per 100 females, or as a simple ratio.

sexology The study of sex and sexual behaviour.

sexual Pertaining to sex, sex-related characters or to the process of sexual reproduction.

sexual dimorphism Marked phenotypic differences between males and females of the same species.

sexual reproduction Reproduction involving the regular alternation of gamete formation by meiosis and gamete fusion (karyogamy) to form a zygote.

sexual selection 1: Choosing a mate; the ability of individuals to differentiate and select a mate. 2: The differential abilities of individuals to acquire mates in competition with other individuals of the same sex.

shad Clupeidae *q.v.*

Shantung soil Non-calcic Brown soil *q.v.*

shark Elasmobranchii *q.v.*

shark-toothed whale Squalodontoidea *q.v.*

shearwater Procellariidae *q.v.*

sheep Bovidae *q.v.*

sheet erosion The removal of a more or less uniform layer of surface soil over a wide area by the action of runoff water.

sheet-web spider Agelenidae *q.v.*

shelf break The outer margin of the continental shelf marked by a pronounced increase in the slope of the sea bed.

she-oak Casuarinales *q.v.*

shield bug Pentatomidae *q.v.*

shingle Rounded rock fragments larger than 16 mm in diameter.

shipworm A member of the family Teredinidae of the order Myoida, containing about 65 species of usually marine bivalve molluscs which burrow into wood.

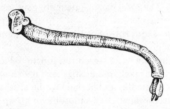

shipworm

shore fly Ephydridae *q.v.*

showy mistletoe Loranthaceae *q.v.*

shrew 1: Soricidae *q.v.* 2: Tupaiidae *q.v.* (tree shrew). 3: Macroscelididae *q.v.* (elephant shrew).

shrike Laniidae *q.v.*

shrimp Dendrobranchiata *q.v.* and Pleocyemata *q.v.*; prawn.

shrimp plant Acanthaceae *q.v.*

shrimpfish Centriscidae *q.v.*

SI Units (Système International d'Unités) An internationally agreed system of metric units; comprising seven base units – metre, kilogram, second, ampere, Kelvin, candela and mole; other metric units may be derived units or supplementary units.

Sialidae Alderflies; family containing about 50 species of primitive neuropteran insects which typically have slow, awkward flight; subaquatic, carnivorous larvae found in well-aerated fresh water.

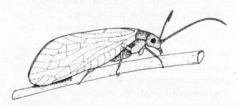

alderfly (Sialidae)

sib mating Mating of progeny derived from the same parents.

sib pollination Pollination of one plant by another plant derived from the same parents.

siblings Organisms sharing close genetic relationship, especially those derived from the same parents; **sibs.**

sibling species Pairs or groups of closely related and frequently sympatric species which are morphologically indistinguishable but which are reproductively isolated.

siccocolus Living in dry arid habitats; **siccocole.**

side-necked turtle Pleurodira *q.v.*

sidereal day The mean time taken for one revolution of the Earth; each year comprises 365.256 sidereal days; *cf.* solar day.

siderophilous Thriving in an iron-rich medium; **siderophile, siderophily.**

siderotrophic Used of water bodies rich in dissolved iron compounds.

sidewinder Crotalidae *q.v.*

Sierozem A zonal soil with pale grey A-horizon, a slightly darker alkaline B-horizon merging into a calcareous lower layer, formed in temperate to cool arid climates; cerozem.

Siganidae Rabbitfishes; family containing 10 species of herbivorous Indo-Pacific marine perciform teleost fishes; body oval, dorsal and anal fins elongate bearing venomous spines.

signal Any behaviour that conveys information from one individual to another.

siliceous ooze A fine-grained pelagic sediment containing more than 30% siliceous skeletal remains of pelagic organisms, primarily radiolarians and diatoms; *cf.* ooze.

silicoflagellate Dictyochophycidae *q.v.*

Silicoflagellida Dictyochophycidae *q.v.* treated as an order of the protozoan class Phytomastigophora.

silicolous 1: Living in soil rich in silica or silicates; **silicole.** 2: Growing on flints.

silicula A dry dehiscent fruit developed from 2 carpels fused to form a short, broad, flattened pod with 2 loculi separated by a false septum.

silification A process of fossilization in which the fossil is encrusted, partially replaced or impregnated with silica derived by the degradation of silicates, from biogenic organic silica solutions or from products of vulcanism.

siliqua A dry dehiscent fruit developed from 2 carpels fused to form a long, narrow pod with 2 loculi separated by a false septum; silique.

silkcotton tree Bombacaceae *q.v.*

silkworm Bombycidae *q.v.*

silky flycatcher Bombycillidae *q.v.*

silky oak Proteaceae *q.v.*

Silphidae Carrion beetles; family of beetles (Coleoptera) most of which are carrion feeders or scavengers although some are predators and plant feeders; contains about 175 species.

silt Cohesionless sediment particles between 0.004 and 0.0625 mm in diameter; coarse mud; see sediment particle sizes.

silt load The quantity of particulate matter carried in suspension by a stream or river; *cf.* bed load.

Silurian A geological period within the Palaeozoic (*c.* 441–413 million years B.P.); see geological time scale.

Siluridae Family of small to large (70 mm – 3 m) Eurasian freshwater catfishes (Siluriformes); body elongate, naked, 2–3 pairs of barbels present; dorsal and pelvic fins small to absent, adipose fin absent, anal fin elongate, in some species continuous with caudal fin.

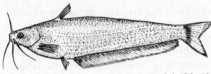

catfish (Siluridae)

Siluriformes Catfishes; cosmopolitan order of small to very large, mostly freshwater, teleost fishes comprising over 2000 species in about 30 families; sexual dimorphism absent; median fins typically with strong spines, pelvics posterior to pectorals; gas bladder connected to inner ear.

silver berry Elaeagnaceae *q.v.*

silver dollar Serrasalmidae *q.v.*

silver-bell tree Styracaceae *q.v.*

silverfish Thysanura *q.v.*

silverside Atherinidae *q.v.*

silvicolous Inhabiting woodland; **silvicole.**

silvics The study of forest trees; **silvical.**

silviculture The management and exploitation of forests.

Simaroubaceae Tree of heaven; family of Sapindales containing about 150 species of trees and shrubs mostly with very bitter bark and seeds; primarily tropical in distribution; flowers small with an intrastamenal disk; fruit often drupe-like.

Simenchelyidae Monotypic family of small (to 600 mm) deep-sea eels (Anguilliformes); snout short and blunt, mouth small; dorsal and anal fins long, continuous with caudal.

Simmondsiaceae Jojoba; family of Euphorbiales containing a single species of

xeromorphic, evergreen dioecious shrub producing large, wax-containing seeds; native to southwestern United States and Mexico.

Simuliidae Black flies, buffalo flies; family containing about 1100 species of small biting flies (Diptera) that feed on the blood of warm-blooded vertebrates and may be serious pests of cattle; one tropical species is a vector of river-blindness caused by a filarial parasite.

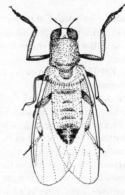

black fly (Simuliidae)

simultaneous hermaphrodite Synchronous hermaphrodite *q.v.*

sink A buffering reservoir; any large reservoir that is capable of absorbing or receiving energy or matter without undergoing significant change.

siotropism An orientation response to shaking; **siotropic.**

Siphonales Bryopsidales *q.v.*; also treated as a distinct class of the protoctistan phylum Chlorophyta.

Siphonaptera Fleas; order containing about 1750 species of small wingless holometabolous insects that as adults are blood-feeding ectoparasites of warm-blooded animals, mainly mammals but including some birds; body compressed and hairy, mouthparts specialized for piercing and sucking; legs

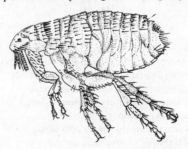

flea (Siphonaptera)

adapted for grasping and jumping; the larvae are blind and limbless, typically non-parasitic feeding on organic matter around domicile of host; fleas are of medical and veterinary importance as disease vectors and intermediate hosts of various parasitic worms.

Siphoniulida Order of minute (to 7 mm) helminthomorphan diplopods (millipedes) comprising only 2 species from Sumatra and Guatamala.

Siphonocladales Small order of tropical marine green algae in which the thallus is non-cellular (vesicular) and attached by rhizoids which are often septate.

siphonogamy Fertilization by means of a pollen tube; **siphonogamous.**

Siphonophora An exclusively marine order of colonial Hydrozoa found worldwide in surface waters; characterized by the lack of a medusa phase and by the polymorphism of the individual zooids; all zooids are in communication via their gastrovascular canals but zooids may be specialized as a gas float (pneumatophore), swimming bell (nectophore), for prey capture (dactylozooid), feeding (gastrozooid) or reproduction (gonozooid); several species, including the Portuguese man o'war, inflict painful stings.

Portuguese man o'war (Siphonophora)

Siphonophorida Order of mainly tropical helminthomorphan diplopods (millipedes) having a slender setose or glabrous body lacking a median dorsal suture; head prolonged anteriorly, bearing reduced suctorial mouthparts.

Siphonostomatoida Order containing about 1500 species of predominantly marine copepod crustaceans which have a tubular sucking mouth and are typically parasites of fishes and marine invertebrates.

Sipuncula Peanut worms; small phylum of unsegmented sedentary coelomate marine worms that burrow into soft sediments, gravel, calcareous rock or coral, inhabit crevices or live beneath stones, and range from the intertidal to abyssal depths; body usually divided into stout trunk and elongate retractable anterior introvert carrying the mouth and often encircled by tentacles; gut U-shaped, opening on dorsal surface at base of introvert; larvae planktonic; contains about 320 species.

peanut worm (Sipuncula)

sire In animal breeding, the male parent; *cf.* dam.

Sirenia Sea cows; order of large aquatic herbivorous mammals (Ferungulata) comprising 2 families, Trichechidae (manatee) and Dugongidae (dugong); head robust bearing large lips, teeth present or absent; body massive, forelimbs modified as flippers, hindlimbs absent, tail forming lateral flukes; eyes small, ear lobes absent.

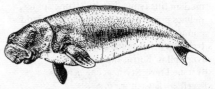

sea cow (Sirenia)

Sirenidae Sirens; family containing 3 species of eel-like aquatic salamanders (Caudata) from southeastern United States; hindlimbs and pelvic girdle absent; metamorphosis incomplete, external gills and gill slits present, eyelids absent; eggs laid singly or in groups on submergent vegetation, fertilization external.

Siricidae Horntails; family containing about 90 species of large wood wasps (Hymenoptera) typically having a long cylindrical body lacking a distinct waist and coloured black banded with red or yellow, or with metallic blue sheen; larvae bore into trees or wood, and may carry a fungus that causes damage to timber.

Sisoridae Family containing about 40 species of freshwater catfishes (Siluriformes) found in fast-flowing montane streams of south and southeast Asia; body depressed, to 2 m length, some forms with ventral adhesive organ; dorsal fin bearing 2 spines; 4 pairs of barbels present.

sitotropism An orientation response to a food stimulus; **sitotropic.**

Sittidae Nuthatches; family containing about 20 species of small passerine birds found primarily in forests and rocky habitats of the northern hemisphere; bill stout and compressed; legs short, toes and claws elongate, strong; habits solitary or weakly gregarious, arboreal, non-migratory, feeding mainly on insects and seeds; nest in tree or rock cavity.

sivapithecines Extinct group of Miocene apes known from fossil fragments found in Europe, Turkey, India, Pakistan and China; closely resemble the ramapithecines and placed by some authorities in the same genus, *Sivapithecus*; their hominid affinities are uncertain, having facial and other features that link them also to chimpanzees and gorillas as well as to orang-utans.

skate Rajiformes *q.v.*

skatobios Those organisms inhabiting detritus or faecal matter; **skatobiont, skatobiontic.**

skeletal soil Lithosol *q.v.*

skeleton shrimp Caprellidea *q.v.*

skilfish Anoplopomatidae *q.v.*

skimmer Rynchopidae *q.v.*

skin beetle Dermestidae *q.v.*

skink Scincidae *q.v.*

skiophilous Thriving in shaded situations, or in habitats of low light intensity; **skiophile, skiophily.**

skiophyllous Sciophyllous *q.v.*

skiophyte A plant growing in shady situations or in habitats of low light intensity; *cf.* heliophyte.

skipper 1: Scomberesocidae *q.v.* 2: Hesperiidae *q.v.*

skotophilic 1: Thriving in darkness or in darkened situations; scotophilic; **skotophile, skotophily.** 2: Pertaining to the dark phase of a light/dark cycle.

skotoplankton Planktonic organisms living in darkness, below the photic zone; *cf.* knephoplankton, phaoplankton.

skototaxis The directed movement of a motile organism towards a dark area, as distinct from negative phototaxis; scototaxis; **skototactic.**

skototropism An orientation movement towards a dark area; scototropism; **skototropic.**

skua Stercorariidae *q.v.*

skunk Mustelidae *q.v.*

skunk cabbage Araceaceae *q.v.*

slack water An interval of low velocity tidal current, usually the period of reversal between ebb and flow.

slater Oniscoidea *q.v.*

slick An aggregation of floating debris resulting in reduced wave activity and a shiny water surface.

slickhead Alepocephalidae *q.v.*

slime mould Myxomycota *q.v.*

slime net Labyrinthulata *q.v.*

slimehead Trachichthyidae *q.v.*

sloth Bradypodidae *q.v.* (tree sloth).

slow worm Anguidae *q.v.*

sludge worm An oligochaete worm of the family Tubificidae (Tubifieina); exhibit extreme tolerance to heavily polluted fresh water.

slug Stylommatophora *q.v.* and Systellommatophora *q.v.*

smelt 1: Osmeridae *q.v.* 2: Argentinidae *q.v.* (herring smelts). 3: Bathylagidae *q.v.* (black smelts).

Smilacaceae Nearly cosmopolitan family of Liliales comprising about 300 species of mostly climbing, herbaceous or woody vines with branched leafy stems frequently bearing recurved prickles; leaves petiolate and often with a pair of tendrils arising from the petiole, blades expanded and with curved convergent veins.

smoke tree Anacardiaceae *q.v.*

smooth hound Triakidae *q.v.*

smut fungus Ustilaginales *q.v.*

snaggletooth Astronesthidae *q.v.*

snail Gastropoda *q.v.*

snailfish Cyclopteridae *q.v.*

snake Serpentes *q.v.*

snake eel Ophichthidae *q.v.*

snake lizard Pygopodidae *q.v.*

snake mackerel Gempylidae *q.v.*

snake plant Agavaceae *q.v.*

snake star Ophiuroidea *q.v.*

snake vine Dilleniaceae *q.v.*

snakefly Raphidiidae *q.v.* (Neuroptera).

snakehead Channidae *q.v.*

snake-necked turtle Chelidae *q.v.*

snapdragon Scrophulariaceae *q.v.*

snapper Lutjanidae *q.v.*

snapping turtle Chelydridae *q.v.*

snipe Scolopacidae *q.v.*

snipe eel 1: Nemichthyidae *q.v.* 2: Cyemidae *q.v.*

snipefish Macroramphosidae *q.v.*

snook Centropomidae *q.v.*

snow flea Mecoptera *q.v.*

snow line The line marking the lower altitudinal limit of perpetual snow.

snowberry Caprifoliaceae *q.v.*

snowdrop tree Styracaceae *q.v.*

soap fish Grammistidae *q.v.*

soapberry Sapindaceae *q.v.*

soari nut Caryocaraceae *q.v.*

social Pertaining to a society, or to the interrelationships between organisms.

social parasitism A symbiosis in which a parasite diverts to its own use the workers of the host-species society.

socialization The modification of the behaviour of an organism as an adaptation to social existence.

society 1: A group of organisms of the same species which has a social structure and consists of repeated members or modular units. 2: A ranked category in the classification of vegetation; a small plant climax community.

sociobiology Study of the biological basis of social behaviour.

sociology The study of communities or of collectively organized groups of animals or plants.

sociotomy The fission of a termite colony.

soft coral Alcyonacea *q.v.*

soft tick Any of the ixodid tick family Argasidae that have a leathery cuticle, a ventrally located head and no head plate.

soft-shelled clam Myoida *q.v.*

soft-shelled turtle Trionychidae *q.v.*

soil The superficial weathered layers of the Earth's crust and any intermixed organic material.

soil evolution The sequence of changes in a soil profile during development towards maturity.

soil granulation scale A scale of soil particle sizes: coarse gravel/rubble (greater than 20 mm diameter), gravel (20-2 mm diameter), coarse sand (2-0.2 mm diameter), fine sand (0.2-0.02 mm diameter), silt/dust (0.02-0.002 mm diameter), fine silt/clay (less than 0.002 mm diameter).

soil horizons The horizontal subdivisions of a soil profile; often given the notation O-, A-, E-, B-, C- and R-horizons. See below.

soil moisture deficit The amount of water required to restore soil to its field capacity *q.v.*

soil profile A vertical section of a soil through the system of horizontal layers (soil horizons)

HORIZON			DESCRIPTION	
O	Ol or L	A_{00}	Litter layer; loose fragmented organic matter in which original plant structures are still readily discernible	Horizons of maximum biological activity and eluviation
	Of or F		Fermentation layer; partially decomposed organic matter	
	Oh or H	A_0	Humified layer; heavily transformed organic matter with no discernible macroscopic plant remains	
A	A_1		Dark-coloured horizon, with high organic content intermixed with minerals. Ap denotes ploughed layer	
E	A_2 or E_a		Light-coloured horizon, with low organic content due to high eluviation	
	A_3		Transitional layer, sometimes absent	
B	B_1		Transitional layer, sometimes absent	Horizons of illuviation; accumulations of clay, humus or iron oxides leached or translocated from the upper layers
	B_2		A dark layer of accumulation of transported silicate, clay, minerals, iron and organic matter showing the maximum development of blocky and/or prismatic structure	
	B/C or B_3		Transitional layer	
C			Weathered parent material, comprising mineral substrate with little or no structure; containing a gleyed layer or layers of accumulated calcium carbonate or sulphate in some soils	
R	0		Underlying rock	

soil horizons

more or less within the depth zone penetrated by plant roots.

soil reaction The measure of acidity or alkalinity of a soil, categorized according to pH value; extremely acid (less than 4.5), very strongly acid (4.5–5.0), strongly acid (5.1–5.5), moderately acid (5.6–6.0), slightly acid (6.1–6.5), neutral (6.6–7.3), slightly alkaline (7.4–7.8), moderately alkaline (7.9–8.4), strongly alkaline (8.5–9.0), very strongly alkaline (greater than 9.1).

soil skeleton The chemically inert quartz fraction of a soil.

soil structure The macrostructure of a soil; the aggregation of soil particles into compound units (or peds), described according to shape of units – crumb (rounded), prismatic (vertically elongated), and platy (flattened).

Solanaceae Potato, tomato, capsicum (peppers), belladonna, Chinese lantern, aubergine, egg plant, thorn apple, tobacco, petunia, deadly nightshade, henbane, jimsonweed; large family containing about 2300 species of herbs, shrubs, vines or trees; often prickly, commonly producing alkaloids especially of the tropane and nicotine groups; cosmopolitan in distribution but best developed in South America; flowers usually regular and bisexual, commonly with 5 sepals and petals, 5 or fewer stamens and a superior ovary that develops into a berry or capsule.

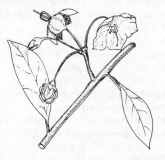

solanum (Solanaceae)

Solanales Order of Asteridae containing 8 families of plants often producing alkaloids but never cardiotonic glycosides; the majority of species are in the families Convolvulaceae and Solanaceae.

solar day The mean time interval between consecutive sunrises, or any other given position of the sun; 24 h; *cf.* lunar day, sidereal day.

solar radiation The radiant energy emitted by the sun; the ecosphere receives solar radiation

at wavelengths ranging from 290 nm to about 3000 nm.

solarization The inhibitory effect of extremely high light intensities on photosynthesis, resulting largely from photo-oxidation.

soldier beetle Cantharidae *q.v.*

soldier fly Stratiomyidae *q.v.*

soldierfish Holocentridae *q.v.*

sole Soleidae *q.v.*

Soleidae Soles; family containing about 120 species of primarily shallow marine soleiform flatfishes (Pleuronectiformes) in which the eyes are on the right side of the head; body length to 500 mm; some species commercially important as food-fishes.

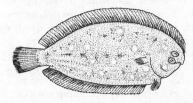

sole (Soleidae)

Solemyoida Order of burrowing, marine protobranch bivalve molluscs characterized by the anterior shell-closing muscle being much larger than the posterior and by a papillate sole on the foot.

Solenochilida Extinct order of globose cephalopod molluscs (Nautiloidea) in which the siphuncle is ventral and the sutures simple; known from Lower Carboniferous to end of Permian.

Solenodontidae Family of primitive, rat-like, terrestrial, nocturnal mammals (Insectivora) found in forests of Cuba and Hispaniola; contains 2 Recent monotypic genera, both rare, one possibly extinct.

solenodon (Solenodontidae)

Solenogastres Class of marine, free-living or epizoic molluscs which feed almost exclusively on cnidarians; characterized by a vermiform, laterally compressed body which is primitively shell-less although the mantle enclosing most of the body comprises one or more layers of embedded calcareous scales or spicules and a covering cuticle; foot represented by a long midventral groove; comprises about 180 species in 4 orders, Cavibelonia, Neomeniamorpha, Pholidoskepia and Sterrofustia; formerly known as the Aplacophora.

solenogastre (Solenogastres)

Solenostomidae Family containing 4 species of small (to 170 mm) tropical Indo-Pacific, shallow marine gasterosteiform pipefishes; body stout, with armour of bony plates, snout elongate; female pelvic fins attached to body distally to form a brood pouch for eggs.

Soleoidei Soles; suborder of primarily marine flatfishes (Pleuronectiformes) comprising 2 families, Soleidae and Cynoglossidae; median fins lacking spine rays, pectorals small or absent; jaws asymmetrical, vomerine and palatine teeth absent; use undulations of body for locomotion.

solifluction The gradual downhill flow of fragmented surface material, typically over a frozen substratum.

Solifugae Solpugida *q.v.*

Solifugida Solpugida *q.v.*

solitary wasp Eumenidae *q.v.*

Solod soil Soloth soil *q.v.*

Solonchak An intrazonal soil with grey shallow salty surface layer over a fine mulch and a greyish salty lower horizon, formed in subhumid to arid climates under conditions of poor drainage and halophytic vegetation.

Solonetz An intrazonal soil with a thin friable surface layer over a dark alkaline horizon, formed in subhumid to arid climates under conditions of moderate drainage and sparse halophytic vegetation; soil formed by leaching of salts from a Solonchak *q.v.* soil.

Soloth An intrazonal soil with a light brown friable leached surface layer over a dark lower horizon, formed as a desalinated, decalcified, degraded Solonetz *q.v.* soil, in subhumid to arid climates under conditions of moderate drainage; Solod soil.

Solpugida Wind scorpions, sun scorpions; order of long-legged predatory arthropods (Arachnida) comprising about 900 species confined mostly to deserts and arid habitats; chelicerae massive, pedipalps elongate and leg-like; first pair of legs small and slender,

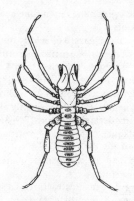

sun scorpion (Solpugida)

used as feelers; abdomen 11-segmented, bearing 3 pairs of tracheae.

solstitial tide The tide occurring near the times of the solstices when the sun reaches its greatest northerly or southerly declination; *cf.* tropic tide.

solubility The extent to which a substance (solute) mixes with a liquid (solvent) to produce a homogeneous mixture (solution).

Somali Current A warm surface ocean current that flows northwards off the coast of Somalia, produced by the southwesterly winds of the summer monsoon; South West Monsoon Drift; see ocean currents.

Somasteroidea Subclass of primitive asterozoan echinoderms comprising a single living species in the order, Goniactinida.

somatic Pertaining to the body or any non-germinal cell, tissue, structure or process.

somatogamy A form of plasmogamy *q.v.* in which somatic cells or hyphae fuse.

somatogenesis The development of somatic structure; **somatogenetic.**

somatropism An orientation response to the substratum; **somatotropic, somatropic.**

sonagram A graphic representation of animal vocalizations.

song An elaborate vocal signal.

soniferous Capable of producing sounds; **sonic.**

Sonneratiaceae Small family of Myrtales containing 8 species of tanniferous trees with internal phloem; all native to the Old World tropics; flowers often nocturnal, sour-smelling and pollinated by bats.

Sooglossidae Small family containing 3 species of terrestrial frogs (Anura) confined to the Seychelle Islands.

sooty mould Chaetothyriales *q.v.*

Sordariales Order of pyrenomycete fungi which

are widely distributed on plant debris, seeds and dung; characterized by a dark, membranous ascocarp wall and mostly cylindrical asci which are deliquescent at maturity.

sorghum Poaceae *q.v.*

Soricidae Shrews; family containing about 250 species of small, active, mouse-like mammals (Insectivora) which may be terrestrial, semiaquatic or burrowing.

shrew (Soricidae)

Soricoidea Superfamily of insectivores (Lipotyphla) comprising the families Soricidae (shrews) and Talpidae (moles), also sometimes including the Solenodontidae (solenodon), Tenrecidae (tenrecs) and Chrysochloridae (golden moles).

sorosis A multiple fruit derived from the ovaries of several flowers.

sorrel Polygonales *q.v.*

sorus A reproductive structure comprising several spore-producing sporangia; found on the underside of fern fronds and in some algae and fungi; **sori.**

souari nut Caryocaraceae *q.v.*

sour gum Nyssaceae *q.v.*

sourwood Ericaceae *q.v.*

South Atlantic Gyre The major anticlockwise circulation of surface ocean water in the South Atlantic; see ocean currents.

South Equatorial Countercurrent A warm surface ocean current that flows eastwards in the equatorial Pacific; see ocean currents.

South Equatorial Current A warm surface ocean current that flows westwards in the equatorial Pacific and forms the northern limb of the South Pacific Gyre; see ocean currents.

South Indian Gyre The major anticlockwise circulation of surface ocean water in the southern Indian Ocean; see ocean currents.

South Pacific Gyre The major anticlockwise circulation of surface ocean water in the South Pacific; see ocean currents.

South West Monsoon Drift Somali Current *q.v.*

sow bug Oniscoidea *q.v.*

soybean Fabaceae *q.v.*

spadefish Ephippidae *q.v.*

spadefoot toad Pelobatidae *q.v.*

spadix A spike *q.v.* of usually unisexual flowers on a fleshy central axis, typically protected by a large petal-like bract, the spathe.

spaghetti eel Moringuidae *q.v.*

Spalacidae Mole rats; family containing 3 species of fossorial myomorph rodents found in southern Europe and around the eastern Mediterranean; body fusiform, length to 300 mm, eyes and tail absent; limbs short; incisor teeth used for digging; feed mainly on tubers and roots.

Spanish moss Bromeliales *q.v.*

Sparganiaceae Bur reeds; family of Typhales containing 13 species of perennial herbs emergent from shallow water or with floating leaves; native chiefly to north temperate regions.

bur reed (Sparganaceae)

Sparidae Porgies, sea breams; family containing about 100 species of primarily bottom-feeding teleost fishes (Perciformes) widespread in tropical and temperate seas; body deep and compressed, to 1.2 m length; head large, mouth weakly protractile; single dorsal fin present; many species hermaphroditic.

sparrow 1: Passeridae *q.v.* 2: Ploceidae *q.v.* 3: Emberizidae *q.v.*

Spatangoida Heart urchins, bottle urchins; cosmopolitan order of littoral to abyssal echinoids (Atelostomata) comprising about 275 species; exhibiting pronounced bilateral symmetry and variously specialized spines and tube-feet; mouth displaced to anterior position; burrow into soft sediment maintaining respiratory passages to surface by means of long spines or tube-feet; fossil record extends from the Cretaceous. (Picture overleaf).

heart urchin (Spatangoida)

Spathebothriidea Order of tapeworms parasitic in freshwater and marine fishes and utilizing crustaceans as intermediate hosts; characterized by a weakly developed scolex and by the absence of external segmentation but possessing several, serially arranged, sets of reproductive organs.

spawn The eggs of certain aquatic organisms; also the act of producing such eggs or egg masses.

spay The removal of, or interference with the activity of, the ovaries of a female; *cf.* geld.

specialization Evolutionary adaptation to a particular mode of life or habitat.

specialized Adapted to perform a particular function; highly modified from the ancestral condition and no longer able to carry out the function performed when in the ancestral state.

speciation The formation of new species; the acquisition of reproductive isolating mechanisms producing discontinuities between populations.

species 1: A group of organisms, minerals or other entities formally recognized as distinct from other groups. 2: A taxon of the rank of species; in the hierarchy of biological classification the category below genus; the basic unit of biological classification; the lowest principal category of botanical and zoological classification.

species odour The particular odour found on the bodies of social insects that is peculiar to a given species.

species richness The absolute number of species in an assemblage or community.

species-flock A group of several ecologically diverse and closely related species that have evolved within a single macrohabitat, such as a particular lake basin.

specific Pertaining to a particular individual, trait, or species.

spectre Phasmoptera *q.v.*

speedwell Scrophulariaceae *q.v.*

Spelaeogriphacea Primitive order of small, blind, shrimp-like peracaridan crustaceans comprising a single freshwater, cavernicolous species found in southern Africa; eggs incubated in ventral brood pouch.

speleology The study of caves; **speleological.**

speleotherm A structure within a cave formed by the deposition of secondary minerals, such as a stalactite or stalagmite.

sperm Male gamete; spermatozoon.

sperm whale Physeteridae *q.v.*

spermatize Fertilize; impregnate.

spermatocyte The cell that produces male gametes by meiotic division: *cf.* oocyte.

spermatogamete A male gamete; sperm.

spermatogenesis The process of sperm formation; spermogenesis; **spermatogenic.**

Spermatophyta The seed-bearing plants; a division of plants comprising the gymnosperms (Pinophyta) and angiosperms (Magnoliophyta); probably arose from pteridophyte ancestors during the Devonian.

spermatozoon A male gamete; sperm; spermatozoan.

spermiogenesis The transformation of a spermatid into a spermatozoon.

spermology The study of seeds.

spermophilic Feeding on seeds; **spermophile, spermophily.**

Sphacelariales Order of widely distributed brown algae with a tufted filamentous body form; growth usually occurring from a large, dense, apical meristematic cell.

Sphaerocarpales Small order of liverworts (Hepaticopsida); plants organized as a small lobate thallus or a stem-like axis with lateral leaves or a dorsal lamina: possessing smooth rhizoids; female sex organs not involving apical cell; produce a sporophyte capsule on a short to obsolescent stalk which liberates spores as its walls decay.

Sphaeropleales Small order of freshwater green algae in which the thallus is an unbranched filament composed of very elongate, banded, multinucleate cells.

Sphaeropsidales Large order of coelomycete fungi containing over 6000 species which are mostly parasitic or saprobic on plant material; produce fruiting bodies containing a conidia-producing layer within enclosing walls of sterile tissues.

Sphaerosepalaceae Small family of Theales comprising 14 species of woody plants with simple unicellular hairs, confined to Madagascar.

Sphaerotheriida Order of oniscomorphan diplopods (millipedes) comprising about 160 species widespread in the southern hemisphere, except South America.

sphagnicolous Living in peat-moss; **sphagnicole.**

sphagnophilous Thriving on *Sphagnum* moss, or in *Sphagnum*-rich habitats; **sphagnophile, sphagnophily.**

Sphagnopsida Peat mosses; class of bryophytes comprising a single genus, *Sphagnum*; typically found in acidic mire habitats and may dominate lowland boreal areas of northern hemisphere; stems mostly erect and sparsely, dichotomously branching with differentiated stem, branch and leaves; stem consisting of central core surrounded by cortical cells and a hyalodermis; sporophytes produced terminally and borne on a stalk (pseudopodia) of gametophytic tissue.

sphagnum moss (Sphagnopsida)

Sphecidae Digger wasps; family of minute to large, winged wasps (Hymenoptera) which dig nests in soil, wood or other plant cavities; adults feeding on nectar, honeydew or occasionally on blood fluids oozing from captured prey; eggs laid one per cell on prey placed in cell by female; sometimes additional prey are brought during larval development; contains nearly 8000 species, distributed worldwide except for extremely cold regions.

sphecology The study of wasps.

Spheniscidae Penguins; family containing 18 species of flightless marine neornithine birds widespread in the cold Southern Ocean extending north to Galapagos Islands; very efficient swimmers using wings modified as flippers; feet positioned posteriorly providing upright stance; move on land by hopping, walking or tobogganing; coloration typically black and white; feed on fishes, crustaceans and squid.

Sphenisciformes Order of birds comprising a single family, Spheniscidae (penguins).

Sphenocleaceae Family of Campanulales containing only 2 species of annual herbs with

penguin (Spheniscidae)

large air passages in the cortex; pantropical in distribution.

Sphenodontidae Tuatara; monotypic family of primitive lizard-like reptiles (Rhyncocephalia) confined to a few small islands off New Zealand; body length to about 0.5 m; inhabit petrel burrows, feeding nocturnally on invertebrates and small vertebrates; estimated to take 20 years to attain sexual maturity.

Sphenomonadales Order of colourless euglenoid flagellates which may be osmotrophic or phagotrophic.

Sphenopsida Horsetails; the Equisetophyta *q.v.* classified as a subdivision of the Pteridophyta.

Sphinctozoa Subclass of calcareous sponges containing a single species found in shaded submarine caves and crevices between 5 and 40 m in the Indo-West Pacific; characterized by the production of a solid aragonite skeleton consisting of a series of chambers.

Sphingidae Hawk moths; family of medium-sized to large moths (Lepidoptera) typically having long triangular wings and an elongate body, adapted for hovering flight; several are mimics of hummingbirds or bumblebees; contains about 850 species, most abundant in tropical regions; larvae of some species are pests of agricultural crops such as tomato, tobacco and sweet potato.

sphingophilous Pollinated by hawk moths or nocturnal Lepidoptera; **sphingophile, sphingophily.**

Sphyraenidae Barracudas; family containing 18 species of large (to 1.8 m) tropical and subtropical, predatory marine teleost fishes (Perciformes) found mainly in inshore waters and estuaries; body slender with pointed snout, mouth large bearing strong sharp teeth;

2 dorsal fins present; popular as sport and food-fishes.

Sphyraenoidei Suborder of predatory marine teleost fishes (Perciformes) comprising a single family, Sphyraenidae (barracudas).

sphyric Pertaining to rock slides.

Sphyrnidae Hammerhead sharks; family containing 8 species of small to large (to 6 m) marine carcharhiniform elasmobranch fishes in which the sides of the head are expanded into flattened blades bearing the eyes; found in shallow waters, to 300 m depth in tropical and warm temperate regions, feeding mainly on other fishes; reproduction viviparous, with yolk sac placenta; the expanded head lobes are thought to enhance manoeuvrability.

hammerhead shark (Sphyrnidae)

spider Araneae *q.v.*
spider beetle Ptinidae *q.v.*
spider monkey Cebidae *q.v.*
spider plant Liliaceae *q.v.*
spider wasp Pompilidae *q.v.*
spiderwort Commelinaceae *q.v.*
spike A racemose inflorescence comprising stalkless flowers on an elongated central axis.

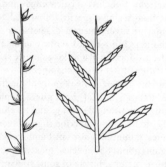

spike

spike moss Selaginellales *q.v.*
spikefish Triacanthidae *q.v.*
spiladophilous Thriving on clay or in clay-rich habitats; **spiladophile, spiladophily.**
spiladophyte A plant living on clay soils.
spinach Chenopodiaceae *q.v.*
spindle tree Celastraceae *q.v.*
Spinosauridae Extinct family of carnivorous dinosaurs (Saurischia) in which the neural spines were elongated, presumably forming a

sail-like dorsal fin; known from the Cretaceous.

Spintherida Order of polychaete worms comprising a single family and about 9 species ectoparasitic on marine sponges; body flattened ventrally, anterior parapodia fused, pharynx large and eversible; parapodia biramous; cirri and branchiae absent.

Spinulosida Large order of intertidal to deep-sea asteroidean echinoderms that includes the notorious crown-of-thorns sea star, *Acanthaster*, which can cause extensive damage to coral reefs when abundant.

spiny anteater Tachyglossidae *q.v.*
spiny dormouse Platacanthomyidae *q.v.*
spiny eel 1: Notacanthidae *q.v.* 2: Mastacembelidae *q.v.*
spiny lobster Palinura *q.v.*
spiny rat Echimyidae *q.v.*
spinyfin Diretmidae *q.v.*
spiny-headed worm Acanthocephala *q.v.*
Spionida Order of burrowing or tubicolous, marine and occasionally freshwater, polychaete worms comprising about 700 species in 8 families, found from intertidal to abyssal depths; body divided into distinct regions; prostomium bearing pair of tentacular palps; pharynx eversible, ciliated, unarmed; parapodia biramous with simple setae.

Spiraculata Extinct order of echinoderms (Blastoidea) with hidden hydrospire slits; known from the Silurian to the Permian.

spirea Verbenaceae *q.v.*

Spiriferida Extinct order of articulate brachiopods with spiral brachidia; known from the Ordovician to the Jurassic.

Spirobolida Order of helminthomorphan diplopods (millipedes) comprising about 450 species, having a largely tropical distribution but extending into temperate North America; body cylindrical, polished, to 180 mm length; some large species can eject an allomone spray over distances of up to 1 m.

Spirochaetes Group of helical, non-flagellated, motile bacteria which are chemoheterotrophic and occur as free-living saprobic forms in aquatic habitats, or as pathogens.

Spirophorida Small order of tetractinomorph sponges usually found on soft bottom sediments to which they are attached by a basal mat of tangled spicules.

Spirostreptida Diverse order containing about 1500 species of helminthomorphan diplopods (millipedes) that includes the largest known species with a body length of 300 mm; body may have up to 90 segments.

Spirotricha Sole subclass of polyhymenophoran ciliates.

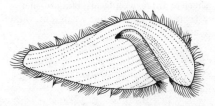

spirotrich (Spirotricha)

Spiruria Subclass of secernentean nematodes that are obligate parasites of vertebrates typically with arthropod intermediate hosts; characterized by 6 apical lobes surrounding the mouth or by 2 lateral lips; males have 2 spicules often unequal in length; comprises 2 orders, Spirurida and Camallanida.

Spirurida Order of spirurian nematodes found as parasites of terrestrial and aquatic vertebrates, and in annelids; involving various invertebrate intermediate hosts in the life cycle; characterized by a cephalic hook and a pore-like phasmid on the tail of the newly hatched larva.

spittlebug Cercopidae *q.v.*

Splachnales Small order of mosses (Bryidae); plants forming mats and tufts on soil or decaying organic matter; stems erect and simple or forked, bearing terminal sporophytes; sometimes classified in the Funariales.

splash erosion The displacement of soil particles by impacting rain droplets.

splash zone The region of the shore immediately above the highest water level that is subject to wetting by splash from breaking waves; see marine depth zones.

sponge Porifera *q.v.*

spongicolous Living on or in sponges; **spongicole.**

spongilla fly Neuroptera *q.v.*

Spongiomorphida Extinct order of cnidarians (Hydroida) exhibiting a structure of stellate, vertical tubules; found in warm shallow seas during the Triassic and Jurassic.

spoonbill Threskiornithidae *q.v.*

spoonworm Echiura *q.v.*

spore A plant reproductive cell capable of developing into a new individual, directly or after fusion with another spore; produced either by meiosis (meiospores) or by mitosis (gonidia).

sporiferous Bearing or producing spores.

Sporochnales Order of brown algae found in warm waters; characterized by a branched body form with the branches terminating in distinctive tufts of filaments.

sporogamy Production of spores by an organism developed from a zygote; **sporogamic.**

sporogenesis The formation of spores; reproduction by means of spores; **sporogenic.**

sporogenous Spore-producing.

sporogony Multiple fission of a cell (sporont) to produce many dormant spores (sporozoites); **sporogonous.**

sporophyte The diploid, spore-producing, asexual generation in the life cycle of a plant; typically formed by fusion of haploid gametes.

Sporozoa Class of apicomplexan protozoans which generally reproduce by alternation of asexual and sexual phases, the latter resulting in the formation of resistant cysts containing infective stages (sporozoites) which commence the asexual phase; contains about 3900 species in 3 subclasses, Gregarinia, Coccidia and Piroplasmia; Telosporidea.

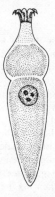

gregarine (Sporozoa)

Sporozoasida A group of unicellular organisms comprising the Gregarinia *q.v.* and the Coccidia *q.v.*; treated as a class of the protoctistan phylum Apicomplexa.

sport Any freak, or somatic mutation, that deviates from the basic type; rogue.

sporulation The production or liberation of spores.

spory The formation of spores; reproduction by means of spores.

spray zone The region of the shore immediately above the splash zone that is subject to wetting by the spray from breaking waves; see marine depth zones.

spring tide A tide of maximum range occurring at the time of new and full moon, when the gravitational attraction of the sun and the moon act together during syzygy *q.v.*; *cf.* neap tide.

springhaa Pedetidae *q.v.*

springtail Collembola *q.v.*

Spumellarida Order of radiolarians (Polycystina) in which the central capsule is uniformly perforated and which produce swarmers containing a crystal of strontium.

spurge laurel Thymelaeaceae *q.v.*

Squalidae Large cosmopolitan family containing about 70 species of small to medium-sized (to 7 m) squaliform sharks; body slender or stout, dorsal fins small, with or without spines, mouth short; mostly benthic, active or sluggish, some species meso- or bathypelagic; reproduction viviparous, without placenta.

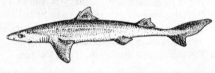

spur dog (Squalidae)

Squaliformes Order of small to medium-sized (0.1–7 m) squalomorph sharks, mostly benthic in deep water to 3000 m; body subcylindrical with 2 dorsal fins, anal fin absent, caudal fin with small lower lobe, rostrum trough-shaped; notochord usually constricted with well calcified vertebral centra.

Squalodontoidea Shark-toothed whales; extinct superfamily of toothed whales (Odontoceta) resembling porpoises, often with an elongate beak carrying triangular teeth proximally; widely distributed during the Miocene.

Squalomorphii Superorder of primarily deep-water elasmobranch fishes comprising about 80 species in 3 orders, Hexanchiformes, Squaliformes and Pristiophoriformes.

Squamata Large and very diverse order of lepidosaurian reptiles comprising 3 suborders, Sauria (lizards), Serpentes (snakes) and Amphisbaenia (worm lizards); habits including terrestrial, aquatic, arboreal and fossorial; reproductive strategies vary from oviparous to viviparous; contains the great majority of living reptiles.

squash Cucurbitaceae *q.v.*

squat lobster Anomura *q.v.*

Squatinidae Angel sharks, monkfish; family containing 10 species of tropical to warm temperate elasmobranch fishes; body strongly depressed, pectoral fins large and demarcated from head, pelvic and caudal fins also well developed; mouth bearing numerous small sharp teeth; feed on fishes, molluscs and crustaceans; reproduction viviparous, without placenta.

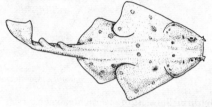

angelshark (Squatinidae)

Squatiniformes Monofamilial order of benthic marine elasmobranch fishes.

Squatinimorphii Superorder of neoselachian elasmobranch fishes comprising a single order, Squatiniformes.

squid Teuthoidea *q.v.*

squirrel Sciuridae *q.v.*

squirrel monkey Cebidae *q.v.*

squirrelfish Holocentridae *q.v.*

SSSI Site of special scientific interest.

stable fly Muscidae *q.v.*

Stachyuraceae Small family of Violales containing about 6 species of tanniferous shrubs or small trees native from the Himalayan region to Japan; typically with waxy yellow flowers in leaf axils, arranged in groups of 12–20 in stiff, catkin-like racemes.

Stackhousiaceae Family of Celastrales containing about 25 species of xerophytic, rhizomatous herbs; native principally to Australia and New Zealand; flowers typically with 5 fused sepals, 5 free or partly fused petals, 5 stamens and a superior ovary; seeds large with fleshy endosperm.

stade A climatic episode within a glacial stage during which a secondary glacial advance occurs; **stadial**; *cf.* interstade.

stag beetle Lucanidae *q.v.*

stagnicolous Living in stagnant water; **stagnicole.**

stagnoplankton Planktonic organisms and floating vegetation of stagnant water bodies; **stagnoplanktonic.**

stalked puffball Tulostomatales *q.v.*

stamen Each of the male reproductive organs in a flower; typically differentiated into a filament bearing a distal anther where the pollen is produced; the whorl of stamens is known as the androecium.

staminode A sterile stamen.

stand A unit of vegetation comprising a single species, homogeneous in composition and age.

standing crop Biomass; the total mass of organisms comprising all or part of a population or other specified group, or within a given area, measured as volume, mass (live, dead, dry, ash-free) or energy (calories); standing stock.

Stannomida Order of Xenophyophores distinguished by the general flexibility of the body and by the presence of organic threads in the test; known only from the Indo-Pacific Ocean.

Staphyleaceae Bladdernut; family of Sapindales containing about 50 species of mucilage-producing shrubs and trees of widespread distribution; flowers arranged in showy panicles, each flower usually with 5 sepals, petals and stamens, and a superior ovary.

Staphylinidae Rove beetles; widely distributed family of often small beetles (Coleoptera) with usually short, truncate elytra; containing about 30 000 species found in a variety of habitats, especially leaf litter; mostly predators but some feed on fungal spores.

rove beetle (Staphylinidae)

star anise Illiciaceae *q.v.*

star apple Sapotaceae *q.v.*

star thistle Asterales *q.v.*

starch Complex carbohydrate composed of D-glucose molecules, occurring in 2 forms, amylase (straight-chained) and amylopectin (branch- chained).

star-eater Astronesthidae *q.v.*

starfish Asteroidea *q.v.*

stargazer Uranoscopidae *q.v.*

starling Sturnidae *q.v.*

stasis Cessation or retardation of growth or movement; static.

stasophilous Thriving in stagnant water; stasophile, stasophily.

stasophyte A stagnant-water plant.

statary Pertaining to a relatively quiescent phase in the activity of a colony of social insects.

statice Plumbaginales *q.v.*

statistics Any kind of numerical data collection, organization and analysis; the branch of mathematics dealing with generalization, estimation and prediction in situations where uncertainty exists.

staurogamy Cross fertilization *q.v.*

Stauromedusae Small order of medusae (Scyphozoa) found primarily in shallow cold temperate and polar seas; characterized by a sessile medusa stage, typically attached to the substratum by an aboral stalk; Lucernariidae.

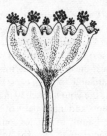

medusa (Stauromedusa)

steady state An apparently unchanging condition of a system maintained at equilibrium by antagonistic processes.

Stegosauridae Stegosaurs; extinct family of quadripedal, ornithischian dinosaurs; typically armoured with a double row of plates and spines along the back and tail; known mainly from the Jurassic.

stegosaur (Stegosauridae)

Stegostomatidae Zebra shark; monotypic family of inoffensive bottom-dwelling orectolobiform elasmobranch fishes (carpet sharks) found in coastal marine and estuarine waters of the Indian Ocean and western Pacific; body length to 4 m, pectoral fins much larger than pelvics, caudal without ventral lobe; reproduction oviparous; young with light markings on dark background, becoming dark spots on light surface in adult.

Steindachneriidae Monotypic family of tropical western Atlantic deep-water gadiform teleost fishes sometimes classified as part of the Macrouridae or Merlucciidae; tail elongate and tapering, single dorsal and anal fins; photophores present.

Stellatosporea Class of Ascetospora *q.v.* parasitic on invertebrates, characterized by the speckled appearance of the cytoplasm due to the presence of minute, electron-dense particles (haplosporosomes).

Stelleroidea Sole extant class of Asterozoa *q.v.*

stem wasp Symphyta *q.v.*

stem-flow Rainwater flowing down a tree trunk or plant stem.

Stemmiulida Order of helminthomorphan diplopods (millipedes) comprising about 70 species found in damp forest litter of the New World tropics; body compressed with marked dorsal suture; these millipedes may be very active and are able to jump clear of the substratum; the young hatch with numerous body segments and many pairs of legs.

Stemonaceae Family of Liliales comprising about 30 species of erect herbs, herbaceous vines or low shrubs; commonly producing lactone alkaloids of a unique type; leaves consisting of a long petiole and broad blade with curved convergent primary veins; fruit is a bivalved capsule; native to eastern Asia, Malesia and northern Australia.

Stemonitida The slime mould subclass Stemonitomycetidae *q.v.*, treated as an order of the protozoan group Myxogastria.

Stemonitomycetidae Subclass of plasmodial slime moulds *q.v.* (Myxomycetes) comprising a single order, Stemonitales, characterized by a plasmodium formed from a flattened network of strands which mass into a fan with trailing veins prior to formation of fruiting body which grows upwards from the underside of the plasmodial thallus; spores carried in dark masses within the fruiting body.

Stenidae Small family of dolphins comprising 4 species which are commonly included in the family Delphinidae *q.v.*

stenobaric Tolerant of a narrow range of atmospheric pressure or hydrostatic pressure; *cf.* eurybaric.

stenobathic Tolerant of a narrow range of depth; *cf.* eurybathic.

stenobenthic Living on the sea or lake bed within a narrow range of depth; *cf.* eurybenthic.

stenobiontic Used of an organism requiring a stable uniform habitat; *cf.* eurybiontic.

stenochoric Having a narrow range of distribution; *cf.* eurychoric.

stenoecious Restricted to a narrow range of habitats and environmental conditions; *cf.* amphioecious, euryoecious.

stenohaline Used of organisms that are tolerant of only a narrow range of salinities; *cf.* euryhaline, holeuryhaline, oligohaline, polystenohaline.

stenohydric Tolerant of a narrow range of moisture levels or humidity; *cf* euryhydric.

stenohygric Tolerant of a narrow range of atmospheric humidity; *cf.* euryhygric.

stenoionic Having or tolerating a narrow range of pH; *cf.* euryionic.

Stenolaemata Class of marine bryozoans comprising a single order, Cyclostomata, in which the zooids are typically cylindrical and calcified; polymorphism limited to autozooids (feeding), gynozooids (female, reproductive) and kenozooids (structural); lophophore circular, epistome absent; reproduction involves polyembryony in which embryo repeatedly subdivides.

Stenolpelmatidae Sand crickets, stone crickets; family of large, wingless, orthopterodean insects (Ensifera); commonly found in rotting logs, under stones or burrowing in sand; antennae typically longer than body.

stenolumic Tolerant of a narrow range of light intensity; *cf.* eurylumic.

stenomorphic Pertaining to organisms that are of small size owing to cramped habitat conditions.

stenophagous Utilizing only a limited variety of foods or food species; **stenophage**, **stenophagy**; *cf.* euryphagous.

stenophotic Tolerant of a narrow range of light intensity; *cf.* euryphotic.

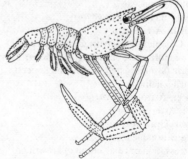

stenopod shrimp (Stenopodidea)

Stenopodidea Shrimps; infraorder of pleocyematan decapod crustaceans; pleura of second abdominal segment not overlapping those of first segment; third thoracic leg chelate.

stenothermic Tolerant of a narrow range of environmental temperatures; **stenotherm**; *cf.* eurythermic.

stenothermophilic Tolerant of only a narrow range of high temperatures; **stenothermophile, stenothermophily**; *cf.* eurythermophilic.

stenotopic 1: Tolerant of a narrow range of habitats; **stenotopy**. 2: Having a narrow geographical distribution; *cf.* amphitopic, eurytopic.

stenotropic Used of organisms exhibiting a limited response or adaptation to changing environmental conditions; **stenotropism**; *cf.* eurytropic.

stenoxenous Tolerating only a narrow range of host species; *cf.* euryxenous.

Stephanoberycidae Pricklefishes; family containing 3 species of small (to 120 mm) tropical abyssobenthic or bathypelagic beryciform teleost fishes; head very large, teeth minute and villiform; caudal fin bearing about 10 procurrent spines.

steppe Semi-arid areas of treeless grassland found in the mid-latitudes of Europe and Asia.

stercoraceous Living on or in dung.

Stercorariidae Skuas; family containing 5 species of large gull-like charadriiform sea birds; bill stout and strongly hooked, wings long and pointed, legs short; habits predacious, gregarious, aggressive, migratory, monogamous, feeding on variety of animals and carrion; cosmopolitan in distribution, breeding in high polar latitudes.

Sterculiaceae Cacao, flame tree, bottletrees; family of Malvales containing about 1000 species of mostly woody tanniferous plants with stellate hairs or peltate scales; commonly producing complex cymes of regular flowers with 3–5 sepals, 5 or no petals, 2 whorls of stamens and superior ovaries.

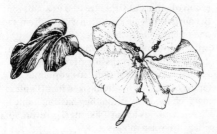

flannel bush (Sterculiaceae)

stereokinesis A change in linear or angular velocity in response to a contact stimulus; **stereokinetic.**

Stereomyxida Order of Acarpomyxa containing 4 species of marine amoebae with branched pseudopodia serving as organs of locomotion or flotation.

stereotaxis A directed response of a motile organism to continuous contact with a solid surface; **stereotactic.**

stereotropism An orientation response to a contact stimulus; **stereotropic.**

sterile 1: Unable to produce viable propagules or to reproduce sexually; **sterility**. 2: Free from contamination by microorganisms.

Sternaspida Order of polychaete worms comprising a single family of 6 species that burrow in littoral to abyssal soft sediments; body short and papillose; prostomium and peristomium naked, pharynx eversible, unarmed.

Sternoptychidae Hatchetfishes; cosmopolitan family containing 35 species of tiny (to 70 mm) luminescent pelagic marine stomiiform teleost fishes found from surface to depths over 3500 m; body very deep and compressed, silvery; mouth almost vertical, eyes occasionally telescopic; photophores present on ventral surface.

hatchetfish (Sternoptychidae)

sterrhophilous Thriving on moorland; **sterrhophile, sterrhophily.**

sterrhophyte A moorland plant.

sterric Pertaining to a heath community or heathland habitat.

Sterrofustia Order of solenogastres characterized by a mantle integument with several layers of solid, mainly needle-like, calcareous bodies, and by the presence of stalked epidermal papillae.

Stichaeidae Pricklebacks; family containing about 55 species of mainly northern Pacific, intertidal and shallow marine, blennioid teleost fishes (Perciformes); body slender, to 500 mm length, naked or with small scales; dorsal fin long and spinose.

stick insect Phasmoptera *q.v.*

stickleback Gasterosteidae *q.v.*

stigma The receptive apex of a carpel in flowering plants upon which the pollen grain germinates.

Stigonematales Order of complex blue-green algae found in freshwater and marine habitats; contains uniseriate or multiseriate filamentous forms that exhibit true branching.

Stilbellales Cosmopolitan order of hyphomycete fungi comprising about 500 species in a single family; includes many saprophytes as well as parasites of flowering plants and insects.

stilt Recurvirostridae *q.v.*

stingray 1: Dasyatidae *q.v.* 2: Urolophidae *q.v.* 3: Myliobatidae *q.v.*

stinkhorn Phallales *q.v.*

stoat Mustelidae *q.v.*

stock 1: The stem of a plant that receives a graft. 2: Brassicaceae *q.v.*

Stolidobranchia Order of littoral to abyssal, solitary or colonial, tunicates (Ascidiacea) having the gonads in the body wall, and the branchial sac with internal longitudinal bars and folding; fertilization often external with free-living larva, but development may be more or less direct and may include asexual budding from atrial wall.

stolon A branch which comes into contact with the ground and may produce a new plant from an axillary bud at a node touching the soil.

Stolonifera Order of primitive colonial octocorals found mainly in shallow tropical seas; characterized by the production of simple cylindrical polyps from a ribbon-like stolon creeping over the substratum.

Stolonoidea Extinct order of encrusting or sessile graptolites known from the Lower Ordovician.

Stomatopoda Mantis shrimps; only extant order of the crustacean subclass Hoplocarida, containing about 350 species, most abundant in shallow tropical seas; all are raptorial predators, typically inhabiting burrows and crevices from where they emerge to spear or smash prey using powerful claws.

mantis shrimp (Stomatopoda)

Stomiatidae Stomiidae *q.v.*

Stomiidae Scaly dragonfishes; family containing 9 species of luminescent deep-sea stomiiform teleost fishes; body elongate, to 400 mm in length, photophores present laterally and ventrally; long barbel below mouth bearing light organ acting as lure; teeth fang-like.

Stomiiformes Order of mainly deep-sea luminescent teleost fishes comprising about 230 species in 8 families and including bristlemouths, lightfishes, hatchetfishes, viperfishes, dragonfishes, snaggletooths, and loosejaws.

Stomochordata Hemichordata *q.v.*

stone A sediment particle larger than 20 mm in diameter; a component of gravel; see sediment particle size.

Stone Age An archaeological period dating from about 10 000 years B.P.; the earliest technological period of human history characterised by the use of stone tools; three periods are recognised, Palaeolithic, Mesolithic and Neolithic *q.v.*

stone cricket Stenolpelmatidae *q.v.*

stone curlew Burhinidae *q.v.*

stone plant Aizoaceae *q.v.*

stonefish Synanceiidae *q.v.*

stonefly Plecoptera *q.v.*

stonewort Charophyceae *q.v.*

stony coral Scleractinia *q.v.*

storey A horizontal layer of a given height within a stratified plant community.

stork Ciconiidae *q.v.*

storm petrel Hydrobatidae *q.v.*

strandline A line on the shore comprising debris deposited by a receding tide; commonly used to denote the line of debris at the level of extreme high water.

stratification Organization into horizontal layers; the structuring of a community or a habitat into superimposed horizontal layers.

stratigraphic range The distribution of a taxon through geological time, determined by its distribution in strata of known geological age.

stratigraphic unit A geological stratum or group of adjacent strata treated as a unit for the purposes of classification with reference to one or more characters.

stratigraphy Study of the origin, composition, distribution and succession of rock strata.

Stratiomyidae Soldier flies; cosmopolitan family of small to medium-sized flies (Diptera) in which the body commonly has a metallic sheen; contains about 1400 species several of which are wasp mimics.

stratosphere 1: The nearly uniform cold ocean

water masses in high latitudes and near-bottom waters of middle and low latitudes; ocean water below the thermocline. 2: The layer of the atmosphere above the troposphere, 15-50 km above the Earth's surface, in which temperature ceases to fall with increasing height, typically without strong convection currents; *cf.* troposphere.

stratum (strata) 1: A layer, as of rock, possessing characters that serve to distinguish it from adjacent layers. 2: A horizontal layer of vegetation within a stratified plant community.

strawberry Rosaceae *q.v.*

Streblidae Batfly; family of small flies (Diptera) that are blood-feeding ectoparasites of bats.

Strelitziaceae Bird-of-paradise, traveller's tree; small family of Zingiberales containing only 7 species of large perennial herbs or banana-like trees native to South Africa; flowers adapted to pollination by birds or insects; each flower usually with 6 perianth segments, 5 or 6 stamens and an inferior ovary, but always grouped into striking inflorescences.

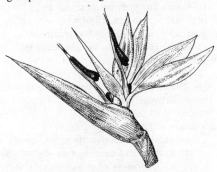

bird-of-paradise flower (Strelitziaceae)

Strepsiptera Small order of holometabolous insects that are endoparasites of other insects (Hymenoptera, Hemiptera, Orthopterodea, Thysanura); sexual dimorphism pronounced, male with broad head, reduced forewings and fan-like hindwings; female typically larviform

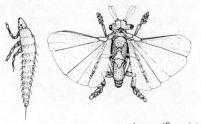

strepsipteran (Strepsiptera)

and limbless; about 400 species in 2 suborders, Mengenillidia and Stylopidia.

Strepsirhini Prosimians; suborder of primates comprising 6 families, Lemuridae and Cheirogaleidae (lemurs), Indriidae (indriid lemurs), Daubentoniidae (aye aye), Galagidae (galagos) and Lorisidae (lorises); nostril with lateral slit; rhinarium moist and glandular; mandible not fully fused medially; distributed through the Ethiopian and Oriental Regions.

stress Any environmental factor that restricts growth and reproduction of an organism or population; any factor acting to disturb the equilibrium of a system.

stridulation Production of sound by rubbing together of specially modified structures or surfaces; **stridulate.**

Strigeidida Order of digenetic trematodes *q.v.* characterised by 2 pairs of flame cells (excretory organs) in the miracidia larva; includes the blood fluke, *Schistosoma*, that causes bilharzia (schistosomiasis).

Strigidae Owls; family containing about 135 species of small to large nocturnal raptors found in grassland and woodland habitats worldwide; head large with round facial disk; habits typically solitary, monogamous, non-migratory, nocturnal or crepuscular, mostly arboreal, feeding on a variety of small vertebrates and insects; nest in tree hole or ground burrow.

owl (Strigidae)

Strigiformes Owls; order of nocturnal raptors comprising about 150 species in 2 families, Tytonidae and Strigidae; head large and mobile, eyes and ears well developed, bill strong and hooked, wings broad, feet short and strong; crop absent; prey on variety of small vertebrates and insects.

strobilation Reproduction by successive budding.

strobilus Cone; a reproductive structure

consisting of closely packed sporophylls (leaves specialised to bear sporangia) in lower vascular plants, or ovuliferous scales (leaves specialised to bear ovules basally) in gymnosperms.

Stromateidae Butterfishes; family containing 13 species of coastal marine perciform teleost fishes; body deep and compressed, to 500 mm length, pelvic fins absent, caudal deeply forked; young live in association with jellyfish.

Stromateoidei Suborder of marine perciform teleost fishes comprising about 60 species in 6 families, including medusa fishes, driftfishes and butterfishes; young typically found in association with medusae and siphonophores.

stromatolite Rock-like sedimentary structure found in warm, shallow marine habitats; built up from mats of blue-green algae and trapped sediment particles; also found as fossils.

Stromatoporoida Order of sclerosponges containing many fossil forms and two extant genera with reticulate skeletons composed of the aragonite form of calcium carbonate.

Strongylida Order of parasitic rhabditian nematodes in which the adults are found in vertebrate hosts and the larval stages are either parasites of annelids or molluscs, or feed on bacteria; characterised by the muscles in the copulatory bursa of the male.

Strophomenida Large order of extinct brachiopods (Articulata) with one valve flat or concave and the other convex; hinge line straight; known from the Lower Ordovician to the Lower Jurassic.

strophotaxis A twisting movement in response to an external stimulus; **strophotactic.**

Struthionidae Ostrich; monotypic family of large ratite birds (Palaeognathae) found in open arid parts of Africa and Arabia; legs long and powerful, bearing 2 digits; neck elongate, head naked, eyes large; feathers plumose; female smaller than male; habits gregarious, cursorial, polygamous, feeding largely on plant material; the ostrich is the largest living bird, up to 2.5 m tall.

Struthioniformes Order of ratite birds comprising the monotypic family, Struthionidae (ostrich).

sturgeon Acipenseridae *q.v.*

Sturnidae Starlings; family containing about 110 species of Old World passerine birds found in variety of forest, grassland and open habitats from Africa to Australia; plumage often dark with metallic sheen; bill slender to robust, straight or curved; habits gregarious, arboreal to terrestrial, mostly non-migratory, feeding on invertebrates, seeds and fruit; nest solitary or colonial, made of twigs or grass, on or off the ground.

stygon The groundwater biotope; *cf.* crenon, thalasson, troglon.

stygophilic Thriving in caves or subterranean passages; **stygophile, stygophily.**

stygoxenous Found only occasionally in caves or subterranean passages; **stygoxene.**

Stylasterina Hydrocorals; order of marine Hydrozoa known from most seas and a range of depths, characterized by a colonial polyp generation which secretes a calcareous skeleton; free medusae not produced but sessile medusoid gonophores are retained in cavities in the colony.

style The sterile, often elongate, portion of a carpel between the ovary and the terminal stigma.

Stylidiaceae Trigger plants; family of Campanulales containing about 150 species of mostly herbs with a basal rosette of grass-like leaves which possess an irritable style column which participates in pollination; widely distributed in southern hemisphere; fruit typically a capsule.

Stylommatophora Amber snails, door snails, edible snail, garden snails, slugs; large cosmopolitan order containing about 15 000 species of terrestrial pulmonate snails; characterized by eyes situated on the tips of the upper pair of retractile tentacles, and with a lower pair of usually tactile tentacles; typically possessing an external shell.

ostrich (Struthionidae)

slug (Stylommatophora)

Stylophoridae Tube-eye; monotypic family of mesopelagic lampridiform teleost fishes widespread in tropical seas; body ribbon-like, to 280 mm in length; head large with tubular eyes, mouth markedly protractile; dorsal fin extending entire length of body, anal fin short, pelvics comprising solitary fin ray, caudal fin bearing 2 long filaments.

Styracaceae Silver-bell tree, snowdrop tree; family of Ebenales containing about 150 species of trees or shrubs often cultivated as ornamentals or as sources of gums; typically with 4 or 5 sepals and petals, and the same number or twice as many stamens.

subaerial Occurring immediately above the surface of the ground, or at the ground/air interface.

Subatlantic period A period of the Blytt–Sernander classification *q.v.* (*c.* 2500 years B.P. to present), characterized by beech and linden vegetation and by mild and wet conditions.

storax (Styracaceae)

subboreal Pertaining to a climate or biogeographic zone approaching frigid or boreal conditions.

Subboreal period A period of the Blytt–Sernander classification *q.v.* (*c.* 4500–2500 years B.P.), characterized by oak, ash and linden vegetation and by cooler and drier conditions than in the Subatlantic period.

subduction zone A zone in which the edge of one tectonic plate descends below an adjacent plate.

subfossil A post-Pleistocene fossil; used of plant and animal remains not strictly Recent but which are not old enough to be regarded as fossil.

subgeocolous Living underground; **subgeocole.**

subhydrophilous Thriving in habitats experiencing periodic inundation by fresh water; **subhydrophile, subhydrophily.**

sublethal Not causing death by direct action, but possibly modifying behaviour, physiology, reproduction, or life cycle, and showing cumulative effects.

subliminal Below threshold *q.v.* level; insufficient to elicit a response.

sublittoral 1: The deeper zone of a lake below the limit of rooted vegetation. 2: The marine zone extending from the lower margin of the intertidal (littoral) to the outer edge of the continental shelf at a depth of about 200 m; sometimes used for the zone between low tide and the greatest depth to which photosynthetic plants can grow; see marine depth zones.

submergent Pertaining to a plant or plant structure growing entirely under water; *cf.* emergent.

submersed Pertaining to a plant or plant structure growing entirely under water; *cf.* emersed.

submission The behaviour exhibited by an animal, defeated in an aggressive contact, to prevent further attack.

subsidence theory That the upward growth of coral atolls and barrier reefs matched the slow subsidence of a volcanic island over a prolonged period of time; postulated to explain the formation of coral atolls.

subsoil The layer of soil beneath the topsoil and overlying the bed rock; the C-horizon of a soil profile.

subspecies A group of interbreeding natural populations differing morphologically and genetically, and often isolated geographically from other such groups within a biological species but interbreeding successfully with them where their ranges overlap.

substrate 1: The substance acted upon by an enzyme, or that utilized by a microorganism as a food source. 2: Substratum *q.v.*

substratohygrophilous Thriving on moist substrata; **substratohygrophile, substratohygrophily.**

substratum The sediment, surface, or medium to which an organism is attached or upon which it grows; substrate.

subterranean Living or occurring beneath the surface of the Earth.

subxerophilous Thriving in moderately dry situations; **subxerophile, subxerophily.**

succession Ecological succession *q.v.*; the chronological distribution of organisms within an area; the geological sequence of species within a habitat.

succulency The condition of having specialized fleshy tissue in a plant root or stem for the conservation of water; **succulent.**

sucker 1: Catostomidae *q.v.* 2: An underground shoot arising adventitiously from the base of the stem or from the roots, and emerging to form a new plant.

sucker-footed bat Thyropteridae *q.v.*

sucking louse Anoplura *q.v.*

Suctoria Suctorians, Tentaculifera; subclass of kinetofragminophoran ciliates characterized by suctorial tentacles used for prey capture and feeding, commonly on other ciliates; widely distributed in marine and freshwater habitats, often attached to other organisms and sometimes endosymbiontic in invertebrates and vertebrates.

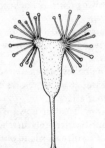

suctorian (Suctoria)

sudd A floating mass of plant material.

suffructicose chamaephyte A chamaephyte *q.v.* subtype in which the upper vegetative and flowering shoots die back leaving only the lower parts to survive unfavourable seasons.

sugarcane Poaceae *q.v.*

Suidae Swine, pigs; family containing 8 species of omnivorous terrestrial ungulates (Artiodactyla) native to the Ethiopian, Oriental and Palaearctic regions but now cosmopolitan through domestication; canine teeth forming tusks, hooves well developed; stomach comprising 2 chambers; typically gregarious, in forest habitats.

wart hog (Suidae)

Suiformes Suborder of mostly omnivorous cloven-hooved ungulates (Artiodactyla) comprising 3 families, Suidae (pigs), Tayassuidae (peccaries) and Hippopotamyidae (hippopotamuses); feet with 4 digits, each of which may be in contact with the ground; hooves well developed or not.

Sulidae Gannets, boobies; cosmopolitan family containing 9 species of medium-sized to large coastal marine birds (Pelecaniformes) that feed on fishes caught by plunging from flight; breed in colonies, nesting in trees or on the ground.

gannet (Sulidae)

sulphur plant Polygonales *q.v.*

sulphuretta The prokaryotic and protistan inhabitants of anaerobic sulphur-rich deposits.

sumach Anacardiaceae *q.v.*

sun scorpion Solpugida *q.v.*

sunbeam snake Xenopeltidae *q.v.*

sunbird Nectariniidae *q.v.*

sunbittern Eurypygidae *q.v.*

sundew Droseraceae *q.v.*

sundrop Onagraceae *q.v.*

sunfish 1: Molidae *q.v.* 2: Centrarchidae *q.v.*

sunflower Asterales *q.v.*

sungrebe Heliornithidae *q.v.*

sunset shell Veneroida *q.v.*

superfoetation The fertilization of a plant ovary by two or more kinds of pollen.

superior ovary Hypogyny *q.v.*

supernatant Floating at the surface.

supernumerary chromosome Any chromosome or chromosome fragment differing from the normal A-chromosomes in structure, genetic effectiveness and pairing behaviour; a B-chromosome.

superorganism Any colony possessing features of social organization analogous to the physiological properties of a single organism; any group of organisms acting as a single functional unit.

superparasitism 1: Parasitism of a host that is itself a parasite; **superparasite**. 2: The multiple infection of a host by several parasites, usually of a similar kind.

superterranean Occurring at or above ground level.

suprafolious Growing on leaves.

supralithion Those aquatic organisms swimming above rock, but dependent upon it as a food source.

supralittoral The region of the shore immediately above the highest water level and subject to wetting by spray or wave splash; splash zone; spray zone; see marine depth zones.

suprapelos Those aquatic organisms swimming above soft mud that are dependent upon it as a food source.

suprapsammon Those aquatic organisms swimming above sand that are dependent upon it as a food source.

supratidal The zone on the shore above mean high tide level.

surf The waves breaking on the shore.

surf clam Veneroida q.v.

surf fly Canaceidae q.v.

surfperch Embiotocidae q.v.

surgeonfish Acanthuridae q.v.

Surianaceae Small family of Rosales containing 6 species of shrubs or trees; widespread in tropics, especially along coasts.

survival of the fittest The differential and greater success of the best adapted genotypes.

survivorship The proportion of individuals from a given cohort q.v. surviving at a given time.

susceptible Prone to influence by a stimulus, or to infection by parasites or pathogens; **susceptibility.**

suspended animation The temporary suspension of life processes.

suspension A dispersion of fine insoluble particulate matter in a fluid.

suspension feeder Any organism that feeds on particulate organic matter suspended in water.

swallow Hirundinidae q.v.

swallower Chiasmodontidae q.v.

swallowtail Papilionidae q.v.

swamp Wet spongy ground, saturated or intermittently inundated by standing water, typically dominated by woody plants but without an accumulation of surface peat.

swamp eel Synbranchidae q.v.

swan Anatidae q.v.

swarm cell An amoeboid cell produced by a germinating spore in the Myxomycota q.v., and characterized by the absence of a cell wall and possession of 2 anterior flagella.

swarm-spore Motile spore in certain algae, fungi and protozoans; zoospore; swarmer.

sweat bee Halictidae q.v.

swede Brassicaceae q.v.

sweet fern Myricales q.v.

sweet gum Hamamelidaceae q.v.

sweet pea Fabaceae q.v.

sweet potato Convolvulaceae q.v.

sweet William Caryophyllaceae q.v.

swell 1: A deep-water, wind-generated, wave system of long periodicity. 2: An area that has subsided less than the surrounding area and is covered by a thinner sequence of sedimentary deposits.

swift Apodidae q.v.

swine Suidae q.v.

Swiss cheese plant Aracaceae q.v.

swordfish Xiphiidae q.v.

sycamore Platanaceae q.v.

Sycettida Order of calcaronian sponges of variable size and shape; canal system consisting of either radially arranged tubes, or scattered spherical chambers; widely distributed in all oceans, from intertidal to depths of about 4000 m.

syconium A type of false fruit (pseudocarp) in which achenes are carried on the inside of a hollow receptacle.

sycophagous Feeding on figs; **sycophage, sycophagy.**

sylvestral Pertaining to, or inhabiting, woods and shady hedgerows; **sylvestrine.**

sylvicolous Inhabiting woodland; **sylvicole.**

Sylviidae Warblers; family containing about 350 species of primarily Old World passerine birds found in forest, grassland and open habitats; habits gregarious to solitary, arboreal or terrestrial, some species migratory; feed on insects and arachnids by gleaning or hawking; nest in trees or reeds.

symbiology The study of symbioses.

symbiont A participant in a symbiosis; a symbiotic organism.

symbiosis The living together of two organisms; the relationship between two interacting organisms or populations, commonly used to describe all relationships between members of two different species, and also to include intraspecific associations; sometimes restricted to those associations that are mutually beneficial; symbioses; **symbiont, symbiotic.**

symbiotrophic Used of an organism obtaining nourishment through a symbiotic relationship; **symbiotrophy.**

Symmetrodonta Extinct order of primitive mammals (Pantotheria) that were probably small predators; molar teeth 3-cusped; known mainly from the late Jurassic and Cretaceous.

sympatric Used of populations, species or taxa

occurring together in the same geographical area; the populations may occupy the same habitat (biotic sympatry) or different habitats (neighbouring sympatry) within the same geographical area; **sympatry**; *cf.* allopatric, dichopatric, parapatric.

symphily An amicable relationship between one organism (the symphile) and its host colony of social insects; **symphilic.**

Symphyacanthida Small order of acantharians in which the bases of 20 radial spines are intimately fused into a solid central body.

Symphyla Class of small blind mandibulate arthropods that inhabit soil and woodland litter; trunk 14-segmented, bearing 12 pairs of legs in the adult, and a pair of posterior cerci with spinning glands; legs may be elongate in active forms; sexes separate with little dimorphism, eggs laid in soil; most are phytophagous, rarely saprophagous; contains about 160 species.

symphylan (Symphyla)

Symphyta Sawflies, horntails, woodwasps, stem wasps; suborder of hymenopteran insects comprising about 4700 species, in which the adults are winged and wasp-like with a broad waist between first and second abdominal segments; female ovipositor rod-like or saw-like; larvae caterpillar-like feeding on, or boring into, leaves, stems or wood; Chalastogastra.

sawfly (Symphyta)

Symplocaceae Family of Ebenales containing about 250 species of usually evergreen, glabrous trees or shrubs; found mostly in tropical and warm temperate regions.

Synanceiidae Stonefishes; family containing 30 species of bizarre tropical Indo-Pacific coastal marine scorpaeniform teleost fishes having powerfully venomous dorsal fin spines; body extremely robust, often warty, scales absent; body length to 600 mm.

Synaphobranchidae Cutthroat eels; family containing about 10 species of predatory deep-sea anguilliform teleost fishes; dorsal and anal fins continuous with caudal; mouth large; gill openings close together or forming single ventral aperture.

synaposematic coloration Protective coloration in which the warning signal is shared with other species, as in Müllerian mimicry; *cf.* aposematic coloration.

Synapsida Subclass of mammal-like reptiles characterized by a single temporal opening in the skull above which the postorbital and squamosal bones meet; includes the therapsids, which are ancestral to the mammals, and the pelycosaurs.

synaptospermy The clumping of seeds by means of cohering or interlocking surface structures.

synaptospory The clumping of spores by means of cohering or interlocking surface structures.

Synbranchidae Swamp eels; family containing 15 species of circumtropical freshwater teleost fishes (Synbranchiformes) with slender eel-like body lacking paired fins; dorsal and anal fins greatly reduced; body length to 1.5 m; atmospheric oxygen is utilized by gulping air into vascularized gill pouch or hind gut; during drought some species burrow into the mud.

Synbranchiformes Order of eel-like, tropical freshwater teleost fishes comprising a single family, Synbranchidae (swamp eels).

Syncarida Superorder of freshwater eumalacostracan crustaceans comprising 2 extant orders, Anaspidacea and Bathynellacea; representing relicts of a diverse and formerly widespread marine group.

synchronic Existing or occurring at the same time; contemporaneous; *cf.* allochronic.

synchronous hermaphrodite Any organism producing mature sperm and ova at the same time; simultaneous hermaphrodite; *cf.* consecutive hermaphrodite.

synclerobiosis Symbiosis between two species of social insects that usually inhabit separate colonies.

synchorology Study of the occurrence, distribution and classification of plant communities.

syncytium A multinucleate protoplasmic mass not differentiated into cells.

syndiacony A mutually beneficial relationship between ants and plants; **syndiaconic.**

Syndiniophycidae Subclass of marine dinoflagellates which are parasitic on various protistan and invertebrate hosts; typically endoparasitic, sometimes intracellular parasites.

synecology Study of the ecology of organisms, populations, communities, or systems; *cf.* autecology.

synecthry Commensalism in which the participants (synecthrans) display mutual dislike.

synergism 1: Cooperative action of two or more agencies such that the total is greater than the sum of the component actions; **synergetic**, **synergistic**. 2: Mutualism *q.v.*

syngamete A diploid cell formed by the fusion of two haploid gametes; zygote.

syngamy Sexual reproduction; fusion of male and female gametes; **syngametic.**

syngenesis Sexual reproduction; **syngenetic.**

syngenic Having the same set of genes; *cf.* allogenic.

Syngnathidae Pipefishes, sea horses; family containing 175 species of tropical and warm temperate, shallow marine or freshwater gasterosteiform teleosts; body elongate with armour of bony rings, length to 500 mm, but usually much smaller; mouth terminal on long snout used as suction pipette to catch small invertebrates; tail prehensile in sea horses; swim using dorsal and pectoral fins.

sea horse (Syngnathidae)

syngonic Producing male and female gametes in the same gonad; **syngony**; *cf.* digonic.

synhesma A swarming society, or a swarm.

Synhymeniida Order of hypostomatan ciliates that have a cylindrical shape with bipolar ciliary rows over the general body surface: all free-living and mostly freshwater forms.

synmorphology The study of floristic composition, minimal area, and distribution of a plant community.

Synodidae Mochokidae *q.v.*

Synodontidae Lizard fishes; family containing 35 species of bottom-living, carnivorous, myctophiform teleost fishes found in tropical and warm temperate shallow waters to 550 m depth; body slender, subcylindrical, bearing heavy shiny scales, pelvic fins abdominal; head depressed, mouth large with long sharp teeth; photophores and gas bladder absent.

synoecy 1: A symbiosis between a colony of social insects and a tolerated guest organism (a synoekete). 2: The production by an organism of both male and female gametes; **synoecious.**

synpiontology The study of ancient patterns of distribution and migration of plant communities.

syntopic Pertaining to populations or species that occupy the same macrohabitat, are observable in close proximity and could thus interbreed; **syntopy**; *cf.* allotopic.

syntrophy The condition exhibited by two organisms mutually dependent for food.

synusia A community of species with similar forms and ecological requirements; a habitat of characteristic and uniform conditions; **synusial.**

synxenic Used of a culture comprising two or more organisms under controlled conditions.

synzoochorous Dispersed by the agency of animals; **synzoochore.**

syringa Oleaceae *q.v.*

Syringocnemida Extinct order of solitary Archaeocyatha having a conical shape and an outer wall with simple pores; known from the Lower Cambrian.

syringograde A form of locomotion in which the organism propels itself with a jet of water; jet propulsion.

Syrphidae Hover flies; cosmopolitan family containing over 5000 species of small to large, colourful flies (Diptera) often found hovering over flowers; adults feed on pollen and nectar, and mimic bees and wasps; larvae are diverse

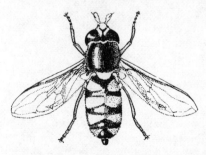

hover fly (Syrphidae)

and feed as phytophages, saprophages, scavengers, or as predators of aphids and caterpillars.

syrtidophilous Living on dry sand-bars; **syrtidophile, syrtidophily.**

syrtidophyte A dry sand-bar plant.

Systellommatophora Slugs; small order of slug-like pulmonate molluscs which lack a shell but possess 2 pairs of tentacles, the upper bearing the eyes; typically with the pulmonary cavity located posteriorly.

systematics The classification of living organisms into hierarchical series of groups emphasizing their phylogenetic interrelationships; often used as equivalent to taxonomy.

systemic Used of a chemical absorbed and transported throughout the tissues of a plant to make them toxic to fungi (systemic fungicide) or insects (systemic insecticide).

syzygy 1: A close association between protozoans prior to gamete formation, but during which nuclear fusion does not occur. 2: The time at which the sun and moon are in line, either in conjunction or in opposition, with respect to the Earth; *cf.* quadrature.

t

2,4,5-T 2,4,5-Trichlorophenoxyacetic acid; a translocated hormone weedkiller used to control scrub and woody vegetation.

Tabanidae Horseflies, clegs, deerflies; cosmopolitan family of robust biting flies (Diptera) especially abundant in marshy habitats, that feed on blood and are of medical and veterinary importance as pests of livestock, as well as vectors of diseases such as loiasis and tularaemia.

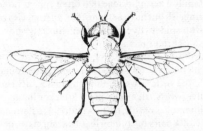

horsefly (Tabanidae)

Tabulata Extinct order of colonial corals (Zoantharia) possessing well developed tabulae; important as reef-building organisms in the Palaeozoic.

Tabulospongida Order of sclerosponges represented by a single living species; characterized by a basal skeleton of the calcite form of calcium carbonate with a lamellar microstructure.

Taccaceae Arrowroot; pantropical family of Liliales comprising about 10 species of perennial herbs with large, long-petiolate, basal leaves having parallel or palmate veins; flowers trimerous, grouped in umbels, and with 6 perianth segments, 6 stamens and inferior ovaries; fruit usually a berry; seeds with copious endosperm containing reserves of protein and oil.

tacheion Actively moving aquatic organisms comprising both crawling (herpon) and free-swimming forms.

tachyauxesis Allometric growth in which a given structure is relatively larger in large individuals than in small ones.

Tachyglossidae Echidnas (spiny anteater); family containing 2 or 3 species of small terrestrial, semifossorial, egg-laying mammals (Monotremata) found in Australia and New Guinea; body covered with coarse hair and spines, tail vestigial; tongue long and sticky,

spiny anteater (Tachyglossidae)

teeth absent; feet with strong claws used to dig out termites and other invertebrates on which they feed, and to escape from predators; eggs incubated in a transitory abdominal pouch.

tachysporous Used of a plant that disperses its seeds quickly.

Tachysuridae Ariidae *q.v.*

tactile Pertaining to the sense of touch.

tadpole The long-tailed larval form of frogs and toads (Anura).

tadpole shrimp Notostraca *q.v.*

tagmosis The functional specialization occurring within metamerically segmented animals, producing a subdivision of the body into distinct regions (tagmata).

taiga Northern coniferous forest biome *q.v.*; the ecosystem adjacent to the Arctic tundra, but used with varying scope to include only the Arctic timberline ecotone through to the entire sub-Arctic subalpine north temperate forest.

tailless whip scorpion Amblypygi *q.v.*

Talpidae Moles, desmans; family containing about 20 species of small fossorial or semiaquatic mammals (Insectivora) in which the eyes are reduced or absent; feet webbed in aquatic species (desmans), or broad with stout claws in burrowing forms (moles); distributed through the Holarctic and Oriental regions.

mole (Talpidae)

Tamaricaceae Tamarisk; family of Violales containing about 100 species of often evergreen shrubs or small trees found in halophytic or xerophytic habitats in Eurasia and Africa; flowers typically bisexual, with 4 or 5 sepals and petals, and 4–5, 8–10 or many stamens, and a superior ovary.

tamarin Callitrichidae *q.v.*

tamarisk Tamaricaceae *q.v.*

tanager Thraupidae *q.v.*

Tanaidacea Order containing about 550 species of small marine peracaridan crustaceans found in benthic habitats from the littoral zone to abyssal depths; often tube-dwelling; body elongate, carapace reduced, eggs retained in brood pouch.

tang line The highest continuous line on the shore along which a particular seaweed grows; has variously been applied to wracks, laminarians and other algal forms; *cf.* algal line.

Tantulocarida Class of microscopic copepod-like crustaceans found as ectoparasites on other marine crustaceans; larval stages and females attached to host by minute adhesive disk, adult males free-swimming, with copulatory stylet on seventh trunk segment; contains about 10 species, mainly from deep oceanic waters.

tantulus larva (Tantulocarida)

tapaculos Rhinocryptidae *q.v.*

tapestry A more or less continuous cover of trees on a steep slope.

tapetail Eutaeniophoridae *q.v.*

tapeworm Eucestoda *q.v.* (Cestoda)

taphocoenosis An assemblage of dead organisms brought together at or after the time of death, or of fossils that were buried together; **taphocoenose**; *cf.* biocoenosis.

taphoglyph An imprint in a sediment of a dead or dying organism.

taphonomy Study of the environmental phenomena and processes that affect organic remains after death, including the processes of fossilization.

Taphrinales Order of hemiascomycete fungi comprising a single worldwide genus of about 100 species, all of which parasitize ferns and higher plants, producing galls or other lesions on leaves, stems and fruits.

taphrophilous Thriving in ditches; **taphrophile**, **taphrophily**.

taphrophyte A ditch plant.

Tapiridae Tapirs; family containing 4 species of forest-dwelling nocturnal mammals (Perissodactyla) in which the nose is

tapir (Tapiridae)

developed into a short, mobile trunk; feed mainly on aquatic plants and fruit; one species is Oriental in distribution, the others Neotropical.

tarantula 1: Any of the large hairy spiders in the family Theraphosidae that often appear rather sluggish but have a strong bite. 2: A species of wolf spider (Lycosidae) found in Italy and southern France.

Tardigrada Water bears; phylum of microscopic (50–1200 μm) coelomates found in the surface water films on mosses, lichens, liverworts, algae and plant litter, or in the interstices of soil; body cylindrical to flattened with 4 pairs of stumpy legs bearing claws or other armature; mouthparts typically styliform for piercing, with muscular sucking pharyx; sexes separate, monomorphic, some species parthenogenetic; comprises about 380 species in 3 orders, Heterotardigrada, Mesotardigrada and Eutardigrada.

water bear (Tardigrada)

Tarphyceratida Extinct order of nautiloids (Cephalopoda) having simple septa; known from the Ordovician and Silurian.

tarpon Megalopidae *q.v.*

Tarsii Monofamilial infraorder of primates (suborder Haplorhini).

Tarsiidae Tarsiers; family of small nocturnal, arboreal primates containing 3 species found in the Oriental and Australian regions; lower jaw not fused medially; eyes and ears very large; limbs slender, hindlimbs longer than forelimbs and used for jumping, toes bearing disk-shaped pads; feed on insects and other small animals.

tarsier (Tarsiidae)

Tarsipedidae Honey possum; monotypic family of small (to 180 mm) arboreal diprotodont marsupials found in southwestern Australia; marsupium opens anteriorly, tail prehensile; tongue long and protrusible for feeding on nectar and pollen.

Tasmanian devil Dasyuridae *q.v.*

Tasmanian wolf Thylacinidae *q.v.*

tassel-finned fish Coelacanthiformes *q.v.*

Taxaceae Yew; family of evergreen conifers (Pinatae); shrubs to trees with needle-like or linear leaves spirally arranged; male (staminate) cones solitary or on spikes, with 2–8 pollen sacs per scale; female (pistillate) cones with a coloured fleshy aril surrounding the hard seed.

taxis A directed reaction or orientation response of a motile organism towards (positive) or away from (negative) the source of the stimulus; **tactism**; **tactic**; *cf.* klinotaxis, telotaxis, tropotaxis.

Taxodiaceae Sequoia: family of tall resiniferous coniferous trees (Pinatae); leaves spirally arranged on small branchlets; male (staminate) cones with 2–9 pollen sacs per scale; female (pistillate) cones solitary with 2–9 ovules per scale; when mature, the cone is leathery to woody.

swamp cypress (Taxodiaceae)

taxon A taxonomic group of any rank, including all the subordinate groups; any group of organisms, populations, or taxa considered to be sufficiently distinct from other such groups to be treated as a separate unit; **taxa.**

taxonomic hierarchy A hierarchical system of taxonomic categories arranged in an ascending series of ranks; in botany the following 12 main ranks are recognized – Kingdom, Division, Class, Order, Family, Tribe, Genus, Section, Series, Species, Variety, Form; in zoology 7 main ranks are recognized – Kingdom, Phylum, Class, Order, Family, Genus, Species; additional categories can be introduced by the use of super- and sub-prefixes.

taxonomy The theory and practice of describing, naming and classifying organisms.

Taxopodida Order of Heliozoa comprising a single species with a skeleton of 14 rosettes of strong hollow spines and many additional rods; abundant at depths of 300 m in the Mediterranean.

Tayassuidae Peccaries; family containing 3 species of omnivorous terrestrial ungulates (Artiodactyla) found in the Neotropical and southern Nearctic regions; 3 digits on the hindlimbs, 4 on the forelimbs, hooves well developed; stomach comprising 3 chambers.

tea Theaceae *q.v.*

teak Verbenaceae *q.v.*

teasel Dipsacaceae *q.v.*

tecnophagous Used of an individual that feeds on its own eggs; **tecnophage, tecnophagy.**

tectology The study of organisms as groups of morphological units or individuals.

tectonic Pertaining to the movement of the rigid plates that comprise the Earth's crust, and to deformation of the crustal plates.

tegulicolous Living on tiles, as of lichens; **tegulicole.**

Teiidae Whiptail; family containing about 200 species of tropical New World lizards (Sauria) that possess a long tail; head plates separated from skull bones; mainly terrestrial, but some semiaquatic or arboreal; habits chiefly diurnal, insectivorous, a few species predatory on vertebrates; reproduction oviparous.

telegamic Pertaining to the process or behaviour of attracting a mate from a distance.

telegenesis Artificial insemination.

teleology The doctrine that natural phenomena result from, or are shaped by, design or purpose; **teleological.**

Teleostei Teleosts; loose assemblage of bony fishes (Osteichthyes) that excludes the primitive orders, Acipenseriformes, Amiiformes and Semionotiformes; sometimes used as an infraclass.

Teleostomi The true bony fish (Osteichthyes) with a terminal mouth.

teleplanic Used of pelagic larvae that have a protracted planktonic existence, with the capacity for widespread dispersal.

Telestacea Order of colonial octocorals found in shallow and deep water in which polyps arise from a creeping stolon and by budding from existing polyps.

Teliomycetes Class of basidiomycotine fungi, all of which parasitize vascular plants; resting spores (teliospores) are produced which germinate with a promycelium or a basidium; the septate mycelium invades the host intercellularly, penetrating cells to obtain nutrients by means of specialized haustoria; contains 2 orders, Uredinales and Ustilaginales.

telmatology The study of wetlands, swamps and marshy areas.

telmatophilous Thriving in wet meadows; **telmatophile, telmatophily.**

telmatophyte A wet-meadow plant.

telmatoplankton Planktonic organisms of freshwater bogs and marshes.

telmicolous Living in freshwater marshes; **telmicole.**

telmophagous Used of an organism, usually an insect, that feeds from a blood pool produced by tissue laceration; **telmophage.**

Telosporidea Sporozoa *q.v.*

telotaxis An orientation response of a motile organism to a single stimulus without reference to other sources of stimulation; *cf.* klinotaxis, tropotaxis.

TEM Transmission electron microscope.

Temnocephala Order of flatworms (Turbellaria) found on the outer surface of various freshwater animals, usually crustaceans; possess a pharynx and bag-shaped intestine; often treated as a suborder of the Neorhabdocoela.

Temnopleuroida Order of mainly sublittoral Indo-West Pacific sea urchins (Echinacea) comprising about 120 extant species; test depressed or spherical, to 150 mm in diameter, with or without sculpturing, dentition camarodont; includes the venomous *Toxopneutes* that has large pedicellariae able to pierce human skin.

temperate 1: Moderate; not excessive or extreme; used of climates with alternating long warm summers and short mild winters. 2: Pertaining to latitudes between the tropics and the polar circles in each hemisphere; *cf.* polar, tropical.

Temperate grassland biome Continental grassland regions such as pampas, prairies, steppes and veld, characterized by a rainfall intermediate between forest and desert, a long dry season, seasonal extremes of temperature; dominated by grasses and large grazing mammals, with burrowing animals frequent.

Temperate forest biome Mixed forest of conifers and broad-leaf deciduous trees, or mixed conifer and broad-leaf evergreen trees, or entirely broad-leaf deciduous, or entirely broad-leaf evergreen trees, found in temperate regions across the world; characterized by high rainfall (temperate rainforests), warm summers, cold winters occasionally subzero, and seasonality; typically with dense canopies, understorey saplings and tall shrubs, large animals, carnivores dominant, and very rich in bird species.

temporary parasite A parasite that makes contact with its host only for feeding.

Tenebrionidae Darkling ground beetles; family containing about 18 000 species of small to medium-sized beetles (Coleoptera) that feed on a wide variety of materials, mostly of plant origin and including stored products such as flour.

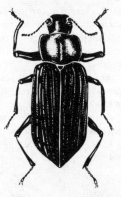

ground beetle (Tenebrionidae)

tenpounder Elopidae *q.v.*

Tenrecidae Tenrecs, otter shrews; diverse family of small (75–600 mm) terrestrial mammals (Insectivora) containing about 35 species found in central and west Africa and Madagascar.

Tentaculata Class of ctenophores possessing tentacles.

Tentaculifera Suctoria *q.v.*

Tentaculitida Extinct order of molluscs (Cricoconarida) known from Lower

Ordovician to Upper Devonian.

tenuis Benthic larval stage of teleost fishes in the family, Carapidae (pearlfishes).

tenrec (Tenrecidae)

tepal Each element in the perianth of a flower which does not have differentiated sepals and petals.

tephra Clastic volcanic materials ejected from a volcano during an eruption and transported through the air.

Tephritidae Fruit flies; cosmopolitan family of small colourful flies (Diptera) that feed on sap and fruit; larvae are phytophagous and are economically important pests of apples, olives, cucumbers, celery and other crops; comprises about 4000 species, most widespread in tropical regions.

tephrochronology Geological dating based on the layering of volcanic ash.

tera- (T) Prefix used to denote unit x 10^{12}; see metric prefixes.

teratogenic Causing developmental malformations or monstrosities; **teratogenesis.**

teratology The study of malformations and monstrosities; **teratologic.**

Terebellida Order of tubicolous polychaete worms comprising about 1000 species in which

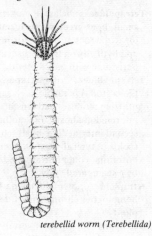
terebellid worm (Terebellida)

body regions are strongly differentiated; prostomium and peristomium fused, branchiae typically present anteriorly; oral tentacles used for deposit feeding and tube construction.

Terebrantia Parasitica *q.v.*

Terebratulida Order of articulate brachiopods comprising about 250 species in which the shell is punctate, often with light coloration; pedicle variable, functional or absent; spicules present or absent in soft parts; epifaunal, typically sessile and attached, intertidal to depth of 5000 m.

termitarium A termites' nest.

termite Isoptera *q.v.*

termiticolous Inhabiting termite nests; **termiticole.**

termitology The study of termites.

termitophilic Having an affinity for termites; inhabiting termite nests; **termitophile, termitophily.**

tern Laridae *q.v.*

terraneous Pertaining to, or living on, the land or ground surface; terrestrial.

terrapin Emydidae *q.v.*

terraqueous Comprising both land and water.

terrestrial Pertaining to, or living habitually on, the land or ground surface.

terrestrial biomes The major biogeographical regions of the world's land surface; see map on page 48.

terricolous Living on or in soil; used of an organism that spends most of its active life on the ground; **terricole.**

terrigenous Derived from the land.

terrigenous mud A marine sediment comrising at least 30% silt and sand derived from the land; *cf.* black mud, blue mud, green mud, red mud, white mud.

territoriality Behaviour related to the defence of a territory.

territory An area within the home range *q.v.* occupied more or less exclusively by an animal or group of animals of the same species, and held through overt defence, display or advertisement; **territorial.**

Tertiary A geological period of the Cenozoic era (*c.* 65–1.6 million years B.P.), comprising the Palaeocene, Eocene, Oligocene, Miocene and Pliocene; see geological time scale.

Testaceafilosida Order of Filosa; testate amoebae with filose pseudopodia (filopodia), the test usually with one opening, sometimes 2; reproduction typically by binary fission; mostly found in fresh water, soil and forest litter, occasionally marine.

Testacealobosa Subclass of amoebae with lobose pseudopodia and tests, comprising 2 orders, Arcellinida and Trichosida.

testate amoeba Arcellinida *q.v.*

Testudines Turtles, terrapins, tortoises; order of aquatic and terrestrial reptiles (Chelonia) characterized by a rigid body shell comprising a dorsal carapace and a ventral plastron; jaws beaked, without teeth; girdles within the rib cage; reproduction oviparous; contains about 220 extant species in 11 families; distribution cosmopolitan, except for high latitudes and high montane regions.

Testudinidae Tortoises; family containing about 40 species of herbivorous, terrestrial reptiles (Testudines: Cryptodira) widespread in grassland and arid habitats; carapace typically domed, limbs stout with reduced digits.

tortoise (Testudinidae)

Tethys The epicontinental sea separating Laurasia from Gondwana following the break up of Pangaea *q.v.* in the Mesozoic.

tetra Alestidae *q.v.*

Tetracentraceae Family containing a single species of deciduous trees (Trochodendrales) known from Nepal, China and Burma.

Tetractinomorpha Subclass of demosponges ranging in form from encrusting to massive and branching, found from the intertidal to the abyssal zone; characterized by a skeleton of tetraxonid and monaxonid megascleres; reproduction oviparous.

Tetragonidiales Small order of freshwater flagellate algae (Cryptophyceae).

Tetrameristaceae Small family of Theales containing 2 species of trees or shrubs from Malaysia and the Guayana highlands.

Tetramerocerata Suborder of pauropods widespread in tropical to cold temperate regions; body with 9–10 pedigerous segments, antennae weakly telescopic, mandibles specialized for piercing, tracheae absent; feed on liquid foods; contains about 500 species in 4 families, the vast majority in one family, Pauropodidae.

Tetraodontidae Puffers; family containing about 120 species of tropical and subtropical, marine, brackish and freshwater teleost fishes found on reefs and sea-grass beds; body robust, to 0.9 m length, naked or minutely spinose; jaws forming parrot-like beak; when threatened pufferfishes inflate the body with water; flesh may be highly toxic.

Tetraodontiformes Order of tropical and temperate marine teleost fishes comprising about 300 species in 8 families, including spikefishes, triggerfishes, filefishes, boxfishes, pufferfishes, porcupinefishes and sunfishes; body usually armoured with plates, shields or spines; many species are venomous.

Tetraonidae Grouse; family containing 16 species of medium-sized gallinaceous birds (Galliformes) widespread in cool temperate forests and grasslands of the northern hemisphere; habits solitary or gregarious, polygamous or monogamous, feeding on berries, buds and insects; nest in a scrape on the ground.

grouse (Tetraonidae)

Tetraphidae Subclass of mosses (Bryopsida); small, light-green plants forming loose mats on acidic rock surfaces, rotting wood, or epiphytically on tree ferns; capsule has a peristome with only 4 teeth.

Tetraphyllidea Order of tapeworms (Eucestoda) parasitic in elasmobranchs, utilizing molluscs and crustaceans as first intermediate hosts, and molluscs or fishes as second intermediate hosts; characterized by a scolex of typically 4 stalked or sessile bothridia, commonly divided into loculi, and by a segmented body.

tetraploid A polyploid having 4 sets of homologous chromosomes; **tetraploidy** *cf.* ploidy.

Tetrapoda Subgroup of vertebrates with 4

limbs; includes all terrestrial vertebrate taxa – amphibians, reptiles, birds and mammals; *cf.* Pisces.

Tetrasporales Order of mostly freshwater or terrestrial green algae in which the cells are usually massed into gelatinous colonies.

Tettigoniidae Grasshoppers, katydids; large family of orthopterodean insects (Ensifera) in which auditory communication is well developed; ovipositor sword-shaped; most of the 5000 species are phytophagous.

Teuthoidea Squid; order of marine carnivorous cephalopod molluscs which are typically streamlined, active swimmers found in shallow coastal and open oceanic waters; characterized by the reduction of the shell to an internal, cartilaginous rod, by 8 arms and 2 tentacles, and by posteriorly developed fins.

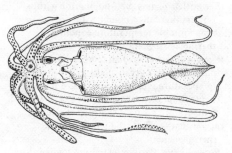

giant squid (Teuthoidea)

thalassic Pertaining to the sea or deep oceanic waters.

Thalassinidea Mud lobsters; infraorder of pleocyematan decapod crustaceans; sometimes classified as a section of the Anomura *q.v.*

Thalassocalycida Small order of pelagic ctenophores found in tropical and subtropical waters; characterized by an expanded body which forms a medusa-like bell around the mouth and by the lack of tentacular sheaths.

Thalassocyathida Extinct order of Archaeocyatha; cup-shaped animals with single walls, known from the Lower Cambrian.

thalasson The marine biotope; *cf.* crenon, stygon, troglon.

Thalassomycetales The single order of elliobiophycid dinoflagellates.

thalassophilous Thriving in the sea; **thalassophile, thalassophily.**

thalassophyte A marine plant.

thalassoplankton Marine planktonic organisms.

Thaliacea Salps; class of solitary or colonial,

pelagic tunicates that have the branchial (anterior, inhalent) and atrial (posterior, exhalent) siphons at opposite ends of body; the water current may be used for propulsion; a ventral stolon is present posterior to the endostyle from which asexual budding occurs; 3 orders recognized, Pyrosomida, Doliolida and Salpida.

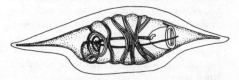

salp (Thaliacea)

Thallobionta Subkingdom of plants comprising eukaryotic algae and fungi; *cf.* Embryobionta.

Thallophyta A group name formerly used for all plant-like organisms lacking differentiation into roots, stems and leaves.

thallus A primitive plant body lacking differentiated true leaves, stems and roots.

thamnobiontic Living in bushes and shrubs; **thamnobiont.**

thamnocolous Living on or in bushes and shrubs; **thamnocole.**

thamnophilic Thriving on, or living in, bushes or shrubs; **thamnophile, thamnophily.**

thanatocoenosis An assemblage of organisms brought together after death; **thanatocenosis;** *cf.* biocoenosis.

thanatogeography Study of the distributions of dead organisms.

thanatology Study of death and dead organisms.

thanatosis Death-feigning behaviour; death feint.

Theaceae Camellia, tea; family containing about 600 species of woody, tanniferous plants with large showy flowers comprising 4–7 sepals and petals, many stamens and usually a

camellia (Theaceae)

superior ovary; widespread in tropical and subtropical regions; the leaves of *Camellia sinensis* are used to make tea.

Theales Order of Dilleniidae containing 18 families of mostly trees and shrubs, sometimes evergreen, very often tanniferous and with mucilage or resin in special cells or ducts.

Thecanephria Order of pogonophoran marine worms possessing a horseshoe-shaped anterior body cavity in the tentacular region with medial ciliated ducts opening to the exterior; spermatophores typically foliate.

Thecina Suborder of Amoebida *q.v.*; also treated as a class of the protoctistan phylum Rhizopoda.

Thecodontia Primitive order of archosaurian reptiles with teeth set in sockets; probably ancestral to dinosaurs, pterosaurs and crocodiles; known from the Upper Permian to the Upper Triassic.

thecodont (Thecondontia)

Thecostomata Pteropods; cosmopolitan order of pelagic marine, opisthobranch molluscs which feed on planktonic organisms carried to the mouth by ciliary tracts; typically with a calcareous shell, and a foot greatly modified for swimming by enlargement into lateral fins.

thegosis The process of tooth sharpening; tooth-grinding behaviour.

Theligonaceae Family of Rubiales containing only 3 species of herbs with wind-pollinated unisexual flowers; found in eastern Asia, the Mediterranean and the Canary Islands.

thelygenic Producing offspring that are entirely, or predominantly, female; **thelygeny**; *cf.* allelogenic, amphogenic, arrhenogenic, monogenic.

Thelyphonida Uropygi *q.v.*

thelytoky Obligatory parthenogenesis in which a female gives rise only to female offspring; **thelytokous**; *cf.* arrhenotoky.

Theophrastaceae Small family of Primulales containing about 100 species of woody, often palm-like, plants native to the tropical New World.

Theraphosidae Tarantulas; family of large, hairy spiders characterized by a group of 8 eyes, and by claw tufts; feed on large insects and small vertebrates which are typically chewed.

Theraponidae Tiger perches, tigerfish; family containing 35 species of predatory Indo-Pacific marine brackish and freshwater teleost fishes (Perciformes); body oblong and compressed, to 500 mm length; dorsal fin deeply notched with 11–14 fin spines; may produce sounds by action of muscles on the swim bladder.

Therapsida Order of synapsid reptiles which were ancestral to the mammals; known from the Permian to the early Jurassic.

Theria The viviparous mammals; a subclass of mammals comprising 2 extant infraclasses. Metatheria (marsupials) and Eutheria (placentals), and a fossil infraclass, Pantotheria.

thermobiology Study of the effect of heat on living organisms and biological processes.

thermocleistogamy Self-pollination within flowers that fail to open because of unfavourable temperature conditions; **thermocleistogamic.**

thermocline A horizonal temperature discontinuity layer in a lake in which the temperature falls by at least 1°C per metre depth; a boundary layer in the sea in which temperature changes sharply with depth.

thermogenesis The production of heat; **thermogenic.**

thermohaline Pertaining to both temperature and salinity.

thermolabile Unstable or destroyed when exposed to high temperatures; *cf.* thermostable.

thermonasty A reponse to a non-directional temperature stimulus; **thermonastic.**

thermopegic Pertaining to hot-water springs.

thermoperiodic Pertaining to the response of an organism to periodic changes in temperature; **thermoperiodicity.**

thermophilic Thriving in warm environmental conditions; used of microorganisms having an optimum for growth above 45°C; **thermophile, thermophily.**

thermophobic Intolerant of high temperatures.

thermophylactic Heat-resistant; tolerant of high temperatures.

thermophyte 1: A plant tolerant of, or thriving at, high temperatures. 2: A hot-spring plant.

Thermosbaenacea Small order of primitive pancaridan crustaceans containing 10 species found in hot springs, freshwater hypogean and marine habitats; carapace reduced but forming a dorsal brood chamber; eyes absent.

thermostable Relatively stable or resistant to heat; *cf.* thermolabile.

thermotaxis The directed response of a motile organism towards (positive) or away from (negative) a temperature stimulus; **thermotactic.**

thermotoxy Death or injury caused by heat; **thermotoxic.**

thermotropism An orientation response to a temperature stimulus: **thermotropic.**

therophyllous Deciduous; used of plants that have leaves only during the warm season.

therophyte An annual plant; a plant that completes its life cycle in a single season, passing unfavourable seasons as a seed; *cf.* Raunkiaerian life forms.

Theropoda Suborder of bipedal, carnivorous dinosaurs (Saurischia) known from the Triassic to the Cretaceous; includes the coelosaurs and carnosaurs.

thigmokinesis A change in the velocity of linear or angular movement of an organism in response to a contact stimulus; **thigmokinetic.**

thigmomorphosis A change in form due to contact; **thigmomorphic.**

thigmotaxis A directed response of a motile organism to a continuous contact with a solid surface; **thigmotactic.**

thigmothermic Used of an animal that draws heat into its body from contact with a warmed object in the environment; **thigmotherm.**

thigmotropism An orientation response to touch or a contact stimulus; **thigmotropic.**

thinicolous Living on sand dunes; **thinicole.**

Thinocoridae Seedsnipe; family containing 4 species of small quail-like charadriiform birds having a short finch-like bill, found in dry open habitats of South America; feed mainly on seeds and other plant material.

thinophilous Thriving on sand dunes; **thinophile, thinophily.**

thinophyte A sand-dune plant.

thiobios Those organisms inhabiting anaerobic sulphur-rich deposits; **thiobiotic.**

thiogenic Sulphur-producing.

thiophilic Thriving in sulphur-rich habitats; **thiophile, thiophily.**

thistle Asterales *q.v.*

thixotropic Used of a colloidal sediment that becomes liquid as a result of agitation or pressure; **thixotropy**; *cf.* dilatent.

tholichthys Larval stage of fishes in the family Chaetodontidae.

Thoracica Acorn barnacles, goose barnacles; largest order of barnacles (Cirripedia) containing about 700 living species; mostly sessile filter feeders with the body enclosed within protective calcareous plates; thoracic

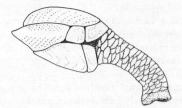

goose barnacle (Thoracica)

legs modified as food-gathering cirri; found attached to hard substrata at all depths from the intertidal to the abyss, also found on whales and other marine animals.

thorn apple Solanaceae *q.v.*

thornbill Maluridae *q.v.*

thorny coral Antipatharia *q.v.*

thorny-headed worm Acanthocephala *q.v.*

Thraupidae Tanagers; family containing about 240 species of New World passerine birds found mostly in tropical forests and woodland; plumage often colourful; bill robust and conical, frequently hooked; habits solitary to gregarious, arboreal, non-migratory, feeding on insects, fruit and flowers; nest in tree or bank.

threadfin Polynemidae *q.v.*

threefin blenny Tripterygidae *q.v.*

thremmatology Study of animal and plant breeding in domestic situations.

thresher shark Alopiidae *q.v.*

threshold The minimum level or value of a stimulus necessary to elicit a response; *cf.* subliminal.

Threskiornithidae Ibises, spoonbills; family containing 33 species of medium-sized wading birds (Ciconiiformes) having long legs, long bills and large broad wings; bill decurved (ibises) or straight and spatulate (spoonbills); feed on fishes and other aquatic organisms; breed in colonies, nesting in trees or on the ground; widespread in tropical and warm temperate freshwater and coastal marine habitats.

ibis (Threskiornithidae)

thrips Thysanoptera *q.v.*

Thripidae Family of thysanopteran insects comprising about 1500 species, including most of the economically important thrips.

throatwort Campanulaceae *q.v.*

thrush Turdidae *q.v.*

Thurniaceae Small family of Juncales comprising only 3 species of coarse perennial herbs with spheroidal silica bodies in shoots and stems; small flowers borne in large, dense terminal heads; native to tropical South America.

thumbless bat Furipteridae *q.v.*

Thylacinidae Tasmanian wolf; monotypic family of dog-like carnivorous marsupials recorded in modern times only from Tasmania but thought to be extinct since 1936; body length to 2 m; habits nocturnal, typically living in a den amongst rocks or logs and feeding on other marsupials and birds.

Thylacomyidae Family of long-eared rabbit-like burrowing bandicoots found in southern Australia; habits nocturnal, feeding on insects and small vertebrates.

thyme Lamiaceae *q.v.*

Thymelaeaceae Spurge laurel, mezereon; cosmopolitan family of Myrtales containing about 500 species of mostly poisonous shrubs; individual flowers bisexual, and regular with 4–5 petaloid sepals, usually 4–5 or 8–10 stamens and a superior ovary; flowers arranged in racemose inflorescences.

Thyropteridae Sucker-footed bats; family containing 2 species of bat (Microchiroptera) having stalked suckers on the wrists and ankles; found in the Neotropical region and West Indies.

Thysanoptera Thrips; order of small (0.5–15 mm) hemipterodean insects comprising about 5000 species that feed on plant juices, fungi, spores, pollen or body fluids of other arthropods such as mites and scale insects; mouthparts specialized for piercing and sucking; mouth cone asymmetrical, right mandible reduced, left mandible modified as

thrips (Thysanoptera)

stylet; wings slender, frequently reduced or absent.

Thysanura Silverfish, firebrats, bristletails, rock jumpers; order of primitive wingless insects (Apterygota) commonly found amongst decaying wood, in leaf litter, caves and human habitations, or in association with ants and termites; feed on fungi, lichens, algae, pollen, or decaying vegetable matter; body length 5–25 mm, abdomen terminating in pair of long cerci and long median filament; sexes similar or weakly dimorphic; development ametabolous; contains about 600 species.

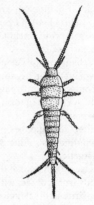

silverfish (Thysanura)

tick Ixodida *q.v.*

tickseed Asterales *q.v.*

tidal current The alternating horizontal movement of water associated with the rise and fall of the tide.

tidal day The period between two consecutive high waters at a given place, averaging 24 h 51 min (24.84 h).

tidal flat An extensive flat tract of land alternately covered and uncovered by the tide, and comprising mostly unconsolidated mud and sand.

tidal marsh A low-elevation marshy coastal area formed of mud and the root mat of salt-loving plants, regularly inundated during high tides.

tide The periodic rise and fall of the ocean water masses and atmosphere, produced by gravitational effects of the moon and sun on the Earth.

tide cycle The duration of a given tidal sequence, as for example a lunar month or a tidal day.

tide range The difference in height between consecutive high and low waters.

tiger Felidae *q.v.*
tiger beetle 1: Carabidae *q.v.* 2: Cicindelidae *q.v.*
tiger moth Arctiidae *q.v.*
tiger perch Theraponidae *q.v.*
tiger shark Carcharhinidae *q.v.*
tigerfish Theraponidae *q.v.*
tilefish Branchiostegidae *q.v.*
Tiliaceae Linden, lime, bass wood; family of mostly tropical and subtropical trees often with stellate hairs or peltate scales; flowers usually with 5 sepals and petals, many free stamens and superior ovaries; phloem fibres from species of the herbaceous genus, *Corchorus* are used as jute.

lime (Tilliaceae)

till Material deposited directly by ice.
Tillodontia Extinct order of mammals known from the northern hemisphere during the Palaeocene and Eocene; some were bear-sized but with rodent-like incisors.
Tilopteridales Small order of sublittoral brown algae with a loosely branched filamentous body form; found in cold waters of the northern hemisphere.
tilth The physical condition of a soil in relation to its fitness for plant growth.
Timaliidae Babblers; diverse family containing about 275 species of Old World passerine birds found in variety of forest and grassland habitats from Africa to Australia; habits gregarious or solitary, mostly arboreal and non-migratory, feeding on insects and fruit; nest composed of sticks and mud, on or off the ground.
timber beetle Cerambycidae *q.v.*
timber line The line marking the latitudinal or upper altitudinal limit of normal dense tree growth.
Tinamidae Tinamous; family containing 47 species of small New World ground-living palaeognathous birds found in variety of forest, brush and grassland habitats; still

tinamou (Tinamidae)

capable of flight; body compact, neck and wings short, tail inconspicuous, legs strong; bill weak, moderately long and weakly decurved; sexes similar although female larger than male; habits mostly crepuscular, secretive, often polygamous, feeding on seeds and fruit; range extends from Mexico to Argentina.
Tinamiformes Order of small New World ratites containing a single family, Tinamidae (tinamous).
tinamou Tinamidae *q.v.*
Tineidae Clothes moths; family of small or minute moths containing about 3000 species some of which are pests of stored products and cause damage to woollen materials.
tintinnid Oligotrichida *q.v.*
tiphicolous Living in ponds; **tiphicole.**
tiphophilous Thriving in ponds; **tiphophile, tiphophily.**
tiphophyte A pond or pool plant.
Tipulidae Craneflies, daddy longlegs, leatherjackets; largest family of dipteran insects containing more than 13 000 species; adult body slender with narrow wings and long fragile legs, antennae 6-segmented; commonly feed on plant juices and nectar.

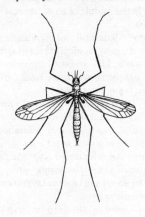

crane fly (Tipulidae)

titmouse 1: Paridae *q.v.* (tits). 2: Aegithalidae (long-tailed tits).

tit warbler Regulidae *q.v.*

toad Bufonidae *q.v.* (Anura).

toadfish Batrachoididae *q.v.*

toadflax Scrophulariaceae *q.v.*

toadstool Informal name for any typical umbrella-shaped fungal fruiting body.

tobacco Solanaceae *q.v.*

tolerance 1: The ability of an organism to endure extreme conditions. 2: The range of an environmental factor within which an organism or population can survive.

tomato Solanaceae *q.v.*

tombolo A sand-bar connecting an island or group of islands with the mainland.

tomiparous Reproducing by fission; **tomiparity.**

tongue sole Cynoglossidae *q.v.*

tonotaxis The directed response of a motile organism to a change in osmotic pressure or to an osmotic stimulus; **tonotactic.**

tonotropism An orientation response to an osmotic stimulus; **tonotropic.**

tooth shell Scaphopoda *q.v.*

toothcarp Cyprinodontidae *q.v.*

top minnow Cyprinodontidae *q.v.*

top shell Archaeogastropoda *q.v.*

topknot Scophthalmidae *q.v.*

topochemotaxis A directed response of a motile organism towards the source of a chemical stimulus; **topochemotactic.**

topodeme A local interbreeding population occurring in a particular geographical area; *cf.* deme.

topogalvanotaxis The directed response of a motile organism towards the source of an electrical stimulus; **topogalvanotactic.**

topographic desert A low-rainfall desert region located in the middle of a continental landmass.

topography All natural and man-made surface features of a geographical area; **topographical.**

topophototaxis The directed response of a motile organism towards the source of a light stimulus; **topophototactic.**

topotaxis The directed response of a motile organism to spatial differences in the intensity of a stimulus, especially with reference to the source of the stimulus; **topotactic.**

topotropism An orientation response towards the source of a stimulus; **topotropic.**

topsoil The upper soil layer or layers containing some organic matter.

tornaria Ciliated planktotrophic larva of enteropneust and planctosphaeroidean hemichordates.

Torpedinidae Torpedo rays, electric rays; family containing 15 species of marine torpediniform fishes; body length to 2 m, disk broader than long, 2 dorsal fins present; mouth large, teeth small and formed of single cusps; depth range from littoral to 500 m.

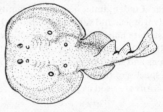

electric ray (Torpedinidae)

Torpediniformes Electric rays, torpedo rays; order comprising about 40 species of mostly small (to 2 m) elasmobranch fishes (Batoidea) in which the body disk is strongly flattened, subcircular, and contains large electric organs; body devoid of denticles; caudal fin large with axis straight; habits benthic, sluggish, using electric organ to stun prey; reproduction viviparous, without placenta; found mainly in the tropics and subtropics, ranging from the littoral to depths over 1000 m.

torpedo ray Torpedinidae *q.v.*

torpid Dormant; lacking vigour.

torpor A dormant state.

torrentfish Cheimarrichthyidae *q.v.*

torrenticolous Living in river torrents.

tortoise Testudinidae *q.v.*

tortoiseshell butterfly Nymphalidae *q.v.*

Tortricidae Large cosmopolitan family of small moths (Lepidoptera) the larvae of which have diverse leaf rolling, mining, boring and gall-forming habits; comprises about 4000 species and includes many serious agricultural pests such as the codling moth, numerous fruit moths, and the spruce and pine budworms.

toucan Ramphastidae *q.v.*

touch-me-not Balsaminaceae *q.v.*

touraco Musophagidae *q.v.*

Tovariaceae Small family of Capparales containing 2 species of coarse herbs or soft shrubs, native to tropical America.

toxic Poisonous.

toxicity The virulence of a poisonous substance (the toxicant).

toxicology The study of poisonous substances and their effects.

toxigenic Producing toxin.

toxin A biogenic poison, usually proteinaceous.

Toxotidae Archerfishes; family containing 6 species of small (to 250 mm) Indo-Pacific, coastal marine to freshwater, perciform teleost fishes frequently found in mangrove habitats; body deep and compressed, dorsal profile somewhat flattened; eyes large, mouth strongly protractile; feed on insects dislodged from vegetation by spitting drops of water over distances up to 3 m; archerfishes are popular amongst aquarists.

trace Any distinct biogenic structure either fossil or recent, such as a track, trail, burrow, boring or faecal cast.

trace element An element essential for normal growth and development of an organism, but required only in minute quantities.

trace fossil A sedimentary structure resulting from the activity of a living animal, such as a track, burrow or tube.

Tracheophyta Vascular plants; division of plants possessing vascular tissues (xylem and phloem) for internal transport of nutrients and water; includes the pteridophytes (Filicophyta), gymnosperms (Pinophyta) and angiosperms (Magnoliophyta).

Trachichthyidae Slimeheads; family containing about 15 species of beryciform teleost fishes widespread in shallow inshore as well as deep oceanic waters; body deep and compressed, to 500 mm in length; head with large mucous cavities.

Trachinidae Weeverfishes; family of venomous, bottom-dwelling, coastal marine teleost fishes (Perciformes) containing 4 species widespread in the eastern Atlantic; body to 400 mm in length, head depressed, eyes dorsal, mouth oblique; first dorsal fin spines and opercular spines with poison glands; live partly buried in sand during the day emerging to feed at night.

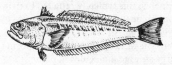

weeverfish (Trachinidae)

Trachipteridae Ribbonfishes; family containing 8 species of mesopelagic lampridiform teleost fishes; body elongate, to 1.7 m in length, compressed, silvery, scales deciduous; caudal and dorsal fins elongate, caudal with upper lobe only, anal fin absent; feed mainly on fishes, squid and crustaceans.

Trachylina Primitive order of Hydrozoa occurring mostly in open water in the warmer oceans, characterized by the reduction or loss of the polyp generation; the medusoid stage is a transformed actinula larva and is not homologous with the medusae of other hydrozoans.

Trachypsammiaceae Extinct order of octocorals possessing a central medullar canal system; known only from the Permian.

trade wind Uniform tropical oceanic wind system blowing towards the equator from the northeast in the northern hemisphere and from the southeast in the southern hemisphere.

tradescantia Commelinaceae *q.v.*

Tragulidae Chevrotains; family containing 4 species of small (up to about 1 m length) forest-dwelling ungulates (Artiodactyla: Ruminantia) found in tropical Ethiopian and Oriental regions; upper canines elongate; horns absent; feet with 4 digits; habits nocturnal or crepuscular, solitary, feeding mainly on fruit and fleshy plant material; these ungulates do not ruminate.

chevrotain (Tragulidae)

trahiras Erythrinidae *q.v.*

trail pheromone A chemical substance laid down as a trail by one individual and followed by another member of the same species.

trait Any character or property of an organism.

tramp A species having a wide geographical range and an efficient means of dispersal.

transcription Synthesis of messenger RNA from a DNA molecule acting as a template; *cf.* translation.

transduction The transfer of genetic material from donor to recipient microorganism, as from one bacterial cell to another by a bacteriophage.

transect A line or narrow belt used to survey the distributions of organisms across a given area.

transfer RNA A relatively small RNA molecule that transfers a given amino acid to a polypeptide chain during translation *q.v.* of messenger RNA at the site of protein synthesis in a ribosome; tRNA.

transgression The spread of the sea over a land area; *cf.* regression.

transient Of short duration; transitory.

translation The sequential reading of the base sequence of messenger RNA and its translation into a specific amino acid chain during protein synthesis in ribosomes; *cf.* transcription.

translocated herbicide A herbicide that is dispersed throughout the plant tissue following absorption through the leaves or roots.

translocation 1: The transport of material within a plant. 2: Movement of a segment of a chromosome to another part of the same chromosome or to a different chromosome.

transpiration Loss of water vapour from an organism through a membrane or through pores.

transpiration efficiency The ratio of dry organic matter produced by a plant to water loss by transpiration.

transplant Transfer of tissue from one part of an organism to another, or to another organism.

Trapaceae Water chestnut; small family of Myrtales containing about 15 species of aquatic annual herbs; native to tropical and subtropical Africa and Eurasia.

traplining A feeding strategy during which an animal visits a sequence of widely dispersed food sources, obtaining a small part of its daily food requirement at each feeding station.

trauma Injury or stress caused by an extrinsic agent; **traumatic, traumatism.**

traumatonasty A growth response to a traumatic stimulus; **traumatonastic.**

traumatotaxis Movement of organisms, cells or organelles in response to injury; **traumatotactic.**

traumatotropism An orientation response to injury or stress; **traumatotropic.**

traveller's tree Strelitziaceae *q.v.*

tree creeper Certhiidae *q.v.*

tree line The line which marks the northerly, southerly or upper altitudinal limit of tree cover.

tree of heaven Simaroubaceae *q.v.*

tree shrew Tupaiidae *q.v.*

tree sloth Bradypodidae *q.v.*

tree swift Hemiprocnidae *q.v.*

treehopper Membracidae *q.v.* (Homoptera).

tree-ring chronology Study of annual growth rings of trees for the purposes of dating.

Tremandraceae Family of Polygalales containing 28 species of heather-like shrubs,

native to Australia and Tasmania; flowers often brightly coloured, with 3–5 sepals and petals, twice as many stamens, and a superior ovary.

Trematoda Class of exclusively parasitic flatworms characterized by a dorsoventrally flattened body covered with a cuticle, one or more attachment organs (holdfasts), a well developed gut consisting of mouth, pharynx, oesophagus and 1 or 2 intestinal caeca, usually with no anus; reproductive system well developed, usually hermaphroditic and producing numerous encapsulated eggs; comprises over 8000 species grouped into 3 subclasses, Monogenea, chiefly parasitizing fish, Aspidogastrea, endoparasites of molluscs, fishes and turtles, and Digenea, parasitizing all groups of vertebrates.

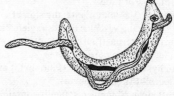

schistosome (Trematoda)

Trentepohliales Small order of green algae in which the thallus forms an erect tuft or prostrate disk of loosely packed, branched, uniseriate filaments; found subaerially and occasionally as parasites, in the tropics and subtropics.

Trepostomata Extinct order of bryozoans known from the Ordovician to the Permian; colonies solidly built of calcite, the individuals forming long tubes.

Triacanthidae Spikefishes, triplespines; family containing 20 species of small (to 70 mm) bottom-living tetraodontiform teleost fishes found in shallow coastal waters and in the deep sea; pelvic fins typically reduced to pair of strong spines.

triacylglycerol An ester of glycerol and carboxylic or fatty acids; often important as energy storage molecules in plants.

Triakidae Hound sharks; widespread family containing about 40 species of small to medium-sized (to 2 m) inoffensive carcharhiniform elasmobranch fishes that are benthic or pelagic in continental seas from the intertidal to 500 m; may be locally common; feed on fishes and a variety of invertebrates, especially crustaceans; reproduction ovoviviparous to viviparous; sometimes treated as part of the Carcharhinidae.

Triassic A geological period of the Mesozoic era (*c.* 245–210 million years B.P.); Trias; see geological time scale.

Tribonematales Order of freshwater and marine algae (Xanthophyceae) characterized by their filamentous organization; cells within filaments mostly uninucleate.

Trichechidae Manatees; family containing 3 species of large nocturnal, herbivorous aquatic mammals (Sirenia) found in tropical freshwater and coastal habitats of the western and eastern Atlantic; incisor and canine teeth absent; tail fluke without median notch.

Trichiales Order of Myxogastromycetidae characterized by fruiting bodies that lack lime but bear brightly coloured spores internally; contains about 80 species typically found on litter and decaying wood.

Trichiida The slime mould order Trichiales *q.v.*, treated as an order of the protozoan group Myxogastria.

Trichiuridae Cutlassfishes; family containing 17 species of scombroid teleost fishes (Perciformes) found mostly in midwater to about 1000 m depth; body elongate and compressed, to 1.5 m in length, jaw teeth fang-like; pelvic and caudal fins small or absent; feed on squid, crustaceans and other fishes.

Trichocephalida Small order of enoplian nematodes found only as parasites of vertebrates; characterized by arrangement of the oesophageal glands in a single row, and by the protrusible axial spear present in early larval stages but lost in the adult.

Trichodontidae Sandfishes; family containing 2 species of northern Pacific, shallow marine perciform teleost fishes, typically found partly buried in sandy sediment; body compressed, deeply convex ventrally, to 300 mm in length, mouth subvertical opening on top of head; anal fin elongate, pectorals broad; exploited as food-fish in Japan.

Trichomonadida Order of mostly parasitic zooflagellates with 4 to 6 flagella, one of which is recurrent and either free or adhered to the body surface forming an undulating membrane.

Trichomycetes Class of zygomycotine fungi comprising obligate gut symbionts of various arthropods that obtain nutrients from material in host's digestive tract; usually attached to chitinous lining of hind gut; 4 orders recognized, Amoebidiales, Asellariales, Eccrinales and Harpellales.

Trichomycteridae Family of small (to 150 mm) Central and South American freshwater catfishes (Siluriformes); body often vermiform, naked, dorsal fin lacking spine; 2–3 pairs of barbels present; contains about 180 species, some parasitic on other fishes, others free-living and nocturnal or crepuscular, burrowing in substratum during the day.

Trichoptera Caddisflies; order of holometabolous insects containing about 7000 species in which the adults are terrestrial and the larvae and pupae are almost exclusively aquatic in fresh water; adults are moth-like with hairy forewings held at an oblique vertical angle at rest, and are liquid feeders; larvae are caterpillar-like and construct a variety of protective cases, silken nets or other domiciles in fresh water, feeding on algae, fungi, or decomposing plant material.

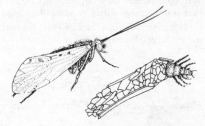

caddisfly (Trichoptera)

Trichosida Order of multinucleate amoeboid protozoans (Testacealobosa) which produce flexible fibrous sheaths with multiple apertures; common in shallow coastal and estuarine waters.

Trichostomatida Primitive order of vestibuliferan ciliates in which the ciliature of

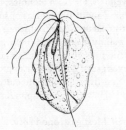

trichomonad (Trichomonadida)

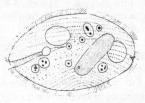

ciliate (Trichostomatida)

the vestibule is derived during fission from the terminal portions of the ciliary rows over the general body surface; occasionally free-living but mostly found as endocommensals in the digestive tracts of ruminants.

Tricladida Large order of typically free-living turbellarians found commonly in marine, freshwater and terrestrial habitats; characterized by a single pair of ovaries, a posteriorly directed plicate pharynx and an intestine divided into an anterior median and 2 posterior branches, all highly diverticulated; often attaining large size, most are predatory; some are resistant to desiccation.

Triconodonta Extinct order of carnivorous, early mammals (Prototheria) known from the late Triassic to the early Cretaceous; molar teeth with row of 3 sharp cusps.

trigenetic Pertaining to a symbiont requiring three different hosts during the life cycle; *cf.* digenetic, monogenetic.

trigger plant Stylidiaceae *q.v.*

triggerfish Balistidae *q.v.*

Triglidae Sea robins, gurnards; family containing 85 species of bottom-living, tropical and temperate, marine scorpaeniform teleost fishes found on soft sediments of inshore and shelf waters; body often armoured with bony plates, head large and ornate; pectoral fins large with a number of isolated fin rays; use muscles acting on swim bladder for sound production.

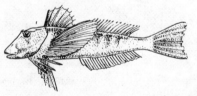

gurnard (Triglidae)

trigoneutic Producing three broods in a single season; **trigoneutism**; *cf.* monogoneutic, digoneutic, polygoneutic.

Trigoniaceae Family of Polygalales containing 24 species of woody plants, widespread in moist forests of tropical America; flowers irregular, with 5 sepals, 3–5 petals, 3–12 stamens on one side, and a superior ovary; fruit usually a capsule.

Trigonioida Small order of marine paleoheterodont bivalves found burrowing in shallow waters of the Indo-Pacific.

Trilobita Trilobites; subphylum of primitive aquatic arthropods known from the Cambrian to the Permian; body divided into anterior

cephalon, thorax and posterior pygidium; trilobed appearance due to pair of longitudinal furrows running length of body; compound eyes often present; limbs typically with ambulatory inner branch and foliaceous, respiratory outer branch.

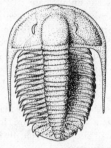

trilobite (Trilobita)

Trimeniaceae Small family of trees or climbing shrubs (Laurales) restricted to New Guinea, New Caledonia, Fiji and eastern Australia.

trimonoecious Used of an individual plant bearing male, female and hermaphrodite flowers.

trimorphic Used of organisms having three distinct forms in the life cycle, or in a population; **trimorphism.**

trioecious Used of a plant species having male, female and hermaphrodite flowers on different individuals; *cf.* dioecious, monoecious.

Trionychidae Soft-shelled turtles; family containing about 20 species of large (to 1 m) predatory freshwater turtles (Testudines: Cryptodira) found in North America, North Africa and the Indo-Pacific region; carapace without horny scutes; snout forming a proboscis bearing fleshy lips; feet webbed.

triplespine Triacanthidae *q.v.*

tripletail Lobotidae *q.v.*

triploid A polyploid having three sets of homologous chromosomes; **triploidy** *cf.* ploidy.

tripod fish Bathypteroidae *q.v.*

Tripterygiidae Threefin blennies; family containing about 100 species of small (to 90 mm) tropical marine perciform teleost fishes found mainly in coral and rocky habitats; 3 dorsal fins present, the first 2 spinose, the third soft-rayed; sexual dimorphism pronounced.

tripton Non-living particulate matter suspended in water; a component of seston *q.v.*

triton shell Mesogastropoda *q.v.*

tritonymph Immature stage, between

protonymph and adult, found in some mites (Acari).

Triuridaceae Family of about 70 species of small mycotrophic herbs (Triuridales) without chlorophyll, with a single ring of vascular bundles in the stem and leaves reduced to scales; widespread in tropical and subtropical regions; flowers regular, borne in racemes.

Triuridales Order of Alismatidae comprising 2 small families of mycotrophic herbs lacking chlorophyll, with endospermous seeds and distinct or nearly distinct carpels, Petrosaviaceae and Triuridaceae.

trivoltine Having three generations or broods per year; *cf.* voltine.

trixenous Used of a parasite utilizing three host species during its life cycle; **trixeny**; *cf.* dixenous, heteroxenous, monoxenous, oligoxenous.

tRNA Transfer RNA *q.v.*

Trochilidae Hummingbirds; family containing about 340 species of small colourful New World birds (Apodiformes) noted for their acrobactic hovering flight; bill slender, tongue extensible, adapted for feeding on nectar, pollen and insects from flowers; sexes dimorphic; habits solitary, mostly monogamous, often migratory.

hummingbird (Trochilidae)

Trochodendraceae Primitive family of Trochodendrales containing a single tree species with tiny green flowers lacking a perianth; fruit is a follicle.

Trochodendrales The most archaic surviving order of the Hamamelidae comprising two monotypic families of trees, the Tetracentraceae and the Trochodendraceae, both native to eastern Asia.

trochophore Free-swimming, unsegmented ciliated larval stage of polychaetes, sipunculans, echiurans, entoprocts, bryozoans, brachiopods, phoronids and molluscs.

trochosphere Trochophore *q.v.*

troglobiont An obligate cavernicole; an

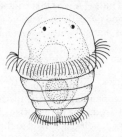

trochophore

organism found only in caves or subterranean passages; *cf.* troglophile, trogloxene.

Troglodytidae Wrens; family containing about 60 species of small active passerine birds found in variety of forest, woodland and open arid habitats of the Old World; body compact, wings short and rounded, tail erect; habits typically solitary, arboreal, some species migratory, monogamous to polygamous; nest in trees, bushes, grass, or banks.

wren (Troglodytidae)

troglon The biotope comprising subterranean water bodies in caves and subterranean passages; *cf.* crenon, stygon, thalasson.

troglophile An animal frequently found in underground caves or passages but not confined to them; *cf.* troglobiont, trogloxene.

troglophilic Thriving in caves and subterranean passages; **troglophile, troglophily.**

trogloxene An organism found only occasionally in caves or subterranean passages; **trogloxenous**; *cf.* troglobiont, troglophile.

Trogoniformes Order of colourful, sedentary, arboreal birds found worldwide in tropical forests; feed on fruit, insects and other small animals; contains about 35 species in a single family, Trogonidae.

Trogonophiidae Family containing 6 species of worm lizards (Amphisbaenia) found in North Africa and southwestern Asia.

Trombidiformes Trombidiid mites; diverse

suborder of mites (Acari) that contains many predatory and parasitic species; includes spider mites, harvest mites, velvet mites and water mites.

troop A group of primates; a flock or assemblage of birds or mammals.

Tropaeolaceae Nasturtium; family of Geraniales containing about 90 species of mustard oil-producing herbs, often climbing with twining petioles; confined to the New World; flowers bisexual, irregular and often showy, with 5 sepals forming a spur below the flower, 5 irregular petals, 8 stamens and a superior ovary.

trophallaxis Mutual or unilateral exchange of food between colony members.

trophic Pertaining to nutrition; **trophism.**

trophic level 1: Each step of a food chain or food pyramid *q.v.*, from producer to primary, secondary or tertiary consumer. 2: The nutrient status of a body of fresh water.

trophic unit A group of individuals of one or more species occupying the same relative position in a food chain.

trophobiosis Symbiosis in insects in which food is obtained from one species (the trophobiont) by another in return for protection.

trophodynamics Study of the energy relationships of feeding strategies and food webs.

trophogenic 1: Produced by, or resulting from, food or feeding behaviour. 2: Used of the illuminated surface layer of a lake in which photosynthesis occurs; *cf.* tropholytic.

tropholytic Pertaining to the deeper part of a lake below the trophogenic *q.v.* zone, in which organic matter is utilized as the energy source.

trophotaxis The directed response of a mobile organism towards (positive) or away from (negative) a food stimulus; **trophotactic.**

trophotropism An orientation response to food; **trophotropic.**

tropic bird Phaethontidae *q.v.*

tropic tide The tide occurring twice monthly when the moon reaches its greatest northerly or southerly declination; *cf.* solstitial tide.

tropical 1: Pertaining to the zone between the Tropic of Cancer (23° 27′ N) and the Tropic of Capricorn (23° 27′ S); *cf.* polar, temperate. 2: Used of a climate characterized by high temperature, humidity and rainfall, and having rare light frosts at night.

tropical desert Hot dry desert situated close to the Tropics of Cancer and Capricorn, where subtropical high pressure meteorological conditions result in very low sporadic rainfall.

Tropical grassland biome The circumtropical savannah belt transitional between equatorial forest and grassland, characterized by constant high temperature, a long dry season, low-fertility soil, frequent fires, dominated by grasses and scattered trees and large grazing mammals; Tropical savannah biome.

tropical lake A lake having a surface temperature that never falls below 4°C at any time of year.

Tropical rainforest biome A circumtropical forest region characterized by constant high humidity and temperature, no frost, little seasonal fluctuation of climate except precipitation, typically with a rich variety of trees, a dense canopy, lianas and epiphytes, little undergrowth, an abundance of birds and arboreal vertebrates and the greatest diversity of animal life in any terrestrial biome.

Tropical savannah biome Tropical grassland biome *q.v.*

tropical zone The latitudinal zone between 15.0° and 23.5° in either hemisphere.

tropicopolitan Cosmopolitan within the tropics.

Tropidophiidae Family containing 20 species of small (to 500 mm) terrestrial, ovoviviparous booid snakes (Serpentes) from West Indies and Mexico through to northern South America.

tropism An orientation or growth response of a non-motile organism or one of its parts towards (positive) or away from (negative) a stimulus; **tropic.**

tropoparasite An organism that lives as an obligate parasite for only part of its life cycle.

tropophilous Thriving in an environment that undergoes marked periodic fluctuations of light, temperature and moisture; **tropophile, tropophily.**

tropophyte A plant that thrives under mesic conditions for part of the year and xeric conditions at other times.

troposphere 1: The warm upper layer of oceanic water at middle and low latitudes, typically having strong currents; ocean waters above the thermocline *q.v.* 2: The layer of the atmosphere below the stratosphere extending from ground level to 10–15 km above the Earth's surface, in which temperature falls rapidly with increasing altitude, typically with active convection currents; *cf.* stratosphere.

tropotaxis An orientation response of a motile organism in which the stimulus is compared simultaneously on each side of the midline by bilateral sense organs, such that no deviations from the line of movement are required to

compare the intensities of the stimuli; *cf.*
klinotaxis, telotaxis.

trout Salmonidae *q.v.*

trout-perch Percopsidae *q.v.*

true fungi Eumycota *q.v.*

truffle The underground fruiting body of some
species of Tuberales *q.v.*; often edible.

trumpet creeper Bignoniaceae *q.v.*

trumpeter 1: Psophiidae *q.v.* 2: Latridae *q.v.*

trumpetfish Aulostomidae *q.v.*

trunkfish Molidae *q.v.*

Tryblidioidea The only order comprising the
class Monoplacophora *q.v.*; contains Recent
forms and fossils from Cambrian onwards.

Trypanorhyncha Order of tapeworms parasitic
in cartilaginous fishes (elasmobranchs);
characterized by a scolex consisting of 2–4
bothridia and 4 eversible, apical tentacles used
for attachment to the host; utilize molluscs and
crustaceans as intermediate hosts.

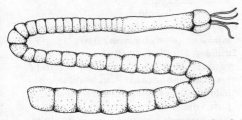

tapeworm (Trypanorhyncha)

tryphobiontic Inhabiting bogs.

tsetse fly Glossinidae *q.v.*

tsunami A large ocean wave caused by
submarine volcanic or earthquake activity.

tuatara Sphenodontidae *q.v.*

tube anemone Ceriantharia *q.v.*

tube-eye Stylophoridae *q.v.*

tubenose Aulorhynchidae *q.v.*

tuber A swollen part of a root or stem, typically
modified for storage and lasting for a single
year only.

Tuberales Order of discomycete fungi that
produce underground fleshy or waxy, fruiting
bodies (truffles) which are highly prized as

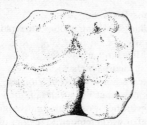

truffle (Tuberales)

food; characterized by underground ascocarps
that remain closed and by asci that have lost
their active spore-discharge mechanism.

Tuberculariales Order of hyphomycete fungi in
which the fertile hyphae are compacted into
false fruiting bodies called sporodochia, which
are highly variable in shape, texture and
colour; most species are saprophytes and
parasites of flowering plants.

tubeshoulder Searsiidae *q.v.*

tubesnout Aulorhynchidae *q.v.*

tubicolous Tube-dwelling; **tubicole.**

Tubificina Sludge worms; large suborder of
oligochaete worms (Haplotaxida) found in
fresh water; chaetae diverse, gills sometimes
present; blood may contain haemoglobin
pigment.

Tuboidea Extinct order of sessile, dendroid
graptolites known from the Lower Ordovician
to Upper Silurian.

Tubulidentata The smallest extant order of
mammals comprising a single species, the
aardvark (Orycteropodidae); characterized by
the absence of tooth enamel, each tooth
having tubular pulp cavities surrounded by
hexagonal dentine prisms.

Tubulina Suborder of Amoebida *q.v.*; also
treated as a class of the protoctistan phylum
Rhizopoda.

tulip Liliaceae *q.v.*

tulip shell Neogastropoda *q.v.*

tulip tree Magnoliaceae *q.v.*

Tulostomatales Stalked puffballs; small order
of gastromycete fungi that produce more or
less globose fruiting bodies supported on a
stalk (stipe).

stalked puffball (Tulostomatales)

tumulus Pertaining to dunes.

tun A resistant cryptobiotic stage in
tardigrades.

tun shell Mesogastropoda *q.v.*

tuna Scombridae *q.v.*

Tundra biome A barren treeless region north of
the Arctic Circle (Arctic tundra), also found

above the tree line of high mountains (alpine tundra) and on some sub-Antarctic islands; characterized by very low winter temperatures, short cool summers, permafrost below a surface layer subject to summer melt; dominated by lichens, mosses, sedges and low shrubs, vast seasonal swarms of insects, and migratory animals.

Tundra soil A zonal soil with abundant litter and duff, a dark brown peaty A-horizon rich in organic material over a greyish B-horizon; formed under conditions of extreme cold, high humidity and poor drainage.

Tungidae Jiggers, sand fleas; family containing 10 species of siphonapteran insects (fleas) in which the females bury themselves beneath the skin of the host and then swell up by engorging blood.

Tunicata Tunicates; subphylum of solitary or colonial, sessile or free-living, marine invertebrate chordates in which the adult body lacks segmentation and body cavity, and is enclosed in a soft, sometimes gelatinous test; water passes through a large branchial sac by ciliary action via inhalent and exhalent siphons, food particles being trapped by mucus secreted by the endostyle; the gut is U-shaped; typically hermaphroditic, oviparous, the larval stage possessing the chordate features of a notochord, dorsal nerve chord and elongate post-anal tail; colonies formed by asexual budding; about 1250 species are recognized in 3 classes, Ascidiacea, Thaliacea and Appendicularia; Urochordata.

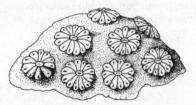
colonial tunicate (Tunicata)

Tupaiidae Tree shrews; family containing 15 species of small arboreal and terrestrial

tree shrew (Tupaiidae)

mammals (Insectivora) found in the Oriental Region; habits diurnal, active, omnivorous, locomotion scansorial.

Turbellaria Class of mostly free-living, predatory flatworms, but including some commensal and parasitic forms, found commonly in freshwater and marine habitats, and in some damp terrestrial situations; characterized by the possession of a sperm with 2 flagella, a blind gut with no anus, an unsegmented body and a cellular ciliated epidermis; most forms have a muscular pharynx and some possess statocysts; comprises 12 orders.

turbid Cloudy; opaque with suspended matter.

turbidite A marine sediment deposited at the base of a submarine slope by a turbidity current.

turbidity current A gravity flow resulting from a mass of unstable sediment sliding down a submarine slope.

turbot Scophthalmidae *q.v.*

turbulence Irregular motion or agitation of liquids or gases; *cf.* laminar flow.

Turdidae Thrushes; diverse family of small passerine birds found in a variety of forest, grassland and open habitats, worldwide; habits solitary to gregarious, arboreal or terrestrial, migratory or non-migratory, feeding mainly on insects, worms, molluscs and fruit; nest typically cup-shaped, on or off the ground; contains about 300 species, including wheatears, chats, bluebirds, nightingale and blackbirds.

blackbird (Turdidae)

turfaceous Pertaining to bogs.

turfophilous Thriving in bogs; **turfophile, turfophily.**

turkey Meleagridae *q.v.*

turkeyfish Scorpaenidae *q.v.*

Turneraceae Small family of Violales containing about 120 species of often cyanogenic herbs or shrubs; native mainly to tropical or warm temperate regions of America and Africa.

Turnicidae Hemipodes; family containing 14 species of small quail-like gruiform birds found in tropical and warm temperate open woodland and grassland of the Old World from Europe to Australia; habits solitary, terrestrial, flight usually short and explosive, female polyandrous; feed on seeds and insects, nest on the ground.

turnip Brassicaceae *q.v.*

turnover 1: The ratio of productive energy flow to standing crop biomass in a community or ecosystem. 2: That fraction of a population which is exchanged (lost by mortality and emigration, replaced by recruitment). 3: Water circulation occurring in deep temperate lakes in spring and autumn, tending to equalize temperature throughout the water column.

turnstone Scolopacidae *q.v.*

turtle Testudines *q.v.*

turtle grass Hydrocharitales *q.v.*

tusk shell Scaphopoda *q.v.*

tussock moth Lymantriidae *q.v.*

tychopelagic Used of organisms that are normally benthic, but which have been carried up into the water column by chance factors.

tychoplankton Organisms occasionally carried into the plankton by chance factors such as turbulence; **tychoplanktont, tychoplanktonic.**

tychopotamic Pertaining to aquatic organisms thriving in the still backwaters of rivers and streams; *cf.* autopotamic, eupotamic.

Tylenchida Order of diplogasterian nematodes containing fungal feeders, and parasites of insects and higher plants; characterized by the presence of an axial spear in the oesophagus.

Tylopoda Suborder of cloven-hooved ungulates (Artiodactyla) comprising a single extant family, Camelidae (camels, llamas), and many fossil families of ruminant-like artiodactyls known from the Eocene to the Pliocene.

Typhaceae Cattails, reedmace; cosmopolitan family of often dense, colonial, perennial herbs growing in marshes or in shallow water; with numerous wind-pollinated flowers borne in dense, cylindrical spikes, female flowers typically lower down spike, male flowers in upper part.

Typhales Order of Commelinidae comprising 2 small families of aquatic or semiaquatic perennial herbs, the bur reeds (Sparganiaceae) and the cattails (Typhaceae).

Typhlonectidae Family containing about 20 species of aquatic South American caecilians (Gymnophiona); body length to about 700 mm, scales and tail absent; reproduction

reedmace (Typhaceae)

oviparous or viviparous, larvae without gill slits; distribution confined to Orinoco, Amazon and Paraná Basins in South America.

Typhlopidae Family of worm-like, insectivorous, burrowing snakes (Serpentes) widespread in tropical and subtropical regions; length up to 1 m; teeth present on upper jaw only; eyes concealed beneath scales; contains about 150 species.

Typhogena Superorder of blind helminthomorphan diplopods (millipedes) comprising 2 orders, Siphonophorida and Platydesmida; mouthparts reduced, sternal and pleural sclerites loosely articulated; 8 pairs of legs present anterior to gonopods.

Tyrannidae Tyrant flycatchers; diverse family containing about 375 species of Neotropical passerine birds found in wide variety of habitats from wet forest to arid open areas; habits arboreal to terrestrial, solitary, monogamous, feeding on insects and small vertebrates; nest on or above the ground.

Tytonidae Barn owls; cosmopolitan family containing 11 species of medium-sized nocturnal raptors (Strigiformes); head large with heart-shaped facial disk; habits solitary, monogamous, non-migratory, feeding on variety of small mammals, birds and insects.

u

ubiquitous Having a worldwide distribution, effect or influence; widespread; cosmopolitan.

Udden grade scale A scale of sediment particle size *q.v.*, categories based on a doubling or halving, above or below, respectively, the fixed reference point of 1 mm.

uliginous Inhabiting swampy soil or wet muddy habitats.

ulluco Basellaceae *q.v.*

Ulmaceae Elm, hackberry; family of Urticales containing about 150 species of woody plants widely distributed in the northern hemisphere; often with mineralization of cell walls, especially in the epidermis; flowers borne in small clusters on twigs, each flower with a 4–8 lobed perianth, the same number of stamens and an ovary of 2 fused carpels.

ulmification The process of peat formation.

Ulotrichales Small order of unbranched filamentous green algae; filaments composed of uninucleate cells each containing a parietal, laminate or cylindrical chloroplast; also treated as a class of the proctoctistan phylum Chlorophyta.

ultrananoplankton Planktonic organisms less than 2 μm in diameter.

ultraplankton Planktonic organisms less than 5 μm in length or diameter.

ultrastructure Cellular structure studied with the aid of a transmission electron microscope; fine structure.

ultraviolet radiation (UV) That part of the solar radiation spectrum below a wavelength of 380 nm.

Ulvales Sea lettuce; order of essentially marine or brackish-water green algae in which the body is a multicellular thallus typically in the form of a monostromatic blade, tube or sac; including *Ulva* and *Enteromorpha*.

umbel 1: A typically umbrella-shaped inflorescence in which all pedicels arise at the apex of an axis; commonly compound. 2: Any member of the family Apiaceae *q.v.*

Umbellales Apiales *q.v.*

Umbelliferae Apiaceae *q.v.*

umbraticolous Living in shaded habitats; **umbraticole**; *cf.* lucicolous.

umbrella shell Notaspidea *q.v.*

Umbridae Mudminnows; family containing 5 species of small (to 200 mm) freshwater salmoniform teleost fishes found in streams

and rivers of northern temperate and Arctic regions; body elongate and weakly compressed; swim bladder may serve accessory respiratory function.

umbrophilic Thriving in shaded habitats; **umbrophile, umbrophily.**

umbel

understorey The vegetation layer between the overstorey or canopy and the ground-storey of a forest community, formed by shade tolerant trees of moderate height.

undulipodia All flagella and cilia sharing a 9+2 microtubular structure.

ungulate Any large hoof-bearing, grazing mammal; extant species typically belonging to the orders Perissodactyla and Artiodactyla.

unguligrade Walking or running on hooves or the modified tips of one or more digits.

unicorn plant Martyniaceae *q.v.*

uniformitarianism The doctrine that natural geological processes affecting the Earth are still operating at essentially the same rate and intensity as they have throughout geological time; actualism; *cf.* catastrophism.

Unionoida Naiads, pearly freshwater mussels; order of freshwater paleoheterodont bivalve molluscs which release a parasitic larval stage (the glochidium) after incubation in a marsupium formed from the gills in the female.

uniparous 1: Producing a single offspring in each brood; *cf.* biparous, multiparous. 2: Having only one brood during the life cycle.

Uniramia Phylum or subphylum of arthropods comprising 2 (or 3) classes, Myriapoda (pauropods, diplopods, chilopods, symphylans), Insecta (insects) and, by some authorities, Onychophora; characterized by a single pair of antennae, 1 or 2 pairs of maxillae, uniramous limbs, and mandibles that bite transversely at the tip.

unisexual 1: Used of a population or generation composed of individuals of one sex only. 2:

Used of an individual having either male or female reproductive organs and producing only male or female gametes. 3: Used of a flower possessing only male or female reproductive organs; imperfect flower; *cf.* bisexual.

unistratal Pertaining to vegetation lacking distinguishable horizontal layering; *cf.* multistratal.

univoltine Having one brood or generation per year; *cf.* voltine.

univorous Feeding on only one type of food; **univore, univory.**

Upupidae Hoopoes; small group of colourful woodland birds (Coraciiformes) found in Africa and in temperate parts of Eurasia; habits monogamous, mostly non-migratory, feeding largely on insects and other invertebrates; nest in hole in tree or bank, nest commonly foul smelling.

hoopoe (Upupidae)

upwelling An upward movement of cold nutrient-rich water from ocean depths.

Uranoscopidae Stargazers; family containing 25 species of bottom-living, tropical and warm temperate, marine teleost fishes (Perciformes) typically found partly buried in sand or sandy mud; body robust and deep, to 600 mm in length, head large, mouth vertical; some species have electric organ derived from modified eye muscles.

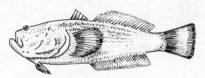

stargazer (Uranoscopidae)

Uredinales Rust fungi; large cosmopolitan order of parasitic teliomycete fungi occurring as an intercellular mycelium within the host plant; producing emergent reproductive units or spores which are wind-disseminated; most cause infections of host organs, often in economically important crops; includes the genus *Puccinia* which has over 4000 species parasitizing cereal crops and many other host plants.

Urkaryota One of the three primary kingdoms (urkingdoms *q.v.*) of living organisms, comprising all eukaryote organisms; *cf.* Eubacteria, Archaebacteria.

urkingdoms The three primary kingdoms proposed recently as a basic tripartite scheme for grouping all living organisms, as an alternative to the widely used five-kingdom *q.v.* scheme of Whittaker; *cf.* Eubacteria, Archaebacteria, Urkaryote.

Urochordata Tunicata *q.v.* (sea squirts).

Urodela Caudata *q.v.*; modern order of tailed amphibians.

Urolophidae Stingrays; family containing about 30 species of mostly small, shallow marine, myliobatiform elasmobranch fishes having a largely tropical and subtropical distribution; body disk oval to subcircular, tail bearing one or more serrated spines, caudal fin well developed; feed on crustaceans and polychaetes; reproduction viviparous.

Uropeltidae Family of small fossorial booid snakes (Serpentes) containing about 50 species from India through southeast Asia to New Guinea; typically possessing an enlarged scale near the tip of the tail.

urophilic Thriving in habitats rich in ammonia; **urophile, urophily.**

Uropodina Large and diverse suborder of free-living mites that derive their name from the curious anal secretion which forms a stalk by which the phoretic deutonymphs of many species attach to insects or other arthropods.

Uropygi Whip scorpions, vinegaroons; order of

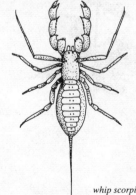

whip scorpion (Uropygi)

nocturnal predatory terrestrial arthropods (Arachnida); body length 15–75 mm, chelicerae with proximal fang, pedipalps powerful and spinose; first pair of legs long and slender; abdomen divided into broad mesosoma bearing book lungs and short metasoma sporting an apical whip-like flagellum; contains about 85 species.

Ursidae Bears; family containing 7 species of large terrestrial to arboreal mammals (Carnivora) found in Holarctic, Oriental and northwestern Neotropical regions; most are omnivorous with diets including terrestrial and aquatic vertebrates, insects and other invertebrates, carrion, fruit and diverse plant material.

Urticales Large and ecologically diverse order of Hamamelidae; woody or herbaceous plants all with reduced flowers which may be wind- or insect-pollinated; often with cystoliths or milky latex; contains 6 families, Barbeyaceae, Ulmaceae, Cannabaceae, Moraceae, Cecropiaceae and Urticaceae.

Ustilaginales Smut fungi; large cosmopolitan order of parasitic teliomycete fungi parasitizing vascular plants, including cereals and sugar cane; mycelia grow within host tissues; sporulation produces soot-like masses of teliospores on the surface of the host plant.

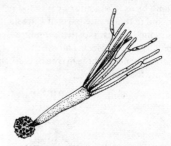

smut fungus (Ustilaginales)

polar bear (Ursidae)

Urticaceae Nettles; family of mostly herbaceous plant species having a widespread distribution; often with stinging hairs, and a tendency towards mineralization of cell walls; flowers unisexual, with a 4 or 5-segmented perianth; 4–5 stamens in male flowers; ovary one-celled.

stinging nettle (Urticaceae)

V

vadal Floating close to the shore.

vagile Wandering; freely motile; mobile.

vagility The tendency of an organism or population to change its location or distribution with time; mobility; **vagile.**

vaginicolous Living in secreted sheaths or cases; **vaginicole.**

vagrant Unattached; wandering; used of unattached wind-blown organisms, and of organisms that move about by their own activity.

Valdian glaciation The most recent glaciation of the Quaternary Ice Age *q.v.* in Russia, with an estimated duration of about 70 000 years.

Valerianaceae Cosmopolitan family of Dipsacales containing about 300 species of mostly herbs with a characteristic rank odour caused by ethereal oils; flowers occur in compound cymose or capitate inflorescences; each flower typically with 1–3 minute sepals, 3–5 petals, 1–4 stamens and an inferior ovary.

valvate dehiscence Spontaneous release of the contents of a ripe fruit through a valve, flap or other large aperture.

Valvatida Large order of often colourful, intertidal to deep-sea, asteroidean echinoderms typically with rigid body, large disk, short to long arms and suckered tube feet; anal pore present.

Valvifera Suborder of marine and brackish-water isopod crustaceans in which the uropods are enlarged and fold towards the midline, covering and protecting the respiratory pleopods.

vampire bat Phyllostomatidae *q.v.*

vampire squid Vampyromorpha *q.v.*

Vampyromorpha Vampire squid; order of cephalopod molluscs comprising a single cosmopolitan, deep-water species resembling both octopuses and squid; characterized by a

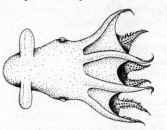

vampire squid (Vampyromorpha)

pair of tendril-like filaments and 8 long arms connected by an extensive web.

Varangian glaciation A widespread glaciation during the Precambrian era, about 670 million years B.P.

Varanidae Monitor lizards, Komodo dragon; small family of medium-sized to very large (200 mm to 3 m) terrestrial lizards (Sauria); habits diurnal, preying on variety of invertebrates and vertebrates; reproduction oviparous; contains 21 species ranging from Africa to Australia, and also on some Pacific islands.

variant Any individual or group showing marked deviation from type, in form, quality or behaviour.

variegation The occurrence within an individual of a mosaic phenotype, typically with respect to coloration; partial albinism in plants.

varve A banded layer of sediment deposited in a lake during the course of a single year; comprising a paler coarse layer deposited in spring and summer and a darker fine layer deposited in autumn and winter.

vascular plant Any plant containing specialized conducting tissues, *i.e.* xylem and phloem, and typically differentiated into roots, stems and leaves; Tracheophyta.

Vaucheriales Order of algae (Xanthophyceae) commonly found forming mats over moist soil and mud; characterized by the lack of cell walls in the vegetative filaments (siphoneous organization).

vector 1: An organism that carries or transmits a pathogenic agent; *cf.* carrier. 2: Any agency responsible for the introduction or dispersal of an animal or plant species.

vegetable sponge Cucurbitaceae *q.v.*

vegetation The total plant life or cover in an area; also used as a general term for plant life; *cf.* faunation.

vegetative Pertaining to assimilative, somatic and trophic growth processes rather than to reproduction.

vegetative reproduction Reproduction by asexual processes such as budding or fragmentation; vegetative propagation.

veil fish Veliferidae *q.v.*

veld The open temperate grassland areas of southern Africa, typically with scattered shrubs or trees; veldt.

Veliferidae Veil fish; family containing 3 species of moderately deep-water, Indo-Pacific lampridiform teleost fishes; body deep and compressed, to 300 mm length; dorsal and anal fins very large with scaly basal sheaths.

veliger Free-swimming larval stage of some molluscs possessing a large ciliated apical lobe used in swimming and feeding, a foot, shell and other adult features.

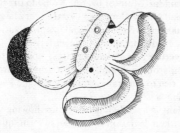

veliger

Velloziaceae Widespread family of Liliales comprising about 250 species of resinous or gum-producing xerophytic shrubs; slender stem appearing stout because of persistent leaf bases on upper part and adventitious roots on lower parts; producing solitary, trimerous flowers.

velvet ant Mutillidae *q.v.*

velvetfish 1: Aploactinidae *q.v.* 2: Caracanthidae *q.v.*

venation The pattern of veins in a leaf of vascular plants, or in an insect's wing.

Veneroida Large order of mainly marine heterodont bivalve molluscs which are generally active burrowers or nestlers, rarely sedentary forms; characterized by more or less equal shell valves, two shell-closing muscles and by the structure of the hinge; includes the heart shells, cockles, giant, surf and fingernail clams, razor shells, sunset shells, Venus shells and wedge shells.

cockle (Veneroida)

venomous Poisonous; noxious; venom-producing.

Venus shell Veneroida *q.v.*

Venus's fly trap Droseraceae *q.v.*

Venus' flower basket Hexasterophora *q.v.*

Verbenaceae Teak, spiraea, pagoda flower;

family of Lamiales containing about 2600 species of sometimes roughly hairy herbs and woody plants, mostly from tropical regions; flowers bisexual and arranged in cymose or racemose heads frequently set off by an involucre of coloured bracts; individual flowers bisexual and commonly with 5 sepals, 5 petals forming a narrow tube basally, and 4 stamens.

vernal Pertaining to the spring; *cf.* aestival, hibernal, hiemal, serotinal.

vernal pool A temporary pool formed during spring from meltwater or flood water.

vernalization A process of thermal induction in plants, in which growth and flowering are promoted by exposure to low temperatures.

vernation The pattern of folding and rolling of leaves in a bud.

Verongiida Order of ceractinomorph sponges found most commonly in tropical and warm temperate waters from the intertidal zone down to 400 m; characterized by a branching or reticulate skeleton of spongin fibres which are not differentiated into size categories.

Verrucariales Order of loculoanoteromycetid fungi which are lichenized or parasitic in lichen thalli; widespread in temperate and cold regions.

versicoloured Exhibiting colour change, or different colours; variegated.

Vertebrata Subphylum of chordates characterized by the presence of a braincase, vertebral column, medial fins, skeleton of bone and/or cartilage, skin comprising dermal and epidermal layers, heart and blood cells, kidney, liver, pancreas and neural crest; higher vertebrates also possess jaws, teeth, paired fins/or limbs having an internal skeleton articulating with girdles, bony scales, feathers or hair, lungs, and in tetrapods a neck formed by specialization of the vertebral column; 8 classes are traditionally recognized, Placodermi (extinct), Agnatha (jawless fishes), Chondrichthyes (cartilaginous fishes), Osteichthyes (bony fishes), Amphibia (amphibians), Reptilia (reptiles), Aves (birds) and Mammalia (mammals).

Vespertilionidae Cosmopolitan family of mainly insectivorous microchiropteran bats containing about 280 species; habits solitary to gregarious, often migratory; some species hibernate.

vespertine Pertaining to the evening.

Vespidae Social wasps; family of truly social or socially parasitic wasps (Hymenoptera) which construct nests of regularly arranged cells made of masticated plant fibres; colonies consist mostly of sterile females; larvae are fed on a masticated paste of arthropod prey; contains about 800 species distributed worldwide.

Vestibulifera Subclass of kinetofragminophoran ciliates characterized by the presence of a usually apical invagination (vestibular cavity) at the base of which lies the cell mouth and pharyngeal apparatus; includes free-living aquatic and soil forms as well as endocommensals in the stomachs of mammals.

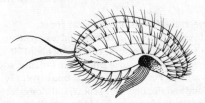

ciliate (Vestibulifera)

vestigial Degenerate or imperfectly developed; used of structures or functions that have become diminished or reduced during the course of evolution or ontogeny; **vestige.**

Vestimentifera Order of pogonophoran worms lacking a mesosomal bridle; body devoid of setae; can attain considerable size (2 m) when living in association with deep-sea hydrothermal vents.

vetch Fabaceae *q.v.*

vexillifer Pelagic larval stage of teleost fishes in the family Carapidae (pearlfishes).

viable Having the capacity to live, grow, germinate or develop; **viability.**

viatical Growing on the roadside or beside paths.

viburnum Caprifoliaceae *q.v.*

vicariance The existence of closely related forms (vicariants) in different geographical areas, which have been separated by the formation of a natural barrier (a vicariance event).

vice-counties Biogeographical subdivisions of former administrative counties in Britain, used as unit areas in distributional studies.

vicine Used of organisms invading or entering from adjacent areas or communities.

vinegar fly Drosophilidae *q.v.*

vinegaroon Uropygi *q.v.*; vinegarone.

Violaceae Viola, violet, pansy; large family of herbaceous or woody plants widespread from northern temperate regions to tropical rainforests; typically with bisexual flowers each comprising 5 sepals, petals and stamens, and a superior ovary.

pansy (Violaceae)

Violales Large order of Dilleniidae containing about 5000 species in 24 families; diverse assemblage of herbs to trees but with the great majority of species contained in the families Begoniaceae, Cucurbitaceae, Flacourtiaceae, Passifloraceae and Violaceae.

violet Violaceae *q.v.*

violet snail Mesogastropoda *q.v.*

viperfish Chauliodontidae *q.v.*

Viperidae Vipers, rattlesnakes, moccasins; cosmopolitan family containing about 180 species of medium-sized highly venomous snakes (Serpentes); typically terrestrial, nocturnal, feeding on small mammals, a few species arboreal or fossorial; reproduction oviparous or ovoviviparous.

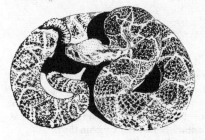

rattlesnake (Viperidae)

Vireonidae Vireos; family containing about 40 species of small passerine birds found in forest habitats of the New World; plumage usually greenish, bill robust and weakly hooked, feet strong; habits solitary, arboreal, sometimes migratory, feeding on insects and fruit; cup-shaped nest placed in tree.

virgin 1: Pertaining to a native habitat, fauna or flora that is essentially unaffected by the activities of man. 2: An animal that has not copulated.

Virginia creeper Vitaceae *q.v.*

Virginian waterleaf Hydrophyllaceae *q.v.*

virginiparous Used of organisms reproducing only by parthenogenesis.

virology The study of viruses.

virulence The capacity of a pathogen to invade host tissue and reproduce; the degree of pathogenicity.

viruliferous Virus-carrying; used of organisms harbouring viruses.

Virus Kingdom of prokaryotes comprising the viruses.

viruses A group of microorganisms traditionally regarded as a kingdom of prokaryotes; consist of a proteinaceous shell (capsid) containing the viral nucleic acid; growth and replication takes place only within a host cell, with the viral nucleic acid causing the host cell to synthesize the materials necessary for making more viral particles; recently regarded by some authorities as being rogue sections of the host nucleic acid which have become self-replicating by diverting the resources of the host cell.

virus

Viscaceae Mistletoes; cosmopolitan family of Santalales containing about 350 species of brittle evergreen shrublets hemiparasitic on tree branches and producing haustoria that penetrate and ramify within the host.

viscacha Chinchillidae *q.v.*

visible light That part of the solar radiation spectrum between the wavelengths 380 and 780 nm.

Vitaceae Grape vine, Virginia creeper, Boston ivy; Vitidaceae; family of Rhamnales containing about 700 species of mostly woody vines, commonly with leaf-opposed tendrils; mainly found in warm regions; flowers small and arranged in cymose inflorescences, each with 4–5 sepals, petals and stamens, and a superior ovary.

vitellogenous Yolk-producing.

viticolous Growing on vines; **viticole.**

Vitidaceae Vitaceae *q.v.*

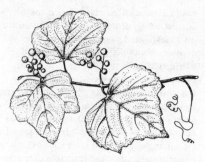

grape vine (Vitaceae)

Viverridae Civets, mongooses, genet; diverse family of terrestrial and arboreal mammals (Carnivora) containing about 70 species widespread in Ethiopian, Oriental and southern Palaearctic regions; body typically long and slender, limbs short, tail elongate; mostly carnivorous, feeding on small vertebrates and invertebrates, some forms omnivorous.

civet cat (Viverridae)

viviparous 1: Producing live offspring from within the body of the parent; **viviparity**, **vivipary**; *cf.* larvipary, oviparous, ovoviviparous. 2: Germinating while still attached to the parent plant.

vixigregarious Sparsely distributed; occurring in small, poorly defined groups.

vocalization Production of songs, calls and other vocal sounds by animals.

Vochysiaceae Family of Polygalales containing about 200 species of mostly woody plants with a resinous juice; native mainly to tropical America.

volant Adapted for flying or gliding.

volcanism Volcanic activity.

volcanogenic Used of a sediment produced as a result of volcanic activity.

vole Cricetidae *q.v.*

voltine Pertaining to the number of broods or generations per year or per season; usually quantified with a prefix as in univoltine, bivoltine, trivoltine, quadrivoltine, multivoltine.

voltinism A behavioural polymorphism in

insects in which some members of a population enter diapause and others do not.

volute Neogastropoda *q.v.*

volvation Enrolment of the body to form a ball.

Volvocales Order of typically flagellate, motile green algae distributed ubiquitously in freshwater, brackish and marine habitats; typically with uninucleate cells which may be solitary or united into colonies of definite structure (coenobia); cells usually containing a single cup-shaped chloroplast; also treated as a class of the protoctistan phylum Chlorophyta, and as an order of the protozoan class Phytomastigophora, under the name Volvocida.

Volvocida Volvocales *q.v.* treated as an order of the protozoan class Phytomastigophora.

Vombatidae Wombats; family containing 2 species of burrowing diprotodont marsupials from Australia and Tasmania; body length to about 1 m, incisor teeth long, limbs subequal, tail vestigial; marsupium opening posteriorly; habits typically nocturnal, herbivorous.

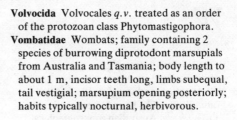

wombat (Vombatidae)

voucher specimen Any specimen identified by a recognized authority for the purposes of forming a reference collection.

vulcanism Volcanic activity.

vulture 1: Cathartidae *q.v.* (New World vultures). 2: Accipitridae *q.v.* (Old World vultures).

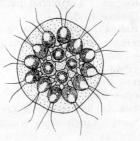

green alga (Volvocales)

W

wagtail Motacillidae *q.v.*

walkingstick Phasmoptera *q.v.*

wallaby Macropodidae *q.v.*

Wallace's line The boundary separating the Oriental and Australian zoogeographical regions drawn to the west of Weber's line *q.v.* and passing between the Philippines and Maluku (Moluccas) in the north, then southwest between Sulawesi and Borneo, and continuing south between Lombok and Bali.

Wallace's realms The major zoogeographical regions – Neotropical, Nearctic, Palaearctic, Oriental, Australian and Ethiopian.

Wallacea The transition zone between the Oriental and Australian zoogeographical regions, bounded to the east by Weber's line *q.v.* and to the west by Wallace's line *q.v.*, and comprising Sulawesi, Lombok, Flores and Timor.

wallflower Brassicaceae *q.v.*

walnut Juglandaceae *q.v.*

walrus Otariidae *q.v.*

wandering jew Commelinaceae *q.v.*

warble fly Oestridae *q.v.*

warbler Sylviidae *q.v.*

warm monomictic lake A lake with a winter overturn in which the water temperature never falls below 4°C; *cf.* cold monomictic lake.

warm temperate zone The latitudinal zone between 34.0° and 45.0° in either hemisphere.

warm-blooded Homoiothermic *q.v.*

warning coloration Conspicuous coloration used to advertize the noxious, unpalatable, or otherwise harmful properties of an organism to a potential predator.

wart snake Acrochordidae *q.v.*

Warthe glaciation A glaciation of the Quaternary Ice Age *q.v.* in northern Germany and Poland with an estimated duration (together with the Saale glaciation) of 100 000 years.

wasp Vespidae *q.v.* (Apocrita, Aculeata).

water bear Tardigrada *q.v.*

water boatman Notonectidae *q.v.*

water capacity The amount of water that a soil can retain against gravity.

water chestnut Trapaceae *q.v.*

water cycle The global biogeochemical cycle of water involving exchange between the hydrosphere, atmosphere, lithosphere and living organisms; hydrologic cycle.

water fern 1: Marsileales *q.v.* 2: Salviniales *q.v.*

water flea Cladocera *q.v.*

water hyacinth Pontederiaceae *q.v.*

water lily Nymphaeales *q.v.*

water louse Asellota *q.v.*

water mass A body of water within an ocean characterized by its physicochemical properties of temperature, salinity, depth and movement.

water measurer Hydrometridae *q.v.*

water mould 1: Leptomitales *q.v.* 2: Saprolegniales *q.v.*

water nymph Najadaceae *q.v.*

water plantain Alismataceae *q.v.*

water poppy Limnocharitaceae *q.v.*

water purslane Onagraceae *q.v.*

water scorpion 1: Eurypterida *q.v.* 2: Nepidae *q.v.* (Heteroptera).

water table The horizontal plane defining the upper limit of the ground layer fully saturated with water.

water willow Lythraceae *q.v.*

watercress Brassicaceae *q.v.*

watermelon Cucurbitaceae *q.v.*

watershed An elevated boundary area separating tributaries draining into different river systems.

wax flower Asclepiadaceae *q.v.*

wax myrtle Myricales *q.v.*

waxbill Estrildidae *q.v.*

waxwing Bombycillidae *q.v.*

W-chromosome The sex chromosome which in cases of female heterogamety is present in the female only; *cf.* X-chromosome, Y-chromosome, Z-chromosome.

weasel Mustelidae *q.v.*

weasel shark Hemigaleidae *q.v.*

weather Local, short-term atmospheric conditions; *cf.* climate.

weathering Physical, chemical and biological changes resulting from exposure to the atmosphere, that accompany soil formation from parent rock.

weaver finch Ploceidae *q.v.*; weavers.

web spinner Embiidina *q.v.*

Weber's line The boundary separating the Oriental and Australian zoogeographical regions drawn to the east of Wallace's line *q.v.* and passing between Maluku (the Moluccas) and Sulawesi to the north, and between Timor and the Kei Islands to the south.

wedge shell Veneroida *q.v.*

weed Any plant growing where it is not wanted.

weeverfish Trachinidae *q.v.*

weevil Curculionidae *q.v.*

Weichselian glaciation The most recent glaciation of the Quaternary Ice Age *q.v.* in northern Germany and Poland with an estimated duration of about 70 000 years; subdivided into Pomeranian, Frankfurt and Brandenburg glaciations.

Welwitschiidae Subclass of gymnosperms (Pinophyta) containing a single species occurring in the arid deserts of Angola and southwestern Africa; stem is an inverted woody cone and bears 2 leaves only, which are very long, strap-like, and grow throughout the life of the the plant.

welwitschia (Welwitschiidae)

wentletrap Mesogastropoda *q.v.*

Wentworth grade scale An extension of the Udden grade scale *q.v.* with descriptive class terms for sediment particles.

West Australia Current A cold surface ocean current that flows northwards off the west coast of Australia, fed from the Antarctic Circumpolar Current and forming the eastern limb of the South Indian Gyre; see ocean currents.

West Greenland Current A cold surface ocean current that flows northwards off the west coast of Greenland, giving rise to the cold Labrador Current; see ocean currents.

West Wind Drift A major cold surface ocean current that flows eastwards through the Southern Ocean producing a circumglobal Antarctic circulation; Antarctic Circumpolar Current; see ocean currents.

wetland An area of low-lying land, submerged or inundated periodically by fresh or saline water.

whale Cetacea *q.v.*

whale louse Cyamidae *q.v.*

whale shark Rhincodontidae *q.v.*

whalefish Small bathypelagic beryciform teleost fishes having flabby body with distensible stomach; fin spines and gas bladder absent; luminous organs often present; used for members of the families Barbourisiidae, Cetomimidae and Rondeletiidae.

wheat Poaceae *q.v.*

wheatear Turdidae *q.v.*

wheel animalcule Rotifera *q.v.*

whelk Neogastropoda *q.v.*

whip scorpion Uropygi *q.v.*

whipspider Amblypygi *q.v.*

whiptail Teiidae *q.v.*

whirligig beetle Gyrinidae *q.v.*

whisk fern Psilotophyta *q.v.*

whistler Pachycephalidae *q.v.*

white ant Isoptera *q.v.*

white butterfly Pieridae *q.v.*; white.

white mangrove Combretaceae *q.v.*

white mud A terrigenous marine sediment derived from coral reef debris.

white rust Peronosporales *q.v.*

white whale Monodontidae *q.v.*; beluga.

white-eye Zosteropidae *q.v.*

whitefish Salmonidae *q.v.*

whitefly Aleyrodidae *q.v.* (Homoptera).

wild bottlebrush Greyiaceae *q.v.*

wild cinnamon Canellaceae *q.v.*

wild type The natural or typical form of an organism, strain or gene, arbitrarily designated as standard or normal for comparison with mutant or aberrant individuals or alleles.

willow Salicales *q.v.*

willow herb Onagraceae *q.v.*

wilting Loss of turgidity in plants caused by an excess of water loss by transpiration over water uptake by absorption from the soil.

wind scorpion Solpugida *q.v.*

windchill The effect of the wind in causing excessive cold penetration into plants.

windkill Plant death resulting from the effects of wind.

window fly Scenopinidae *q.v.*

window-pane shell Ostreoida *q.v.*

winkle Mesogastropoda *q.v.*

winnowing Grading and separation of the fine fractions of particulate matter by wind or water currents.

Winteraceae One of the most archaic families

drimys (Winteraceae)

of flowering plants (Magnoliales), containing about 100 species of trees and shrubs widely distributed in the southern hemisphere; characterised by regular and bisexual flowers with 2–6 sepals, 2 to many petals, 15 to many stamens and a superior ovary.

wireworm Elateridae *q.v.*

Wisconsin glaciation The most recent glaciation of the Quaternary Ice Age *q.v.* in North America, with an estimated duration (together with the Iowan glaciation) of about 70 000 years.

wisteria Fabaceae *q.v.*

witch hazel Hamamelidaceae *q.v.*

wobbegong Orectolobidae *q.v.*

wolf Canidae *q.v.*

wolf herring Chirocentridae *q.v.*

wolf spider Lycosidae *q.v.*

wolffish Anarhichatidae *q.v.*

Wolstonian glaciation A glaciation of the Quaternary Ice Age *q.v.* in the British Isles with an estimated duration of 100 thousand years; Gipping glaciation.

wolverine Mustelidae *q.v.*

wombat Vombatidae *q.v.*

wood sorrel Oxalidaceae *q.v.*

wood swallow Artamidae *q.v.*

wood warbler Parulidae *q.v.*

woodcreeper Dendrocolaptidae *q.v.*

woodland An area of vegetation dominated by a more or less closed stand of short trees.

woodlouse Oniscoidea *q.v.*

woodlouse fly Rhinophoridae *q.v.*

woodpecker Picidae *q.v.*

woodwasp Xiphydriidae *q.v.* (Symphyta).

woodworm Anobiidae *q.v.*

woody plant A perennial plant having a secondarily thickened lignified stem.

woolly bear The hairy, caterpillar-like larval stage of beetles of the family Dermestidae and of moths of the family Arctiidae.

woolly monkey Cebidae *q.v.*

work Any structure or impression resulting from the activity of an animal, including tubes, burrows, nests and tracks.

worm eel Ophichthidae *q.v.*

worm lizard Amphisbaenia *q.v.*

wormfish Microdesmidae *q.v.*

wrack Fucales *q.v.*

wrasse Labridae *q.v.*

wren 1: Troglodytidae *q.v.* 2: Maluridae *q.v.* (Australian wrens). 3: Xenicidae *q.v.* (New Zealand wrens).

wren-thrush Zeledoniidae *q.v.*

wryneck Picidae *q.v.*

Würm glaciation The most recent glaciation of the Quaternary Ice Age *q.v.* in the Alpine area with an estimated duration of about 70 000 years; subdivided into Würm 3, Würm 2 and Würm 1 glaciations.

X

Xanthophyceae Class of chromophycote algae found in freshwater and marine habitats and varying in organization from unicellular to filamentous or siphoneous; characterized by the possession of chlorophylls *a* and *c*, and various carotenoids, and by the production of motile zoospores which are typically pear-shaped with a hairy, anteriorly directed flagellum and a shorter, naked, posteriorly directed flagellum; also classified as a separate phylum of Protoctista under the name Xanthophyta.

xanthophycean alga (Xanthophyceae)

xanthophyll A type of photosynthetic pigment; including the primary light-absorbing pigments of brown algae, fucoxanthin and peridinin.

Xanthophyllaceae Family of Polygalales containing about 40 species of small trees native to the Indo-Malaysian region; flowers pentamerous, one petal usually hood-like; fruit indehiscent, woody or somewhat fleshy.

Xanthophyta The Xanthophyceae *q.v.* treated as a phylum of Protoctista; comprises 4 classes, Heterochloridales, Heterococcales, Heterosiphonales and Heterotrichales.

Xanthorrhoeaceae Grass trees; small family of Liliales comprising about 55 species of stout, often arborescent shrubs or coarse short-stemmed herbs; often with persistent old leaf bases and with perennial, parallel-veined and often pungent leaves; restricted to Australia, New Caledonia and New Guinea.

Xantusiidae Night lizards; family containing 14 species of viviparous terrestrial lizards (Sauria) found in Central and North America; habits mostly crepuscular, insectivorous.

X-chromosome The sex chromosome which in cases of male heterogamety is present in both sexes; *cf.* W-chromosome, Y-chromosome, Z-chromosome.

Xenarthra Suborder of mammals (Edentata) characterized by additional articulations in the lumbar vertebrae; comprises the South American edentates known since the Eocene including the anteaters and sloths (Pilosa) and the armadillos (Cingulata).

xenautogamous Used of organisms in which cross fertilization normally occurs, but in which self-fertilization is possible; **xenautogamy.**

Xenicidae New Zealand wrens; family containing 4 species of small insectivorous passerine birds with short rounded wings and short tail; habits solitary or gregarious, arboreal or terrestrial, flight weak; nest dome-shaped, in tree or rock crevice.

xenobiosis Symbiosis in which one species lives freely within the colony of another species whilst maintaining broods separately.

xenobiotic A foreign (allochthonous) organic chemical; used of environmental pollutants such as pesticides in runoff water.

Xenocongridae False moray eels; family containing 15 species of little-known tropical and warm-temperate anguilliform teleost fishes; body smooth, dorsal and anal fins elongate and continuous with caudal, pectorals present or absent.

xenodeme A local interbreeding population of a parasite that differs from other demes in its host specificity.

xenoecic Inhabiting the empty domicile or shell of another organism; **xenoecy.**

xenogamy Cross fertilization; fertilization between flowers on different plants; *cf.* geitonogamy.

xenogenous Originating from outside the organism or system.

xenology The study of host–parasite relationships.

xenoparasite A parasite infesting an organism that is not its normal host; **xenoparasitism.**

Xenopeltidae Sunbeam snake; family containing a single species of burrowing snakes (Serpentes) found in southeastern Asia; scales iridescent; feeds mainly on small vertebrates, including other snakes.

Xenophyophorea Class of giant marine rhizopod protozoans found primarily in the deep-sea benthos; body consists of a multinucleate plasmodium enclosed within a branching organic tube and a massive test composed of organic and inorganic foreign matter (xenophyae); comprises 2 orders, Psamminida and Stannomida.

Xenopneusta Small order of echiuran marine worms in which the longitudinal muscles of the trunk body wall are between the outer circular and inner oblique layers, the blood system is open, there are 2 or 3 pairs of nephridia and the hind gut is modified as a respiratory chamber; feed on suspended particles and bacteria using mucus net placed across opening of burrow.

Xenosauridae Small family of robust, diurnal lizards (Sauria), comprising 3 terrestrial, insectivorous species from Central America, and a single semiaquatic piscivorous species from China.

Xenungulata Extinct order of mammals known from the Cenozoic of South America; separated from the Pyrotheria by differences in premolars and canine teeth.

xerantic Becoming parched or dried up; withering.

xerarch succession An ecological succession beginning in a dry habitat; *cf.* hydrarch succession, mesarch succession.

xeric Having very little moisture; tolerating or adapted to dry conditions.

xerochastic Used of a fruit in which dehiscence is induced by desiccation; **xerochasy**; *cf.* hygrochastic.

xerochore That region of the Earth's surface covered by dry desert.

xerocleistogamy Self-pollination within flowers that remain unopened because of inadequate moisture; **xerocleistogamic.**

xerocolous Living under dry conditions; **xerocole.**

xerogeophyte A plant which enters a resting stage during periods of drought.

xerohylophilous Thriving in dry forests; **xerohylophile, xerohylophily.**

xerohylophyte A dry-forest plant.

xeromorphic Pertaining to plants having structural or functional adaptations to prevent water loss by evaporation; **xeromorphism, xeromorphy.**

xerophilous Thriving in dry habitats; **xerophile, xerophily.**

xerophobous Intolerant of dry conditions; **xerophobe, xerophoby.**

xerophyte A plant living in a dry habitat, typically showing xeromorphic or succulent adaptations and able to tolerate long periods of drought; **xerophytic**; *cf.* hydrophyte, hygrophyte, mesophyte.

xeropoophilous Thriving in heathland; **xeropoophile, xeropoophily.**

xeropoophyte A heath plant.

xerosere An ecological succession commencing on a dry rock surface.

Xerothermal period A postglacial interval of warmer and drier climate; approximately equivalent to the Altithermal period *q.v.* or the Subboreal period *q.v.*; Long Drought; *cf.* Hypsithermal period.

xerothermic Used of organisms tolerating or thriving in hot and dry environments; **xerotherm.**

xerotherous Used of organisms adapted to dry summer conditions.

xerotropism An orientation response of plants to desiccation; **xerotropic.**

Xiphiidae Swordfish; monotypic family of large (to 4.5 m) predatory scombroid teleost fishes (Perciformes) characterized by a long pointed rostrum; jaw teeth and pelvic fins absent; feeds on squid and other fishes.

swordfish (Xiphiidae)

Xiphodontidae Extinct family of primitive artiodactyls known from the Eocene and Oligocene of Europe.

Xiphosura King crabs, horseshoe crabs; order of large benthic marine arthropods (Chelicerata) comprising a single extant family with 4 species; carapace expansive covering all appendages; prosomal appendages consisting of chelicerae, pedipalps and 4 pairs of legs, 3 of which have pincers; opisthosoma with 6 pairs of limbs and 5 pairs of gills.

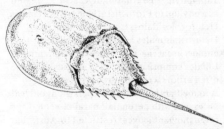

horseshoe crab (Xiphosura)

Xiphydriidae Woodwasps; cosmopolitan family containing about 90 species of wasps (Hymenoptera) typically having a slender cylindrical body lacking a distinct waist, and an awl-like ovipositor in the female; coloration black, banded with red, yellow or white; larvae commonly bore into dead branches of trees and shrubs.

Xylariales Large order of pyrenomycete fungi

candle snuff fungus (Xylariales)

containing thousands of species, mostly saprobic on wood, soil, plant debris and dung but occasionally parasitic.

xylophagous Feeding on wood; **xylophage, xylophagy.**

xylophilous Thriving on or in wood; **xylophile, xylophily.**

xylophyte 1: A woody plant. 2: A plant living in or on wood.

xylotomous Used of organisms able to cut or bore into wood.

Xyridaceae Yellow-eyed grass; family of Commelinales containing more than 200 species of herbs, usually having basal leaves and flowers with 3 sepals, petals, stamens and carpels; petals united into corolla tube, usually yellow in colour; widespread in tropical and subtropical regions.

yam Dioscoreaceae *q.v.*

Yarmouthian interglacial An interglacial period in the middle of the Quaternary Ice Age *q.v.* in North America.

Y-chromosome The sex chromosome which in cases of male heterogamety is present in the male only; *cf.* W-chromosome, X-chromosome, Z-chromosome.

yeast Endomycetales *q.v.*; a fungus found in the form of single cells and reproducing by budding or by fission.

yellow-eyed grass Xyridaceae *q.v.*

yew Taxaceae *q.v.*; an evergreen coniferous shrub or tree with needle-like or linear leaves, spirally arranged.

yew (Taxaceae)

yield That part of production utilized by a consumer species or by a group belonging to a higher trophic level.

yucca Agavaceae *q.v.*

Z

Zamiaceae Cycads; family of gymnosperms (Pinophyta) containing about 80 species of cycads; pinnules with straight venation; widely distributed in tropical and temperate America, Australia and Africa.

Zaniolepididae Combfishes; family containing 2 species of small (to 300 mm) marine scorpaeniform teleost fishes found in shallow waters along the North American west coast; body elongate, anteriorly compressed, body scales ctenoid.

Zannichelliaceae Horned pondweed; family of Najadales comprising a few species of rhizomatous herbs with thread-like stems; growing submerged in fresh, alkaline or brackish water; flowers unisexual, usually lacking perianth or with 3 reduced scales; female flowers with up to 9 free carpels each terminating in a long horn bearing the stigma; fruits are achenes.

Zapodidae Jumping mice; family containing 10 species of small myomorph rodents found in Palaearctic and Nearctic regions; hindlimbs longer than forelimbs; typically nocturnal, feeding on fruit, seeds and insects; most species hibernate during winter.

Z-chromosome The sex chromosome which in cases of female heterogamety is present in both sexes; *cf.* W-chromosome, X-chromosome, Y-chromosome.

zebra Equidae *q.v.*

zebra plant Acanthaceae *q.v.*

zebra shark Stegostomatidae *q.v.*

Zeidae Dories; family of mostly shallow marine zeiform teleost fishes, cosmopolitan in temperate waters; body deep and strongly compressed, often silvery, to 600 mm length, naked or with tiny scales; typically having large mouth and protrusible eyes; dorsal, anal, and pelvic fins spinose.

Zeiformes Small order of deep-bodied marine teleost fishes found in tropical or temperate waters to about 600 m depth; body strongly compressed, mouth protractile, pelvic fins thoracic or jugular; comprises 6 families, including dories, boarfishes and oreos.

zeitgeber Any external stimulus that acts to trigger or phase a biological rhythm.

Zeledoniidae Wren-thrush; family containing a single species of passerine birds found in Central America; usually included in the Turdidae.

Zeugloptera Suborder of primitive lepidopteran insects comprising about 100 species in which the adults have toothed functional mandibles, unspecialized maxillae, and feed on pollen; larvae found in soil feeding on liverworts and detritus.

zincophyte A plant adapted to, or tolerating, high zinc levels in the soil.

Zingiberaceae Ginger; family of about 1000 species of leafy-stemmed herbs with silica cells and scattered ethereal oil-secreting cells; native to tropical regions, especially in southern Asia; flowers trimerous, with 3 sepals and petals, 6 stamens and an inferior ovary, bilaterally symmetrical; fruit usually a capsule.

ginger (Zingiberaceae)

Zingiberales Order of the subclass Zingiberidae consisting of 8 families of almost exclusively tropical herbs or small trees with an unbranched trunk; broad, fragile leaf blade having a prominent midrib and numerous lateral veins in a pinnate-parallel arrangement; flowers basically trimerous, but often bilaterally symmetrical.

Zingiberidae Subclass of monocotyledons (Liliopsida) consisting of 2 orders of terrestrial or epiphytic herbs that lack secondary growth; leaves are alternate with a sheathing base and a narrow, parallel-veined blade; flowers

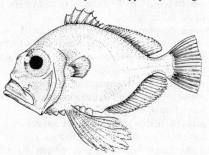

dory (Zeidae)

showy and adapted for pollination by nectar-gathering insects and other animals; fruit usually a capsule or berry, seeds with copious endosperm storing food as starch.

Ziphiidae Beaked whales; family containing 18 species of little-known marine mammals (Odontoceta) widespread in deeper waters of cool temperate to tropical oceans and seas; teeth typically absent in upper jaw, sparse or absent in lower jaw; thought to feed mainly on squid.

Zoantharia Subclass of Anthozoa containing solitary and colonial representatives which may be naked or enclosed within a calcareous exoskeleton; characterized by paired mesenteries in the gastrovascular cavity of the polyp, typically arranged in multiples of 6; Hexacorallia; comprises 5 extant orders, Actiniaria, Corallimorpharia, Scleractinia, Ptychodactiaria and Zoanthinaria.

Zoanthinaria Order of zoantharians found in abundance in warm shallow seas but also known from deeper and colder waters; typically resembling sea anemones but usually colonial or social with polyps arising from a common basal stolon or mat; lacking an intrinsic skeleton; Zoanthidea.

Zoarcidae Eelpouts; family containing 65 species of mainly shallow marine gadiform teleost fishes widespread in cold waters; body length to 1 m, tail tapering, caudal fin continuous with dorsal or anal, pelvics reduced or absent; sexual dimorphism marked in some species.

zoea A larval stage found in higher crustaceans; at least one pair of biramous locomotory thoracic limbs is present; usually with compound eyes.

zoea

zoidiogamy Fertilization by a motile male gamete; **zoidiogamic.**

zoidiophilous Pollinated by animals; **zoidiophily.**

zonal soil A deep soil having moderate internal and surface drainage with a well defined profile comprising several distinct horizons; used of mature soils with well developed characteristics that reflect the influence of vegetation and climate as soil-forming factors; *cf.* azonal soil, intrazonal soil.

zonation The distribution of organisms in distinctive areas, layers or zones.

zone 1: An area, or subdivision of a biogeographical region that has a characteristic biota. 2: A stratum or series of strata distinguished by characteristic fossils; **zonal.**

zoobenthos Those animals living in or on the sea bed or lake floor.

zoochorous Dispersed by the agency of animals; **zoochore, zoochory.**

zoocoenosis An animal community.

zoodomatia Plant structures acting as shelters for animals.

zooecology The study of relationships between animals and their environment; animal ecology.

zooflagellate Zoomastigophora *q.v.*

zoogamete A motile gamete.

zoogamous 1: Used of animals that reproduce sexually; **zoogamy.** 2: Used of plants having motile gametes.

zoogenic Produced by or associated with the activity of animals.

zoogeographical regions The major geographical divisions of the Earth's land surface characterized by a particular faunal composition; the 6 original regions were Australian, Ethiopian, Nearctic, Neotropical, Oriental and Palaearctic; see map on page 412.

zoogeography Study of the geographical distribution of animals and animal communities.

zoolith An animal fossil.

zoology The study of animals; **zoological.**

Zoomastigina Zooflagellates; phylum of Protoctista comprising unicellular heterotrophic organisms possessing at least one flagellum, often more; may be free-living or parasitic.

Zoomastigophora Zooflagellates; class of Mastigophora comprising mostly parasitic but sometimes free-living protozoans, with one to several hundred flagella, all of which are obligate heterotrophs lacking plastids; pseudopodia as well as flagella are present in some forms; reproduction by binary or multiple fission although sexual processes have been confirmed in some species; equivalent in part to the protoctistan phylum Zoomastigina.

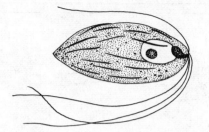

zooflagellate (Zoomastigophora)

zoometry The application of statistical methods to the study of animals; **zoometrics.**

zoomorphic 1: Pertaining to or produced by the activity of animals; **zoomorphosis.** 2: Having the form of an animal.

zooneuston The animals of the neuston; **zooneustonic.**

zoonosis A disease transmitted from animals to man under natural conditions; also used more loosely for any disease of animals; **zoonoses.**

Zoopagales Order of zygomycete fungi which are haustorial ectoparasites or endoparasites of other fungi, protistans and animals; also treated as a class of the phylum Zygomycota.

zooparasitic Used of a parasite having an animal host.

zoophagous Feeding on animals or animal matter; **zoophage, zoophagy.**

zoophilous Pollinated by animals; having an affinity for animals; **zoophile, zoophily.**

zoophyte Any animal that resembles a plant in morphology or mode of life; **zoophytic.**

zooplankton The animals of the plankton; **zooplanktont;** *cf.* phytoplankton.

zoosaprophagous Feeding on decaying animal matter; **zoosaprophage, zoosaprophagy.**

zoosemiotics The study of animal communication.

zoosis Any disease caused by an animal.

zoospore An independently motile spore in protistans, some fungi and algae.

zoosuccivorous Used of organisms that feed on liquid secretions of animals, or on decaying animal matter; **zoosuccivore, zoosuccivory.**

zootoxin A poison produced by an animal.

Zooxanthellales Order of photosynthetic marine dinoflagellates that are intracellular symbionts in foraminiferans, radiolarians, acantharians, cnidarians and molluscs; their presence in corals greatly enhances the calcification of reef-building corals.

Zoraptera Order of tiny (2–3 mm) orthopterodean insects somewhat resembling slender termites, comprising about 20 species,

most widespread under bark and logs in wet tropical habitats; there are 2 adult forms, one white, blind and wingless, the other pigmented, with eyes, and bearing 2 pairs of slender membranous deciduous wings.

Zosteraceae Sea grass, eelgrass; family of Najadales comprising about 12 species of rhizomatous marine herbs growing intertidally or submerged at depths down to 50 m; with slender parallel-veined leaves and small unisexual, water-pollinated flowers.

Zosteropidae White-eyes; family containing about 80 species of small Old World passerine birds widespread in forest habitats from Africa to Japan and New Zealand; bill short and slender, tongue grooved and frilled distally; habits gregarious, arboreal, active, feeding on insects, nectar and fruit; nest in trees.

zorapteran (Zoraptera)

Zygnematales Large order of essentially freshwater or subaerial green algae (Chlorophyceae) containing two body forms, either uniseriate filaments, like *Spirogyra*, or unicellular as in the desmids; characterized by sexual reproduction involving conjugation and the formation of a thick-walled, resistant zygote; equivalent in composition to the protoctistan phylum Gamophyta.

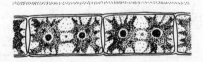

green alga (Zygnematales)

zygogenesis Reproduction during which male and female nuclei fuse; **zygogenetic.**

zygogenic Resulting from fertilization.

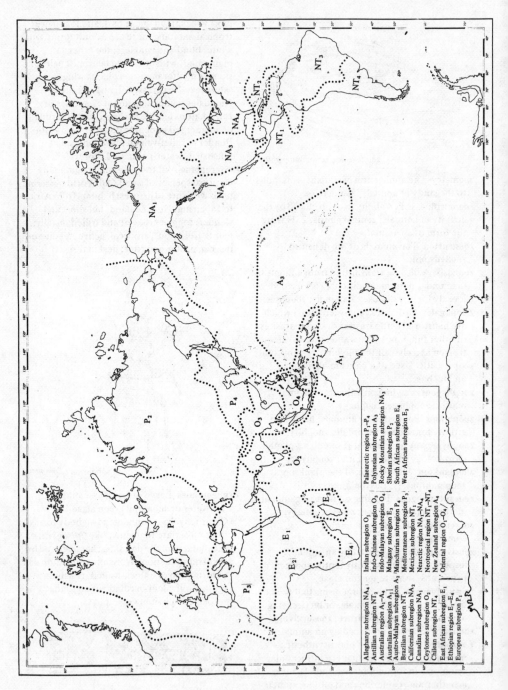

Alleghany subregion NA₄
Antillian subregion NT₂
Australian region A₁–A₄
Australian subregion A₁
Austro-Malayan subregion A₂
Brazilian subregion NT₃
Californian subregion NA₂
Canadian subregion NA₁
Ceyloness subregion O₂
Chilean subregion NT₄
East African subregion E₁
Ethiopian region E₁–E₄
European subregion P₁

Indian subregion O₁
Indo-Chinese subregion O₃
Indo-Malayan subregion O₄
Malagasy subregion E₃
Manchurian subregion P₄
Mediterranean subregion P₃
Nearctic region NA₁–NA₄
Neotropical region NT₁–NT₄
New Zealand subregion A₄
Oriental region O₁–O₄

Palaearctic region P₁–P₄
Polynesian subregion A₃
Rocky Mountain subregion NA₃
Siberian subregion P₂
South African subregion E₄
West African subregion E₂

Zoogeographical regions

zygomorphic Irregular; used of bilaterally symmetrical flowers.

Zygomycetes Class of zygomycotine fungi characterized by a type of asexual reproduction involving the endogenous production of non-motile sporangiospores; comprises 6 orders, Dimargaritales, Endogonales, Entomophthorales, Kickxellales, Mucorales and Zoopagales.

Zygomycota Zygomycetes *q.v.*; treated as a distinct phylum of the kingdom Fungi.

Zygomycotina Cosmopolitan subdivision of true fungi (Eumycota) containing many free-living, saprophytic and parasitic forms, characterized by sexual reproduction involving the fusion of 2 gametangia to form a zygospore; also reproducing asexually by sporangia or conidia formation; hyphae typically lacking septa; comprises 2 classes, Zygomycetes and Trichomycetes.

zygophase The diploid phase of a life cycle; *cf.* gamophase.

Zygophyllaceae Lignum vitae, creosote bush; family of Sapindales containing about 250 species of shrubs, trees and herbs; mostly from warm arid regions; flowers regular, usually with 4 or 5 overlapping sepals and petals; stamens arranged in whorls of 5, ovary superior and often winged; fruit variable.

zygophyte A plant produced by sexual reproduction.

Zygoptera Damselflies; suborder containing about 3000 species of paleopterous insects (Odonata) in which the forewings and hindwings are of similar shape, typically held together above the abdomen at rest; larvae aquatic, body slender with caudal gills.

damselfly (Zygoptera)

zygosis Union of gametes; conjugation; **zygotic.**

zygospore A thick-walled, usually dark-pigmented resting spore formed by zygomycotine fungi.

zygotaxis The mutual attraction between male and female gametes; **zygotactic.**

zygote A fertilized gamete; a diploid cell formed by the fusion of two haploid gametes.

zymogenic Causing fermentation.